Thanks for contribution Ji
best wishes

Mark

Thermochemical Conversion of Biomass to Liquid Fuels and Chemicals

RSC Energy and Environment Series

Editor-in-Chief:
Professor Laurence Peter, University of Bath, UK

Series Editors:
Professor Heinz Frei, Lawrence Berkeley National Laboratory, USA
Professor Ferdi Schüth, Max Planck Institute for Coal Research, Germany
Professor Tim S. Zhao, The Hong Kong University of Science and Technology, Hong Kong

Titles in this Series:
1: Thermochemical Conversion of Biomass to Liquid Fuels and Chemicals

How to obtain future titles on publication:
A standing order plan is available for this series. A standing order will bring delivery of each new volume immediately on publication.

For further information please contact:
Book Sales Department, Royal Society of Chemistry, Thomas Graham House, Science Park, Milton Road, Cambridge, CB4 0WF, UK
Telephone: +44 (0)1223 420066, Fax: +44 (0)1223 420247,
Email: books@ rsc.org
Visit our website at http://www.rsc.org/Shop/Books/

Thermochemical Conversion of Biomass to Liquid Fuels and Chemicals

Edited by

Mark Crocker
Center for Applied Energy Research, University of Kentucky, Lexington, KY 40511, USA

RSC Publishing

RSC Energy and Environment Series No. 1

ISBN: 978-1-84973-035-8
ISSN: 2044-0774

A catalogue record for this book is available from the British Library

Published by The Royal Society of Chemistry,
Thomas Graham House, Science Park, Milton Road,
Cambridge CB4 0WF, UK

Registered Charity Number 207890

For further information see our web site at www.rsc.org

Foreword

The thrust for alternative, renewable sources of liquid transportation fuels is driven by two different, but related occurrences. One of these factors is the global warming issue, where a wealth of scientific data points to greenhouse-gas (GHG) emissions as being a major contributor to this burgeoning problem. Of these GHGs, the biggest in terms of volume and concentration is carbon dioxide, which is released when fossil-based fuels are burnt for energy generation. The second factor concerns the long-term availability of the fossil-based crude oil that has powered the world's economic growth over the last century or so. "Peak Oil" or "Hubbert's Peak", devised by the late Shell scientist M. King Hubbert, predicted in 1956 that US oil production would peak in about 1970 (history proved this prediction to be correct) and global production would peak some 40 years later (today). The ramifications of this latter point will be enormous as production rates start to decline and demand worldwide continues to increase.

The combined pressures of rising fuel prices, diminishing global supplies of crude oil and legislation to control climate change by reducing greenhouse-gas emissions has resulted in both aggressive renewable fuel policies and a rapid growth in the emerging biofuels industry. Surplus agricultural feedstocks in the US and Europe are quickly being exhausted, contributing to commodity price increases, including corn, soybean and rapeseed and other biofuel feedstocks. It is widely argued that the continued commercialization and development of first-generation fuels such as corn ethanol and soy biodiesel will put additional pressure on food supplies during a time of rapid growth in demand from developing countries.

As a result of the foregoing there has been a growing acknowledgement of, and emphasis on, next-generation (some refer to these as second-generation, some as third-generation) biofuels such as cellulosic ethanol, algae-derived biodiesel and pyrolysis oil (bio-oil) from woody biomass. More recently, attention has turned to more advanced biofuels made from renewable sources. These advanced biofuels are often referred to as "drop-in fuels" since these final fuels (renewable diesel, gasoline and jet fuel) are literally drop-in replacements for today's diesel, gasoline and jet fuel: like fossil-derived fuels they are hydrocarbons (devoid of oxygen), with the same high energy content and are very similar in terms of molecular structures. The fact that these advanced biofuels are drop-in replacements is a critical factor in their anticipated commercial success because they can be readily used in the existing infrastructure (from pipelines to refineries, from automobiles to military aircraft, *etc.*). The importance of this latter point cannot be overstated – in the United States alone this infrastructure represents an investment of more than 12 trillion dollars.

This timely book covers many different routes for utilizing a wide variety of biomass sources *via* a plethora of conversion technologies to afford several families of biofuels. Given the virtually unlimited thirst for energy (the world's largest industry by far) it is likely that many of these approaches will lead to commercial implementation, with the most efficient, scaleable and cost-effective ones having global impact and with others having impact in certain regions by virtue of geographical and/or political considerations. Among the former there is growing consensus that the most promising global-scale biomass solution is represented by microalgae since they are Mother Nature's most efficient practitioners of photosynthesis (the fixation of carbon dioxide), resulting in the highest yields of biomass and oils among all aquatic species, which are in turn an order of magnitude more efficient than terrestrial plants.

Topics included in this book range from biodiesel, made by transesterification of triglycerides (vegetable oils) using different catalysts and processes and utilization of the coproduced glycerol, to the hydrotreating of vegetable oils to afford renewable diesel and renewable jet fuel (more correctly HRJ or hydrotreated renewable jet).

Considerable attention is given to pyrolysis oil that is typically generated by rapid pyrolysis (residence times of a couple of seconds) at high temperature ($>450\,^{\circ}C$) in the absence of oxygen. Many feedstocks have been explored, from woody biomass, to microalgae to vegetable oils. The products of pyrolysis are pyrolysis oil (also known as bio-oil) and char (in addition to some lights). The char can be used as a form of "green coal", while pyrolysis oil can be used in boilers, furnaces and power plants. The oil can also be upgraded (typically using hydrotreating technology) to renewable diesel and possibly renewable jet fuel, although further development is envisaged before commercialization can be commenced.

The use of lignocellulosic feedstocks to generate fuels directly, or *via* the intermediate formation of sugars, is also presented.

Another area receiving significant attention in this work is the development of Fischer–Tropsch (FT) syngas (synthesis gas – a mixture of carbon monoxide and hydrogen) conversion processes. FT technology has long been promoted by the prospect of "stranded" natural gas, predominantly methane, being available at remote global locations at near-zero cost. More recently, increasing attention to low-cost, potentially abundant biomass sources, such as algae, switchgrass, corn stover and the like have added a new stimulus to FT technology development. It is well known that the economic viability of gas-conversion processes is largely determined by the capital cost and the market value of the final hydrocarbon products. It is also a widely accepted fact that the manufacture of syngas is by far the most capital-intensive part of the process, and so a competitive technology must combine low-cost feedstocks (stranded gas or cheap, abundant biomass), and cost-effective syngas generation (*via* gasification of the biomass) with efficient syngas conversion. Clearly, selectivity and efficiency are paramount in the design of a viable FT process, as is catalyst activity, which determines the reactor size, and catalyst lifetime, which directly influences the choice of reactor.

Finally, an emerging trend in the biomass to liquid transportation fuels (and chemicals) arena is the concept of an integrated biorefinery in which both the biology and the chemical engineering are brought together in one locale to complete the process – all the way from agriculture to fuel and chemical feedstocks, and in some cases to final consumer products. This trend is also reviewed in this timely work.

Brian L. Goodall
Sapphire Energy, Inc., San Diego, CA 92121, USA

Preface

Given the increasing world-wide demand for energy, coupled with concerns surrounding global warming and the instability of current petroleum supplies, there is a pressing need for low-cost, high-volume processes for the conversion of biomass to fuels and chemicals. Against this backdrop, this book aims to provide an up to date overview of the thermochemical methods available for biomass conversion to liquid (transport) fuels and chemicals. In so doing, this book is intended to provide researchers and students with a comprehensive introduction to this emerging field, while also functioning as a reference for those already active in the area.

In order to provide the reader with sufficient background information, the first three chapters deal with general aspects of the production of biofuels and chemicals from biomass, focusing on the rationale for biofuels, considerations governing the selection and cultivation of energy crops, and the concept of biorefineries. After this, pathways for the thermochemical valorization of biomass are considered in detail, starting with biomass gasification and subsequent conversion of the obtained syngas to fuels and chemicals via Fischer-Tropsch and alcohol synthesis routes. Biomass pyrolysis, representing an important technology for the direct conversion of biomass to liquid fuels, is then discussed, with emphasis on fast pyrolysis and hydrothermal processing; a discussion of lignin utilization is also included, representing a biomass constituent that to date has received relatively little attention in the context of bio-derived fuels and chemicals. Given that the crude bio-oils resulting from pyrolysis processes require upgrading before they can be refined to transport fuels, bio-oil upgrading via *in situ* (catalytic biomass pyrolysis) and *ex situ* (hydrotreating) approaches is also considered in detail. The conversion of cellulose to sugars and utilization of carbohydrates for the production of fuels and chemicals is next discussed, including a review of recent work employing ionic liquids for the utilization of lignocellulosics. The last five chapters concern

RSC Energy and Environment Series No. 1
Thermochemical Conversion of Biomass to Liquid Fuels and Chemicals
Edited by Mark Crocker

Published by the Royal Society of Chemistry, www.rsc.org

the upgrading of vegetable oils and fats for the production of liquid fuels, covering both existing and emerging processes.

As editor of this book, I am indebted to the participating authors who responded in such a positive fashion to my requests for contributions and who (mainly) adhered to their deadlines. Thanks are also due to the staff at the RSC, and in particular to Gwen Jones for her unflagging support of this project and to Sue Humphreys for her production expertise. At the Center for Applied Energy Research, Leslie Hughes provided invaluable assistance and somehow remained good natured even when reformatting references by the hundreds. Finally, as ever, my thanks go to Terri and Saskia.

Mark Crocker

Contents

RSC Energy and Environment Series No. 1
Thermochemical Conversion of Biomass to Liquid Fuels and Chemicals
Edited by Mark Crocker

Published by the Royal Society of Chemistry, www.rsc.org

CHAPTER 1
The Rationale for Biofuels

MARK CROCKER AND RODNEY ANDREWS

Center for Applied Energy Research, University of Kentucky, Lexington, KY 40511, USA

1.1 The Rise of Petroleum

Biomass, for most of history, has been the primary energy source powering human development. This energy supply has taken various forms, including wood and dung for cooking and heating, charcoal for metallurgy, and animal feeds for food and transportation. With increasing concerns regarding human impacts on the environment, humanity is once again looking towards biomass resources to meet a significant portion of our energy needs. The challenges today in using biomass are many, but can best be related to scale and density. The scale of energy needed far exceeds all past demands; both the increasing world population and the energy intensity of modern life compound the need for energy as never before. Similarly, the distances over which energy is moved and the concentration of population into dense urban centers results in the need for fuels with high energy density to insure overall efficiency of use.

Over the past century, the developed world has enjoyed cheap and abundant energy supplies through the adoption of a fossil energy economy. The 1900s have been declared the "Petroleum Century", with both positive and negative connotations. The widespread use of petroleum allowed rapid economic expansion throughout the industrialized world, increasing national and personal affluence, and enabled the modern ideal of personal automobile ownership. With expanded automobile ownership came an increasing demand for liquid transportation fuels, a demand that led to a shift in primary production

RSC Energy and Environment Series No. 1
Thermochemical Conversion of Biomass to Liquid Fuels and Chemicals
Edited by Mark Crocker

Published by the Royal Society of Chemistry, www.rsc.org

away from the consumer nations (chiefly the United States and Great Britain) to majority importation of fuels from more petroleum-rich areas of the world such as the Middle East and South America. As this shift from self-sufficiency to foreign dependence evolved over thc latter half of the 20th century, the formation of the Organization of Petroleum Exporting Countries (OPEC) led to price stabilization and a nearly centralized control in oil output. However, in recent years, OPEC's market control has begun to wane, with expanded production in non-OPEC regions (including Central Asian Republics, Africa and South America) increasing worldwide supply of petroleum and limiting pricing pressures.

This trend away from OPEC dominance in production has been mirrored by a broadened international competition for access to energy, especially petroleum. The United States, consuming more than 25% of worldwide petroleum production by midcentury, was viewed as the dominant consumer to be appeased (or not) in price setting with the international standard for pricing being set in US dollars. However, as the 21st century opened, expanding energy demand in the developing nations, especially China and India with their very large populations, resulted in significant competition for access to petroleum internationally. This increased competition has led to a general upward trend in price and increased market volatility. As non-OPEC production has begun to decline, cooperation between international oil companies (who traditionally supply North America and Europe) and nationalized oil companies in the developing world, along with a renewed fear of OPEC dominance in controlling worldwide supply, has resulted in a desire for decreased dependence on imported sources of energy.[1]

Along with the increased affluence that resulted from the widespread use of fossil energy to power the economies of the western nations came undesirable environmental impacts related to the gaseous emissions from fossil-fuel combustion. At the turn of the century a general upward trend in global carbon-dioxide levels in the atmosphere was recognized. This led to international concerns over global climate change and other devastating impacts that might result from an increase in "greenhouse-gas" levels, chiefly carbon dioxide and methane, in the Earth's atmosphere. This concern is heightened when taking into account the scale and speed with which the developing nations of Asia are expanding fuel use, particularly petroleum and coal. As these nations adopt western ideals in vehicle ownership, petroleum usage will continue to rise, along with concomitant emissions from fossil-fuel combustion, if alternative fuels are not developed on a large scale.

1.2 Worldwide Demand for Liquid Fuels and the Impact of Non-OECD Asia

Worldwide, liquid fuels represent the most heavily utilized source of energy, as shown in Figure 1.1. The dominance of liquid fuels in the overall marketed energy portfolio is expected to continue well beyond 2030.[2] This demand for

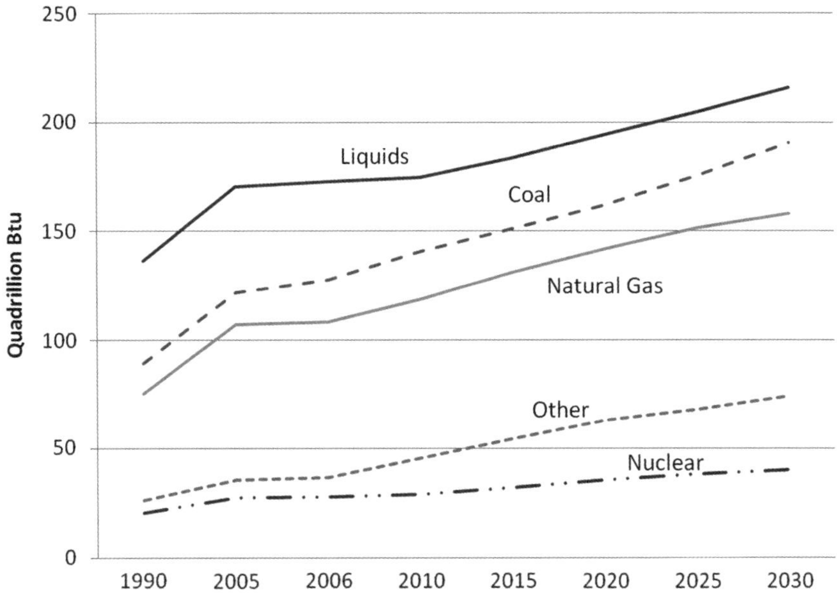

Figure 1.1 World marketed energy use by type.[2]

liquids can be attributed primarily to transportation uses (see Figure 1.2), including automotive, marine bunker, and aviation fuels. Transportation demand for liquid fuels is unique in that demand is relatively unaffected by increases in price.[2] The desire for personal transportation, Figure 1.3, well established in the Organization for Economic Cooperation and Development (OECD) member nations, is increasingly playing a significant role in determining worldwide demand for oil as the non-OECD nations begin to catch up in personal automobile ownership and industrialization of transportation, agriculture and manufacturing. Overall, crude oil demand is projected to grow by 1% per year on average over the next 2 decades, from 85 million barrels per day in 2008 to 105 Mbbl/d in 2030 (or, 4100 Mtoe to 5000 Mtoe; Mtoe = million metric tons of oil equivalent).[3] Similar to overall energy demand, this growth is expected primarily in non-OECD countries, particularly China, India, and non-OECD South Asia (Figure 1.4), while the developed world will see an overall decrease in demand for traditional petroleum feedstocks. The transportation sector is expected to account for 97% of this increase.[3]

To meet this demand, especially in the United States (the largest consumer of transportation fuels), biofuels are expected to play an increasing role in the liquid fuels mix. Biofuels production – biodiesel, ethanol, other alcohol fuels, and bio-oils – has grown substantially within the last decade, and is expected to accelerate over the next several decades from current levels of roughly 4 Mtoe to 133 Mtoe by 2030.[3] This increase in demand is the result of many factors, including the increasing price of traditional petroleum fuels, increasing demand for fuels in the non-OECD world, the implementation of national policies

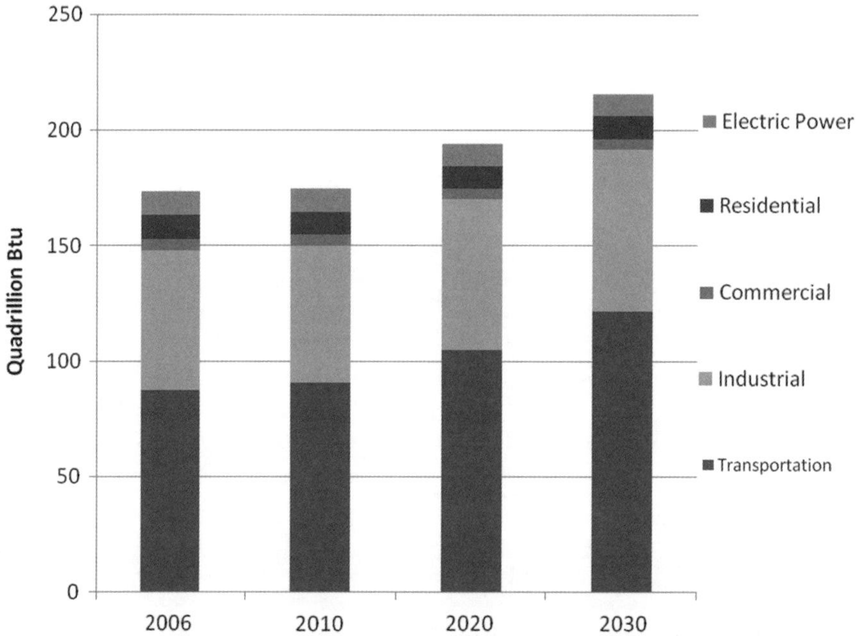

Figure 1.2 World liquid fuel consumption by sector.[2]

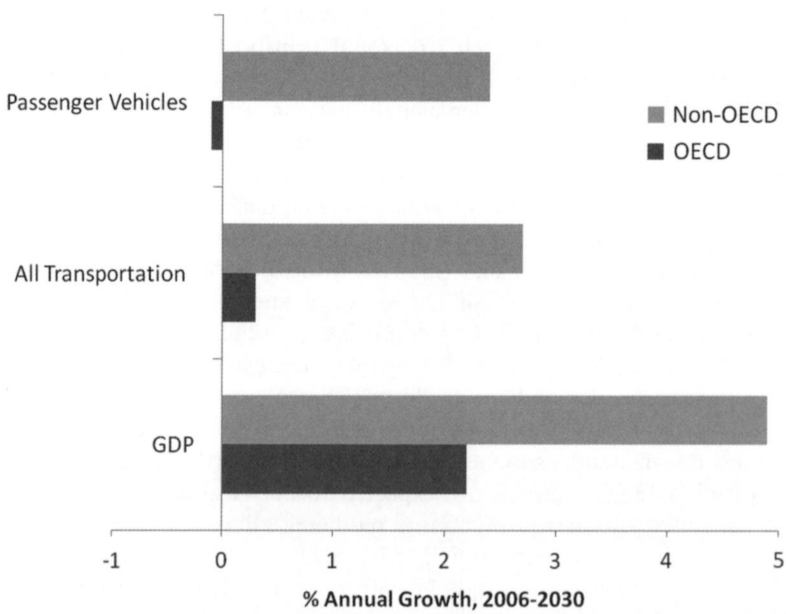

Figure 1.3 Economic growth and demand for passenger vehicles in non-OECD countries push demand for transportation fuels, 2006–2030.[2]

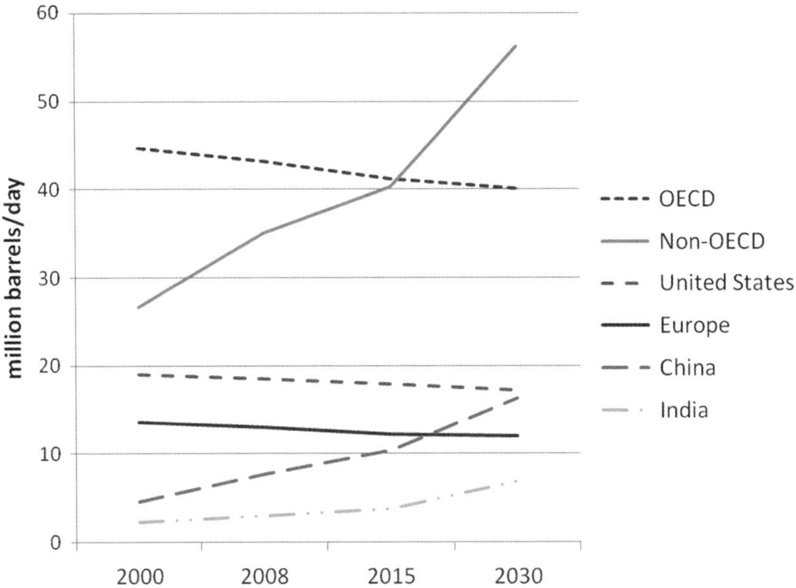

Figure 1.4 China and India drive new demand for oil.[3]

directed at reducing importation of fuels, global climate change legislation favoring renewable over traditional fossil resources, and (within the developed world) a willingness among segments of the public to pay a premium for renewable or green fuels.

1.2.1 Increasing Price and Decreasing Supply of Petroleum

Non-OECD Asia currently dominates the growth in liquid fuels demand over the near term with the remainder of the non-OECD world beginning to similarly exert pressure on demand growth in the next few decades. The traditional markets for petroleum in OECD Europe and, especially North America, will experience significant upward pressures with respect to the cost of petroleum.[2,3] While the average price for crude (as West Texas Intermediate, WTI) dropped significantly in late 2008 and again through 2009, the price is forecast to quickly rebound. Indeed, despite this drop from the all-time high of $147/bbl in June 2008 to approximately $40/bbl, within a few months the price quickly rebounded upward to end 2009 between $75 and $80/bbl. Longer term, the crude petroleum price will continue a steady increase to $130–140/bbl as a base price over the next several decades[2] (Figure 1.5), with some estimates raising the projected price to as high as $200/bbl.

The International Energy Agency (IEA) has recently completed an analysis of world ultimately recoverable petroleum reserves suitable for liquid fuels production.[3] The results of this analysis should give pause to both those

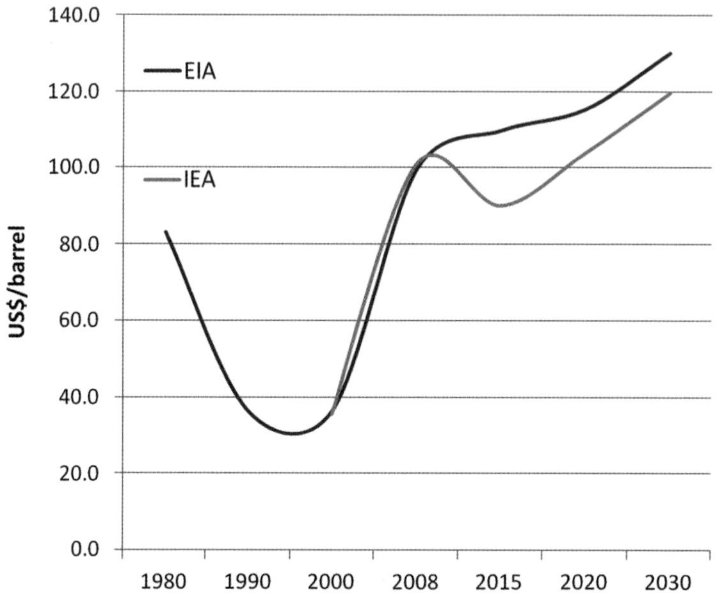

Figure 1.5 World crude oil price estimates from Energy Information Administration[2] and International Energy Agency,[3] adjusted to 2007 index.

predicting a near term "Peak Oil" event and those who believe sufficient reserves exist indefinitely. Essentially, IEA concluded that while nearly two-thirds of reserves remain in place, there will be an accelerated drop toward only one-half remaining by 2030. However, reserves of unconventional oil sources, such as tar sands and heavy oils, are very large and have only begun to be exploited. Assuming that current environmental and logistical constraints in their development can be overcome, these unconventional resources may play a significant role in meeting oil supply going forward, especially in North America. However, these new sources of oil will result in a sustained higher price for fuels,[3] perhaps helping to accelerate development of the more cost competitive biofuels while possibly reducing overall demand for oil importation in the United States.[1]

Overall, dependence on imports is forecast to grow for all consumer sectors except North America, where expansion of nonconventional oil resources (Canada) and biofuels (US) is projected to more than offset demand growth and reduce import dependence by 2030 back below 60%.[3] Projections for world production of unconventional resources are seen to expand from well below 5 Mbbl/d in 2006, to as high as 18 Mbbl/d in 2030 depending on the price of petroleum (Figure 1.6). Even with an unexpectedly slow growth in the price of oil, unconventional production, including biofuels, is expected to grow to well over 12 Mbbl/d.[3] Looking at projected North American production, it is of note that biofuels are expected to surpass the production from Canadian oil sands over the next two decades.

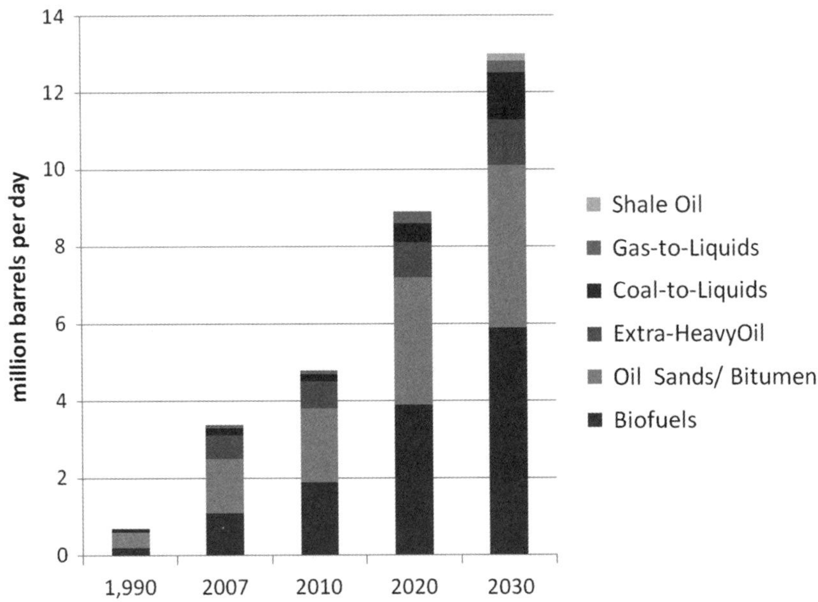

Figure 1.6 Biofuels will become largest source of unconventional liquid fuels.[2]

1.2.2 Instability in Supply and Production of Petroleum

A further pressure on worldwide supply of petroleum is the significant trend towards the nationalization of oil-production companies over the last decade, which is expected to continue. While some nationalized oil companies (NOCs) are effective, many, especially in the developing world, suffer from civil unrest, corruption, inefficiency and diversion of capital from the company to support social programs.[1] As such, this trend toward NOC control of worldwide oil production tends to introduce an unacceptable level of production uncertainty, resulting in price volatility.

More significant has been the emergence of non-OPEC countries as the primary source to meet new demand growth over the last decade. Middle East OPEC producers have seen an overall decline in surplus production capacity, and the need to supply growing demand within their own region is eroding their ability to meet new demand growth in existing markets, as well as the emerging Asian demand. Non-OPEC producers were able to begin making considerable strides in cutting into the OPEC monopoly and by 2000 were producing more than 60% of world oil (Figure 1.7).[3] Meanwhile, the continued decline in the number and size of new discoveries has driven up marginal development costs.

However, this reliance on non-OPEC oil production to meet new demand growth has been destabilizing overall, especially as these fields have begun to decline in productivity much more rapidly than the Middle East OPEC fields.[3] Non-OPEC conventional production (crude oil and natural gas liquids) has

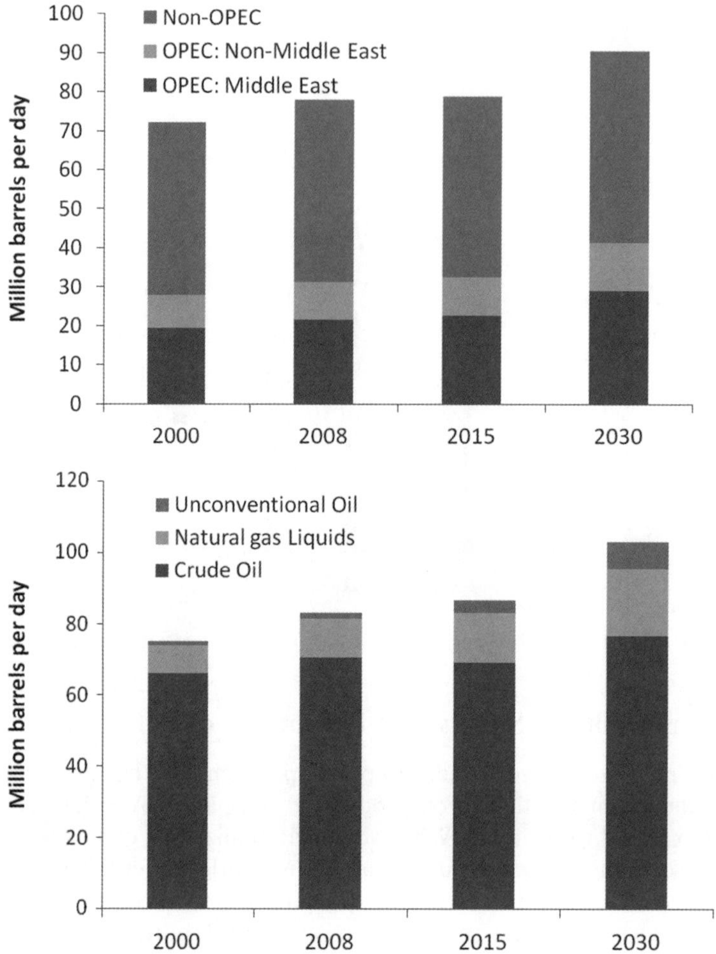

Figure 1.7 Oil production by region and source.[3]

been projected by IEA to peak around 2010 and then begin to decline slowly through 2030. Kazakhstan, Azerbaijan and Brazil are the only non-OPEC producing countries to see any significant increase in output. Non-OPEC conventional oil production is expected to drop by 330 thousand barrels per day (kb/d) between 2008 and 2011.[3]

As the non-OPEC producers begin to lose production capacity (or, at best, the ability to meet new demand growth), the Middle East OPEC countries are expected to see a rapid expansion in exports through the next two decades (see Figure 1.7). This return to the majority of world oil being produced within the OPEC countries will likely result in an overall tightening of oil supply on the world market as competition for new supplies increases to meet demand growth in the non-OECD countries, particularly Asia, and OPEC is once again able to determine worldwide supply levels and pricing.

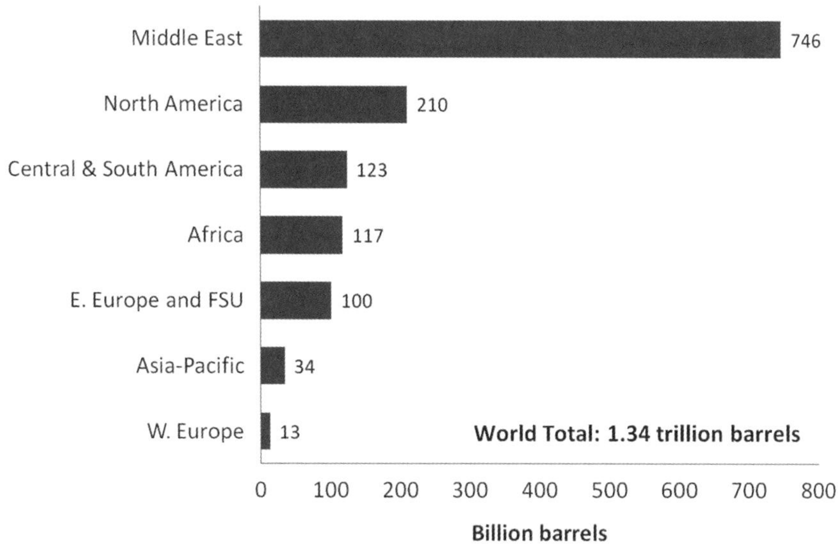

Figure 1.8 World estimated proved oil reserves (January 1, 2009).[4]

Of greater concern for OECD nations wishing to avoid dependence on foreign oil imports, and the associated volatility in consumer prices for transportation fuels, is the location of the Earth's remaining proven reserves of oil (Figure 1.8). Total proven world reserves of oil are up over the last few years, to 1.34 trillion bbl.[4] However, the majority of these reserves are in politically unstable regions of the world, including the Middle East, Africa, and Eurasia, with the increase of 10.5 billion bbl in total oil reserves between 2007 and 2008 coming in large part as a result of higher estimates for Libya and Venezuela, as reported by OPEC. Further exasperating fears over political instability, the Central and South American regional reserve value (123 billion bbl) is dominated by the 99.4 billion bbl reported by Venezuela.[4] While North America does have the second largest regional reserve base (see Figure 1.8), the majority of the endowment consists of Canadian oil sands (173 billion bbl). Further, it is unclear if environmental restrictions on the development of both conventional and unconventional reserves will limit access to development in North America over the long term. Such restrictions, which currently limit access to the Arctic and Coastal areas, may reduce the realization of full development of unconventional sources such as tar sands and oil shales as well.

At the rate of 2007 worldwide production, the world's proven reserves of conventional and nonconventional oil at current estimates would last about 51 years. As the economies of the world recover and the demand for oil increases, the reserve base becomes shorter if there are no new discoveries of significant fields. At the slightly increased 2008 production rate of 73 Mbbl/d, these reserves would last about 50 years.[4]

1.3 Forecast for Biofuels

As world demand for liquid transportation fuels grows, the surplus production capacity of traditional petroleum reservoirs is approaching a vanishingly small margin, resulting in upward pressure on the global price of petroleum. Coupling this with increasing volatility in the price and uncertainty of supply, many national programs encouraging domestic production of fuels have arisen, typically focused on the use of domestic biomass resources to meet demand for liquids.

Global biofuels production expanded 37% between 2006 and 2007, to 0.7 Mbbl/d (34.1 Mtoe) in 2007, and reached 0.8 Mbbl/d in 2008,[3] roughly accounting for 1.5% of total transportation fuels. Aggressive growth is expected to continue, driven primarily by new regulations and investment subsidies in the United States and, to a lesser extent, Europe. Despite the recent economic downturn, world use of biofuels is projected by IEA to recover in the longer term, reaching 1.6 Mbbl/d in 2015 and 2.7 Mbbl/d in 2030 (Figure 1.9). If these growth projections are met, biofuels would account for more than 5% of road transportation fuels by 2030, as well as meeting 1% of aviation energy demand. It should be noted that the Energy Information Agency (EIA) estimates for biofuels production are considerably higher than those of the IEA, primarily due to the prediction of a more rapid expansion of biofuels in North America.[2]

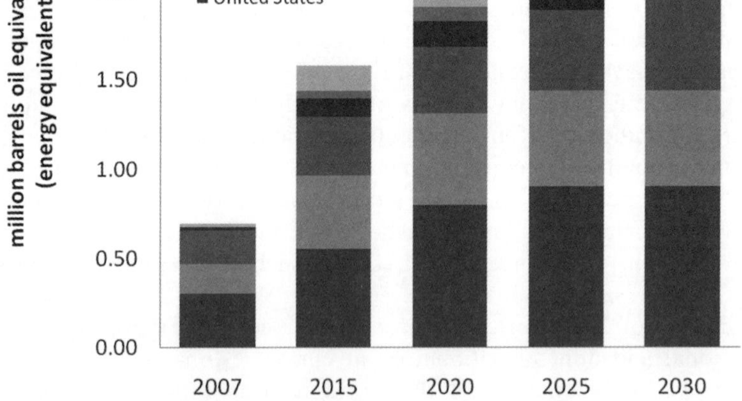

Figure 1.9 Biofuels demand by region.[3]

1.4 Biomass as a Renewable Source of Energy

Biomass can be utilized for the production of process heat, steam, motive power, and electricity, and can be converted by thermal or biological routes into a range of useful energy carriers such as liquid fuels and synthesis gas. The term biomass is used to describe any material of recent biological origin and includes plant materials such as trees, grasses and agricultural crops, as well as animal manure and municipal biosolids (sewage). As a raw material, biomass is a nearly universal feedstock due to its domestic availability and renewability. Indeed, until the widespread utilization of crude oil as an energy source in the 19th century, biomass supplied the majority of the world's energy needs. In one sense, the situation has now come full circle: as outlined above, concern over the environmental effects of fossil-fuel combustion, as well as disquiet about dwindling petroleum reserves – coupled with increasing global energy demand – have brought about a resurgence of interest in the utilization of biomass as an energy source.

The potential of biomass is significant. According to the European Biomass Industry Association, Europe, Latin America and Africa have the potential to produce respectively 8.9, 19.9 and 21.4 EJ of biomass per year, with a corresponding energy content equivalent to 1.4×10^9, 3.2×10^9 and 3.5×10^9 barrels of oil.[5] A joint study by the US Department of Agriculture and the Department of Energy has estimated that 1.3×10^9 metric tons of dry biomass could be sustainably produced in the US on an annual basis[6] (with an energy equivalence of 3.8×10^9 barrels of oil). For comparison, current US oil consumption amounts to ca. 8×10^9 barrels of oil per year. In some countries, biomass utilization as an energy source is already significant. Biomass has traditionally supplied a large fraction of the energy needs in many developing countries, while its use is well advanced in some Western countries, including Finland (where it supplies 20% of energy needs), Sweden (16%), Austria (13%) and Brazil (23%).[7,8]

As for any form of large-scale agriculture, the production of biomass as an energy source can have both positive and negative consequences. The benefits most frequently cited for biomass utilization as an energy source can be summarized as follows:[9,10]

Fuel-supply diversification: biomass is a widespread resource, the utilization of which can diversify the fuel supply and in turn lead to an energy supply that is more secure, *i.e.* less subject to geopolitical constraints.

Reduction of greenhouse gases: substituting fossil fuels with biofuels can lead to a reduction in greenhouse gases, principally CO_2. This stems from the fact that the amount of CO_2 released from combustion of the biomass is inherently equal to that fixed by the plant during its growth. However, the extent of greenhouse-gas reduction is dependent on a number of factors, including the degree to which fossil fuels are used in the production and distribution of biofuels. Another consideration is the extent to which stored CO_2 is released to the atmosphere during the clearing and tilling of grassland and forests that are to be used for biomass cultivation.[11]

Increased rural income: the increased agricultural activity associated with the production of energy crops can generate employment in rural areas and result in increased farm income, thereby reversing the trend of rural depopulation prevalent in many countries.

Restoration of degraded land: If appropriate crops are selected, degraded land (currently unsuited for the production of food crops) can be utilized for biofuels production. In the US, for instance, several projects are underway to utilize former mine-land for the production of energy crops such as miscanthus.

While the foregoing benefits are considerable, it must be appreciated that many factors must be taken into consideration when evaluating the merits of a particular biofuel. Only through a full life-cycle analysis can the sustainability of biofuel production be assessed. Further, as implied above, there is real uncertainty over the degree to which biofuels actually contribute to greenhouse-gas reduction. Although a detailed discussion of these issues is beyond the scope of this book, it is illustrative to consider the net energy ratio (NER) of common biofuels, that is, the energy contained in the fuel, along with possible coproducts, divided by the energy required to produce them. Figure 1.10 summarizes the NER values reported for some representative biofuels.[12–16] Most estimates[12–14] suggest that for every unit of energy expended in ethanol production from corn starch, approximately 1.2–1.5 units are returned (although some ethanol critics have argued that it actually has a negative energy balance[17]). The energy balance of sugarcane-derived ethanol is more favorable, since the bagasse produced can be burned to provide process energy,

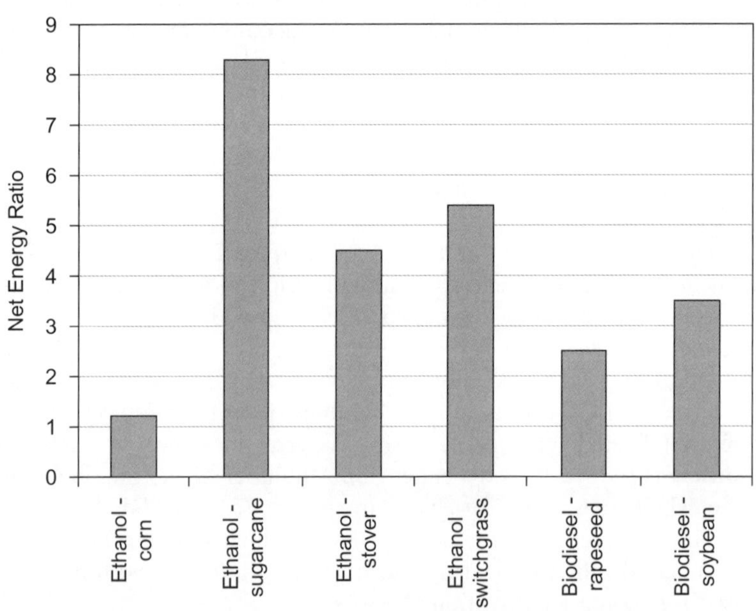

Figure 1.10 Net energy ratio of different biofuels.

thereby obviating the need for additional energy. In this case, the NER is calculated to be 8.3.[14] Preliminary estimates also suggest a NER significantly greater than 1 for cellulosic ethanol, provided that the lignin coproduct is burned to provide necessary energy for the process.[15] In the case of biodiesel, the NER is also generally favorable, but varies with the feedstock employed; for biodiesel derived from soybean oil, a NER of 3.5 has been reported.[16] Given the wide range of net energy ratios for these biofuels, it follows that there is an equally wide range of carbon and greenhouse gas benefits that these biofuels provide.[14]

1.4.1 Biomass Composition

Plants capture solar energy as fixed carbon, converting CO_2 and water to sugars, $(CH_2O)_x$:

$$CO_2 + H_2O + \text{light} \rightarrow (CH_2O)_x + O_2 \tag{1.1}$$

The sugars thus produced are stored in three different types of polymers: cellulose, hemicellulose and starch. Biomass is typically composed of 65–85 wt% sugar polymers (principally cellulose and hemicellulose), with another 10–25 wt% corresponding to lignin. Other biomass components that are generally present in minor amounts include triglycerides, sterols, alkaloids, resins, terpenes, terpenoids and waxes (often collectively referred to as lipids), as well as inorganic minerals. In the case of seeds and certain algae strains, significant amounts of oil can be present, corresponding mainly to triglycerides. This is exemplified by soybeans (ca. 20 wt% oil), rapeseed (ca. 40 wt% oil) and oil-palm fruit (ca. 50 wt% oil), which together account for the majority of the feedstock currently used in biodiesel production.

Together, cellulose, hemicellulose and lignin constitute lignocellulose, the fibrous material that forms the cell walls of plants and trees. The cellulose forms bundles of fibers that provide strength. The chemical structure of cellulose (Figure 1.11) corresponds to a linear polymer of D-glucopyranose monomers containing β-1,4 linkages.[18] Typically, each polysaccharide chain contains between 5000 and 10 000 glucose units. Intra- and intermolecular hydrogen bonds allow the hydrophobic ribbon faces to stack, forming a rigid, flat network.[19] The crystalline nature of this structure renders cellulose completely

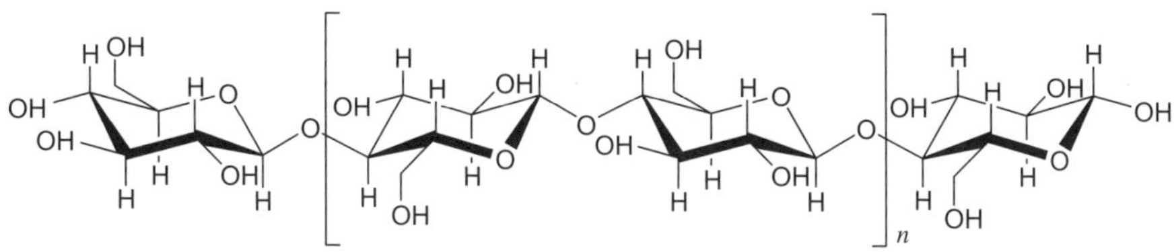

Figure 1.11 Simplified chemical structure of cellulose. See Chapter 13 for a more detailed description.

insoluble in water, although it can be broken down by acid hydrolysis; further details are provided in Chapter 13 of this book. Hemicellulose is also a sugar polymer, which occurs in close association with cellulose. Unlike cellulose, which is a polymer of glucose, hemicellulose contains a variety of hexoses (*e.g.*, glucose, galactose, mannose) and pentoses (usually xylose and arabinose), all of which are highly substituted with acetic acid. Other differences are the relatively low number of saccharide monomers present (~ 150) and the fact that short side chains can occur along the main polymer chain. This latter feature renders hemicellulose amorphous, with the consequence that it is more readily hydrolyzed to its constituent sugars than cellulose.

Lignin, which constitutes the third main component of biomass, is an amorphous, three-dimensional polyphenolic material with no exact structure.[20] Lignin fills the spaces in the cell wall between the cellulose and hemicellulose. It is covalently linked to hemicellulose and confers mechanical strength to the cell wall and by extension the plant as a whole. As shown in Figure 1.12, the three main structural units occurring in lignin correspond to coniferyl, sinapyl and *p*-coumaryl alcohol.

During the biosynthesis of lignin, these monomers undergo enzyme-initiated free-radical polymerization and eventual crosslinking. The complexity of the lignin structure derives from the fact that bonding can occur at many different sites in the monomer units due to delocalization of the radical's electron in the aromatic ring, the double-bond-containing side chain and the oxygen functionalities.[22] Further, the relative distributions of the three main monomer types can vary with the plant type. Softwood lignin is formed mainly from coniferyl alcohol, while hardwood lignin contains coniferyl and sinapyl alcohol as monomer units. Grass lignin typically contains coniferyl, sinapyl and *p*-coumaryl alcohol.[23] For a more detailed discussion of the structure and formation of lignin the reader is referred to Chapter 9.

The van Krevelen plot in Figure 1.13 compares the elemental composition of sugars, lignin and lipids (for which triglycerides are taken to be representative). The energy content of the different biomass constituents increases with decreasing oxygen content and also tends to increase with increasing H:C ratio. Hence, energy content per unit mass follows the order lipids > lignin > sugars. Although lipids are the ideal starting material for the production of biofuels, future large-scale production of biofuels will have to be based on the utilization of lignocellulose as the principal feedstock, owing to the latter's relative abundance.

In general, sugar polymers such as cellulose and starches can be readily broken down to their constituent monomers by hydrolysis, preparatory to conversion to ethanol or other chemicals. In contrast, lignin is less readily degraded. Indeed, the relative chemical inertness of lignin has profound consequences: since lignin occurs principally in the walls of lignocellulosic materials, where it provides mechanical rigidity and protection from chemical or biological attack, extremely forcing conditions are required in order to degrade it and render the cellulose and hemicellulose inside the plant accessible to acid hydrolysis. For this reason, current bioethanol production is based almost

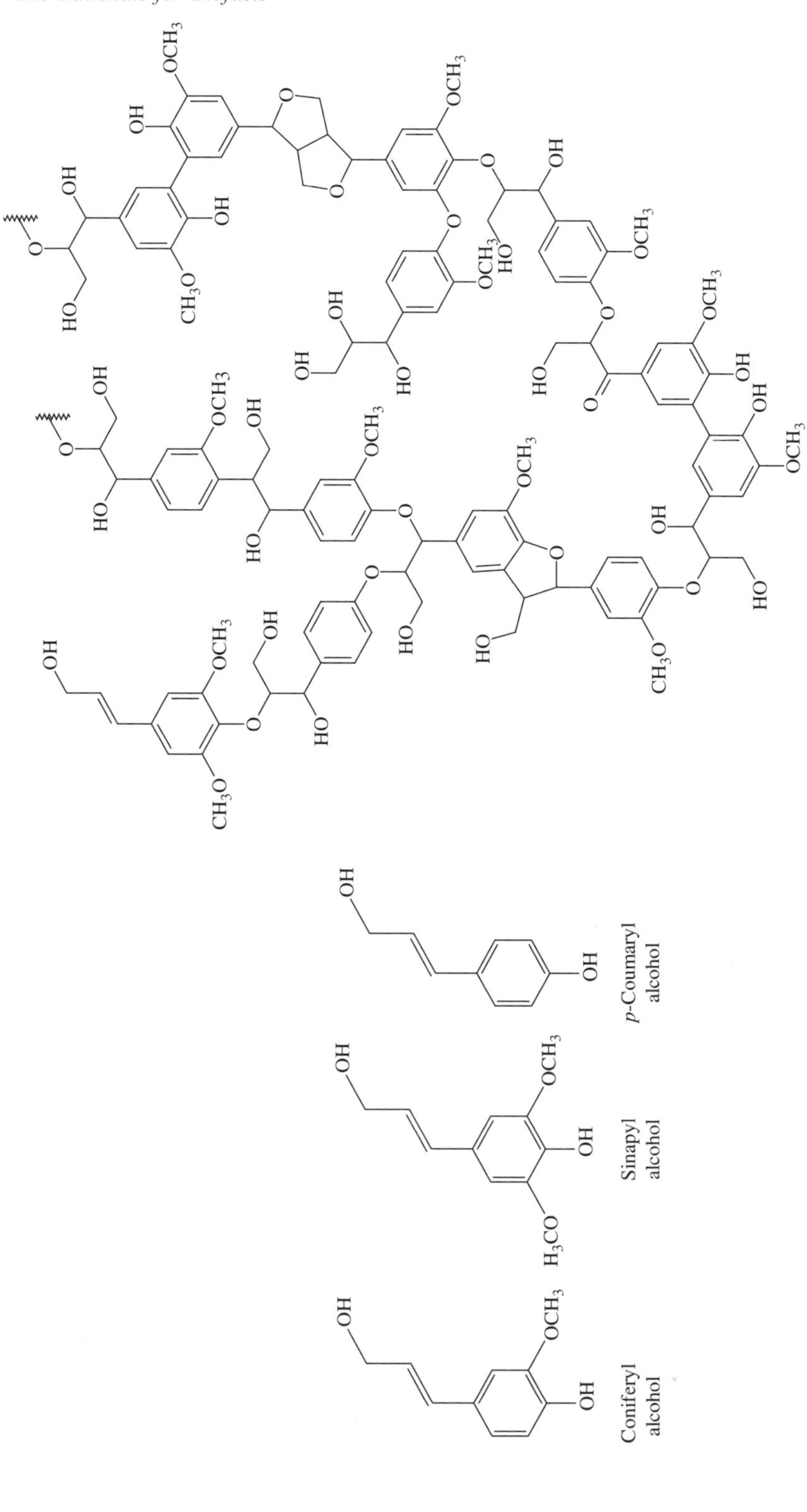

Figure 1.12 Coniferyl, sinapyl and p-coumaryl alcohol structures and partial structure of a typical softwood lignin molecule (after ref. 21). See Chapter 9 for a more detailed description.

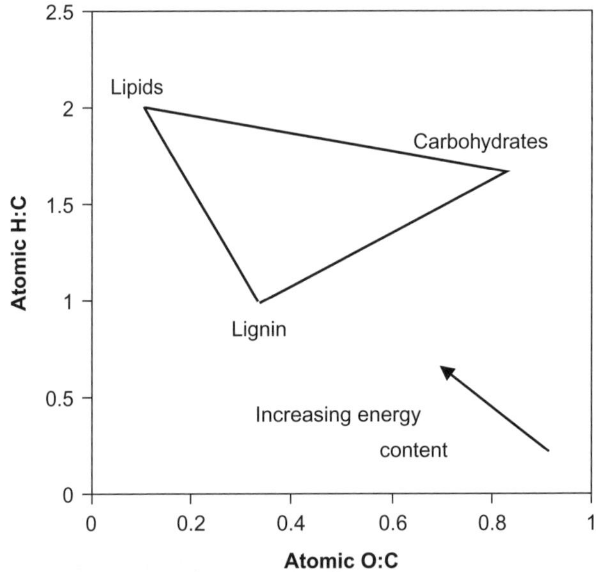

Figure 1.13 Van Krevelen plot for principal biomass constituents (adapted from Sun *et al.*[24]).

entirely on the fermentation of sugars that are readily obtained from the starch in corn grain, in addition to the sugar in sugarcane and sugar beets. In the long term, however, it is clear that future biofuels production will have to be based mainly on lignocellulosic feedstocks owing to their relative abundance compared to available feedstocks that are rich in starches or simple sugars. A further advantage of biofuels production based on lignocelluosic crops is the avoidance of the "food *versus* fuel" conundrum, given that the crops in question (*e.g.*, switch grass, miscanthus, willow, poplar, *etc.*) can typically be grown on marginal land that is unsuited for the production of food crops.[25]

1.4.2 Energy Density of Biomass

Although the utilization of biomass as an energy source offers a number of advantages, there are also limitations. The low density and high water content of biomass makes shipping costs prohibitive in many cases, yet most subsequent refining processes require centralized facilities, where large-scale operations greatly improve process efficiencies and economics. As shown in Table 1.1, the energy density of biomass is low compared to that of most fossil fuels. Indeed, the heat content of biomass, on a dry basis, is at best comparable with that of a low-rank coal or lignite, and substantially lower than that of anthracites, most bituminous coals and petroleum.[26–29] Furthermore, most biomass, as harvested, contains significant amounts of physically adsorbed moisture, up to 50% by weight. Thus, without substantial drying, the energy content of a biomass feed per unit mass is even less.

From the foregoing it follows that the transportation costs of biomass constitute a key component of the overall cost of recovering energy from biomass. Traditional technologies to increase biomass bulk density include baling, cubing and pelletizing. In addition, onsite torrefaction has been proposed,[30] as well as biomass conversion to a crude bio-oil *via* fast pyrolysis.[26] Torrefaction involves heating biomass in the absence of oxygen to a maximum temperature of 300 °C, resulting in what is essentially mild pyrolysis. The treatment yields a solid, uniform product with greatly reduced moisture content and reduced O/C ratio (on a dry basis), resulting in increased energy content relative to the starting biomass.[31,32] By combining torrefaction with pelletization, very energy dense fuel pellets can be produced.[30]

Similar to torrefaction, fast pyrolysis involves heating biomass in the absence of air, although higher temperatures (450–550 °C) and much shorter contact times (typically 1–2 s) are used. During this process the cellulose, hemicellulose and lignin present are depolymerized, while the subsequent rapid quenching "freezes in" the intermediate products of these processes before they can degrade further or condense with other molecules.[33] The product bio-oil is a mobile liquid that comprises a complex mixture of oxygenated compounds. Based on the heating value of bio-oil (16–17 MJ/kg) and density (1.2 kg/L),[34] its volumetric energy density is typically at least six times that of raw biomass such as chipped wood, as shown in Table 1.1.

Table 1.1 Energy density of selected biomass types and fossil fuels.

Fuel	*Moisture content (%)*	*Energy density by mass, GJ/tonne*	*Bulk density, kg/m³*	*Energy density by volume, GJ/m³*
Miscanthus	10	15.8	140	2.2
Wheat straw	15	14.4	100	1.4
Sawdust	6	15.2	160	2.4
Forest wood chip (dry)	40	10.5	240	2.5
Forest wood chip (fresh)	55	7.2	310	2.2
Log wood (ash, air dry)	20	14.7	400	5.9
Cubes (grasses)	8	17.8	450	8.0
Wood pellets (dry)	10	17.5	650	11.4
Torrified wood pellets	3	22	650	14.3
Pyrolysis oil		16–17	1200	19.2–20.4
Anthracite coal		>32.5	800–929	>26
Bituminous coal		>30.2	673–913	>20
Lignite coal		<19.3	641–865	<16.7
Petroleum		42	870	36.5

Based on the foregoing, it can be concluded than on the basis of energy density alone (*i.e.* neglecting likely differences in handling costs), the cost of transporting crude bio-oil should be approximately six times less than that of transporting wood chips or other types of raw biomass (at equivalent energy content). This suggests that biomass transportation costs could be reduced by converting biomass to bio-oil at the point of harvest, and then transporting the bio-oil to a centralized biorefinery for further processing into fuels and chemicals or to a location where it can be directly utilized (such as a power plant). Indeed, in one study,[35] in which the economics of transporting bio-oil to a central power plant was investigated and compared to the corresponding costs for raw biomass (miscanthus), biomass conversion to bio-oil was found to be cost effective when the generation plant was sufficiently large.

A further attribute of bio-oil is that it can be used as a feedstock for many of the same processes as chipped wood and can substitute for fuel oil or diesel in many static applications, including boilers, furnaces, engines and turbines for electricity generation. Bio-oil is particularly attractive for cofiring because it is easier to burn than raw biomass, and its higher energy density means that a greater amount of bio-oil can be used as a fraction of the total fuel; for wood chips, a maximum of 20 wt% can be used (and in practice 10 wt% is typically used) on the basis that higher amounts would require the boiler to be deregulated.[36]

1.4.3 Overview of Pathways for Biomass Conversion to Fuels

As indicated above, although biomass can be utilized in its raw state as an energy source, for example, by cocombustion with fossil fuels for electricity generation, there are compelling reasons to find ways of converting biomass into liquid fuels. At this juncture, it should be noted that in terms of efficiency, a strong case can be made for confining biomass use to electricity generation. For example, a recent study comparing the efficiency of using biomass to power vehicles through either cellulosic ethanol production or bioelectricity production concluded that bioelectricity outperforms ethanol across a range of feedstocks, conversion technologies and vehicle classes.[37] However, owing to their high energy density and ease of shipping and distribution, liquid fuels are far more versatile than solid fuels and are utilized in a far wider range of applications, including transportation, domestic heating, electricity generation, and potentially, as a hydrogen carrier.[38] Furthermore, some energy uses, such those employing diesel and jet engines, require fuel possessing high energy density, which current and foreseeable batteries cannot achieve.[39]

As shown in Figures 1.14 and 1.15, biomass can be converted to liquid fuels through a range of pathways. In the case of lignocellulosic biomass (Figure 1.14), there are three main approaches that can be considered.[5,40] In the first of these, biomass is gasified to produce syngas, which after treatment to remove tars and other impurities, can be catalytically converted to hydrocarbons (*via* the Fischer–Tropsch synthesis) or to alcohols. Other options

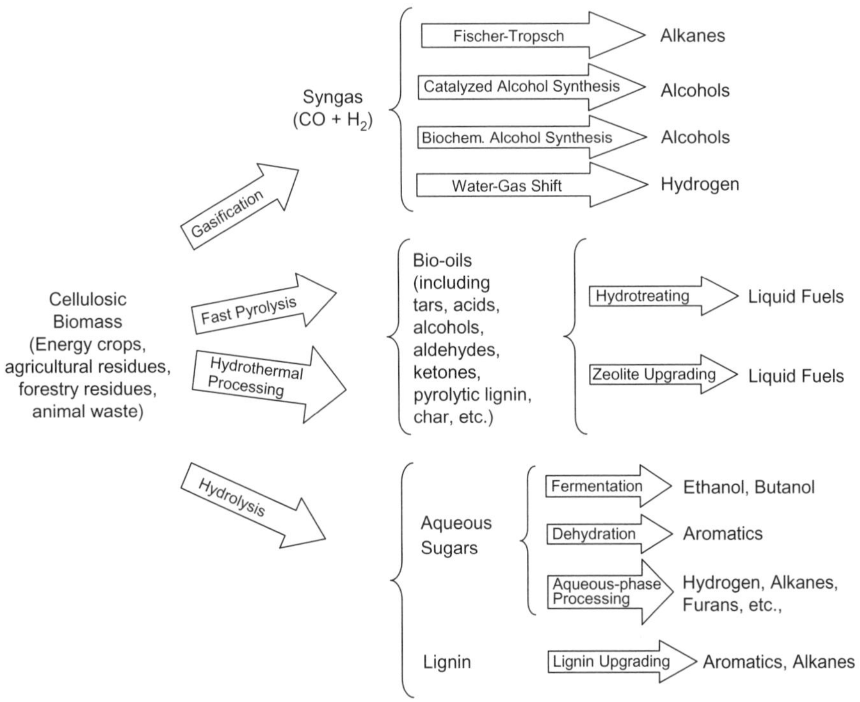

Figure 1.14 Summary of pathways for cellulosic biomass conversion to liquid fuels (adapted from Huber and Dumesic[40]).

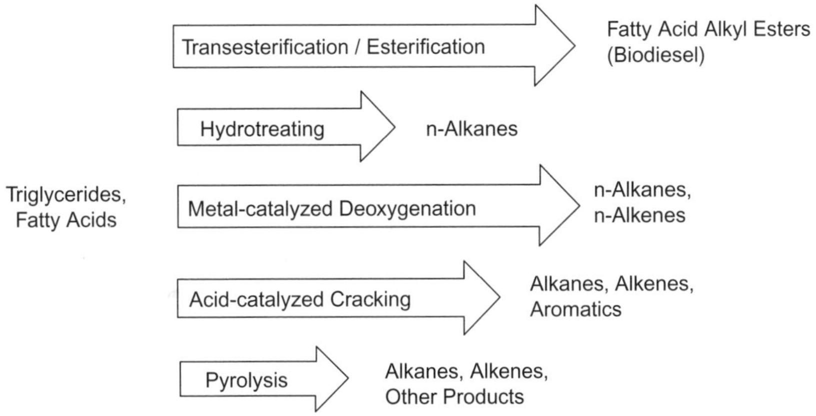

Figure 1.15 Technologies for conversion of bioderived fats and oils to liquid fuels.

include syngas conversion to ethanol using certain types of anaerobic bacteria, or treatment with water-gas shift catalysts to generate additional hydrogen from the water and carbon monoxide present. The resulting hydrogen can be used in a variety of refinery and chemical processes, as a fuel for PEM fuel cells, *etc.*

A second approach involves the use of direct thermochemical conversion processes, in which biomass is heated in the absence of air to produce a crude bio-oil. One such process is fast pyrolysis, in which the production of liquids is maximized by the use of short residence times (1–2 s) and high processing temperatures (*ca.* 500 °C).[41] In solvolysis, which is often referred to as hydrothermal upgrading when the solvent used is water, biomass is typically treated for 5–20 min with water under subcritical conditions (300–350 °C, 10–18 MPa).[42] A third option is high-pressure liquefaction. This is also a single-step process, which utilizes longer residence times and lower temperatures than fast pyrolysis (*ca.* 300–400 °C), hydrogen pressures of up to 20 MPa, and typically requires the addition of a catalyst.[43] Due to the hydrogen requirement, the economics of this process are generally considered to be inferior to those of fast pyrolysis or solvolysis.[44] A commonality of the crude bio-oils resulting from these direct conversion processes is that they cannot be used as transportation fuels without subsequent upgrading due to their high oxygen content (up to 40 wt% on a dry basis) and water content (up to 25 wt%). Upgrading typically requires either hydrotreating or catalytic cracking to eliminate the oxygen functionalities present in the oil.[45,46]

The third approach to lignocellulosic biomass conversion to liquid fuels involves the use of enzymes or acid- or base-catalyzed hydrolysis to decompose the sugar polymers present into their constituent monomers. These sugars can then be fermented to alcohols, dehydrated over acid catalysts to give aromatics,[47] subjected to catalytic aqueous phase processing to give a variety of fuel-type products such as furans, alkanes and hydrogen (depending on the process),[40,48,49] or converted to a variety of chemicals.[49] Efficient processes for utilization of the lignin remaining after removal of the sugars have yet to be found, although catalytic conversion of lignin to aromatics and alkanes (by hydrotreating) or aromatics and coke (by catalytic cracking) has been demonstrated.[5,50]

Finally, as shown in Figure 1.15, a number of routes exist for the conversion of triglycerides and fatty acids to liquid fuels and chemicals. Transesterification of triglycerides and esterification of fatty acids to fatty-acid methyl esters (biodiesel) is currently the most widely used method for their conversion to transportation fuels,[51] although hydrotreating is gaining in commercial importance.[52] Metal-catalyzed deoxygenation (*via* decarboxylation and/or decarbonylation), acid-catalyzed cracking and thermal cracking (pyrolysis) represent additional options.[53,54]

1.4.4 Comparison of Thermochemical and Biological Processes for Biomass Conversion to Fuels

Routes for the conversion of biomass to liquid fuels can be classified into two main types: those based on thermochemical processes and those utilizing biological means. Thermochemical processes use heat, frequently in combination with heterogeneous catalysts, to break biomass down into smaller, constituent units, which are then further processed into useful molecules.[55] Biological processes, in contrast, are typically based on the use of enzymatic catalysts under mild conditions, the most obvious example being the fermentation of sugars to ethanol. Much of the current research in this field is focused on the development of biological routes for the depolymerization of cellulose and hemicelluloses and conversion of the resulting sugars (hexoses and pentoses) to alcohols.

Table 1.2 summarizes the main characteristics inherent to these two approaches, as applied to the conversion of lignocellulose to liquid biofuels. As a generalization, biological processes are more selective than thermochemical approaches, although the former are (to date) unable to utilize the lignin portion of biomass. From this it follows that future biorefineries are likely to use a combination of biological and thermochemical processes to make biofuels. This is implicitly recognized in the US Department of Energy's biorefinery concept (Figure 1.16).[56]

This conceptual biorefinery is built on two different "platforms" to promote different product slates. The "sugar platform" is based on the breakdown of cellulosic biomass into raw component sugars using chemical and biological means. The objective is to produce inexpensive sugar streams that can be used to make a range of fuels (*e.g.*, alcohols), chemicals and other materials that are cost competitive with conventional commodities. The residues can be used for power generation (cofiring) or to make other products (*e.g.*, *via* gasification). The "thermochemical platform" aims at converting biomass or biorefinery residues, *e.g.*, lignin, to intermediates such as pyrolysis oil and syngas. These intermediates can be used directly as raw fuels, or may be upgraded to produce fuels and chemicals that are interchangeable with existing commodities such as liquid transportation fuels and hydrogen.

Table 1.2 Characteristics of thermochemical and biological processes for lignocellulose conversion to liquid fuels (adapted from ref. 55).

	Thermochemical pathways	*Biological pathways*
Feed	Lignocellulose	Cellulose-derived sugars or syngas
Products	Hydrocarbons (many types) or alcohols	Alcohols
Selectivity	Varies, depending on specific pathway	Typically very selective
Reaction conditions	300–1200 °C, 1–250 atm	$<$70 °C, 1 atm
Residence time	0.01 s–30 min	$>$48 h
Reuse of catalysts	Possible for heterogeneous catalysts	Not possible for biological catalysts
Technical/economic hurdles	High energy input (high temperatures required); selectivity to desired products often low	Cellulose to sugar conversion difficult/expensive (high cost of enzymes); difficult to utilize pentoses; large plants required (due to long residence times)

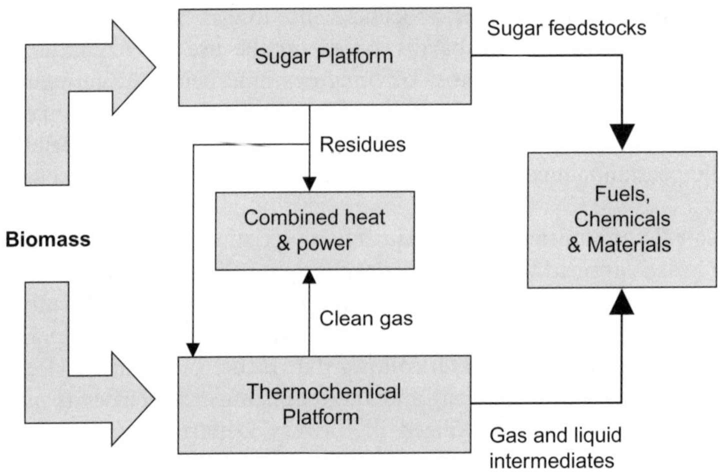

Figure 1.16 Simplified schematic illustrating the biorefinery concept (adapted from ref. 56).

Another important difference between thermochemical and biological conversion processes is the scale at which they are economically viable. Thermochemical processes, with their short residence times and relatively small reactors, are easily scaled, whereas the long residence times and large reactors of biological processes dictate that cellulosic ethanol plants based on fermentation are uneconomic at small scale. Consequently, thermochemical processes provide the greatest opportunity for small-scale distributed biomass processing, *i.e.* close to the point of biomass collection.

1.4.5 Outlook

The utilization of biomass represents one approach, among many that will be required, for meeting future energy needs and reducing greenhouse-gas emissions. Although biofuels offer great promise, their production and use is not without potential drawbacks. Currently, there is real uncertainty over the extent – if any – to which some biofuels contribute to greenhouse-gas reduction and the degree to which their production is truly sustainable. Although biofuels rightly have their critics, at the same time it must appreciated that it is unlikely there will ever be a "perfect" energy source; each potential alternative to the use of fossil fuels, including other renewable-energy sources such as solar, wind and tidal energy, suffers from both economic and environmental drawbacks.

Although biomass can be utilized in its raw state as an energy source, there are compelling reasons to find ways of converting biomass into liquid fuels. Owing to their high energy density and ease of shipping, transport and distribution, liquid fuels are far more versatile than solid fuels and consequently are more widely applied. As outlined above, multiple options exist for the production of liquid fuels from raw biomass, utilizing either thermochemical or biological approaches. In the following chapters of this book, the main thermochemical routes to liquid biofuels are considered in detail.

References

1. P. Dittrick, *Oil and Gas J.*, 2008, **106**(48), 24.
2. Energy Information Agency, International Energy Outlook, 2009, United States Department of Energy.
3. International Energy Agency, World Energy Outlook, 2009, IEA.
4. M. Radler, *Oil and Gas J.*, 2008, **106**(48), 20.
5. G. W. Huber, S. Iborra and A. Corma, *Chem Rev.*, 2006, **106**, 4044.
6. R.D. Perlack, L.L. Wright, A. Turhollow, R.L. Graham, B. Stokes, D.C. Erbach, "Biomass as Feedstock for a Bioenergy and Bioproducts Industry: The Technical Feasibility of a Billion-Ton Annual Supply," Report No. DOE/GO-102995-2135, Oak Ridge National Laboratory, 2005.
7. European Biomass Industry Associations, http://www.eubia.org
8. D. L. Klass, in: *Encyclopedia of Energy*, C. J. Cleveland, (ed.), Elsevier, London, 2004, Vol. 1.
9. E. C. Petrou and C. P. Pappis, *Energy & Fuels*, 2009, **23**, 1055.
10. P. McKendry, *Bioresource Technol.*, 2002, **83**, 37.
11. T. Searchinger, R. Heimlich, R. A. Houghton, F. Dong, A. Elobeid, J. Fabiosa, S. Tokgoz, D. Hayes and T.-H. Yu, *Science*, 2008, **319**, 1238.
12. B. Childs and R. Bradley, *Plants at the Pump. Biofuels, Climate Change, and Sustainability,* World Resources Institute, Washington DC, December 2007.
13. J. Hill, *Argon. Sustain. Dev.*, 2007, **27**, 1.
14. A. E. Farrell, R. J. Plevin, B. T. Turner, A. D. Jones, M. O'Hare and D. M. Kammen, *Science*, 2006, **311**, 506.
15. M. R. Schmer, K. P. Vogel, R. B. Mitchell and R. K. Perrin, *Proc. Nat. Acad. Sci.*, 2008, **105**, 464.
16. J. Sheenan, V. Camobreco, J. Duffield, M. Graboski, H. Shapori, "Life Cycle Inventory of Biodiesel and Renewable Petroleum Diesel for Use in an Urban Bus", Final Report, National Renewable Energy Laboratory, NREL/SR-580-24089 UC Category 1503, May 1998.
17. D. Pimentel and T. Patzek, *Natural Resources Research,* Springer, The Netherlands, 2005.
18. C. E. Wyman, S. R. Decker, M. E. Himmel, J. W. Brady, C. E. Skopec and L. Viikari, in: *Polysaccharides,* 2nd edn, S. Dumitriu, (ed.), Marcell Dekker, New York, 2005.
19. P. Zugemaier, *Prog. Polym. Sci.*, 2001, **26**, 1341.
20. J. McCarthy and A. Islam, in: *Lignin: Historical, Biological and Materials Perspectives,* 2nd edn, W. G. Glasser, R. A. Northey, T. P. Schultz, (ed.), ACS Symp. Series 742, 2000, 2.
21. A. Streitwieser and C. H. Heathcock, *Introduction to Organic Chemistry,* MacMillan, New York, 1981.

22. E. Adler, *Wood Sci. Technol.*, 1977, **11**, 169.
23. R. J. Evans, T. A. Milne and M. N. Soltys, *J. Anal. Appl. Pyrolysis*, 1985, **9**, 207.
24. L. Sun, E. M. Perdue, J. L. Meyer and J. Weis, *Limnol. Oceanography*, 1997, **42**, 714.
25. L. K. Paine, T. L. Peterson, D. J. Undersander, K. C. Rineer, G. A. Bartelt, S. A. Temple, D. W. Sample and R. M. Klemme, *Biomass Bioenergy*, 1996, **10**, 231.
26. P. C. Badger and P. Fransham, *Biomass Bioenergy*, 2006, **30**, 321.
27. Biofuels B2B website, http://www.biofuelsb2b.com/useful_info.php?page=Typic
28. Biomass Energy Foundation, http://www.woodgas.com/fuel_densities.htm
29. The Physics Factbook™, ed. G. Elert, http://hypertextbook.com/facts/2003/JuliyaFisher.shtml
30. P. C. A. Bergman and J. H. A. Kiel, Energy Centre of the Netherlands Report ECN-RX–05-180, 2005.
31. M. J. Prins, K. J. Ptasinski and F. J. J. G. Jansen, *Energy*, 2006, **31**, 3458.
32. M. Pach, R. Zanzi and E. Björnbom, Proc. 6th Asia-Pacific Intl. Symp. on Combustion and Energy Utilization, May 2002, Kuala Lumpur.
33. D. Mohan, C. U. Pittman Jr. and P. H. Steele, *Energy & Fuels*, 2006, **20**, 848.
34. A.V. Bridgwater, Biomass Pyrolysis, IEA Bioenergy report: T34:2007:01.
35. J. G. Rogers and J. G. Brammer, *Biomass Bioenergy*, 2009, **33**, 1367.
36. S. Hayter, S. Tanner, K. Comer, C. Demeter, Biomass Cofiring in Coal-Fired Boilers, US DOE, Energy Efficiency and Renewable Energy, Federal Energy Management Program, Federal Technology Alert, 2004, available at: www.eere.energy.gov/femp/
37. J. E. Campbell, D. B. Lobell and C. B. Field, *Science*, 2009, **324**, 1055.
38. M. M. Mench, *Fuel Cell Engines,* Wiley, Hoboken, New Jersey, 2008, Chapter 8.
39. M. S. Briggs, *Science*, 2009, **325**, 812.
40. G. W. Huber and J. A. Dumesic, *Catal. Today*, 2006, **111**, 119.
41. A. V. Bridgwater, *Chem. Eng. J.*, 2003, **91**, 87.
42. Z. Srokol, A.-G. Bouche, A. van Estrik, R. C. J. Strik, T. Maschmeyer and J. A. Peters, *Carb. Res.*, 2004, **339**, 1717.
43. F. Behrendt, Y. Neubauer, M. Oevermann, B. Wilmes and N. Zobel, *Chem. Eng. Technol.*, 2008, **31**, 667.
44. D. C. Elliot, D. Beckman, A. V. Bridgewater, J. P. Diebold, S. B. Gevert and Y. Solantausta, *Energy & Fuels*, 1991, **5**, 399.
45. S. Czernik, R. Maggi and G. V. C. Peacocke, in: *Fast Pyrolysis of Biomass: A Handbook, Vol. 2,* A. V. Bridgwater, (ed.), CPL Press, Newbury, UK, 2002, Chapter 8.
46. D. C. Elliott, *Energy & Fuels*, 2007, **21**, 1792.
47. T. R. Carlson, T. P. Vispute and G. W. Huber, *ChemSusChem.*, 2008, **1**, 397.

48. J. N. Chheda, G. W. Huber and J. A. Dumesic, *Angew. Chem. Int.*, 2007, **46**, 7164.
49. R. M. West, E. Kunkes, D. A. Simonetti and J. A. Dumesic, *Catal. Today*, 2009, **147**, 115.
50. L. Montague, Lignin Process Design Confirmation and Capital Cost Evaluation, Report No. NREL/SR-510-31579, National Renewable Energy Laboratory, Golden CO, 2003.
51. G. Knothe and J. van Gerpen, *The Biodiesel Handbook,* AOCS Publishing, Urbana, IL, 2005.
52. J. Holmgren, R. Marinangeli, T. Marker, M. McCall, J. Petri, S. Czernik, D. Elliott and D. Shonnard, *Hydrocarbon Eng.*, 2007, **12**, 75.
53. K. D. Maher and D. C. Bressler, *Bioresource Technol.*, 2007, **98**, 2351.
54. P. Mäki-Arvela, I. Kubickova, M. Snåre, K. Eränen and D. Y. Murzin, *Energy & Fuels*, 2007, **21**, 30.
55. National Science Foundation, "Breaking the Chemical and Engineering Barriers to Lignocellulosic Biofuels: Next Generation Hydrocarbon Biorefineries", G. W. Huber, ed., University of Massachusetts Amherst, Washington, D.C., March 2008.
56. Biomass Multi-Year Program Plan, Office of the Biomass Program, Energy Efficiency and Renewable energy, US Department of Energy, May 2009; available at: htttp://www1.eere.energy.gov/biomass/publications.html

CHAPTER 2

Energy Crops for the Production of Biofuels

MICHAEL MONTROSS AND CZARENA CROFCHECK

University of Kentucky, Biosystems and Agricultural Engineering Department, Lexington, KY 40546-0276, USA

2.1 Biorenewable Resources

2.1.1 Definition of Biomass

The term biomass is used to describe any material of recent biological origin and includes plant materials such as trees, grasses, and agricultural crops. The composition and agronomic requirements for each type of biomass crop will affect the handling and processing operations. Even varietal or hybrid differences within the same crop can also influence processing characteristics.[1] This chapter provides an overview of various considerations that need to be weighed before selecting a biomass supply and logistics system for a biorefinery.

2.1.2 Dedicated Energy Crops

Dedicated energy crops are planted and harvested periodically (annually or on a 3- to 10-year rotation) to be used for energy production. These crops are composed of carbohydrates, oils, lipids, protein, and fiber, where the proportion depends on the crop selected, regional variations, and differences between growing seasons. Plants that are high in fiber (lignocellulosic), are better suited for thermochemical conversion, such as switchgrass or poplar. In these plants,

RSC Energy and Environment Series No. 1
Thermochemical Conversion of Biomass to Liquid Fuels and Chemicals
Edited by Mark Crocker

Published by the Royal Society of Chemistry, www.rsc.org

metabolic energy is utilized to produce the fibrous material (roots and stems) resulting in high energy yields compared to the plants producing carbohydrates (sugar and starches). Biomass energy crops include herbaceous energy crops (switchgrass (*Panicum virgatum)*, miscanthus (*Miscanthus giganteus)*, reed canary grass (*Phalaris arundinacea*), *etc.*) and short-rotation woody crops (willow (*Salix* species), eucalyptus, hybrid poplar (*Populus* species), *etc.*).

2.1.2.1 Herbaceous Energy Crops

Herbaceous energy crops are plants made up of roots, leaves and stems and contain little woody material. These crops can be annuals or perennials and are harvested annually or sometimes multiple harvests each year are possible. The average biomass yield is between 5.5 and 11 Mg/ha/yr with a maximum yield up to 25 Mg/ha/yr in temperate regions.[2] Grasses are herbaceous plants that contain a high percentage of lignocellulosic material, and hence are of particular interest for thermochemical conversion. Grasses can be classified as either annuals or perennials, thin-stemmed or thick-stemmed, and warm-season or cool-season. Perennials are preferred as an energy crop because they are more drought resistant, require less weed control, are less likely to erode soils, and only require infrequent planting (possibly once every 20 years) compared to annual grass plants. Thick-stemmed grasses, such as sugarcane (*Saccharum poaceae*) and napier grass (*Pennisetum purpureum*), are indigenous to subtropical or tropical regions and have high labor and equipment requirements for establishment and harvesting. In contrast, thin-stemmed grasses, such as switchgrass, can be harvested with conventional hay equipment currently available on most farms. Warm-season grasses, such as switchgrass and gamma grass (*Tripsacum dactyloides*), grow better in the summer, while cool-season grasses, such as fescue and reed canary grass, grow better in the spring and the fall. In general, warm-season grasses are higher yielding due to their more efficient use of nutrients and water. However, cool-season grasses are usually better adapted to colder climates and may produce more biomass than warm-season grasses. Regional climate and soil differences will help determine which biomass crops are produced in a given region.

Herbaceous energy crops can be further classified by their metabolic pathways used for carbon fixation during photosynthesis. In C4 plants the primary products of photosynthesis are four carbon atoms, while C3 plants produce three carbon atoms. The enzyme that helps the photosynthetic reaction is phosphoenolpyruvate carboxylase (PEP carboxylase) in C4 plants and (ribulose-1,5-bisphosphate carboxylase oxygenase) RuBisCO in C3 plants. Due to the higher efficiency of PEP carboxylase, C4 plants can produce more biomass in a shorter time at higher temperatures and greater light intensities. Most herbaceous energy crops will likely be perennial C4 plants due to their higher water and nitrogen efficiency for producing biomass.[3] In addition, the maximum conversion of solar energy to biomass is 4.6% for C3 plants and 6.0% for C4 plants at a temperature of 30 °C and 380 ppm of atmospheric carbon dioxide.[4] In terms of water-use efficiency, C4 plants require approximately 300 g of water per g of dry matter, while C3 plants require over 600 g water/g dry matter.[5]

2.1.2.2 Short-Rotation Woody Crops

Short-rotation woody crops are fast-growing trees suitable for energy production, harvested on a 3- to 10-year rotation. These crops include both hardwoods and softwoods. Hardwood trees, such as poplar, willow, and oak, are angiosperms, also known as flowering trees. Hardwoods can sprout from stumps after harvest (coppice), which makes the propagation of these trees attractive as an energy crop. Hardwoods also have a lower level of ash, compared to softwoods. Softwoods generally have higher carbon content and higher heating values than hardwood species due to higher concentrations of lignin and resins relative to hardwoods.[6] Hardwood lignin becomes plastic at lower temperatures than softwood lignin that could influence the performance of thermochemical processes.[7] Softwoods are classified as gymonsperms, a term that covers most of the trees commonly referred to as evergreens (pine, spruce, and cedar). A considerable quantity of softwoods is used for commercial purposes such as construction lumber and pulpwood. As a result, there is a considerable quantity of logging and manufacturing residues that could be used as a feedstock for energy.

2.1.3 Waste Biomass Resources

While dedicated energy crops are grown purposely to be harvested as an energy feedstock, there are several options for collecting biomass feedstocks that are byproducts of manufacturing (sawdust or food-processing wastes) or residuals remaining after collection of high value products (collection of wheat straw after grain harvest or small limbs/branches and bark remaining after logging). These waste biomass resources include agricultural crop residues, forestry residues, commercial tree thinnings, fuel reduction in forests, yard waste, food-processing waste, and animal manure. The availability of waste materials varies from region to region and sometimes season to season. Food-processing wastes and manure tend to be centrally located, but highly complex and high moisture content. In comparison, yard wastes are highly complex and high moisture, but are widely distributed, although they are frequently collected by municipalities. The resulting variability, complexity, and availability of wastes are the chief disadvantages of using them as an energy feedstock. Often wastes can be obtained for as little as the cost of transportation. In some cases, the originator of the waste may be willing to pay a fee to facilitate the removal of the waste (in the case of many municipal wastes). Currently, agricultural and forestry residues are likely candidates for use as a thermochemical feedstock due to the predicted quantity available and cost. However, if a market develops for agricultural and forestry residues the costs will likely rise, although improvements in biomass logistics systems can be expected to help offset part of the potential increased feedstock costs.

2.1.4 Biomass Characterization

Biomass can come from a multitude of sources with high variability. Proper characterization of the biomass in order to determine the best possible processing strategy becomes of the upmost importance. In general, the important characteristics include what it is made up of (proximate and ultimate analysis), how much energy it contains (higher heating value), and how easy is it to produce, collect, store, transport, and process (crop yield, economics, equipment availability, grinding performance, *etc.*).

2.1.4.1 Proximate and Ultimate Analysis

Proximate analysis is based on the products that evolve upon controlled heating, in the following order: moisture, volatile matter, fixed carbon, and ash. Moisture content is determined by weighing the biomass sample, drying between 45 and 105 °C for 1 to 3 days depending on the product and reweighing the sample, while it is also determined in the first stage of proximate analysis. Volatile matter is the material that will convert to gases and escape at approximately 400 °C under an inert environment. The fixed carbon will then be driven off in the presence of oxygen. The remaining material at the end is ash, which contains a variety of mineral phases depending on the local soil chemistry, species of plant, and other growth factors controlling incorporation of macro- and micrometerutrients. Ash composition is particularly important to consider if the biomass is to be used as fuel for a combustion process, where differences can result in higher or lower fusion temperatures, clinker formation and undesirable fouling of the boiler tubes and radiative surfaces within the system.

Ultimate analysis is based on the elements that make up the biomass, typically including carbon, hydrogen, oxygen, nitrogen, sulfur, and chlorine. This analysis also quantifies the moisture content and ash. The percentages of the elements are sometimes presented on a dry, ash-free (daf) basis. Both ultimate and proximate analysis are important considerations in thermochemical processes.[8]

2.1.4.2 Heating Value

In addition to characterizing biomass based on composition, biomass can be categorized by how much energy is contained within the chemical bonds in the biomass. The higher heating value (HHV) and the lower heating value (LHV) are used to do just that. The HHV is also referred to as the gross calorific value, while the LHV is also referred to as the net calorific value. HHV is the amount of heat released from the combustion including the latent heat of vaporization of water from the sample. The LHV simply does not include the contribution of the latent heat of vaporization. The HHV is typically used when condensation of the reaction products is practical, such that all of the heat used to vaporize the water in the original sample can be used. HHV and LHV are often used when describing a fuel to be used in a combustion or gasification process, allowing comparison between various fuels and calculation of required air or oxygen consumption. HHV tends to be used by the natural gas and coal industries, while engine and turbine companies tend to rely on LHV.

The HHV for biomass can be measured using bomb calorimetric experiments. However, these experiments tend to be time consuming. Consequently, several attempts have been made to correlate HHV to proximate, ultimate, or biochemical compositions. The most reliable correlation to determine HHV is based on ultimate analysis, where HHV (MJ/kg) $= -1.3675 + 0.3137C + 0.7009H + 0.0318O^{*}$ (C = carbon, H = hydrogen, $O^{*} = 100$-C-H-ash, compositions are weight per cent on dry basis).[9]

2.1.4.3 Additional Processing Characteristics

Beyond the constituents and energy contained in biomass, it is important to consider other biomass characteristics that will affect the ease with which the biomass can be produced, harvested, stored, and processed. Additional parameters that can be anticipated to influence the performance of a biorefinery include grinding performance,[10] flowability,[11] angle of repose,[12] shear strength and other properties that influence the preprocessing and handling of biomass within a biorefinery.

2.2 Available Land Resources

Competition for farm and forestland will likely limit the quantity of biomass that can be produced for energy production. Farmland resources include cropland, idle cropland, and pastureland. Most cropland is under cultivation with traditional agricultural crops (maize, soybeans, rice, wheat, *etc.*) that are used for animal feed and/or food production. Converting cropland to dedicated energy crops will be difficult due to the growing world population, demand for animal protein, economics, and numerous other factors. Idle cropland is land that could be employed for crop production but often has limiting factors such as slope, soil type, rainfall, or other factors that limit its productivity. Pastureland is often in forage production with ruminant animals that could potentially be used for energy crops but will likely have competition with increased demand for animal protein.

Converting farm, forest, and pastureland to energy-crop production will have a significant impact on the sustainability, net energy value, quantity of renewable energy produced, and public acceptance. A large portion of the world is covered with forests and woodland resources that could be used as energy crops. Recent publications have concluded that converting forestland into dedicated bioenergy crops may not be sustainable due to accelerated emissions of carbon stored in the forest that would offset any gains from biofuel production.[13] Underutilized agricultural land is a potentially large source of land (385 to 472 million hectares) that could potentially produce 4.3 t/ha/yr, or about 10% of the primary energy in most countries in North America, Europe, and Asia.[13] Other studies have indicated that the potential energy supplied from biomass could constitute 15% of the global primary energy supply by 2050 after considering changes in arable, grassland, and forestland.[14] Much debate has occurred over the net effects (energy balance and carbon cycling) of converting agricultural and forestry products to energy.[15,16] Some biomass crops could sequester carbon within the soil during production. For example, carbon sequestration benefits from switchgrass production would depend on the crop system replaced (higher C sequestration than annual crops, but no larger than grazed pastures).[17] Switchgrass also reduces carbon emissions and has an improved net energy gain relative to ethanol production from maize grain.[18] Short-rotation woody crops produced on cropland would result in a net sequestration of carbon when all fluxes are considered relative to traditional oak-beech forests.[19]

Biomass is often considered as a method to reduce greenhouse-gas (GHG) emissions relative to fossil energy. Work in Europe has indicated that miscanthus and switchgrass would be competitive with natural gas and crude oil for electricity generation,[20] although coal remains a lower cost energy source. GHG emissions were reduced when either switchgrass or miscanthus were grown on agricultural lands (both cropland and pastureland) for electricity production relative to oil, natural gas, and coal combustion.[20] Switchgrass produced 540% more renewable energy than nonrenewable energy consumed. Further, if the switchgrass was converted to cellulosic ethanol the greenhouse-gas emissions were 94% lower than emissions from gasoline.[21] Land-use decisions, nutrient cycling between the soil and environment, economics, and net emission changes will likely have a significant influence on biomass crop selection and the long-term feasibility of biomass production.

2.3 Biomass Composition and its Effect on Processing

Ideal energy crops would require minimal inputs (low nitrogen and pesticide requirements), grow in areas unsuitable for food crop production (arid, eroded or saline soils), and be high yielding (t/ha or MJ/ha) with a composition well suited for the desired conversion process (cellulose, hemicellulose and lignin content or higher heating value). If an energy crop is used for biochemical conversion to ethanol, a crop with a high cellulose and hemicellulose content relative to the lignin content would be desirable. However, for thermochemical conversion the proximate analysis and heating value will have a greater influence on the product yields. Lignin and extractives from biomass have a higher heating value than the cellulose and hemicellulose in biomass. For thermochemical conversion, an ideal energy crop would therefore have higher concentrations of extractives and lignin that would increase the higher heating value.[9,22,23] Table 2.1 summarizes the composition of typical feedstocks and their higher heating value.[24–31] Numerous other sources are available to evaluate feedstock composition.[32,33] In general, most of the herbaceous energy crops have higher ash contents on average than forestry products,[34] although this is heavily influenced by the collection strategies employed. The lignin content of agricultural residues and herbaceous energy crops will likely be lower than found in forestry products. Trends in cellulose and hemicellulose contents are more difficult to discern between the different crops.

Table 2.2 lists a number of prospective energy crops and their ultimate and proximate analyses. There is a large variation in the proximate analysis between the various crops. The lowest fixed carbon content was 10.6% (miscanthus), while the highest fixed carbon content was 22.0% for corn stover. Large differences in the volatile matter and ash content are observed for the other crops. However, the ultimate analysis is more consistent than the proximate analysis. The range in carbon content was between 43.9 and 49.9% while the nitrogen content varied between 0.05 and 1.1%. Desirable crops would have a low nitrogen, potassium, and phosphorous content since those macronutrients are required for crop production.

2.4 Energy Consumption During Establishment and Production

Equipment selection and performance is important for establishing, maintaining, and harvesting biomass energy crops and may be critical for the economic success of biorefineries. Considerable data exist that allow for the estimation of energy consumption, field capacity, and cost of use for equipment with bias-ply tires used for most agricultural, earthmoving, and forestry applications.[35,36] It should be noted that most of the data have an uncertainty of 50% for the majority of the parameters. This is due to the considerable variation in equipment performance resulting from differences in machine design, machine adjustments, age of machine, operator experience, and site-specific conditions (soil moisture, slope, soil type, *etc.*). Similar time and motion data have been developed for dedicated forestry machinery that allow for the evaluation of optimal equipment performance for specific regions.[37–39] These data can be used to estimate costs, labor requirements, and energy used for biomass crops.

Most energy crops will require fertilization at establishment or as part of an annual maintenance program. Fertilizers are a significant source of energy that are used in the production of biomass crops. The energy required (lower heating value basis) for the manufacture and delivery per kilogram of nitrogen fertilizer is estimated at 57.0 MJ/kg.[40] Phosphate and potash are estimated at approximately 9.3 and 7.0 MJ/kg of P_2O_5 and K_2O, respectively.[40] Furthermore, many crops require the addition of fungicides, herbicides, or pesticides during establishment and production. Each active ingredient requires fossil

Table 2.1 Typical composition of biomass feedstocks (expressed as percentage of bone dry material).

Type	*Extractives*	*Cellulose*	*Hemicellulose*	*Lignin*	*Ash*	*HHV (kJ/kg)*	*Ref.*
			Agricultural Residues				
Maize stover	5.9	35.5	22.8	18.7	11.6	18 300	25
Sugarcane bagasse	4.5	39.1	22.6	24.3	5.2	19 100	25
Wheat straw	13.0	32.6	22.6	16.9	10.2	17 400	26
			Forestry Products				
American Sycamore	3.0	39.7	18.5	25.8	1.2	19 600	25
Black locust	4.7	40.8	18.1	26.2	1.6	19 700	25
Eucalyptus	1.9	46.3	14.9	27.2	1.1	19 600	25
Hybrid Poplar	4.2	41.5	17.9	25.7	1.8	19 600	25
Pine	2.7	41.7	20.5	25.9	0.3	19 600	26
Willow	**	48.5	13.9	19.7	1.7	20 000	27
			Herbaceous Crops				
Switchgrass Alamo	11.1	33.1	26.3	18.0	5.4	18 900	25
Miscanthus x giganteus		45.0	30.0	21.0	2.3	17 700	28
Sericea Lespedeza	7.7	36.6	16.9	26.2	3.0	19 400	25
Reed canary grass	**	42.6	29.7	7.6	**	17 900	27, 29
Timothy grass	**	28.8	27.2	4.8	**	**	30
Alfalfa	**	27.4	11.7	4.8	**	18 500	24, 30
Tall fescue	20.3	24.5	19.5	14.7	11.8	**	31

** Not evaluated

Table 2.2 Proximate and ultimate analysis of some typical biomass feedstocks.

	Proximate Analysis (wt% dry basis)			*Ultimate Analysis (wt% by dry basis)*				
Biomass	*Fixed Carbon*	*Volatile Matter*	*Ash*	*C*	*H*	*N*	*O*	*Reference*
Corn Stover	22.0	73.1	10.2	47.04	5.47	0.68	41.10	32
Eucalyptus	18.4	81.2	1.2	49.89	5.71	0.05	42.29	32
Hybrid Poplar	20.0	78.7	2.0	49.75	5.52	0.52	42.42	32
Miscanthus	10.6	87.6	3.0	43.7	5.7	1.1	44.8	8
Switchgrass	21.57	73.75	5.76	47.27	5.31	0.51	41.59	32
Wheat Straw	21.54	69.38	10.22	43.88	5.26	0.63	38.75	32
Willow	10.7	87.6	1.7	49.9	6.5	0.2	39.9	27

energy for manufacturing that varies between 223 and 328 MJ/kg of product. Applying 100 kg of nitrogen/ha of a biomass crop yielding 15 t/ha with a heating value of 18 000 kJ/kg represents 2% of the energy within the biomass crop. It is obvious that fertility requirements of biomass crops could be significant on the overall energy balance.

Fresh-water supplies are limited in most of the world, while irrigation is capital and energy intensive and probably not feasible for energy crops. Electricity, propane, and natural gas are often used as power sources for irrigation pumps. Studies have indicated that the energy required to pump 1 ha cm of water requires between 38 and 43 kWh.[41] In one study, maximum biomass production was achieved with 116.5 cm of water in Texas with switchgrass, although the maximum water-use efficiency was obtained with 85.5 cm of water (combination of rainfall, irrigation, and soil water).[42] If 50 cm of water was supplied with irrigation pumps, almost 3% of the energy contained within the biomass would be consumed.

Miscanthus and switchgrass use water and nitrogen differently. Miscanthus produced over double the amount of biomass as switchgrass in one study.[43] The authors concluded that miscanthus produces more biomass over a range of variables than switchgrass, but switchgrass was less affected by water availability and more affected by nitrogen fertilization than miscanthus. This would have important consequences for deciding on which energy crop should be produced and the long-term economics and net energy production. Establishing switchgrass with a lower yield potential than miscanthus in areas with shallow, droughty soils or marginal rainfall could be advantageous.

Sustainable energy-crop production will require efficient equipment and methods to establish energy crops, provide nutrients for crop maintenance, and appropriate rainfall quantities related to the soil characteristics. Estimates of the potential labor requirements, energy consumption, yields, composition, and cost to grow biomass energy crops can be obtained using the data presented above. An example is the predicted cost of switchgrass and miscanthus produced in Illinois, US.[44] Considering only fertilizer, chemicals, seed, and planting operations, establishment costs for switchgrass and miscanthus were estimated to be \$271 and \$613/ha, respectively.[44] The establishment cost spread over the expected life of the crops (10 years for switchgrass and 20 years for miscanthus) and the expected yield (7.0 t/ha for switchgrass and 22.3 t/ha for miscanthus), resulted in \$4.76/t and \$2.02/t of the switchgrass and miscanthus cost being due to establishment, respectively. Although miscanthus was over twice as expensive to establish as switchgrass, the total establishment cost of miscanthus was only half the cost of switchgrass. This demonstrates the influence of establishment cost and yield on the total biomass cost.

2.5 Biomass Logistics

2.5.1 Biomass Crop Harvesting

Biomass harvesting can be split into two basic categories – woody and herbaceous crops. Woody biomass is harvested using equipment found within the forestry industry with minor modifications. Likewise, equipment to harvest herbaceous biomass is modified hay and forage equipment using agricultural machinery.

2.5.1.1 Woody Crop Harvest

Several scenarios exist for the collection of currently unused logging residues. The majority of trees would be harvested using either a feller/buncher, shovel logger, or harvester and skidded using a grapple skidder (a grapple skidder is used with all three mechanized fellers) or forwarder (used with a harvester or shovel logger).[37,39,45] The trees would be transported to forwarding areas for further operations.

Whole-tree skidding or forwarding can be used to move the total tree biomass to the landing where the traditional forest products are harvested and the residual biomass can be used for energy production. The residual biomass can be chipped directly into trucks for over-the-road transport. This option is the most time efficient but results in a higher chipped biomass moisture content and presents logistical challenges at the landing and processing site. Other disadvantages include increased landing size and the potentially high degree of soil disturbance associated with whole-tree skidding that can be somewhat reduced using forwarders. Costs to collect the wood chips from commercial thinnings have been estimated to be \$31/green tonne.[46]

Instead of immediate chipping of the residual biomass, the residual biomass could be hauled into the woods and piled in windrows. The piled windrows are allowed to passively dry and are hauled back into the forwarding area and chipped. While not as efficient from a time standpoint, passive moisture loss reduces chipped biomass moisture content and can significantly reduce transportation costs and increase the lower heating value of the fuel. This scenario would require two passes along skid trails, would likely delay the overall harvest operation, and interfere with postlogging best-management practices. Another option would be the bundling of the residual biomass at the landing and the bundles stacked for future chipping. Bundled forestry residue has been estimated to cost \$14/green tonne (43% wet basis).[47] This would allow passive moisture loss but would require a stationary bundler.

A harvester can be used to fell and delimb all trees in the woods followed by in-woods bundling using a mobile bundler. Bundles are then transported immediately to the landing for storage or left in the woods prior to hauling to the landing for chipping. This scenario also has the ability to passively reduce the moisture content. However, it requires the use of harvester (feller/buncher with a processor head) and a mobile bundler.

In all cases, chipped biomass is transported over the road in chip vans at the maximum allowable truck weights.[48] Harvesting that includes whole-tree skidding can have dramatic impacts on soil and duff layers, which is the layer of decomposing leaves and twigs that covers the uppermost soil layer. Further nutrient cycling effects, sustainability and other environmental impacts need further investigation. Chipping is an energy and costly operation. Chipping of small diameter trees has been estimated to cost \$9.18 per green tonne with a moisture content of 94.1% dry basis.[48] Whole-tree chippers could process up to 400 green tonnes per day but require diesel engines up to 400 kW.[38]

Fewer options exist for the harvest of short-rotation woody biomass crops. If coppice harvest systems (self-propelled chippers) were employed chips would be produced using self-propelled harvesters that conveyed chips directly into vehicles for transport out of the forest. Minimal opportunities exist to effectively reduce the moisture content prior to transportation and processing. Specific equipment must be purchased to mechanically coppice the biomass. Species that stump sprout should be used and harvesting should be avoided during the summer months to ensure adequate and vigorous coppice development and minimize nutrient removal. Collection costs from short-rotation poplar stands have been estimated at between \$8 and \$11 per green tonne (58% moisture content, wet basis or \$19 to \$26 per dry tonne).[38]

Urban woody residues are also available in numerous municipalities. Urban residues are diverse and include green and partially dry biomass. Collection sites, developed within or adjacent to urban centres, would accept unprocessed biomass moved to these sites using relatively small to medium capacity hauling vehicles. At the collection site, biomass is piled and allowed to passively dry prior to chipping and transportation.

2.5.1.2 Harvesting of Herbaceous Energy Crops

Most farms have access to equipment required to establish and maintain herbaceous energy crops (switchgrass, Indian grass, Miscanthus, reed canary grass, *etc.*). For harvest, most farms have access to mowers to cut and windrow the crop. After the crop has dried within windrows, balers (round or midsize rectangular balers) or choppers are used to collect the biomass. Handling equipment is utilized to load bales on trucks for transportation or chopped directly into trucks for transportation to storage locations.[49,50] Problems associated with round bales are low density (overloading tractor-trailers is difficult) and handling concerns (securing bales on trailer). However, round bales have the advantage of shedding water when stored outside uncovered, thereby minimizing storage costs relative to rectangular bales.

Midsize rectangular bales (usually with nominal dimensions of around $0.9 \times 1.2 \times 2.4$ m) likely will be the preferred package to handle herbaceous biomass. They have a higher bulk density than round bales, are easier to handle and tie down on a trailer.[51] Midsize rectangular balers will be easier to increase the bulk density to minimize transportation costs. Experiences at the Chariton Valley Cofiring Project in Iowa indicated that one fork truck was required to move 14 t/hr of midsize rectangular bales.[52] Round bales were not accepted at the facility due to the difficulty of handling with fork trucks and feeding into the hammer mill. One significant disadvantage of rectangular bales is the larger tractor size required compared to round bales.

Herbaceous biomass could also be chopped (for example with a forage harvester) directly from the field and stored in bulk piles.[53] Chopped material would then be handled using front-end loaders and live bottom trucks. Forage harvesters typically require high horsepower machines and would be difficult to operate in fields with slopes greater than 15%. Custom harvesters could fill the gap with midsize rectangular balers and forage harvesters not typically found on many farms.

Some facilities may not desire to handle bales. In this case, bales could be ground on-farm prior to transportation to users. Bulk density is increased as the particle size is decreased. In addition, some of the issues associated with bale handling and dust are reduced since the size reduction is performed on-farm. The chopped biomass would then be moved using live bottom trailers.

A large potential source of biomass is crop residues (maize stover and wheat straw) generated during grain harvest. The exact quantity of maize stover and wheat straw that can be sustainably removed is unknown, but is likely in the range of 2 to 4 t/ha.[54] Maize stover will be difficult to handle as a dry system in many areas because of the moisture content of the stover at grain harvest. Multipass systems that would allow the stover to dry prior to baling or chopping would be required. Single-pass collection of stover would reduce collection costs but would require development of additional equipment and possibly wet storage systems.[55–57] Collection of wheat straw using a dry system should not create significant problems. The equipment for herbaceous and crop residues would likely be modifications of existing equipment and ideally would be utilized for hay and biomass collection.

A major concern with the utilization of biomass crops is the production cost. The farm gate price of maize stover has been estimated to be $33 per dry tonne[58] and the farm gate cost of switchgrass has been estimated to be $65 per dry tonne.[59] The difference between the cost of agricultural crop residues (maize stover) and dedicated herbaceous energy crops (switchgrass) corresponds to the difference in the land cost. The land cost is not a significant factor with maize-stover production because the cost of land is considered as part of the grain production. The cost per unit of energy or mass of lignin, cellulose, or hemicellulose can be determined at the farm gate. Additional cost is required to handle and transport the biomass to a biorefinery. There are wide ranges in the estimated price at the farm gate depending on the assumptions made, such as land cost, fuel cost, labor cost, crop yield, moisture content, nutrient removal rates, and machine productivity. It is clear that the cost of feedstocks will have a significant effect on the feasibility of renewable energy production.

2.5.2 Biomass Transportation

Wide-scale utilization of biomass is hindered by its low energy density and low bulk density.[34] Biomass bulk density can vary considerably depending on the format. Rectangular bales of switchgrass are close to the desired bulk density (210 kg/m^3) required to minimize transportation costs.[60] However, switchgrass that is chopped to a particle length between 20–40 mm and loosely filled results in a bulk density between 60–80 kg/m^3.[60] However, switchgrass with a mean particle length of 51.8 and 14.7 mm has a loose density of 52 and 120 kg/m^3, respectively, which is increased to 69 and 147 kg/m^3 when compacted with vibration.[61] Pellets with a diameter of 6.24 mm result in a bulk density between 500-700 kg/m^3.[60] The bulk density of biomass can be increased to minimize transportation costs, but is expensive and energy intensive.

Collection, storage, handling, and transportation costs are a major hurdle to the large-scale consumption of biomass. Other options to reduce transportation costs could include the densificiation of biomass into pellets, briquettes, or cubes prior to transportation to a biorefinery. Briquettes produced from wheat straw and used waste paper as a binder have achieved bulk densities 7 to 10 times greater than the bulk density of baled wheat straw using a temperature of 100 °C and a pressure between 400 and 800 MPa.[62] The cubing cost of corn stover has been estimated to be $20.90/t, the corresponding bulk density being up to 500 kg/m^3.[63] However, in the study cited the use of binders that would reduce the energy consumption and capital cost of the cuber required was not investigated. Other research has estimated energy requirements of 12 to 30 MJ/t to briquette corn.[64] Pelleting cereal straws (barley, canola, and wheat) required between 3 and 10 MJ/t.[65] Drying wood chips to 10% moisture content, grinding, and pelleting required approximately 3 GJ/t.[66] The large difference in energy required to densify agricultural residues is due to differences between the mechanics of pelleting and briquetting and the type of biomass. The higher energy required to pellet wood products relative to cereal straw is due to a number of factors. Biomass densification is highly sensitive to moisture content and the type of densification process. The maximum durability of wafers was found at a moisture content of 11%,[67] sawdust and wheat straw briquettes were strongest when the moisture content was between 12 and 20%,[68] while the optimal moisture content for alfalfa pellets was 19%.[69] In addition to moisture content numerous additional factors influence the properties of densified biomass (durability, bulk density, flowability, *etc*.), such as protein content, starch content, type of densified product (pellet, briquette, cube, or wafer), biomass type and many other factors.[70]

Densification of biomass into briquettes, cubes, or pellets would significantly improve the large-scale handling of biomass. Most briquettes, cubes, or pellets would be free flowing and approach bulk densities of 400 to 800 kg/m^3,[70] which may be required for large facilities. The fixed cost ($/t) for transporting biomass is 5.70, 17.10, and 34.01 for truck, rail, and barge, respectively. However, the variable cost ($/t km) is 0.14, 0.03, and 0.01 for truck, rail, and barge, respectively.[60] Rail is the cheapest transportation mode, but for the most cost-effective freight rates, unit trains with 100 rail cars and up to 10 000 t of material need to be loaded in a 12-h period. Rail transportation, and to a lesser extent transportation by barge, requires dense, free flowing material. Barge and rail transportation would allow biomass to be produced in areas it was most cost effective, and then be transported to other regions. Indeed, grain, coal, and other raw materials are frequently shipped in bulk over long distances by rail and barge.

One of the primary variables that affect the collection, handling, storage, and transportation cost is the bulk density of the biomass. Transportation costs can account for 30% of the feedstock value with the current biomass handling systems.[71] Increasing the bulk density of bales or wood chips will decrease the collection and transportation cost, and if the material is in the form of pellets, transportation costs to a biorefinery will be the lowest.[60] In addition, trucking costs are minimized when trucks can be loaded with biomass at a density sufficient to allow the vehicle to meet the maximum gross vehicle weight. The maximum gross vehicle weight varies considerably with location and trailer specifications. For example, most US locations have a maximum gross vehicle weight of 36 000 kg, the UK allows maximum gross weights of 44 000 kg, and continental Europe allows gross vehicle weights of 40 000 kg. Depending on the configuration of the trailer, a bulk density between 190 and 240 kg/m^3 is required to achieve the maximum gross vehicle weight. It should also be noted that according to a recent study, the primary variable that determined the cost of biomass transportation was the loading and unloading time rather than the cost of travelling for bales and wood chips.[72]

Moisture content plays a significant role in the collection, storage, and transportation cost of biomass. Only the dry-matter fraction of biomass is important for energy production. Collecting, storing, and transporting moisture within biomass has significant negative effects, especially with thermochemical processes. Moisture decreases the heating value, requires additional fuel for transportation, additional storage area, and increases preprocessing costs (drying and grinding).

2.5.3 Storage

Biomass harvesting occurs seasonally over the year, while energy production must proceed daily throughout the year. As a result, the biomass must be stored, sometimes for months at a time. Traditionally, crops can be dried and then stored or ensiled. Without proper storage, biomass will start to degrade within days, or sometimes months. Degradation can occur due to weathering and/or microbial contamination. Degradation can occur in the form of dry-matter loss, reduction in lower heating value, increase in ash content, and can cause difficulty in handling the biomass after storage. The optimal storage system will depend on the initial moisture content of the biomass at harvest, the environmental conditions, and the desired conversion process.

Recently harvested biomass can have a moisture content of 60% for woody biomass or 70–85% for herbaceous biomass. In order to prevent microbial degradation, the biomass can be dried to a moisture content that suppresses microbial growth, typically less than 20% wet basis. If the moisture content is greater than 25%, biomass is aerobically unstable and will have high respiration losses during transportation and handling.[73] Drying biomass using fossil energy is not feasible due to the high energy requirement, typically requiring a minimum of 2442 kJ/kg H_2O removed. Considering the significant amount of water that needs to be evaporated, natural drying is required to ensure a positive energy balance.

Biomass can be collected and stored at the farm in four common formats: round bales, rectangular bales, bundled forestry residues, and chopped/chipped bulk piles. Each form has its unique advantages and disadvantages. Round and rectangular bales are frequently used with agricultural residues or dedicated herbaceous energy crops. Round bales shed some water when stored outdoors, but are inefficient when loaded on trucks or stored inside of buildings. Rectangular bales do not shed water, but are efficient to load on a truck or store in a building. Bundled forestry products would allow for passive drying if appropriate weather conditions existed during storage. Ground bulk piles are partially preprocessed, but require specialized handling equipment and would require packing for efficient storage. A fifth form could be included that involves densification systems to produce pellets, cubes, or briquettes; these densified forms of biomass easily fit into bulk handling and storage systems. Densification systems could be positioned in a regional system that would provide shorter transportation distances from the farm/forest to regional densification systems. These would then provide pellets, cubes, or briquettes for long-distance and high-volume shipping to biorefineries or power plants.

Herbaceous energy crops in temperate climates will dry in the field after the killing frost. Moisture contents during switchgrass harvest should be under 20% since the crop is typically harvested in the winter to early spring. Switchgrass moisture contents were approximately 35% during fall harvest (October 15 to November 15 in Pennsylvania) but decreased to around 7% during spring harvest (April 6 to May 4).[74] The same study also noted a large amount of biomass losses in the spring primarily due to inefficiencies associated with the mower and baler not being able to pick up down crop. Experiences with the Chariton Valley project indicated that switchgrass delivered to the plant was consistently less than 15% moisture content with an average of 11%.[52] Miscanthus appears to have a higher moisture content relative to switchgrass at harvest.[75] However, most of the miscanthus data are from Europe and differences in weather would significantly affect the crop dry down rate and moisture content at harvest.

Moisture content is of great concern when considering agricultural residues as feedstocks. During maize harvest, the material other than grain passing through a combine will frequently have a moisture content greater than the 20% that is typically recommended for hay and forage storage. In Tennessee, a recent study showed that when the corn moisture content was 20%, the maize stalks were approximately 30%, while the leaves and husks were approximately 10%.[76] In a study in Wisconsin, the lowest stover moisture observed was 50% when the corn moisture content was 24%.[77] In Kentucky, during harvest with corn at an average moisture content of 18%, the average moisture content of the cobs, husks, leaves, stalks, and whole stover were 32%, 37%, 19%, 51%, and 43%, respectively.[57] Clearly, there is considerable variation in the moisture content of maize stover both within the plant's fractions and across climatic regions. It would be expected that large differences would also exist in maize stover across the world. These moisture levels would indicate that maize stover needs to be left in the field after grain harvest for ambient drying or stored as an ensiled product. Either storage method is effective, but allowing maize stover to field dry requires additional machine operations that increases the cost and increases soil contamination.

Collecting material other than grain during small grain harvest (wheat, barley, oats, *etc.*) should create fewer moisture problems than found with maize harvest. The moisture content of wheat straw at grain harvest in the Eastern US and Europe could be slightly elevated and will result in moldy straw if immediately baled after wheat harvest[78] but is substantially lower than corn stover. In the Western US, during wheat harvest, straw is frequently at a moisture content of 15% or below.[79]

Management of dry biomass (herbaceous energy crops and agricultural residues) will require storage buildings or stacking bales with tarpaulins or plastic to keep water from rewetting the crop. Additional concerns are fire issues with large quantities of dry biomass. Miscanthus bales stored outside increased by up to 20 percentage points of moisture in Ireland, while the same bales stored indoors or covered outdoors had a moisture decrease.[80] Dry-matter loss from round bales of switchgrass stored on crushed limestone was estimated to be around 5%.[81] Dry-matter loss from large round and large square switchgrass bales stored 200 days was 17 and 55%, respectively, when stored uncovered on a gravel pad; this was the worst-case scenario investigated.[82] Rectangular bales of wheat straw in Idaho had a significant increase in moisture content in the top 30 cm and the bottom 15 cm of bales stored outdoors after 144 days of storage.[79] In chopped piles, the increase in moisture content was significant throughout the pile. Limiting moisture increases in chopped piles relative to bales would appear to be difficult especially considering the dry conditions in Idaho. Similar results have been observed showing the need to store biomass in covered structures to limit moisture increases and dry-matter losses.

Anaerobic storage (ensiling) of agricultural materials is commonly implemented to limit respiration losses in products with a high moisture content. In one study, the dry-matter loss from dry maize-stover bales stored inside was approximately 3.3%, while this increased to 10.0% for round bales of corn stover stored outside with net wrap, and decreased to 2.4% for wet (38% moisture content) round bales of corn stover wrapped in plastic.[56] Some studies have investigated the handling of switchgrass as an ensiled product,[49] although the likely handling and storage scenario for herbaceous energy crops would be dry systems. Anaerobic storage is effective at limiting storage losses, but this comes at a loss of aerobic stability and adds the cost of transporting water.

Forestry products will be stored in the form of chips, windrows, or bundled residuals. Many forest products at harvest will have a wet basis moisture content of between 25 and 60% depending on the tree species, tree maturity, and numerous other factors. Chipping produces an acceptable fuel for biorefineries and increases the bulk volume, homogeneity, and improves the handling of the wood products. However, wood chips are difficult to store due to their high moisture content, microbial respiration, and reduction in heating value. For these reasons, systems have been developed to dry wood chips to increase the storability and heating value and reduce the risk of self-ignition.[83,84] Wood-chip piles have relatively low airflow resistance and can be ventilated with ambient air to reduce the moisture content and increase the storability.[85]

Windrows of forestry residues can have a significant reduction in moisture content, up to 20 percentage points.[84] Bundled forestry residues can also be stored in the forest and allowed to passively dry prior to chipping. In one study, bundles decreased in moisture from 43% (wet basis) to 26% and 30% in covered and uncovered bundles, respectively.[84] However, the windrows and bundles would need to be collected from the forest in a second operation and chipped prior to utilization.

2.6 Summary

There are a wide variety of biomass types available to be used as an energy feedstock. The ideal feedstock will depend on several factors, including growth requirements, growth amenability to the climate, irrigation requirements, efficiency of solar-energy capture, biomass composition, harvesting and collection requirements, storage options, transportation cost, and other costs. In addition, several of these requirements depend on the type of biomass conversion to be used, be it pyrolysis, gasification, hydrothermal processing, *etc.*

References

1. D. W. Templeton, A. D. Sluiter, T. K. Hayward, B. R. Hames and S. R. Thomas, *Cellulose*, 2009, **16**, 621.
2. R. C. Brown, *Biorenewable Resources-Engineering New Products from Agriculture,* Blackwell Publishing, Ames, IA, 2003.

3. H. Marschner, in *Mineral Nutrition of Higher Plants*, H. Marschner, (ed.), Academic Press, London, 1986, pp. 116–123.
4. X. G. Zhu, S. P. Long and D. R. Ort, *Curr. Opin. Biotechnol.*, 2008, **19**, 153.
5. H. W. Woolhouse, *Endeavour*, 1978, **2**, 35.
6. T. Demirbas and C. Demirbas, *Energy Sources Part A-Recovery Utiliz. Environ. Effects*, 2009, **31**, 1464.
7. S. Kubo and J. F. Kadla, *J. Wood Chem. Technol.*, 2008, **28**, 106.
8. M. Jeguirim, S. Dorge and G. Trouve, *Biores. Technol.*, 2010, **101**, 788.
9. C. Sheng and J. L. T. Azevedo, *Biomass Bioenergy*, 2005, **28**, 499–507.
10. V. S. P. Bitra, A. R. Womac, C. Igathinathane, P. I. Miu, Y. C. T. Yang, D. R. Smith, N. Chevanan and S. Sokhansanj, *Biores. Technol.*, 2009, **100**, 6578.
11. S. Mani, L. G. J. Tabil and S. Sokhansanj, *Trans. ASAE*, 2004, **47**, 1983.
12. K. E. Ileleji and B. Zhou, *Powder Technol.*, 2008, **187**, 110.
13. J. E. Campbell, D. B. Lobell, R. C. Genova and C. B. Field, *Environmental Sci. Technol.*, 2008, **42**, 5791.
14. G. Fischer and L. Schrattenholzer, *Biomass Bioenergy*, 2001, **20**, 151.
15. T. Searchinger, R. Heimlich, R. A. Houghton, F. X. Dong, A. Elobeid, J. Fabiosa, S. Tokgoz, D. Hayes and T. H. Yu, *Science*, 2008, **319**, 1238.
16. J. Fargione, J. Hill, D. Tilman, S. Polasky and P. Hawthorne, *Science*, 2008, **319**, 1235.
17. D. I. Bransby, S. B. McLaughlin and D. J. Parrish, *Biomass Bioenergy*, 1998, **14**, 379.
18. S. B. McLaughlin and M. E. Walsh, *Biomass Bioenergy*, 1998, **14**, 317.
19. G. Deckmyn, B. Muys, J. G. Quijano and R. Ceulemans, *Glob. Change Biol.*, 2004, **10**, 1482.
20. E. M. W. Smeets, I. M. Lewandowski and A. P. C. Faaij, *Renewable & Sustainable Energy Rev.*, 2009, **13**, 1230.
21. M. R. Schmer, K. P. Vogel, R. B. Mitchell and R. K. Perrin, *Proc. Nat. Acad. Sci. USA*, 2008, **105**, 464.
22. R. H. White, *Wood Fiber Sci.*, 1987, **19**, 446.
23. F. Shajizadeh and W. G. Degroot, *Thermal Uses and Properties of Carbohydrates and Lignins,* Academic Press, New York, 1976.
24. A. A. Boateng, H. G. Jung and P. R. Adler, *Fuel*, 2006, **85**, 2450.
25. D. K. Johnson, P. Adam, P. Ashley, H. Chum, S. Deutsch, D. J. Fennell and A. Wiselogel, *IEA/BA Task VI Activity 5,* New Brunswick, Canada, 1993.
26. F. Agblevor, H. L. Chum and D. K. Johnson, Energy from Biomass and Wastes XVI: Proceedings of the Institute of Gas Technology Conference, Orlando, FL, 1992.
27. T. G. Bridgeman, J. M. Jones, I. Shield and P. T. Williams, *Fuel*, 2008, **87**, 844.
28. S. Collura, B. Azambre, G. Finqueneisel, T. Zimny and J. V. Weber, *Environ. Chem. Lett.*, 2006, **4**, 75.
29. J. Burvall, *Biomass Bioenergy*, 1997, **12**, 149.

30. P. Alvo and K. Belkacemi, *Biores. Technol.*, 1997, **61**, 185.
31. K. Belkacemi, G. Turcotte, P. Savoie and E. Chornet, *Ind. Eng. Chem. Res.*, 1997, **36**, 4572.
32. US Department of Energy, *Biomass Feedstock Composition and Property Database*, http://www1.eere.energy.gov/biomass/feedstock_databases.html, Accessed July 20, 2009.
33. IEA Bioenergy Task 32, Biomass combustion and cofiring, http://www.ieabcc.nl/, Accessed July 20, 2009.
34. P. McKendry, *Biores. Technol.*, 2002, **83**, 37.
35. ASABE, in "Standards", American Society of Agricultural and Biological Engineers, St. Joseph, MI., 2006.
36. ASABE, in "ASABE Standards, American Society of Agricultural and Biological Engineers, St. Joseph, MI, Editon edn., 2006.
37. D. Culshaw and B. Stokes, *Biomass Bioenergy*, 1995, **9**, 127.
38. R. Spinelli and B. R. Hartsough, *Biomass Bioenergy*, 2006, **30**, 439.
39. R. Spinelli and B. R. Hartsough, *Biomass Bioenergy*, 2001, **21**, 425.
40. H. Shapori and P. Gallagher, ed., U.A.E. Service, 2002.
41. S. H. Amosson, L. New, L. Almas, F. Bretz and T. Marek, *Economics of irrigation systems,* College Station, TX, US, 2001.
42. P. T. Koshi, J. Stubbendieck, H. V. Eck and W. G. McCully, *J. Range Manage.*, 1982, **35**, 623.
43. E. Heaton, T. Voigt and S. P. Long, *Biomass Bioenergy*, 2004, **27**, 21.
44. M. Khanna, B. Dhungana and J. Clifton-Brown, *Biomass Bioenergy*, 2008, **32**, 482.
45. P. Borjesson, *Biomass Bioenergy*, 2000, **19**, 137.
46. M. C. Bolding, L. D. Kellogg and C. T. Davis, *For. Prod. J.*, 2009, **59**, 35.
47. K. Karha and T. Vartiamaki, *Biomass Bioenergy*, 2006, **30**, 1043.
48. D. Mitchell and T. Gallagher, *South. J. Appl. For.*, 2007, **31**, 176.
49. A. Kumar and S. Sokhansanj, *Biores. Technol.*, 2007, **98**, 1033.
50. S. Sokhansanj, A. Kumar and A. F. Turhollow, *Biomass Bioenergy*, 2006, **30**, 838.
51. K. J. Shinners, R. J. Straub, R. L. Huhnke and D. J. Undersander, *Appl. Eng. Ag.*, 1996, **12**, 167.
52. W. Amos, ed., DOE/NREL, Golden, CO, 2002.
53. D. Styles, F. Thorne and M. B. Jones, *Biomass Bioenergy*, 2008, **32**, 407.
54. W. W. Wilhelm, J. M. F. Johnson, J. L. Hatfield, W. B. Voorhees and D. R. Linden, *Agron. Jrnl.*, 2004, **96**, 1.
55. J. R. Hess, C. T. Wright and K. L. Kenney, *Biofuels Bioprod. Biorefin.-Biofpr.*, 2007, **1**, 181.
56. K. J. Shinners, B. N. Binversie, R. E. Muck and P. J. Weimer, *Biomass Bioenergy*, 2007, **31**, 211.
57. R. M. Prewitt, M. D. Montross, S. A. Shearer, T. S. Stombaugh, S. F. Higgins, S. G. McNeill and S. Sokhansanj, *Trans. ASABE*, 2007, **50**, 705.
58. R. L. Graham, R. Nelson, J. Sheehan, R. D. Perlack and L. L. Wright, *Agron. Jrnl.*, 2007, **99**, 1.
59. R. Perrin, K. Vogel, M. Schmer and R. Mitchell, *BioEn. Res.*, 2008, **1**, 91.

60. S. Sokhansanj, S. Mani, A. Turhollow, A. Kumar, D. Bransby, L. Lynd and M. Laser, *Biofuels Bioprod. Biorefining*, 2009, **3**, 124.
61. P. S. Lam, S. Sokhansanj, X. Bi, C. J. Lim, L. J. Naimi, M. Hoque, S. Mani, A. R. Womac, X. P. Ye and S. Narayan, *Appl. Eng. Agric.*, 2008, **24**, 351.
62. A. Demirbas and A. Sahin, *Fuel Process. Technol.*, 1998, **55**, 175.
63. S. Sokhansanj and A. F. Turhollow, *Appl. Eng. Agric.*, 2004, **20**, 495.
64. S. Mani, L. G. Tabil and S. Sokhansanj, *Biores. Technol.*, 2005, **97**, 1420.
65. P. Adapa, L. Tabil and G. Schoenau, *Biosystems Eng.*, 2009, **104**, 335.
66. F. Magelli, K. Boucher, H. T. Bi, S. Melin and A. Bonoli, *Biomass Bioenergy*, 2009, **33**, 434.
67. A. C. Srivastava, W. K. Bilanski and V. A. Graham, *Can. Agri. Eng.*, 1981, **23**, 109.
68. L. Wamukonya and B. Jenkins, *Biomass Bioenergy*, 1995, **8**, 175.
69. N. Mohsenin and J. Zaske, *J. Agr. Eng. Res.*, 1976, **21**, 193.
70. N. Kaliyan and R. V. Morey, *Biomass Bioenergy*, 2009, **33**, 337.
71. J. R. Hess, C. T. Wright and K. L. Kenney, *Biofuels Bioprod. Biorefining*, 2007, **1**, 181.
72. J. G. Rogers and J. G. Brammer, *Biomass Bioenergy*, 2009, **33**, 1367.
73. J. R. Hess, K. L. Kenney, C. T. Wright, R. Perlack and A. Turhollow, *Cellulose*, 2009, **16**, 599.
74. P. R. Adler, M. A. Sanderson, A. A. Boateng, P. I. Weimer and H. J. G. Jung, *Agron. Jrnl.*, 2006, **98**, 1518.
75. I. Lewandowski, J. M. O. Scurlock, E. Lindvall and M. Christou, *Biomass Bioenergy*, 2003, **25**, 335.
76. C. Igathinathane, A. R. Womac, S. Sokhansanj and L. O. Pordesimo, *Trans. ASABE*, 2006, **49**, 97.
77. K. J. Shinners and B. N. Binversie, *Biomass Bioenergy*, 2007, **31**, 576.
78. H. Ohm and J. Camberato, in *Corny News Network,* Purdue University, W. Lafayette, IN, 2008, p. 2.
79. C. W. Radtke, *Post harvest physiology of biomass storage,* Idaho National Lab, 2005.
80. A. Nolan, K. M. C. Donnell, M. M. C. Siurtain, J. P. Carroll, J. Finnan and B. Rice, *Biosystems Eng.*, 2009, **104**, 345.
81. J. S. Cundiff and L. S. Marsh, *Biores. Technol.*, 1996, **56**, 95.
82. C. Wang, J. A. Larson, B. C. English and K. Jensen, in Southern Agricultural Economics Association Annual Meeting, Atlanta, GA, 2009.
83. J. K. Gigler, W. K. P. van Loon, I. Seres, G. Meerdink and W. J. Coumans, *J. Agr. Eng. Res.*, 2000, **77**, 391.
84. R. Jirjis, *Biomass Bioenergy*, 1995, **9**, 181.
85. E. F. Kristensen and P. D. Kofman, *Biomass Bioenergy*, 2000, **18**, 175.

CHAPTER 3

The Biorefinery Concept – Thermochemical Production of Building Blocks and Syngas

BIRGIT KAMM, MIRKO GERHARDT AND SEBASTIAN LEIß

Research Institute Bioactive Polymer Systems (Biopos) e.V. and BTU Cottbus, Teltow-Seehof, Germany

3.1 Introduction

One hundred and fifty years after the beginning of coal-based chemistry and 50 years after the beginning of petroleum-based chemistry, industrial chemistry is now entering a new era. In the 21 st century utilization of renewable raw materials will gain importance in the chemical conversion of substances in industry. Partial or even complete readjustment of whole economics to renewable raw materials will require completely new approaches in research, development and production.

One approach is the development of biorefinery technologies and systems. Biorefining is the transfer of the efficiency and logic of fossil-based chemistry and the substance-conversion industry, as well as energy production, to the biomass industry.[1]

Plant biomass always consists of the basic precursors carbohydrates, lignin, proteins and fats, beside various substances such as vitamins, dyes, flavors, and aromatic essences. Biorefineries combine the essential technologies between biological raw materials and the industrial intermediates and final products.

RSC Energy and Environment Series No. 1
Thermochemical Conversion of Biomass to Liquid Fuels and Chemicals
Edited by Mark Crocker

Published by the Royal Society of Chemistry, www.rsc.org

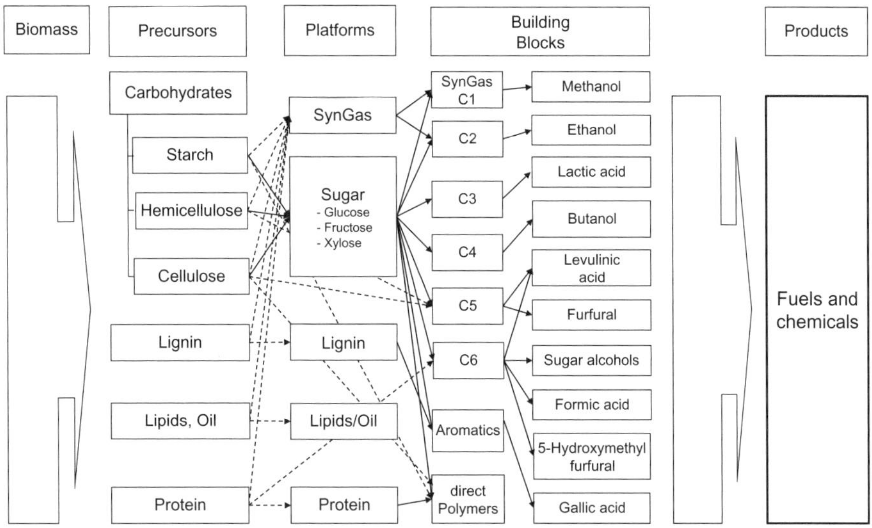

Figure 3.1 Model of a biobased product flowchart for biomass feedstocks (according to ref. 1, adapted from ref. 2).

In this chapter, the focus is directed to the lignocellulosic feedstock biorefinery system, technologies and products, and in particular to platform chemicals (building blocks) (Figure 3.1).[2]

3.2 The Lignocellulosic Feedstock Biorefinery

Among the potential large-scale industrial biorefineries, the lignocellulose feedstock (LCF)-biorefinery is probably poised for the greatest success. On the one hand, the raw material situation is optimal (straw, reed, grass, wood, paper-waste, *etc.*), on the other hand the conversion products are well positioned with respect to the traditional petrochemical as well as the future biobased product market. An important point for utilization of biomass as a chemical feedstock is the cost of the raw material.

Lignocellulose materials consist of three primary chemical fractions or precursors: (a) hemicellulose/polyoses, a sugar-polymer of predominantly pentoses; (b) cellulose, a glucose-polymer; and (c) lignin, a polymer of phenols (Figure 3.2).

The lignocellulosic biorefinery system has a distinct ability for creating genealogical trees. The main advantages of this method are that the natural structures and structure elements are preserved, the raw materials also have low price, and many product varieties are possible (Figure 3.3). Nevertheless, there is still a requirement for development and optimization of these technologies, *e.g.*, in the field of separating cellulose, hemicellulose and lignin, as well as the use of lignin in the chemical industry. The sugar polymers hemicellulose and

$$\text{Lignocellulose} + H_2O \rightarrow \text{Lignin} + \text{Cellulose} + \text{Hemicellulose}$$

$$\text{Hemicellulose} + H_2O \rightarrow \text{Xylose}$$

$$\text{Xylose } (C_5H_{10}O_5) + \text{acid Catalyst} \rightarrow \text{Furfural } (C_5H_4O_2) + 3H_2O$$

$$\text{Cellulose } (C_6H_{10}O_5) + H_2O \rightarrow \text{Glucose } (C_6H_{12}O_6)`$$

Figure 3.2 Generalized equations for biomass conversion at the LCF-biorefinery.

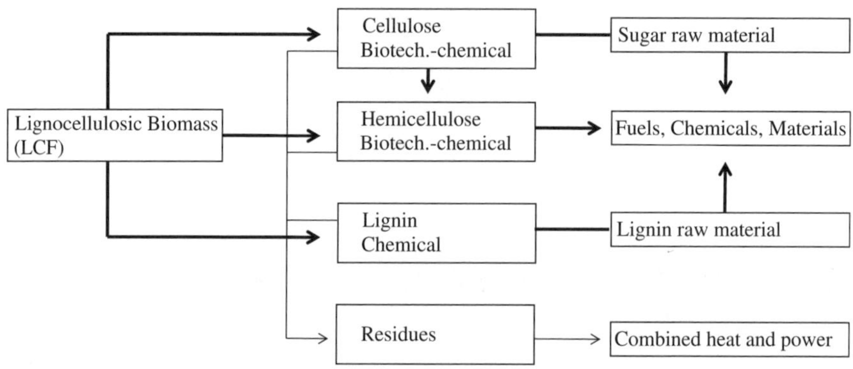

Figure 3.3 Lignocellulosic feedstock biorefinery.

cellulose could be starting materials for C2–C6 building blocks (sugar platform), whereas lignin could be a starting material for the production of syngas (syngas platform) (see Figure 3.1).

Lignocellulosic material shows a high resistance to enzymatic methods. Therefore, whereas acid hydrolysis of lignocelullosics requires much energy, enzymatic hydrolysis can only achieve acceptable carbohydrate yields when using enormous quantities of enzymes. The reason is the specific structure of plant biomass. Lignocellulose consists of the three fractions cellulose, hemicellulose and lignin. Cellulose is an unbranched glucose-polymer[3] that forms microfibrils because of the linear β-1,4-glycosidic bond, which on their part are bundled to crystalline, water-insoluble macrofibrils. Crosslinking these fibrils *via* hemicellulose provides the structural backbone of the plant cell wall.[4,5] Unlike cellulose, hemicellulose is an amorphous, branched heteropolymer that mainly consists of mannose in soft wood and of xylose in hard wood. In addition, these heteropolymers contain galactose, glucose and arabinose. This compound of cellulose and hemicellulose is enclosed by lignin, a three-dimensional linked polymer of phenylpropane units, and so it forms a very resistant complex for hydrolytic or bacterial attack. The acid hydrolysis of LCF raw material can be performed with concentrated hydrochloric acid. Advantages of this process are the exothermic nature of the hydrolysis and the possibility of acid recovery. The process is divided into prehydrolysis (32% w/w HCl) and main hydrolysis (41% w/w HCl). In this manner, the highest yields of fermentable hydrolysates are obtained. In enzymatic hydrolysis high yields of fermentable carbohydrates are also achieved (with respect to cellulose input) with high selectivity. However, in this procedure, the generation of cellulose from LCF raw material by chemical pulping is first required. The disadvantages of chemical pulping of LCF raw material are high energy consumption and low yields of cellulose. For the production of fermentable carbohydrates, acid saccharification is the preferred route. In the future, efforts should be focused on the improvement of separation of hemicellulose, cellulose and lignin, and the enhancement of hydrolysis product yields by acid saccharification.[6]

3.3 Building Blocks by Thermochemical Methods

3.3.1 Furfural

Furfural is produced by acidic hydrolysis of hemicellulose on a scale of approximately 142 000 t/a.[7] Furfural is a starting material for the production of Nylon 6,6 and Nylon 6. Indeed, the original process for the production of Nylon-6,6 was based on furfural. The last of these production plants was closed in 1961 in the USA for economic reasons (the artificially low price of petroleum). Nevertheless, the market for Nylon 6 is huge.[1]

3.3.2 Levulinic acid

A major platform chemical obtained from biomass is levulinic acid, which can be obtained from cellulose and lignocellulose. During the biomass conversion, 5-hydroxymethylfurfural (HMF) is formed *via* acid-catalyzed dehydration, which can be split *via* dehydration into levulinic acid and methanoic acid (formic acid). Even with raw materials that strongly vary in their quality from batch to batch, the yields that can be obtained are very high. The formic acid can be removed *via* distillation for further use. Today, hydrolysis processes of lignocellulosic material are carried out as thermochemical (acid catalyzed/steam assisted) or fungal (enzymatic) catalyzed reactions. Acid catalysis encounters the disadvantage of the corrosive nature of the reaction media, which must be recovered before further processing, while enzymatic reaction leads to longer residence times.

Levulinic acid is a versatile chemical intermediate (Figure 3.4).[8–10]

To decrease waste problems, the State of New York (USA) has built two levulinic acid pilot plants (1 t per day). In these pilot plants, different possibilities will be tested for the exploitation of the whole variety of carbohydrate-rich and humid waste materials (waste paper, sewage sludge) for levulinic acid production. Thermochemical processes tolerate fluctuations in feedstock composition. In the future, decentralized facilities are planned with volumes of 50 to 1000 t/day and more. The high price of levulinic acid has inhibited large-scale use. It currently has a worldwide market of about one million pounds per year at a price of $4 to $6 per pound. The New York *biofine*-process is projected

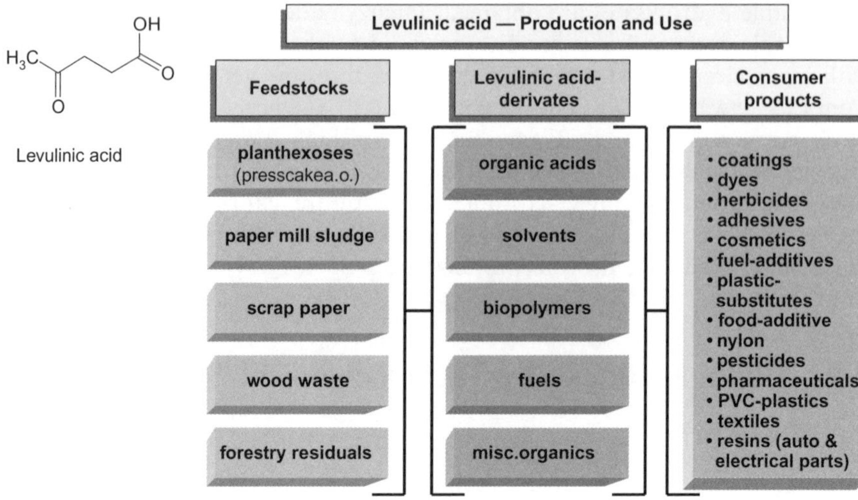

Figure 3.4 Levulinic acid – production and use.

to be capable of producing levulinic acid at \$0.04 to \$0.32 per pound, depending on the scale of operation.[4,11]

A further commercial plant for levulinic acid production has been built in Caserta, Italy and converts 3000 t/day of raw material (waste sludge from the tobacco and paper industries). The process employs a two-stage reaction scheme: continual prehydrolysis at higher temperatures (210–230 °C) with short residence time (13–25 s) in dilute acids and hydrolysis of the first stage product in a second batch reactor at temperatures of 195–215 °C and longer residence times (15–30 min). The yield is between 60–70% of the theoretical yield based on the hexose content of the biomass.

Recently, research has been carried out to develop new methods of pretreating biomass. Sub- and supercritical water have been gaining increasing attention as attractive solvent and reaction media for a variety of applications. Water is cheap, nontoxic and nonflammable and offers some other advantages in the field of "green chemistry". At high temperatures water develops acidic characteristics and accelerates cleavage of chemical bonds and lignocellulose decomposition (see Chapter 8 for further details). One approach to the hydrolysis of polymeric sugars in hemicellulose and cellulose is the implementation of sub- and supercritical water as an environmentally friendly solvent in the pretreatment stage.

3.3.3 Hydroxymethylfurfural

One method for degradation of hyperfunctionality in carbohydrates is the use of acid-catalyzed processes for selective removal of functional groups and formation of defined building blocks. 5-Hydroxymethylfurfural (HMF) is a bifunctional aromatic building block, which can be produced preferentially from inulin (fructose polymer) by acid-catalyzed intramolecular dehydration (Figures 3.5 and 3.6).[9]

Hydroxymethylfurfural, a furane derivative, has been known for over 100 years. Accordingly, numerous publications concerning HMF exist. As early as 1895, Düll,[12] as well as Kiermayer,[13] synthesized HMF *via* heating saccharose with an 0.3% oxalic acid solution (catalyst) under pressure. In the beginning of the 20 th century Ekenstein and Blanksma[14] elucidated the composition of 5-hydroxymethylfurfural. Fenton *et al.*[15–17] were the first to deal with its derivatization. In general, hydroxymethylfurfural is obtained by the catalytic dehydration of hexoses (Figures 3.5 and 3.6).

Historically, oxalic acid was applied as a dehydration catalyst for the first synthesis. However, HMF was obtained in only moderate yields. Haworth and Jones[18] obtained 5-hydroxymethylfurfural in a 54% yield by acid-catalyzed dehydration of saccharose using oxalic acid. Tables 3.1 and 3.2 give an overview of the used catalysts and resulting yields from the dehydration. According to Cottier,[19] the dehydration can be classified into five groups depending on the solvent used or the reaction conditions:

3.3.3.1 Dehydration in Aqueous Milieu

Dehydration in an aqueous milieu has the advantage that water is a sustainable and cheap solvent. However, the yields of HMF are very low. Under drastic conditions (high temperature and low pH value) there are numerous reversible and irreversible reactions in the aqueous phase such as the acid-catalyzed hydrolysis of HMF to levulinic acid and formic acid or the polymerization of HMF to water-soluble and insoluble humic substances; see, for example, Lewkowski.[20]

3.3.3.2 Dehydration in Organic Polar Solvents

Upon implementation of dehydration in organic-polar solvents such as dimethyl sulfoxide (DMSO) or polyethylene glycol, HMF is gained in very high yields of up to 90%. Because of the low concentration of water, the hydrolysis to levulinic acid and formic acid does not happen and the tautomeric equilibrium between fructofuranose and fructopyranose is on the side of the favorable pyranose form. The disadvantage is, on the one hand, the high dilution, due to the fact that the reactants (*e.g.*, hexoses) possess low solubility in organic solvents. On the other hand, it is difficult to separate the HMF from DMSO. Furthermore, there are the high costs of the organic solvents to consider compared to water.

3.3.3.3 Dehydration in the Two-Phase System Water/Methyl Isobutyl Ketone

The use of a two-phase system of water and methyl isobutyl ketone enables continuous removal of the generated HMF and therewith a displacement of the

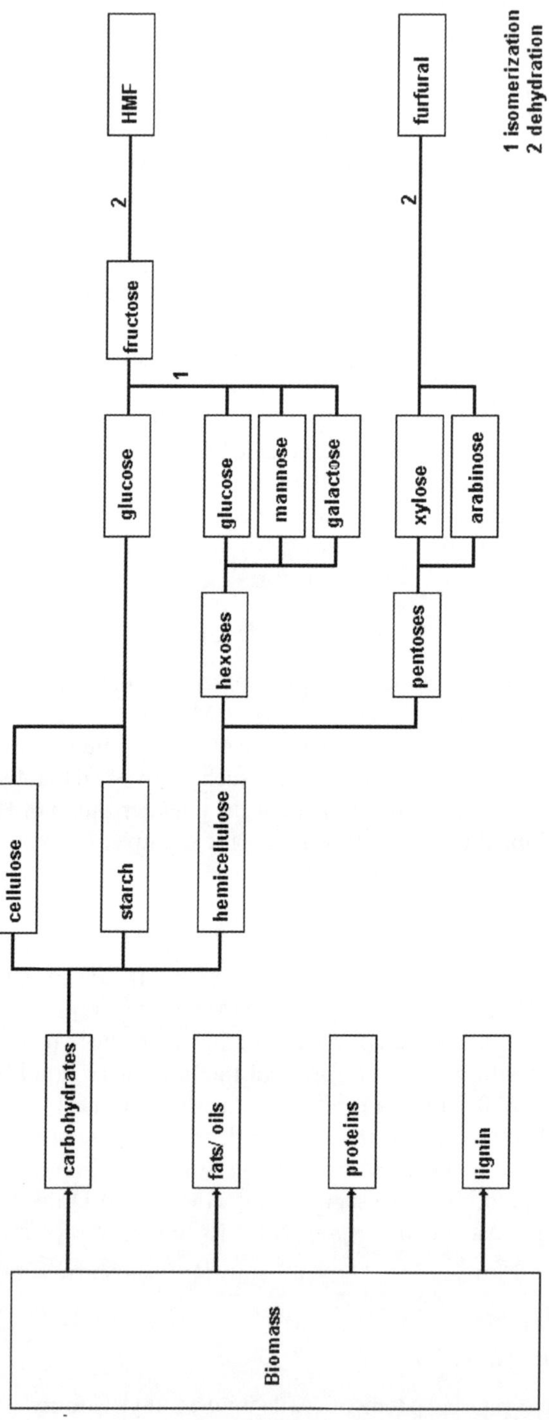

Figure 3.5 Flowchart for HMF production.

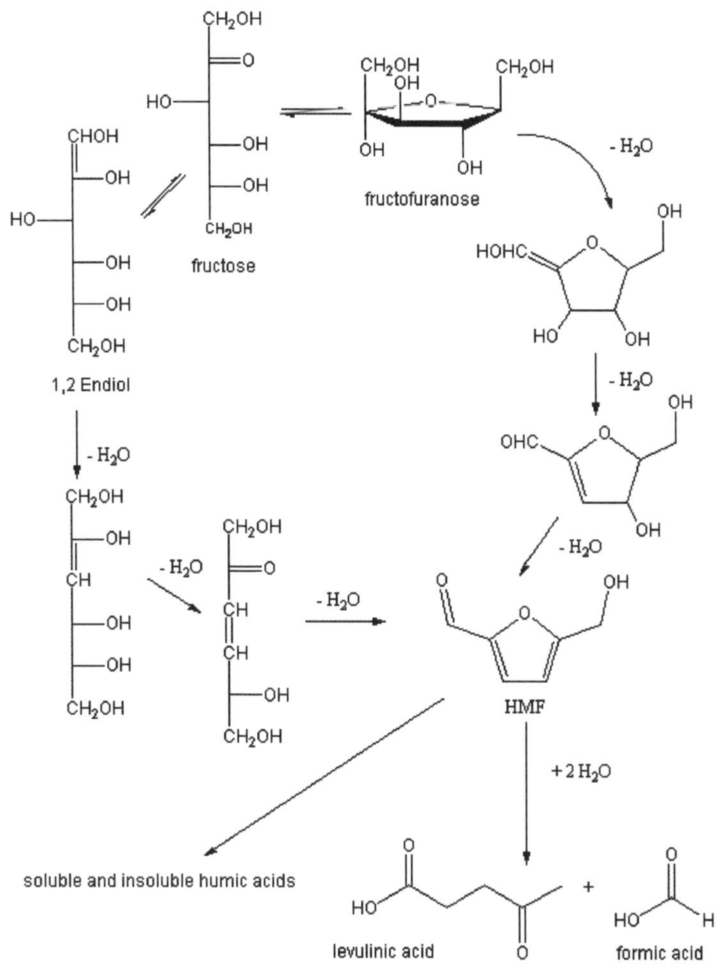

Figure 3.6 Reaction scheme for fructose dehydration.

chemical equilibrium in favor of the generation of HMF. Thereby, yields of up to 90% can be obtained.[21–31] The disadvantage is the high solubility of HMF in water, which requires a large excess of the extracting agent.

3.3.3.4 Dehydration in Ionic Liquids

Further interesting reaction media for the dehydration of fructose are ionic liquids. In 2003 Moreau *et al.*[32] realized the dehydration of fructose in 1-H-3-methylimidazolium chloride, the ionic liquid acting as a solvent as well as a catalyst. HMF was obtained in yields of 92% at 90 °C in 15–45 min. The separation of HMF occurs continuously with diethyl ether, so that the ionic liquid could be reused. Yong *et al.*[33] obtained HMF in a yield of up to 99%

Table 3.1 Overview of catalysts used for the dehydration of fructose.

organic acids	*inorganic acids*	*salts*	*Lewis acids*	*Others*
oxalic acid[12,13,18,35,36]	H_2SO_4, H_3PO_4, HCl[39–50]	$(NH_4)_2SO_4/SO_3$[51]	$ZnCl_2$[67]	iodine[69–71,72]
levulinic acid[37]		pyridine/HCl[52]	$AlCl_3$[67]	acidic ion-exchanger-resins[73–82]
p-toluenesulfonic acid[38,39]		$ZrOCl_2$[53,54] $VOSO_4$, TiO_2[55,56]	BF_3[68]	zeolite[21]
		Zr-phosphate[53,54]		ionic liquids[32,62,83–85]
		vanadyl phosphate[57]		
		Cr-, Al-, Ti-, Ga-, In-salts[58]		
		vanadium porphyrin[59]		
		Zr-, Cr-, Ti-porphyrin[60,61]		
		$CuCl_2/CrCl_2$[62]		
		$(NH_4)_3PO_4$[63,64]		
		$(HNEt_3)_3PO_4$[63,64]		
		pyridinium phosphate[63–66]		

Table 3.2 Overview of yields obtained in the dehydration of fructose to HMF[a]

solvents	*catalysts*	*T in °C*	*t in min*	*X*	*Y*	*S*	$wt\%_{fructose}$	*Ref.*
DMSO	BF_3-etherate	0	30	97%	>90%		25%	106
DMSO	Diaion PK-216	80	500		90%		7.5%	107
DMF	Amberlyst 15	100	180	100%	73%		3.4%	108
IL	$CrCl_2$	100	360		99%		10%	33
IL	$PtCl_2$	80	180	98%	83%		10%	84
IL/THF	WCl_6	50	240		72%		20%	85
H_2O	TiO_2	200	5	73%	35%		2%	109
H_2O	H_3PO_4	240	2	98%	65%		1%	110
H_2O	$NbOPO_4$	100	120	61%	21%		6%	111
H_2O saturated with inorg. salts/butanol	HCl	180	3	74%	66%		30%	112
H_2O/MIBK	zeolite	165	60	76%	69%		1.7%	113
H_2O/MIBK	H-mordenite	165	60	76%		>90–95%	2%	114
H_2O/DMSO/PVP	ion exchanger	90	480	76%	58%		10%	87
acetone/water	H_2SO_4	180	2	98%		77%	0.9%	115
acetone/water	ion exchanger	150	15	95%	73%		2%	116
THF/water	HCl	150	35	53%		83%	30%	117

[a]X – conversion, Y – yield, S – selectivity, MIBK – methyl isobutyl ketone, IL – ionic liquid, PVP – poly(vinyl pyrrolidine)

from the $CrCl_2$-catalyzed dehydration of fructose in 1-butyl-3-methylimidazolium chloride. It is noteworthy that the generated HMF is completely stable in the ionic liquid and that there are no degradative reactions to levulinic acid and formic acid. A disadvantage compared to water as a reaction medium is the high cost of the ionic liquid.

3.3.3.5 Dehydration Without Solvents

If the dehydration is performed without solvents, HMF is obtained in high purity, because secondary reactions like polymerization reactions to humic acids and hydrolysis of the HMF to levulinic acid do not happen. Fayet and Gelas[34] worked with equimolar quantities of hexoses and pyridinium salts and obtained HMF with a yield of 70%.

Schiwek *et al.* suggested that HMF can be converted into furan-2,5-dicarboxylic acid (FDCA) by selective oxidation.[86] FDCA can be used as a replacement for terephthalic acid in the production of polyethyleneterephthalate and polybutyleneterephthalate. Futhermore, the reduction of HMF can lead to products such as 2,5-dihydroxymethylfuran and 2,5-bis(hydroxymethyl) tetrahydrofuran, which can serve as alcohol components in the production of polyesters, providing completely biomass-derived polymers when combined with FDCA. In addition, HMF can serve as a precursor in the synthesis of liquid alkanes to be used, for example, in diesel fuel.[5]

Thus, it would be desirable to develop an efficient industrial process for the production of HMF. A disadvantage of the known HMF production technologies (starting with fructose in homogeneous aqueous sulfuric acid solution) is the formation of unwanted byproducts caused by fragmentation, condensation, additional dehydration, and reversion. Using acidic ion-exchangers and dimethylsulfoxide as the reaction medium it is possible to suppress side reactions. However, the product recovery *via* distillation of DMSO is very expensive.

Román-Leshkov *et al.*[87] recently reported selective intramolecular dehydration at fructose concentrations in the range of 10–50 wt% leading to 5-hydroxymethylfurfural in high yields (80% HMF selectivity, 90% fructose conversion) in a biphasic reaction, where both the organic and aqueous phase are modified. The acid catalyst (hydrogen chloride or an acidic ion-exchanger) modified with a polar aprotic molecule (dimethylsulfoxide) and/or 1-methyl-2-pyrrolidone (NMP) is located in the aqueous phase together with a hydrophilic polymer, poly(1-vinyl-2-pyrrolidone) (PVP), for suppression of side products. The formed HMF was continuously extracted into an organic phase of methyl isobutyl ketone (MIBK) modified with 2-butanol to support HMF separation. Usually, the separation works at a ratio of $[HMF]_{aq} : [HMF]_{org} = 1 : 10$, depending on the composition of the single phases. It was determined by simulation that the separation of the methyl isobutyl ketone solvent *via* vacuum distillation at a concentration of 10 wt% HMF needs 40% less energy compared to the conventional process using DMSO.

Thus, it is possible to produce HMF in high yields from carbohydrates. However, further experimental work should be directed at the conversion of low-cost carbohydrate sources, such as glucose and xylose, accessible from lignocellulosic biomass.[1,2]

As a bifunctional molecule, HMF possesses a rich chemistry and so it opens up access to a multitude of chemical compounds that can be used in different fields of the chemical and pharmaceutical industries (Figure 3.7). Thus, for example, the aldehyde group can be converted by diverse reactions such as nucleophilic addition or the Wittig reaction. The hydroxyl group also allows a versatile consecutive chemistry. For example, it can be etherified, esterified or halogenated. Access to valuable monomer molecules can be obtained by chemical reduction of the aldehyde function to 2,5-bis(hydroxymethyl)furan. By oxidation of the aldehyde and the hydroxymethyl group, furandicarboxylic acid can be obtained, which may be able to replace terephthalic acid, used in the manufacture of PET (polyethylene terephthalate), in the future. Because of this multitude of secondary products, HMF offers a key position for the development of new product trees in the chemical industry. A future challenge is to scale up existing processes to industrial scale. Towards this goal, attention has to be focused on processes that use water as a reaction agent, because this is a cheap and an especially environment friendly solvent. However, for this purpose further research is necessary towards more effective and efficient catalyst systems that increase HMF yields during synthesis.

3.3.4 Sugar Alcohols

The synthesis of sugar alcohol platform chemicals from cellulose as starting material is already a subject of investigation. Due to its crystallinity, degree of polymerization, and availability of chain ends, the hydrolysis of cellulose is carried out under drastic reaction conditions. For the degradation of cellulose mineral acids, bases, supercritical water, and enzymes are used.[88] Problems associated with these methods are low activity and/or selectivity, separation of products and catalysts, corrosion, and generation of a large amount of neutralization waste.

Consequently, heterogeneous catalysis (Pt and Ru catalysts) should represent a more modern, environmentally friendly method of cellulose degradation.[89] Recently, Fukuoka and Dhepe have shown the possibility of direct cellulose conversion into sugar alcohols using a series of heterogeneous catalysts, such as $Pt/\gamma\text{-}Al_2O_3$, $Pt/SiO_2\text{-}Al_2O_3$, and ultrastable Y zeolite in water. Yields of sugar alcohols of approximately 30% have been achieved at an initial H_2 pressure of 5 MPa, reaction temperature of 463 K and a reaction time of 24 h.

The space-time-yield of the reaction – mass of cellulose converted per mass of catalyst per hour – of 0.09 is very low. However, catalysts can be reused after washing with water for up to three cycles.[90] Sorbitol and mannitol are formed in a molar ratio of 4:1 or higher. For the formation of mannitol an epimerization reaction of sorbitol has been proposed and proved experimentally

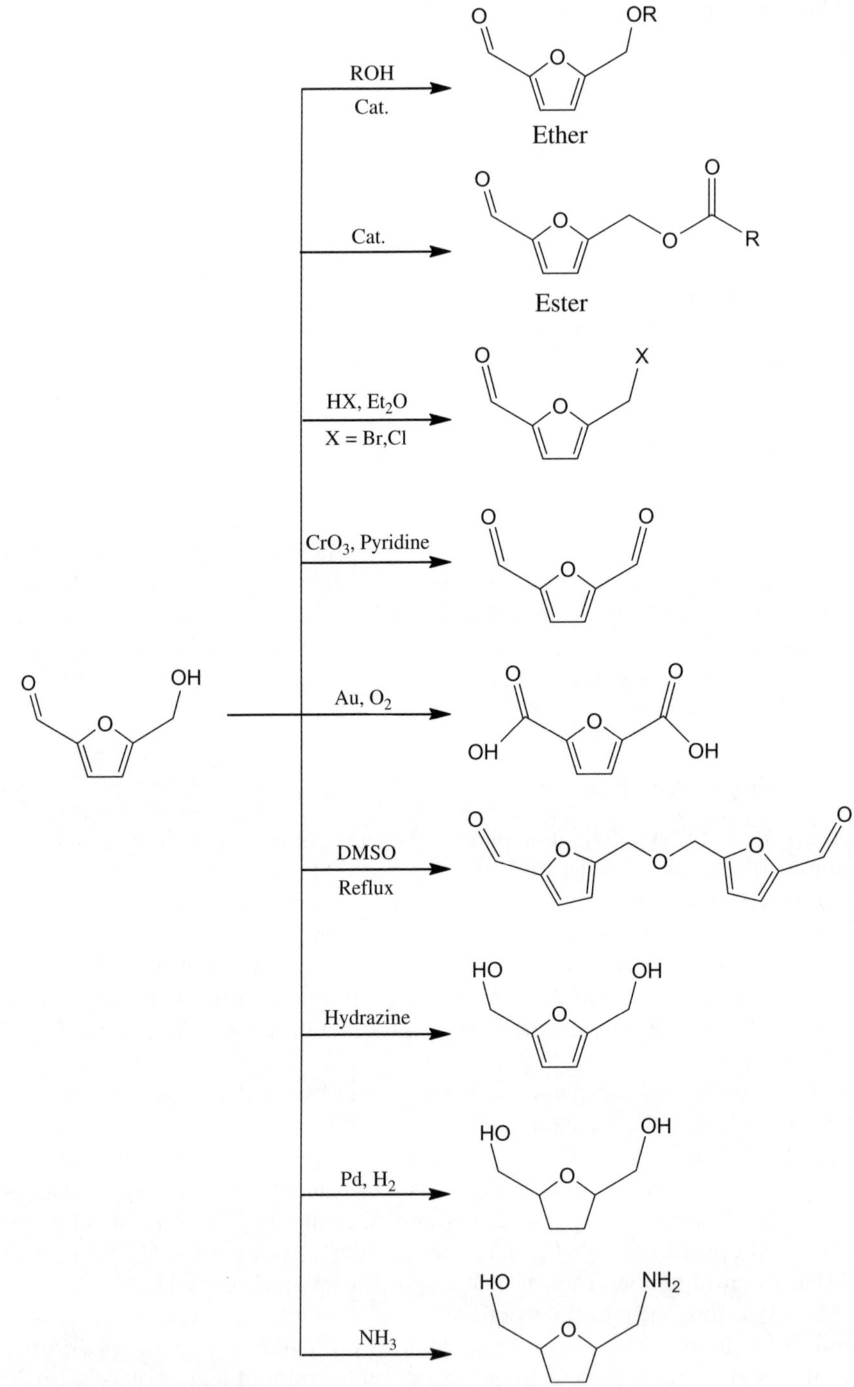

Figure 3.7 Secondary products from HMF.

(Figure 3.8). For the overall reaction the hydrolysis is discussed as the rate-determining step, because the reduction of glucose gave an almost stoichiometric amount of sorbitol (using Pt or Ru catalysts).

Using the support materials as catalysts, such as SiO_2, the H-form of Y zeolite, H-ZSM-5, or H-Mordenite, the corresponding glucose yields are reduced to less than 4%, indicating that the metal promotes the hydrolysis of

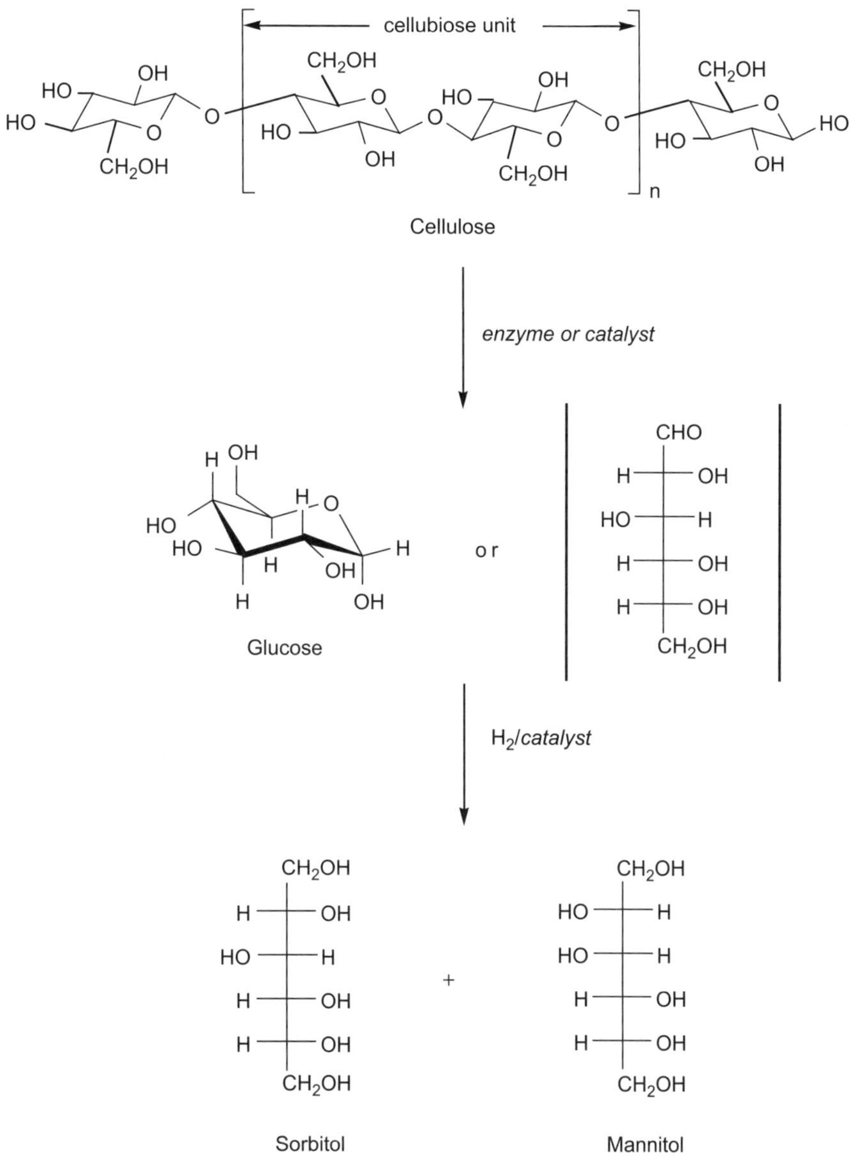

Figure 3.8 Catalytic conversion of cellulose into sugar alcohols.

cellulose. Hence, it is suggested that the acid sites necessary are generated *in situ* from H_2 (*via* dissociative adsorption on the metal and spillover onto the support) in addition to the acidic sites intrinsic to the support.

Sorbitol is applied not only in sweeteners, but also as a platform chemical for the synthesis of isosorbide, 1,4-sorbitan, glycerol, glycols and lactic acid,[2] as well as a potential platform chemical for the production of liquid hydrocarbons.

3.3.5 Syngas

The *two-platform* concept considers both the synthesis of defined functional platform chemicals (sugar platform) and the production of synthesis gas (syngas platform) from biomass.[1] However, there are still some unsatisfactory areas within the LCF system, such as the utilization of lignin as a starting material for syngas.

Generally, syngas can be converted into methanol and subsequently into a variety of sequential products. Furthermore, hydrocarbons that are used as fuels, are accessible *via* Fischer–Tropsch technology.[91]

Regarding economic aspects, Fischer–Tropsch liquids from biomass are not yet available, which means a challenge for research and development.[9] Synthesis gas production methods in conventional Fischer–Tropsch plants require either an O_2-plant or a large Fischer–Tropsch reactor to process the synthesis gas stream that is diluted with air/nitrogen, thereby increasing the capital costs of such facilities. About 50% of the costs of Fischer–Tropsch liquid production from biomass are associated with capital costs, of which 50% are caused by biomass gasification, gas cleaning, and synthesis-gas processing.[92] The company CHOREN in Freiberg (Germany) is producing fuels from renewable resources *via* syngas and Fischer–Tropsch synthesis (10 000 t p.a.).[93]

Soares *et al.* propose a new approach that may allow the operation of a small-scale Fischer–Tropsch reactor producing/utilizing undiluted H_2/CO gas mixtures.[94] They use glycerin as a starting material, which currently is produced as a coproduct in biodiesel manufacturing, and of which there is a large surplus on the market. The combination of both processes, (1) endothermic (350 kJ mol^{-1}) synthesis gas production and (2) exothermic (412 kJ mol^{-1}) Fischer–Tropsch conversion, results in an exothermic process overall with an enthalpy of -63 kJ mol^{-1}. This corresponds approximately to the calorific value of glycerin (Figure 3.9).

Synthesis-gas production was carried out using 30 wt% glycerol solution and Pt catalysts on various carrier materials, such as Al_2O_3, CeO_2/ZrO_2, and MgO/ZrO_2. Pt-Al_2O_3 showed the most stable catalyst performance for this oxidation. Using this catalyst, the process conditions were precisely investigated, the temperature being varied from 573 to 623 K and the pressure from 1 to 20 bar. In addition to the primary reaction (synthesis of CO/H_2), the secondary reaction (synthesis of CO/CO_2) was also investigated. Finally, an overall carbon balance of 10% was achieved regarding the conversion of

$$\underset{\text{glycerol}}{C_3O_3H_8} \longrightarrow 3CO + 4H_2 \quad (1)$$

$$\underset{\text{glycerol}}{C_3O_3H_8} \longrightarrow \frac{7}{25}\,C_8H_{18} + \frac{19}{25}\,CO_2 + \frac{37}{25}\,H_2O \quad (2)$$

Figure 3.9 Production of octane from glycerol *via* synthesis gas.

glycerol into synthesis gas. The final product stream contained unconverted glycerol and small amounts of ethylene glycol, methanol, hydroxypropanone, and ethanol. The conclusion of the authors to closely combine the investigation of syngas production with the corresponding *consequence chemistry* should expressly be supported.

3.3.6 Lactic Acid

Lactic acid (2-hydroxypropionic acid) is the most widely occurring hydroxycarboxylic acid. Due to the two reactive functional groups it displays a rich chemistry and can undergo a variety of chemical conversions.[95] Because of its considerable potential as a feedstock monomer for oxygenated chemicals and the biodegradable plastic PLA (polylactic acid), its worldwide production of about 250 000 t/year (2004) is expected to increase rapidly.[96] Today, fermentation of glucose (conversion 100%; selectivity >90%; 47 °C; pH = 5.5; space–time yield = 12 g/L h) with lime neutralization is state-of-the-art. However, because of the pH optimum of the applied homofermentative lactic acid strains (>5), the produced lactic acid ($pK_a = 3.86$) has to be neutralized to the same extent as it is formed. To release the lactic acid from the lactic acid salt sulfuric acid is added, thus causing an accumulation of 0.8 t gypsum/t lactic acid as a byproduct.[97] Furthermore, lactic acid bacteria need complex fermentation media that directly affect the production costs.[98]

To avoid these disadvantages and to develop a more sustainable process, the chemocatalytical conversion of carbohydrates to lactic acid is under investigation. Bicker *et al.* have postulated the mechanism of the conversion of fructose and glucose in subcritical and supercritical water.[99] Under acid catalysis, mainly HMF is formed and only very small amounts of lactic acid, whereas alkali catalysis provides organic acids and their derivatives. In this case, lactic acid is not selectively produced but it represents the main product (25% mol/mol).[100] However, these yields are not sufficient for process development. Lactic acid yield can be increased when adding $ZnSO_4$ as catalyst. Bing *et al.* showed 48% and 42% yields (mol/mol) from fructose and glucose at 300 °C, 25 MPa, 400 ppm (g/g) $ZnSO_4$ and 30 s residence time. $ZnSO_4$ catalyzes this reaction by forming a pyruvaldehyde/zinc complex that rearranges *via* a hydride shift to lactic acid, thus affecting all of the prior equilibrium reactions.[99] Under the aforementioned conditions, the carbohydrates are fully converted and the produced lactic acid is not obtained as a salt (Figure 3.10).

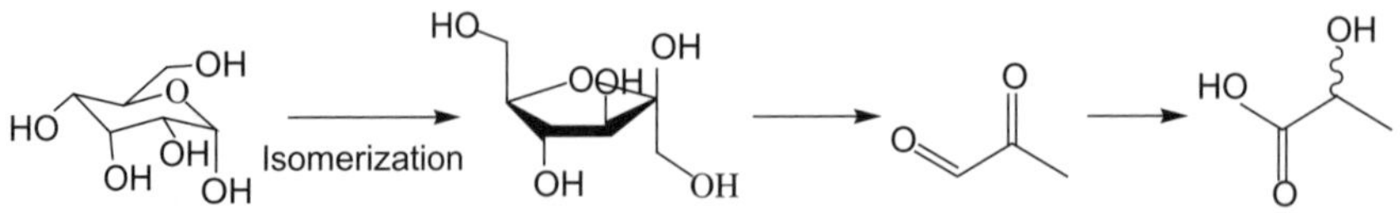

Figure 3.10 Simplified pathway for the conversion of fructose and glucose to lactic acid.

Besides carbohydrates, other starting materials for chemocatalytic production of lactic acid are being investigated. Shen *et al.* used glycerin as a feedstock and achieved a lactic acid yield of 90%,[101] whereas Kishida *et al.* used glycolaldehyde resulting in a yield of 28%.[102] Kong *et al.* used certain types of biomass directly, such as maize straw. However, when using maize straw for catalytic conversion, yields were only about 9% (calculated as mass balance).[103]

Although there have been recent improvements in the field of chemical production of lactic acid, the new processes cannot compete with the established biotechnological conversion. On the one hand, the chemocatalysis has the major advantage that there is no formation of gypsum during the separation step, but on the other hand optical purity cannot be provided. Especially for the production of PLA the optical purity is crucial. Furthermore, the yields described by Bicker *et al.* can only be achieved with low feed concentration, thus causing low space–time yields.

Although the fermentative production of lactic acid is already established, there is still considerable potential for further improvements. Kamm *et al.* showed that it is possible to use organic amines for the production of aminium lactates and high-purity lactic acid without byproducts.[104] Furthermore, reactive extraction systems that remove lactic acid *in situ* or *ex situ* are also under investigation.[105] If it is possible to develop lactic acid bacteria that accept pH values under 4 and cheap nutrient sources, the biotechnological production of lactic acid will become more economic and sustainable.

3.4 Outlook

An important challenge for industrially relevant biorefinery technologies is the efficient synthesis of chemically unified base chemicals with the ability for family tree formation. It is necessary to combine biomass-degradation processes closely with building processes of the intermediates and final products. Therefore, biological and chemical processes have to be combined and new catalysts (heterogeneous, enzymatical) for these processes have to be developed.

References

1. B. Kamm, P. Gruber and M. Kamm, *Biorefineries, Industrial Processes and Products, in Ullmann's Encyclopedia of Industrial Chemistry,* Wiley-VCH, 2007.

2. Top Value Added Chemicals from Biomass, T. Werpy, G. Petersen (eds.), US Department of Energy, Office of Scientific and Technical Information, DOE/GO-102004-1992, 2004.
3. M. J. Taherzadeh and K. Karimi, *Int. J. Mol. Sci.*, 2008, **9**, 1621.
4. NYSERDA (New York State Energy Research and Development Authority), Commercializing Biomass Technologies in New York State. Producing a High-Value Chemical from Biomass (Levulinic acid), project paper, NYSERDA, Albany, New York, 1998.
5. J. O. Metzger, *Angew. Chem.*, 2006, **118**, 710; *Angew. Chem. Int. ed.* 2005, **45**, 696.
6. B. Kamm, M. Kamm, M. Schmidt, I. Starke and E. Kleinpeter, *Chemosphere*, 2006, **62**(1), 97.
7. G. Agricola, De natura fossilium libri X, Basel 1546, Reprint: Dover Publ., Mineola, N.Y., 2004, ISBN 0486-49591-4.
8. J. Dahlmann, *Chem. Ber.*, 1968, **101**, 4251.
9. B. F. M. Kuster, *Starch/Stärke*, 1990, **42**, 314.
10. E. S. Olson, R. K. Micelle, A. J. Schlag and R. K. Sharma, *ACS Symp. Ser.*, 2001, **794**, 51.
11. S. Fitzpatrick, *Process for Conversion of Cellulosic Biomass to Chemicals, in Green Chemistry and Engineering, Proc. 3rd Ann. Conf.,* American Chemical Society, Washington, D.C., USA, June 1999.
12. G. Düll, *Chem. Ztg.*, 1895, **19**, 216.
13. J. Kiermayer, *Chem. Ztg.*, 1895, **19**, 1003.
14. W. A. Ekenstein and J. J. Blanksma, *Chem. Ber.*, 1919, **43**, 2355.
15. H. J. H. Fenton and M. J. Gostling, *J. Chem. Soc.*, 1899, **75**, 423.
16. H. J. H. Fenton and M. J. Gostling, *J. Chem. Soc.*, 1901, **79**, 807.
17. H. J. H. Fenton and F. Robinson, *J. Chem. Soc.*, 1909, **95**, 1334.
18. W. N. Haworth and W. G. M. Jones, *J. Chem. Soc.*, 1944, 667.
19. L. Cottier and G. Descotes, *Trends Heterocycl. Chem.*, 1991, **2**, 233.
20. J. Lewkowski, *Arkivoc*, 2001, **I**, 17.
21. D. Mercadier, L. Rigal, A. Gaset and J. P. Gorrichon, *J. Chem. Technol. Biotechnol.*, 1981, **31**, 489.
22. D. Mercadier, L. Rigal, A. Gaset and J. P. Gorrichon, *J. Chem. Technol. Biotechnol.*, 1981, **31**, 497.
23. D. Mercadier, L. Rigal, A. Gaset and J. P. Gorrichon, *J. Chem. Technol. Biotechnol.*, 1981, **31**, 503.
24. D. Mercadier, Dissertation, Polytechnikum Tolouse, 1980.
25. L. Rigal, A. Gaset and J. P. Gorrichon, *Ind. Eng. Chem. Prod. Res. Dev.*, 1981, **20**, 719.
26. L. Rigal, J. P. Gorrichon and A. Gaset, *Biomass*, 1985, **7**, 27.
27. J. Jow, G. L. Rorrer, M. C. Hawley and D. T. A. Lamport, *Biomass*, 1987, **14**, 185.
28. C. Moreau, R. Durand, C. Pourcheron and S. Razigade, *Ind. Crops Prod.*, 1994, **3**, 85.
29. P. Rivalier, J. Duhamet, C. Moreau and R. Durand, *Catal. Today*, 1995, **24**, 165.

30. C. Moreau, R. Durand, S. Razigade, J. Duhamet, P. Faugeras, P. Rivalier, P. Ros and G. Avignon, *Appl. Cat. A: Gen.*, 1996, **145**, 211.
31. C. Moreau, R. Durand, J. Duhamet and P. Rivalier, *J. Carbohydr. Chem.*, 1997, **16**, 709.
32. C. Moreau, A. Finiels and L. Vanoye, *J. Mol. Catal. A: Chem.*, 2006, **253**, 165.
33. G. Yong, Y. Zhang and J. Y. Ying, *Angew. Chem. Int. ed.*, 2008, **47**, 9345.
34. C. Fayet and J. Gelas, *Carbohydr. Res.*, 1983, **122**, 59.
35. J.J. Breen, Ph.D. Thesis, Delft University of Technology, July 1964.
36. J. A. Middendorp, *Rec. Trav. Chim.*, 1919, **38**, 1.
37. J.D. Garber, R.E. Jones, US patent 2,929,823 (1960).
38. H. E. Van Dam, A. P. G. Kieboom and H. van Bekkum, *Starch*, 1986, **38**, 95.
39. C. J. Moye, *Aust. J. Chem.*, 1966, **19**, 2317.
40. S.W. McKibhins, Ph.D. Thesis, University of Wisconsin, USA, 1958.
41. B.F.M. Kuster, Ph.D. Thesis, Eindhoven University of Technology, June 1975.
42. K. Hamada, H. Yoshihara, G. Suzukamo and O. Hiroaki, *Bull. Chem. Soc. Jpn.*, 1984, **57**, 307.
43. Q.P. Peniston, US patent 2,750,394 (1956).
44. F.H. Snyder, US patent 2,851,468 (1958).
45. J. F. Hurris, J. F. Saeman and L. L. Zoch, *Forest Prods. J.*, 1960, **10**, 125.
46. R.A. Hales, J.W. Le Maistre, G.O. Orth, US patent 3,071,599 (1963).
47. C. J. Moye and Z. S. Krzeminski, *Aust. J. Chem.*, 1963, **16**, 258.
48. C. J. Moye and R. J. Goldsack, *J. Appl. Chem.*, 1966, **16**, 206.
49. B.M. Smythe, C.J. Moye, US patent 3,219,484 (1965).
50. K. Rapp, DOS 3,601,281 (1987).
51. N.H. Smith, US patent 3,118,912 (1964).
52. C. Fayet and J. Gelas, *Carbohydr. Res.*, 1983, **122**, 59.
53. Y. Nakamura, *Jpn. Kokai Tokkyo Koho*, 8,013,243 (1980).
54. Y. Nakamura, *Noguchi Kenkyusho Jiho*, 1980, **23**, 25.
55. V. I. Krupenskii, *Khim. Drev.*, 1983, **2**, 86.
56. V.I. Krupenskii, USSR SU patent 1,054,349 (1983).
57. Y. Nakamura and S. Morikawa, *Bull. Chem. Soc. Jpn.*, 1980, **53**, 3705.
58. V. I. Krupenskii, *lsv. Vyssh. Uchebn. Zaved., Khim. Khim. Tekhnol.*, 1981, **24**, 1081.
59. V.I. Krupenskii, USSR SU patent 1,351,932 (1987).
60. G. P. Potapov, V. I. Krupenskii and M. I. Alieva, *React. Kinet. Catal. Lett.*, 1985, **28**, 331.
61. G. P. Potapov, V. I. Krupenskii and M. I. Alieva, *Kinet. Catal.*, 1987, **28**, 205.
62. Y. Su, H. M. Brown, X. Hunang, X. Zhou, J. E. Amonette and Z. C. Zhang, *Appl. Catal. A: Gen.*, 2009, **361**, 117.
63. M. L. Mednick, *Chem. Eng. News*, 1961, **11**, 75.
64. M. L. Mednick, *J. Org. Chem.*, 1962, **27**, 398.

65. H. D. Stenzenberger, M. Herzog, R. Scheiblich and E. Haberbosch, *Papier (Darmstadt)*, 1986, **40**, 233.
66. H. D. Stenzenberger, M. Herzog, R. Scheiblich and E. Haberbosch, *Papier (Darmstadt)*, 1986, **40**, 323.
67. G. Bolcs, *Hung.*, 1944, 132.
68. H. H. Szmant and D. D. Chundury, *J. Chem. Technol. Biotechnol.*, 1981, **31**, 135.
69. T. G. Bonner, E. J. Bourne and M. Ruszkiewicz, *J. Chem. Soc.*, 1960, 787.
70. A. M. Shur, G. V. Roitburd and I. G. Yazlovetskii, *Tr. Kishinev. Politekh, Inst.*, 1966, **5**, 67.
71. S. Morikawa, *Noguchi Kenkyusho Jiho*, 1978, **21**, 25.
72. W. A. Bonner and M. R. Roth, *J. Am. Chem. Soc.*, 1959, **81**, 5454.
73. P. Vinke and H. van Bekkum, *Proc. 3rd Symposium on Inulin,* Wageningen, 1 March 1989.
74. D. Mercadier, L. Rigal, A. Gaset and J. P. Gorrichon, *J. Chem. Technol. Biotechnol.*, 1981, **31**, 489.
75. R. A. Schraufnagel and H. F. Rase, *Ind. Eng. Chem., Prod. Res. Dev.*, 1975, **14**, 40.
76. Y. Nakamura, S. Morikawa, *Jpn. Kokai Tokkyo Koho,* 7,9154,757 (1979).
77. G. Fléche, A. Gaset, J.P. Gorrichon, E. Truchot, P. Sicard, *Fr. Demande FR* 2,464,260 (1982).
78. A. Gaset, L. Rigal, G. Paillassa, J.P. Salome, G. Fléche, *Fr. Demande FR* 2,551,754 (1985).
79. L. R. Bicker, DOS 3,309,564 (1984).
80. Rigal, A. Gaset and J. P. Gorrichon, *Eng. Chem., Prod. Res. Dev.*, 1981, **20**, 719.
81. D. W. Brown, A. J. Floyd, R. G. Kinsmann and Y. Roshan-Ali, *J. Chem. Technol. Biotechnol.*, 1982, **32**, 920.
82. T. El Hajj, A. Masroua, J. C. Martin and G. Descotes, *Bull. Soc. Chim. Fr.*, 1987, **5**, 855.
83. C. Lansalot-Matras and C. Moreau, *Catal. Comm.*, 2003, **4**, 517.
84. H. B. Zhao, J. E. Holladay, H. Brown and Z. C. Zhang, *Science*, 2007, **316**, 1597.
85. J. Y. G. Chan and Y. Zhang, *ChemSusChem*, 2009, **2**, 731.
86. H. Schiwek, M. Munir, K. M. Rapp, B. Schneider and M. Vogel, New Developments in the Use of Sucrose as an Industrial Bulk Chemical, in *Carbohydrates as Organic Raw Materials,* F.W. Lichtenthaler (ed.), VCH, 1991, pp. 57–94.
87. Y. Román-Leshkov, J. N. Chheda and J. A. Dumesic, *Science*, 2006, **312**, 1933.
88. H. A. Krässig, *Cellulose: Structure, Accessibility and Reactivity,* Gordon and Breach Science Publishers, Amsterdam, 1993.
89. P. L. Dhepe, M. Ohashi, S. Inagaki, M. Ichikawa and A. Fukuoka, *Catal. Lett.*, 2005, **102**, 163.
90. A. Fukuoka and P. L. Dhepe, *Angew. Chem. Int. ed.*, 2006, **45**, 5161; *Angew. Chem.* 2006, **118**, 5285.

91. M. Lancaster, The Syngas Economy, in *Green Chemistry,* The Royal Society of Chemistry, Cambridge, UK, 2002, p. 205.
92. C. N. Hamelinck, A. P. C. Faaij, H. den Uil and H. Boerrigter, *Energy*, 2004, **29**, 1743.
93. http://www.fv-sonnenenergie.de/publikationen/05_b_kraftstoffe_01.pdf.
94. R. R. Soares, D. A. Simonetti and J. A. Dumesic, *Angew. Chem. Int. ed.*, 2006, **45**, 3982; *Angew. Chem.* 2006, **118**, 4086.
95. Y. Wee, J. Kim and H. Ryu, *Food Technol. Biotechnol.*, 2006, **44**, 163.
96. P. Branduardi, C. Smeraldi and D. Porro, *J. Mol. Microbiol. Biotechnol.*, 2008, **15**, 31.
97. P. Claus and H. Vogel, *Chem. Eng. Technol.*, 2008, **31**, 678.
98. K. Hofvendahl and B. Hahn-Hägerdal, *Enzyme Microbial Technol.*, 2000, **26**, 87.
99. M. Bicker, S. Endres, L. Ott and H. Vogel, *J. Mol. Catal. A: Chem.*, 2005, **239**, 151.
100. B. Y. Yang and R. Montgomery, *Carbohydr. Res.*, 1996, **280**, 27.
101. Z. Shen, *Ind. Eng. Chem. Res.*, 2009, **48**, 8920.
102. H. Kishida, F. Jin, X. Yan, T. Moriya and H. Enomoto, *Carbohydr. Res.*, 2006, **341**, 2619.
103. L. Kong, G. Li, H. Wang, W. He and F. Ling, *J. Chem. Technol. Biotechnol.*, 2008, **83**, 383.
104. B. Kamm and M. Kamm, *Chem. Biochem. Eng. Quarterly*, 2004, **18**, 1.
105. P. Claus and H. Vogel, *Chem. Eng. Technol.*, 2008, **31**, 678.
106. H. H. Szmant and D. D. Chundury, *J. Chem. Technol. Biotechnol.*, 1981, **31**, 135.
107. Y. Nakamura and S. Morikawa, *Bull. Chem. Soc. Jpn.*, 1980, **53**, 3705.
108. A. Takagaki, M. Ohara, S. Nishimura and K. Ebitani, *Chem. Commun.*, 2009, 6276.
109. X. Qi, M. Watanabe, T. M. Aida and R. L. Smith Jr, *Catal. Comm.*, 2008, **9**, 2244.
110. F. S. Asghari and H. Yoshida, *Ind. Eng. Chem. Res.*, 2006, **45**, 2163.
111. T. Armaroli, G. Busca, C. Carlini, M. Giuttari, A. M. R. Galletti and G. Sbrana, *J. Mol. Catal. A: Chem.*, 2000, **151**, 233.
112. Y. Roman-Leshkov, C. J. Barrett, Z. Y. Liu and J. A. Dumesic, *Nature*, 2007, **447**, 982.
113. C. Moreau, R. Durand, S. Razigade, J. Duhamet, P. Faugeras, P. Rivalier, P. Ros and G. Avignon, *Appl. Catal. A*, 1996, **145**, 211.
114. C. Moreau, R. Durand, C. Pourcheron and S. Razigade, *Ind. Crops Prod.*, 1994, **3**, 85.
115. M. Bicker, J. Hirth and H. Vogel, *Green Chem.*, 2003, **5**, 280.
116. X. Qi, M. Watanabe, T. M. Aida and R. L. Smith Jr, *Green Chem.*, 2008, **10**, 799.
117. Y. Roman-Leshkov and J. A. Dumesic, *Top. Catal.*, 2009, **52**, 297.

CHAPTER 4

Biomass Gasification

JAMES K. NEATHERY

University of Kentucky, Center for Applied Energy Research, 2540 Research Park Drive, Lexington, KY 40511-8479, USA

The practice of gasification involves as much art as science. There are many papers available in the literature regarding modeling of gasifier reactors that provide an abundance of information that is helpful for reactor design.[1–9] However, there are limited resources available to those individuals with the goal of successfully *operating* a gasifier system with biomass fuels. Therefore, the intent of this chapter is to give an overview of the gasification processes used to convert biomass fuels into a producer gas. The focus here will be on the practical aspects of gasification for small to medium-scale systems.

4.1 Historical Background

Small to medium-scale biomass gasification technologies have been used for making low heating value gas known as "producer gas" for nearly 70 years.[10] The majority of the applications have found utility during times of petroleum fuels shortage. Perhaps the most notable time was in Europe during and shortly after WWII. Currently, the majority of the small-scale gasification applications are in rural areas and in developing countries where power distribution is not widespread. However, interest in the gasification of biomass, such as wood waste and agricultural residues, has been building over the last decade with global-warming concerns regarding fossil-based fuels[11,12] and with increasing fossil energy prices.

Utilization of biomass materials such as crop residues, animal wastes, and construction waste for energy production can benefit the environment since it is a

RSC Energy and Environment Series No. 1
Thermochemical Conversion of Biomass to Liquid Fuels and Chemicals
Edited by Mark Crocker

Published by the Royal Society of Chemistry, www.rsc.org

renewable resource. Additionally, landfill space could be conserved by diverting 245 million tons of relatively clean biomass waste each year in the US.[13] According to the EIA,[14] the rate of biomass utilization for energy production has risen to over 712 trillion BTUs in 2006 from 313 trillion BTUs in 2002 due to ethanol production alone. Utilization of food crops such as corn for ethanol has caused fluctuations in food prices and will likely result in food shortages around the world. Accordingly, it is logical to pursue all available biomass wastes such as crop residues and other agricultural wastes for energy production.

The chemical makeup of biomass wastes varies widely and depends on the starting species. In general, these waste materials consist of high and variable moisture content.[12] This variability can create a multitude of problems related to energy conversion for power or heat, either with gasification or combustion plants. Wood-based materials are the common fuel feedstock for biomass energy systems. Typically, wood-waste feedstock will consist of discarded wood products and nonhazardous wood debris from construction sites.

An expanding biomass waste resource is from animal wastes. According to the USDA,[15] over 45 billion pounds of poultry (or approximately 8.7 billion birds) are produced annually in the US. A majority of this production is from concentrated animal feeding operations (CAFOs). In these facilities, private farmers contracted by large poultry companies raise flocks of birds in broiler houses that contain from 15 000 to 30 000 birds per house. Each house, enclosing over 20 000 ft^2, can grow-out up to 6 or 7 flocks per year of broiler chickens. Starting with each grow-out period, a fresh bedding material, consisting of 6–8″ wood mulch, is spread over the house floor. Over the course of the grow-out phase, the bedding material absorbs manure and provides insulation from the cold. Once the birds are mature and harvested, the bedding material has become poultry litter (PL) and is removed. Assuming 200 tons of is litter removed per flock, up to 87 million tons of poultry litter is produced annually in the US.[15]

A major impediment to utilizing biomass and related wastes is their relatively low energy density. Shown in Table 4.1, the ultimate analyses and respective higher heating values (HHV) of various biomass fuels are displayed with those of bituminous and sub-bituminous coals for comparison. Typically, biomass waste materials have energy densities less than a third to a half that of coal.

Compounding this issue is the fact that biomass residues are typically spread over large distances, making their affordable range of transportation limited to only a few miles. Therefore, if these wastes are to be economically utilized for energy production, small-scale units will be needed to consume the feedstock near the source.

4.2 Gasification Chemistry and Reaction Stages

4.2.1 Gasifier Stages

The chemical and physical changes that take place within a gasifier are quite intricate. Adding further to this complexity is the variability of biomass

Table 4.1 Fuel analyses of potential biomass feedstocks.

	Ultimate analysis wt% dry basis							
Agricultural Residues	*C*	*H*	*N*	*O*	*S*	*Ash*	*Mois. %*	*HHV MJ/kg*
Sawdust	50	6.3	0.8	42.5	0.03	0.3	7.8	19.3
Bagasse	48	6		42	–	4	1	17
Corn Cob	49	5.4	0.4	44.6	–	1	5.8	17
Straw	43.5	4.2	0.6	40.3	0.2	10.1	7.6	17
Poultry Litter	37.4	5.1	3.7	31.6	0.4	21.6	27.4	10.8
Municipal Solid waste (dry sewage)	20.5	3.2	2.3	17.5	0.6	56	4.7	8
Coal, bituminous	61.5	4.2	1.2	6	5.1	21.9	8.7	27
Coal sub-bituminous	67.8	4.7	0.9	17.2	0.6	8.7	43.6	24.6

Source: Ciferno *et al.*[5]

feedstock make-up as mentioned in the previous section. In addition to carbon, hydrogen and oxygen in the biomass fuel, we must often consider the presence of nitrogen (often present as ammonia), sulfur, and ash. For simplicity, we can consider the biomass fuel as a pseudo-compound: $CH_wO_xN_yS_z$, where the subscripts w, x, y and z are the relative ratios of hydrogen, oxygen, nitrogen, and sulfur relative to carbon, respectively. In general, the following stages are usually considered to take place: drying, pyrolysis, combustion (often referred to as flaming pyrolysis), and reduction. These stages may take place on macroscopic (*e.g.*, in distinct zones with a gasifier vessel) or microscopic (*e.g.*, with a fuel particle) scales.

Drying occurs as the biomass fuel approaches the flaming zone where the temperature of the surrounding gas approaches the range of 100–200 °C:

$$\text{Wet Biomass} + \text{Heat} \rightarrow CH_wO_xN_yS_z + H_2O \tag{4.1}$$

The resulting heat transfer to the fuel particles evaporates both the surface and inherent moisture. The rate of moisture drying strongly depends on the type of gasification reactor employed. Drying of the fuel within the gasifier reactor can have a major impact on the other gasification steps since much energy is required for the latent heat of vaporization. Consequently, experienced gasifier operators will typically specify that the entering moisture content of the biomass feedstock be well below 25 wt%. In general, the dryer the feedstock, the higher the thermal efficiency that can be obtained.

As the fuel temperature increases to the range of 200–700 °C, pyrolysis of the biomass occurs in the absence of air or oxygen:

$$\begin{aligned} CH_wO_xN_yS_z + \text{Heat} \rightarrow\ & \text{C (char)} + \text{tars/oils} + aNH_3 + bN_2 + cH_2S \\ & + dH_2O + eH_2 + fCH_4 \end{aligned} \tag{4.2}$$

Here, a portion of the biomass is transformed into condensable hydrocarbon tars, oils, methane, and solid carbon residue or char. As the bonds of the fuel matrix break apart, hydrogen is formed and subsequently reacts with nascent fuel oxygen to form water. The pyrolysis step is endothermic and consequently will tend to suppress temperatures in this zone of the reactor vessel. Energy to drive pyrolysis reactions is derived from an external heating source or from partial combustion of the biomass fuel within the gasifier reactor.

For dense-phase gasifier reactors, air or oxygen is introduced *via* a distributor to create a combustion or "flaming pyrolysis" zone.[10] Here, the temperature of the gas and fuel char particles will increase up to 1800 °C, depending on mass transfer and quantity of oxidizer added. By the time the fuel reaches the combustion zone, most of the volatile matter has been liberated, thus heterogeneous reactions take place between oxygen in the air and solid carbonized fuel as shown in eqn 4.3. Additionally, a portion of the H_2, CO, NH_3, and H_2S in the surrounding gas reacts with the available oxygen in this zone (see reactions 4.4–4.8). Each of the combustion reactions are exothermic and will supply the energy required for all of the endothermic reactions occurring in the remaining gasifier zones. The mode of heat transfer to the other reaction zones will depend strongly on the type of the gasifier process employed. For example, most of the heat transfer could be from the sensible heat of the producer gas or radiant heat from the combustion zone.

Combustion-Zone Reactions:

$$C + \tfrac{1}{2}O_2 \rightarrow CO \qquad \Delta H_r^o = -111\,kJ\,mol^{-1} \tag{4.3}$$

$$H_2 + \tfrac{1}{2}O_2 \rightarrow H_2O \qquad \Delta H_r^o = -242\,kJ\,mol^{-1} \tag{4.4}$$

$$CO + \tfrac{1}{2}O_2 \rightarrow CO_2 \qquad \Delta H_r^o = -283\,kJ\,mol^{-1} \tag{4.5}$$

$$NH_3 + \tfrac{3}{4}O_2 \rightarrow \tfrac{1}{2}N_2 + 3/2H_2O \qquad \Delta H_r^o = -383\,kJ\,mol^{-1} \tag{4.6}$$

$$H_2S + 3/2O_2 \rightleftharpoons SO_2 + H_2O \qquad \Delta H_r^o = -563\,kJ\,mol^{-1} \tag{4.7}$$

The main reactions taking place in the reduction zone are shown in eqns 4.8–4.13. These are mostly endothermic and will require energy from either an external heat source or from the combustion reactions mentioned above. Consequently, the gas temperature in the reduction section will decrease. Typical temperatures in this section range from 800 to 1100 °C. If complete gasification takes place, all the carbon is converted to carbon monoxide and CO_2. The remaining solid residue exiting this section consists of ash and unburned carbon.

Reduction-Zone Reactions:

$$C + CO_2 \rightleftharpoons 2CO \qquad \Delta H_r^o = +172\,kJ\,mol^{-1} \tag{4.8}$$

$$C + H_2O \rightleftharpoons CO + H_2 \qquad \Delta H_r^o = +131\,kJ\,mol^{-1} \tag{4.9}$$

$$CO + H_2O \rightleftharpoons H_2 + CO_2 \qquad \Delta H_r^o = -41\,kJ\,mol^{-1} \tag{4.10}$$

$$CH_4 + H_2O \rightleftharpoons 3H_2 + CO \qquad \Delta H_r^o = +210\,kJ\,mol^{-1} \tag{4.11}$$

$$NH_3 \rightleftharpoons \tfrac{1}{2}N_2 + 3/2H_2 \qquad \Delta H_r^o = +46\,kJ\,mol^{-1} \tag{4.12}$$

$$SO_2 + 3H_2 \rightleftharpoons H_2S + 2H_2O \qquad \Delta H_r^o = -207\,kJ\,mol^{-1} \tag{4.13}$$

The gas-phase concentrations of H_2 and CO increase substantially in the reduction zone of the gasifier due to reactions 4.8 and 4.9. Depending on the catalytic properties of the char present and the reaction temperature, some additional H_2 may be formed by means of the water-gas shift (WGS) reaction (eqn 4.10). Since the WGS reaction is reversible and exothermic, it is less thermodynamically favorable at elevated temperatures.

4.2.2 Overall Reaction Simplification and Air-Equivalence Ratio

If we assume the formation of char and hydrocarbons are minimal, the above-mentioned reaction steps can be simplified to estimate the gas-phase products (ignoring inlet moisture) in the following equation assuming a one mole basis of carbon:

$$CH_wO_xN_yS_z + (O_2 + 3.76N_2) \rightarrow aCO_2 + (1-a)CO + b(H_2O) + cH_2 + dNH_3 + zH_2S \tag{4.14}$$

where the subscripts w, x, y, and z are the molar ratios of hydrogen, oxygen, nitrogen, and sulfur, respectively, with respect to carbon in the biomass. These values can be determined by converting the respective mass percentages of each element reported in the ash-free ultimate analysis to moles on a per mole of carbon basis. The total moles of air fed is denoted by α. The magnitude of the stoichiometric coefficients for the product gases will be determined by both the reactor configuration and operating conditions. Typically, the ratio of CO_2, CO, H_2, and H_2S in the product gas may be estimated from Gibbs free-energy minimization methods. However, if there is sufficient ammonia formed in the gasifier, its pyrolysis to nitrogen and hydrogen can be kinetically limited depending on the process conditions.

The moles of air fed to the process is often expressed as the dimensionless air equivalence ratio or ϕ:

$$\phi = \frac{\alpha}{\alpha_{Stoich}} \tag{4.15}$$

where α is the molar rate of air supplied and α_{Stoich} is the theoretical stoichiometric air coefficient for complete combustion. α_{Stoich} may be calculated by modifying eqn 4.14 to account for complete oxidation of the carbon, hydrogen, and sulfur to CO_2, H_2O and SO_2, respectively and performing a molar balance of C and H:

$$CH_wO_xN_yS_z + \alpha(O_2 + 3.76N_2) \rightarrow CO_2 + b(H_2O) + zSO_2 \qquad (4.16)$$

H Balance: $w = 2b$, solving for b yields $b = w/2$.
O Balance: $x + 2\alpha = 2 + b + 2z$
Therefore, we can solve for the stoichiometric α to obtain:

$$\alpha_{\text{Stoich}} = 1 + \frac{w}{4} + z + \frac{x}{2} \qquad (4.17)$$

Ex. 4.2-1: Calculation of the air equivalence ratio. Calculate the molar rate of air supply required to achieve an air equivalence ratio of 0.3 for a wood mulch/animal waste residue having the following simplified formula of $CH_{1.6}O_{0.62}N_{0.1}S_{0.001}$.

In this example, substitute the values of w, x and z into eqn. 4.16 to calculate α_{Stoich}:

$$\alpha_{\text{Stoich}} = 1 + \frac{1.6}{4} + 0.001 - \frac{0.62}{2} = 1.09$$

Now solve for α in eqn 4.15 and substitute $\phi = 0.3$ to obtain:

$$\alpha = \alpha_{\text{Stoich}}\phi = (1.09 \times 0.3) = 0.33 \frac{\text{mole air}}{\text{mole C in fuel}}.$$

The air equivalence ratio is perhaps the most important parameter for efficient gasifier operation, being one of the few process input variables available to the operator for controlling the gasification zone temperature. Depending on the value of ϕ, the caloric value of the producer gas can vary widely due to changes in equilibrium (with the corresponding changes in temperature) and dilution with nitrogen from the incoming air. The yields and distribution of CO_2, CO, and H_2 within the gasifier depend on numerous operational and design parameters.[16] In Figure 4.1 the calculated equilibrium composition (with associated equilibrium temperature) from an air-blown gasifier utilizing a wood-based biomass fuel is plotted *versus* ϕ. For low values of ϕ (*i.e.* less than 0.2), the resulting low temperature favors the formation of methane. With additional air (*i.e.* $\phi > 0.2$), more fuel carbon is converted to CO_2 and CO. For most fuels, the hydrogen content will peak with ϕ in the range of 0.3 to 0.4.

4.2.3 Quantifying Gasifier Performance

When utilizing a gasifier for energy production, it becomes necessary to gauge the performance of the system such that process variables may be manipulated to optimize the economics. The first impulse for anyone operating a gasifier is to maximize the caloric value of the producer gas; however, this more than

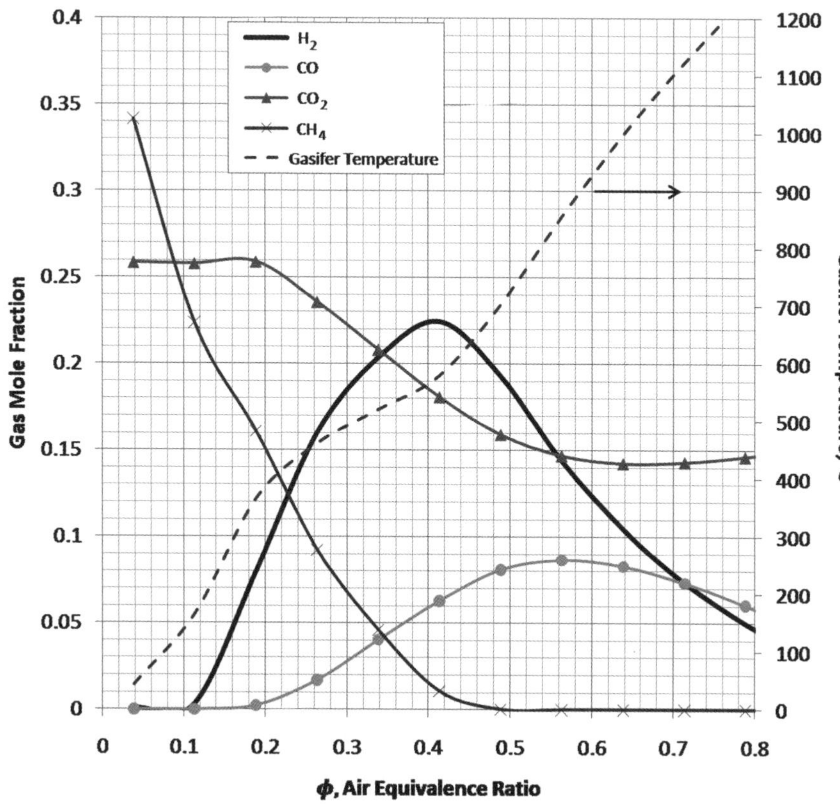

Figure 4.1 Producer-gas composition with variable stoichiometric conditions.

likely does not yield the highest financial return from the equipment. A simple measure of thermal efficiency is the so-called cold gas efficiency defined in eqn 4.18. In this parameter, the product of the gas heating value and the volumetric quantity produced are considered as the thermal output of the gasifier. It should be noted that the cold gas efficiency may be also be reported in terms of a lower heating value basis.

$$\%\,\text{Cold Gas Efficiency} = \frac{\text{HHV}_{\text{gas}} \times v_{\text{gas}}}{\text{HHV}_{\text{fuel}} \times m_{\text{fuel}}} \times 100 \qquad (4.18)$$

where, HHV_{gas} is the higher heating value of the producer gas, kJ m^{-3}; v_{gas} = volumetric rate of producer gas, m^3 h^{-1}; HHV_{fuel} = higher heating value of the biomass feedstock, kJ kg^{-1}; m_{fuel} = input mass rate of the biomass fuel, kg h^{-1}.

In Figure 4.2, the caloric content of a producer gas and associated cold gas efficiency for the same biomass fuel considered in Figure 4.1 is plotted *versus* ϕ.

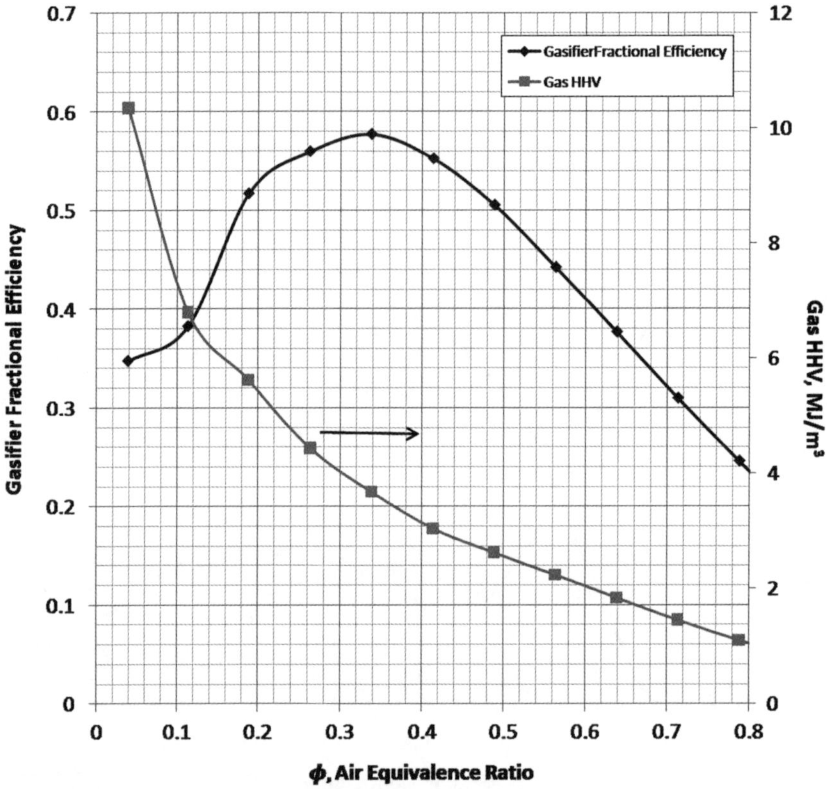

Figure 4.2 Gasifier efficiency and producer-gas heating value *vs.* air equivalence ratio for a wood-mulch gasifier fuel.

For values of ϕ less than 0.2, the caloric heating value is high due to the favorable thermodynamics of methane formation. As more air is added (*i.e.* increasing ϕ), the gas heating value decreases as the formation of CO_2 and CO becomes more predominant and the concentration of the diluent N_2 from the combustion increases.

The thermal losses from gasifier systems can also lower gasification efficiency. These losses are consistent with those observed in furnace/boiler systems: radiant and convective heat loss from the reactor vessel; fan pressure losses; sensible heat loss from nitrogen in the gas; and poor fuel utilization (*e.g.*, unburned carbon in the ash residue). One measure of the fuel utilization is the %Carbon Utilization as defined in eqn 4.19, based on the unburned carbon content of the ash:

$$\text{Carbon Conversion}\,\% = \left\{1 - \frac{m_{\text{ash}} \times \frac{\%C_{\text{ash}}}{100}}{m_{\text{fuel}} \times \frac{\%C_{\text{fuel}}}{100}}\right\} \times 100 \qquad (4.19)$$

where, m_{ash} = mass rate of ash residue exiting the gasifier, kg h^{-1}; m_{fuel} = mass rate of biomass fuel, kg h^{-1}; $\%C_{ash}$ = weight per cent of carbon in the ash residue, %; $\%C_{fuel}$ = weight per cent of carbon in the biomass fuel, %.

Unfortunately, operating a gasifier with 100% carbon utilization while simultaneously maximizing cold gas efficiency is not possible. For example, low carbon content in the ash residue can be achieved by operating with a ϕ approaching unity (near-stoichiometric combustion); however, the cold gas efficiency would approach zero since no combustible gases would be present. Thus, the operator should always sacrifice some carbon efficiency to achieve the desired producer gas specifications for the particular application.

4.3 Gasification Processes

Gasifier reactors are relatively simple in construction; however, the chemistry and physics involved in their operation is still not well understood. Gasifier designs are generally categorized into three types: downdraft, updraft, and fluidized beds.[17] All of the aforementioned gasification processes oxidize a portion of a fuel feedstock (containing predominantly carbon, hydrogen, and in some cases oxygen) at an elevated temperature to provide heat energy in order to drive chemical reactions that convert carbon and hydrogen-containing compounds to carbon monoxide, hydrogen, carbon dioxide, and methane gas.

The selection of the gasifier type will depend on the end use and the quantity of the producer gas required. Gasification with air is the more widely used technology for small-scale operations since there is neither the cost nor hazard of oxygen production. With air gasification, the cold gas efficiency, describing the heating value of the gas stream in relation to that of the biomass stream, is in the order of 55 to 85%, typically less than 65% for wood-based biomass.

4.3.1 Downdraft Gasification

In the downdraft gasifier, fuel is fed near the top of the reactor. A schematic of the process is illustrated in Figure 4.3. Air can be introduced along with the fuel or at some intermediate level below. Hot producer gas along with char and ash exit the bottom of the reactor vessel. A slight vacuum is usually applied at the gasifier exit to withdraw the producer gas.

The four distinct zones of drying, pyrolysis, combustion and reduction occur sequentially from the top to the bottom of the gasifier, respectively. The top section of the gasifier acts as a surge vessel for fuel to be dried before entering the pyrolysis zone. Since the top of the downdraft gasifier is at ambient temperature and pressure, fuel feeding is quite simple. Air in-leakage from the top of the vessel into the gasifier is acceptable as the fuel is added; therefore, elaborate gas-sealing devices are not needed. Radiant and conductive heat transfer from the lower pyrolysis and combustion sections provide the heat for drying of the biomass fuel. Water vapor from the drying fuel flows downward towards the pyrolysis and combustion sections where a portion is reduced to

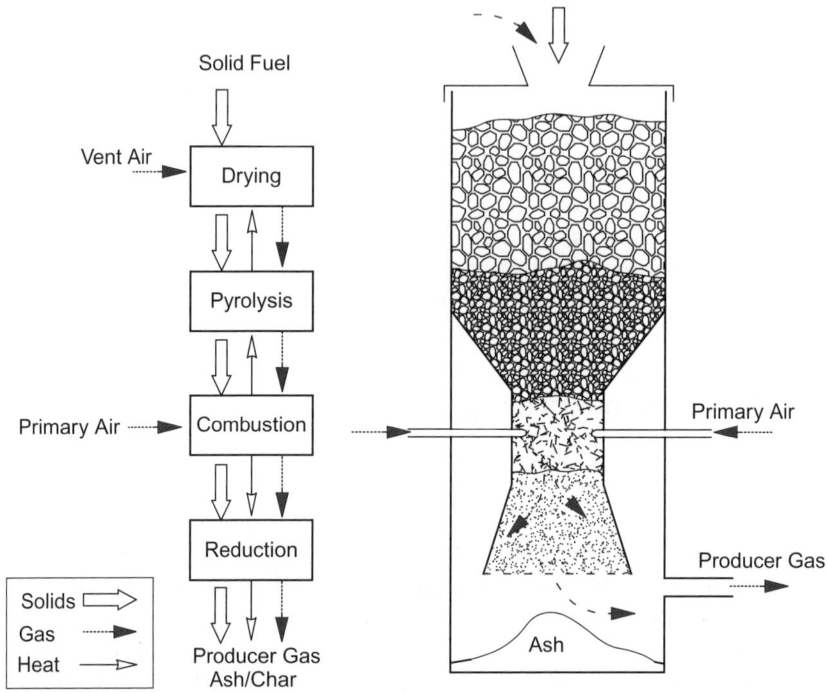

Figure 4.3 Downdraft gasifier.

hydrogen. However, a majority will remain in the final producer gas. A small amount of air may be added in the drying section to facilitate the downward flow of gas, thereby sweeping out excess moisture. If the flow in this area becomes stagnant, water vapor will tend to rise from natural convective forces and will condense onto the entering cold fuel particles. This phenomenon can cause severe bridging within the gasifier hopper.

The dried solid fuel moves downward by gravity into the pyrolysis zone and is heated such that volatile matter is evolved. The resulting gas flows through a reduction zone on an ash support grate before exiting the reactor. Gaseous products of the pyrolysis reactions are drawn downwards along with the char into the combustion zone of the gasifier. A portion of the pyrolysis products and char are burned in the combustion zone as free oxygen becomes available. At this point, an oxidizer gas such as air or oxygen is injected. Reactions with oxygen result in a steep rise of the temperature up to 1200–1800 °C, depending on the moisture content of the fuel. Some downdraft reactor designs employ a restriction, commonly known as a throat section, to increase the velocity of the gas to promote heat and mass transfer. Here, a plurality of air nozzles is included to promote good mixing. Proper arrangement of the throat section is probably the most crucial point of a good downdraft gasifier design. If the restriction is too narrow, proper downward flow of the solids may be a limiting factor. Additionally, nozzles should be properly specified such that sufficient air flowrate is maintained to oxidize the char. If this section is overly starved of oxygen, the combustion zone will tend to move upward and expand into the intended pyrolysis and drying zones. Conversely, if too much air is added, the combustion zone will lengthen into the reduction zone below, resulting in an increase in an excess of CO_2 formation.

The hot gases and char exit the combustion section and continue to move downward into the reduction zone. Here, the hot glowing char in the absence of oxygen converts CO_2 and water vapor into H_2 and CO, the main constituents of producer gas. Additionally, higher molecular weight tars and oils have an opportunity to thermally crack to smaller gas constituents in this section. As most of the carbon is converted, only a char residue with inert ashes remains on a support grate near the bottom of the gasifier vessel. These ashes are removed periodically by vibrating or shaking the support grate to prevent blockages which can obstruct the free flow of the producer gas.

Since the producer gas contacts the hot char before exiting the downdraft reactor, the concentrations of tar and oils in the producer gas are considerably lower than those of other gasification reactors such as the updraft and fluidized bed types that are discussed below. The presence of tar and oil aerosols can cause severe problems in downstream equipment. Consequently, the downdraft-type is the most widely used gasifier for small-scale engine and heating applications.

4.3.2 Updraft Gasification

In updraft gasification, shown schematically in Figure 4.4, air is injected near the bottom of the reactor (in the combustion zone) with the resulting producer-gas flowing up through the interstitial spaces in the solid fuel. The producer gas exits the top of the reactor where incoming biomass fuel is added. Hence, the updraft producer gas exits at a lower temperature (130–150 °C) than that of the downdraft reactor due to the drying of the solid fuel. However, the gas is heavily contaminated with pyrolysis products (primarily tars, oils, and particulate) from the incoming fuel.[18] Since the producer gas exits the top of the reactor vessel near the fuel entry area, special precautions should be taken to avoid air in-leakage with the incoming biomass. Lock hoppers and/or rotary seal valves are typically used to secure fuel introduction into the top of the updraft reactor.

The updraft gasification process is similar to that of the downdraft mode in that the major reaction stages are present: drying, pyrolysis, combustion, and reduction. However, with respect to the solids, the ordering of these stages is changed. First, the incoming fuel is dried from the forced convective heat transfer of the pyrolysis gases. Most of the moisture evaporated from the fuel solids and products of the pyrolysis reactions exit the top of the reactor with the producer gas. The charred solids from the pyrolysis zone migrate below to the reduction zone. Heated combustion gases enter the reduction zone providing the required energy for the endothermic reactions. Finally, the remaining char

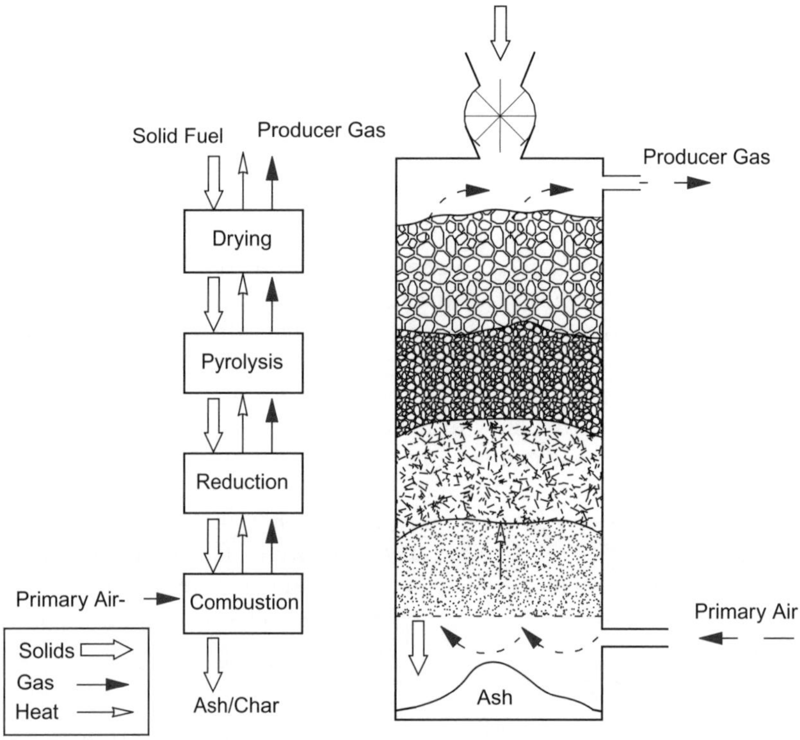

Figure 4.4 Updraft gasifier.

reacts with oxygen or air injected into the bottom of the reactor vessel. Ash is extracted through the bottom of the reactor.

Overall, the updraft gasifier is more forgiving with respect to fuel moisture since the heat transfer is enhanced with the counter flow arrangement. Exotic materials are not required to handle the low-temperature producer gas exiting the reactor. However, the resulting producer gas is heavily loaded with water vapor and condensable hydrocarbons. For engine applications, the producer gas must be treated since the combustion kinetics of the heavy tars are rather slow. Consequently, updraft gasifiers are best suited to direct heating applications where the tars are easily oxidized in a furnace with high residence time.

4.3.3 Fluidized-Bed and Transport-Reactor Gasification

A fluidized-bed gasifier, shown schematically in Figure 4.5, makes use of an inert medium such as sand to mix the solid fuel with the gas phase. Depending on the specific reactor design and average bed media particle size, the superficial gas velocity typically ranges from 0.5 to 1.0 $m\,s^{-1}$. Air is injected in the bottom of the reactor vessel through a distributor consisting of a perforated plate or series of nozzles such that the bed material is suspended in a fluid-like state.

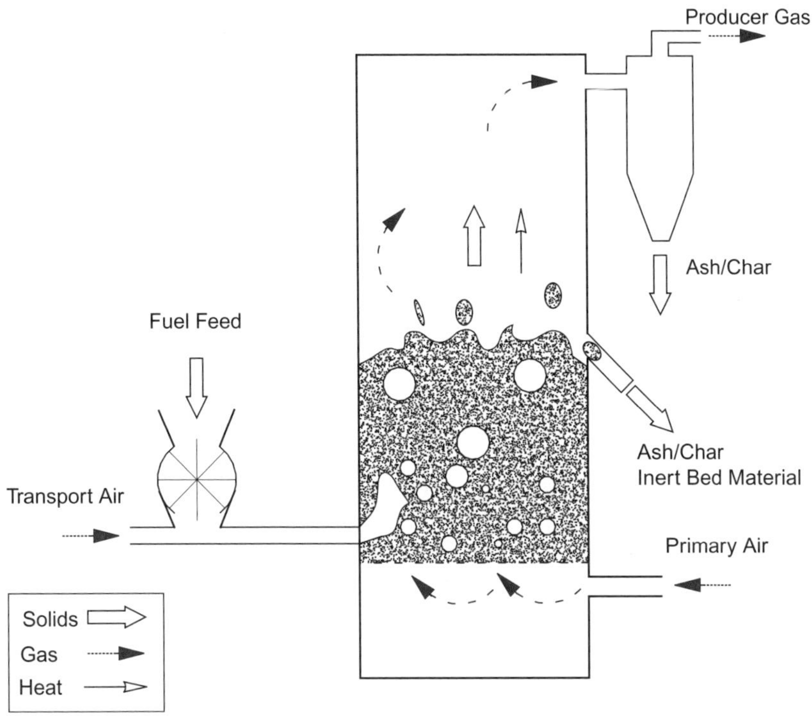

Figure 4.5 Fluidized-bed gasifier.

Gas bubbles form at the distributor, increasing in size as they rise upward toward the bed surface due to the decreasing hydrostatic pressure. Air within the bubbles exchanges with combustion and pyrolysis gases formed in the dense phase as the fuel is converted to producer gas. Fly ash, char, and inert bed material are ejected from the dense-phase bed as gas bubbles erupt on the surface; therefore, it is desirable to size an appropriate freeboard height to allow larger particles to return back to the bed zone.

The turbulent fluidized medium promotes high heat-transfer rates and as a result, the bed temperature is maintained in an essentially isothermal state. Since the dense-phase zone is well mixed, the maximum temperature obtained (only 800–900 °C) during gasification is substantially lower than that of the fixed bed designs. This, combined with the relatively low gas residence time, increases the yield of tars, oils, and light hydrocarbons. As a result, the tar/oil concentration in a fluid bed derived producer gas (ranging from 2–20 g/N m^3 gas), is greater than that derived from the downdraft gasifier.[19]

Fuel is pneumatically conveyed into the bottom of the dense-phase bed. The fuel port is typically under pressure due to the column of fluid bed above the injection port. Consequently, a rotary seal valve or lock hopper system is needed to isolate the fuel storage hopper from hot producer gases. Particles injected into the reactor are mixed with the hot bed material and rapidly heated to the bed temperature. Thus, drying, pyrolysis, combustion and reduction steps proceed in nearly parallel paths.

Fluidized beds require some additional gas-cleaning steps due to the inherent high particle loading in the producer gas. The degree of particle loading will increase with gasifier load since the superficial velocity will increase proportionally. Fluidized-bed systems are usually equipped with cyclones to recover char, ash, and undersized inert bed particles. A portion of the recovered particle stream can be injected back into the bed to increase the carbon conversion. The cyclone surfaces must be maintained at a temperature greater than 500 °C to prevent troublesome tar deposition and fouling. One practical option is to mount the cyclone device inside the freeboard to minimize heat losses.

The inventory of solids within the fluid bed must be controlled to maintain an efficient operation. If the bed level exceeds that necessary to properly convert the fuel into producer gas, a high pressure drop across the vessel will result in excessive fan power being consumed. On the other hand, if the bed inventory is too low, the gas residence time will not be sufficient to convert the fuel and the thermal mass will be unable to sustain a stable bed temperature. The bed level is usually monitored *via* differential pressure measurements from the top of the distributor plate to the freeboard. The average bed density will decrease slightly with increasing superficial velocity. However, the differential pressure measured will be proportional to the mass of solids inventory within the bed. Control of the bed inventory can be as simple as utilizing an overflow pipe connected to the reactor body at a predetermined height. However, as the load changes, the bed density will decrease as the bed expands with increasing superficial velocity such that bed material will be bumped into the overflow pipe (this method is shown in Figure 4.5). Another option is to extract bed material *via* an auger located near the bottom of the reactor. As the differential pressure increases, a controller will vary the auger speed proportionally to maintain the setpoint inventory level.

The ability to match the gasifier demand with the producer gas output is a challenge for all gasifier systems. It is particularly so for fluidized-bed systems, where maintaining a critical range of superficial gas velocity is important for controlling the bed material inventory. If the load demand increases, the control system increases both the fuel and air flows, while keeping the fuel-to-air ratio relatively constant. The fluidized-bed cross-sectional area should be sized such that the maximum allowable velocity within the reactor is below the terminal velocity of the smallest particle in the fuel to minimize char carry over. Additionally, for the lowest load level, the gas velocity should be well above the minimum fluidization velocity for the largest bed particle. Otherwise, the bed could become defluidized or "slumped", leading to hotspots and possible agglomeration of ash particles.

One design option to solve the bed-slumping issue is to decrease the size of the reactor cross-sectional area to intentionally exceed the terminal velocity of all the particles in the bed inventory. This is a hybrid fluidized-bed gasifier also known as a transport reactor (see Figure 4.6). All the elutriated solids

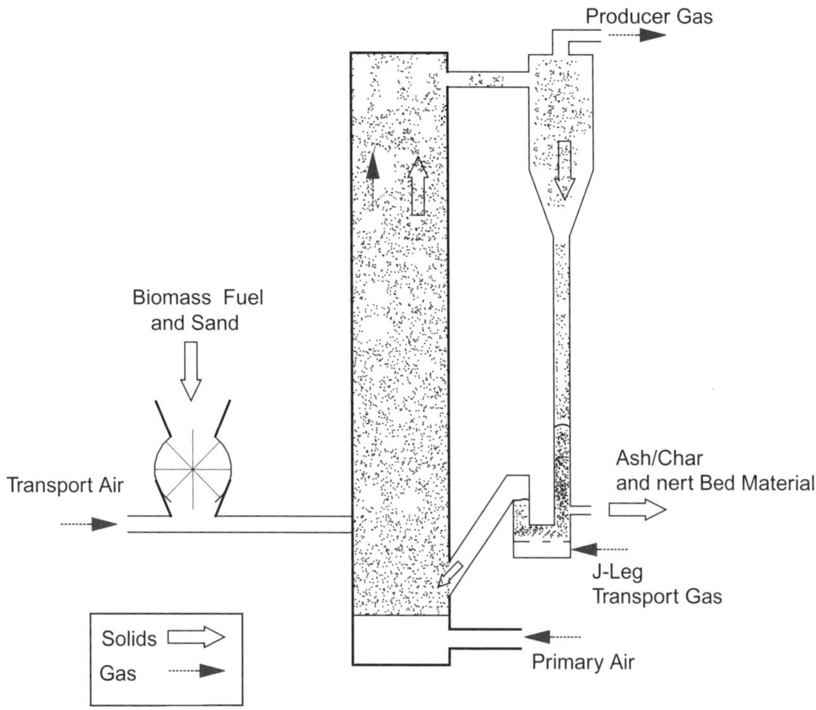

Figure 4.6 Transport reactor or circulating-bed gasifier.

are captured in a cyclone and continually injected into the bottom of the reactor. A column of solids is maintained in the cyclone down-comer *via* a fluidized j-leg. A small purge stream of air or nitrogen fluidizes the ash in the j-leg such that the column of solids acts as a pressure manometer. As the solids in the column increase to the point where the differential pressure of the reactor is overcome, the solids will flow back to the bottom of the reactor through an overflow tube. The solids inventory circulating within the gasifier is controlled by monitoring the differential pressure along the axis of the reactor and removing solids though a flow splitter by the j-leg located in the down-comer of the cyclone.

Unlike fluidized-bed reactors, fixed/moving-bed gasifiers require fuel specifications with larger particles (> 1–1.5 cm) that create sufficient void spaces that provide little resistance to flow. The turbulent medium within the fluidized bed is able to process a wide range of particle sizes due to the void space created by the gas bubbles. However, one limitation is that fine particles in the fuel feedstock will tend to be entrained within the wakes of the gas bubbles and are not easily recirculated within the dense phase of the fluidizing medium. This phenomenon can cause low carbon conversion efficiency, especially for increased loads where higher superficial velocities are encountered.

4.4 Practical Issues of Gasification

4.4.1 Fuel Particle Size

Fixed-Bed Gasifier Systems. Fixed- or moving-bed gasifiers have limitations regarding the range of fuel particle size and shape. Fine-grained feedstocks are prone to bridging in the drying and pyrolysis sections within the gasifier vessel. This phenomenon is exacerbated with high moisture fuels. Additionally, fine and irregular-shaped feed particles may impede gas flow through the bed and result in a large pressure drop.[20] Large pressure drops will result in irregular axial temperature profiles and "rat holes" or channeling of the pyrolysis and combustion zones.

If the fuel particles are too large, fuel reactivity will be an issue. This problem will manifest itself with regard to slow gasifier start-ups and poor gas quality. Additionally, the void spaces between large chunks of biomass can cause channeling of combustion air into the fuel gas. The ideal fuel for a moving-bed system would have a uniform or narrow size distribution of particles ranging from 1 to 8 cm with short aspect ratios. Herbaceous biomass feedstocks such as switch grass, for example, would not be suitable for fixed-bed gasification.

Fluidized-bed gasification. Due to the high degree of turbulence and good heat-transfer characteristics, fluidized beds tend to be more forgiving with respect to utilizing smaller-sized fuel particles. Fluidized-bed gasifiers are able to process fuels with particle diameters down to 0.1 cm. Ideally, the maximum aerodynamic should be limited to that which will easily fluidize for the minimum design velocity within the bed region. Normally, this will be approximately less than 3 cm in diameter for particle densities similar to wood.

The feed systems for delivering fuel into the fluid-bed region will generally dictate the type of biomass that can be used effectively. It is desirable to feed the fuel near the bottom of the fluidized bed in order to maximize the residence time of the solids to achieve high carbon conversion. Pneumatic conveying systems are often specified for these gasifiers. Consequently, large or long fuel particles need to be chopped such that air transport in a pipe is possible.

4.4.2 Fuel Moisture Content

Most raw biomass fuels that are newly harvested have inherently high moisture contents. Sawdust from hardwoods for example, typically ranges from 40–60 wt% moisture depending on the exposure to rain and humid conditions at the mill. Gasification of high moisture fuels will result in more tar formation and low gasification temperatures. Processing wet fuels also can result in erratic gasifier operation and slow start-up times.

The extent of drying required for a stable operation will depend on the type of gasifier used. Updraft and fluidized-type gasifiers generally are more tolerant of fuel moisture due to their internal high heat-transfer rates. A general rule-of-thumb for biomass gasification is the fuel should be dried to 15–25 wt% moisture.[12] If sufficient covered and ventilated space is available, biomass can be effectively air-dried by spreading the material for several days in ambient conditions. However, the risk of spontaneous combustion is always present due to composting or biological degradation.[21] Unit operations such as rotary kilns or fluidized-bed dryers provide for more rapid biomass drying. Quite often, waste process heat can be utilized to provide the necessary energy for these drying operations. In the end, the costs for fuel drying need to be evaluated against the gains in producer gas quality and the capital costs of downstream gas-cleaning equipment.

4.4.3 Process Control and Automation of Gasifiers

The chemistry and physics involved in gasifier reactors are quite complex and sometimes counterintuitive. Hence, the control and automation of gasifier systems is quite challenging. Conventional control systems, such as simple feedback loops with proportional-integral-derivative (PID) control schemes are difficult to tune due to the nonlinear nature of the gasification process.[21] This aspect of the technology alone has inhibited widespread commercialization of the technology for small to medium-scale industrial markets.

For systems supplying producer gas to process heaters or engines for power applications, the consistency of the gas quality is crucial. Therefore, analysis of the product gas will also be required to implement an effective feedback process control system. Additionally, the fuel gas that is noncombustible will need to be flared outside the normal heating system or engine generator-set to account for any transients that may occur.

In the context of producer gas, the concentration of the respective compounds depends mainly on the gasifier temperature and ϕ (or the air supplied to the gasifier).[22] Accordingly, modulation of the combustion air is the main parameter used to control the fuel gas quality and gasification temperature. However, the response of gasification temperature to changing air flowrate is often unpredictable. Therefore, an effective control system should be adaptable with respect to being either positive or negative acting to changes in bed temperature.

One proposed gasifier control concept is based on optimizing gasifier cold gas efficiency by controlling the thermal conductivity of the producer gas.[23] This method relies on the gasifier efficiency being proportional to the hydrogen content of the gas. As shown in Figure 4.7, this is a good assumption with wood-based fuels, depending on the moisture content of the biomass feedstock. The producer gas thermal conductivity (TC) closely follows that of hydrogen, which is nearly an order of magnitude higher than that of the other major producer-gas constituents (CO, CO_2, N_2, CH_4, and H_2O). The thermal conductivities of the major gas compounds that make up a typical producer gas along with air (used as a reference gas) are displayed in Table 4.2. The total thermal conductivity of the gas produced will therefore vary proportionally with the hydrogen content. The art and science of thermal-conductivity detectors (TCD) and thermal-conductivity measurement of gases are well

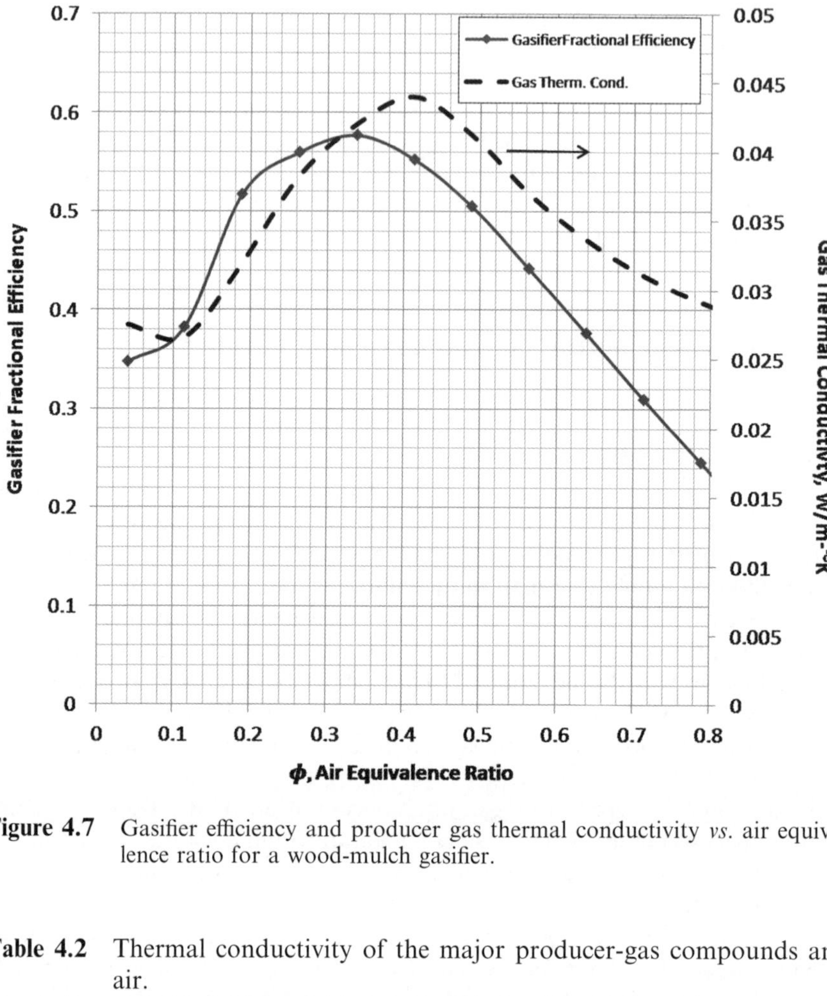

Figure 4.7 Gasifier efficiency and producer gas thermal conductivity *vs.* air equivalence ratio for a wood-mulch gasifier.

Table 4.2 Thermal conductivity of the major producer-gas compounds and air.

Gas thermal conductivity, W/m K					
H_2	CO	CO_2	CH_4	N_2	Air
0.1718	0.0251	0.0166	0.0337	0.0253	0.0259

known.[24,25] Typically, TCDs are used in conjunction with separation devices such as a gas chromatograph column to detect and quantify individual gas compounds.

Consider for example, the air-blown adiabatic gasifier firing a wood-based fuel considered in Figures 4.1 and 4.2. As shown in Figure 4.1, the calculated equilibrium mole fractions of the main gas constituents such as hydrogen, carbon dioxide, and carbon monoxide, with a balance of nitrogen, will vary according to the air or oxidizer equivalence ratio (in this case the producer gas has been dried to a 50 °C dew point to remove water). The calculated thermal conductivity (TC) based on the weighted mean TC of each gas component, is plotted *vs.* ϕ in Figure 4.7. Since hydrogen's TC is an order of magnitude higher than the other gas constituents, the total TC signal varies proportionally to the hydrogen mole fraction.

While the heating value of the gas can be an important control parameter for engine applications, it is not an indicator of the hydrogen content or gasifier efficiency. In the example detailed in Figure 4.2, the producer-gas heating value decreases in proportion to the quantity of air added, thereby steadily increasing the concentration of nitrogen and diluting the energetic compounds such as methane, carbon monoxide, and hydrogen. However, in this example the overall thermal efficiency exhibits a maximum at an air equivalence ratio of 0.34.

In Figure 4.7, the gasifier fractional efficiency is plotted with the theoretical gas thermal conductivity *versus* the air equivalence ratio (same case as considered for Figures 4.1 and 4.2). Here, we see the gasifier efficiency tracks closely with the hydrogen content and thus thermal conductivity. The maximum efficiency and thermal conductivity, in this example, occurs at an air equivalence ratio between 0.36 and 0.41 in proportion to the gas hydrogen mole fraction. Thus, in addition to the thermal-conductivity signal indicating the hydrogen content of the fuel gas, it also provides useful information regarding the efficiency level of the gasifier.

Based upon the strong correlation between hydrogen content, TC and gasification efficiency, a control scheme can be envisaged to optimize gasifier reactor systems. This method can make use of commercially available TC cells containing thermistor sets arranged in a Wheatstone bridge configuration. One side of the thermistor cell would be exposed to the sampled producer gas, while a reference thermistor is immersed in an air stream. Both the sample and reference streams would be maintained at near identical temperatures and flowrates through the TC cell. As the hydrogen content in the producer gas stream increases, it will dissipate more heat from the sample thermistor than that of the air with the reference thermistor. The resulting resistance difference in the bridge is converted to a voltage signal.

In Figure 4.8, a TC instrument is shown integrated within a process control scheme for a typical gasifier system. A small slipstream sample is extracted at a fixed rate from the dried producer gas and heated above its dew point to avoid the formation of water droplets. The heated sample enters the thermal-conductivity detector where an output voltage signal is generated that is proportional to the hydrogen content of the producer gas. The control signal is then processed by a control system computer along with the gasification temperature information. The process controller will maintain a gasifier exit temperature between an acceptable low and high setpoint interval by varying air and fuel flowrates. The low-temperature setpoint is set such that excessive tar and oil formation are avoided in the producer gas (*e.g.*, a minimum of 750 °C to 800 °C is generally chosen depending on the fuel feedstock used). A high-temperature setpoint is chosen to avoid excessive materials wear within

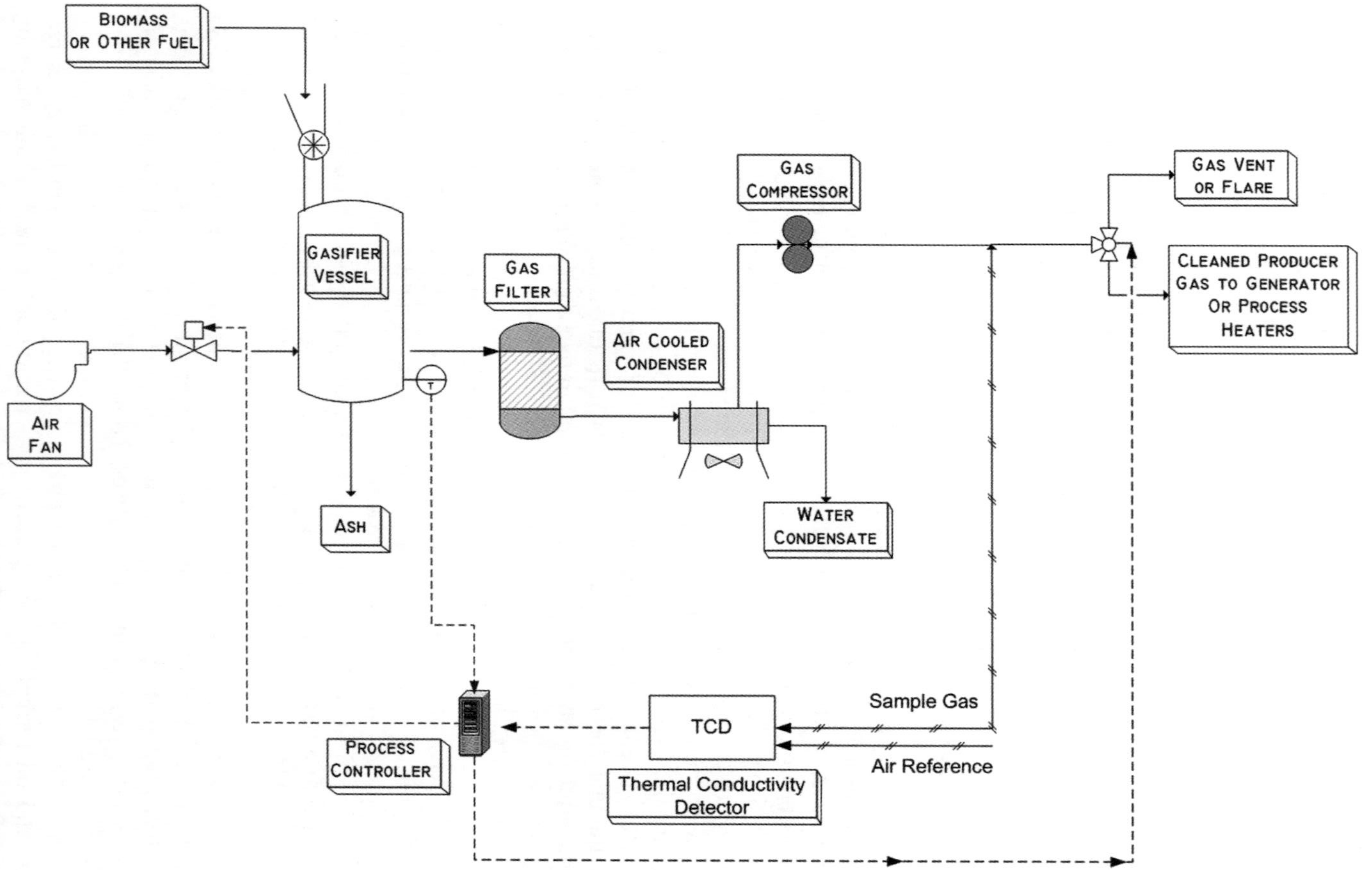

Figure 4.8 A control system based on optimizing producer-gas thermal conductivity.

the gasifier reactor. The control system varies the gasifier temperature setpoint by trimming the air flowrate such that the producer gas thermal conductivity (*i.e.* hydrogen content) is maximized within the desired temperature setpoint range. For power applications, the gasifier control system could be programmed to switch a valve such that the producer gas is sent to a flare or vent system in the event the desired caloric value specification (as detected by a low thermal conductivity) has not been met.

In practice, the control system should maintain the exit gasifier temperature within a bounded range in order to minimize tar and oil formation at some low setpoint temperature, and to minimize high-temperature corrosion and wear (the maximum temperature setpoint). Given a specific fuel composition, we can calculate equilibrium gasifier exit temperature and theoretical gas thermal conductivity curves *vs.* the air equivalence ratio (ϕ), as shown in Figure 4.9. Here, the projected dotted lines to the *y*-axis from the low and high setpoint temperatures define the bounds of the allowable air input in terms of ϕ. Since the composition and moisture content of biomass feedstocks can vary substantially over time, the relationship between the temperature and conductivity curves will also vary significantly with fuel changes. Therefore the control system will need to continuously seek the highest attainable thermal conductivity, while maintaining the gasifier exit temperature within the desired setpoint range.

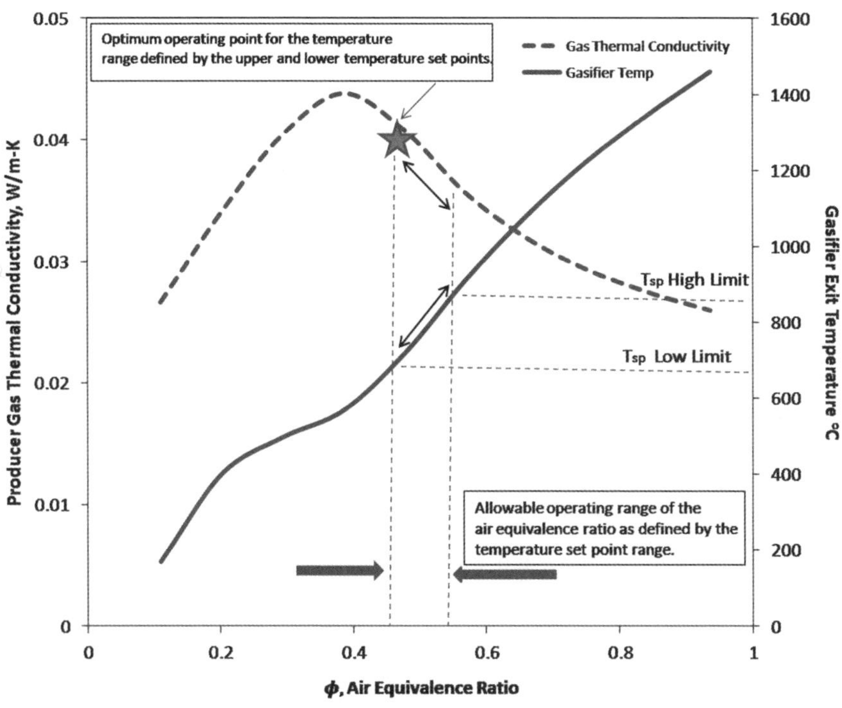

Figure 4.9 Thermal-conductivity control diagram.

In the example displayed in Figure 4.9, the allowable ϕ is in the range of 0.45 to 0.55. Although the thermal conductivity peaks at an AER of 0.38, the resulting equilibrium temperature would fall well below the lower temperature setpoint. Consequently, the optimum TC for control purposes occurs while operating the gasifier at the ϕ of 0.45 (see the star denoting the optimum in Figure 4.9).

The challenge of implementing this type of control method is to adapt to changes in fuel composition. For example, if the fuel in the case of Figure 4.9 were to suddenly become drier, the temperature curve would be shifted to the left, thereby necessitating lowering of the ϕ for the given high- and low-temperature setpoints. Likewise, an increase in moisture would have the opposite effect. Consequently, the control system must constantly modulate the air or oxidizer flowrate to seek out the TC optimum. It should also be able to determine if the control action should be negative or positive. For example, when the gasifier is operating to the left of the TC signal maximum the control action should be to increase the air flowrate. On the other hand, if the operating point is to the right of the TC maximum, the control action should be negative (*i.e.* act to decrease the air flowrate).

4.5 Producer-Gas Conditioning

Producer gases from air-blown gasification processes have low calorific value and high levels of unwanted components, such as particulate and tar, which should be eliminated prior to utilization in heating, chemical synthesis and engine applications. Additionally, the producer gases need to be dried to some extent if transported for short distances in a pipeline. The degree of gas treatment needed will depend strongly on its desired end use.

4.5.1 Particulate Removal

As shown in Table 4.3, the quantity of dust generated largely depends on the type of gasifier employed.[26] Fluidized beds generate the most particulate and have mechanical devices such as cyclones to capture the coarser particles. The particle loading in the producer gas for fixed- or moving-bed gasification systems is dependent on the load; higher velocities experienced at high load levels will entrain more fine particulates. Fuels containing a large quantity of fines will also tend to create a proportional amount of fines in the producer gas.

Table 4.3 Producer gas particulate concentrations in various gasifier designs.

	Fixed bed updraft gasifier	*Fixed-bed downdraft gasifier*	*Fluidized-bed dense phase*
Particle concentration ($mg/N\,m^3$)	100–8000	100–3000	8000–100 000

Source: Pathak *et al.*[26]

The removal of ash, char, oil aerosols, and soot from producer gas is a major issue for engine applications. Ash particulate matter is rather abrasive and can cause excessive engine wear and fouling. Aerosols of oils in the producer gas can further aggravate fouling of engine parts. Although oil aerosols contain considerable heating value, their combustion reactivity is rather limited and can lead to further soot formation in the combustion chamber of the engine. To a limited degree, internal combustion (IC) engines can tolerate particles less than 5 μm. Mechanical capture devices such as cyclones are quite effective in removing particles greater than 50 μm. However, tars will condense on the cyclone internals if the surface temperature is below about 300 °C. Consequently, the cyclone should be situated inside the gasifier as previously mentioned to avoid heat losses.

Capture options for particles less than 50 μm are limited to filter media and gas scrubbing with water.[20] Wet scrubbers or wash towers (shown in Figure 4.10) have been demonstrated to remove 60 to 90 + % of ≥ 2 μm

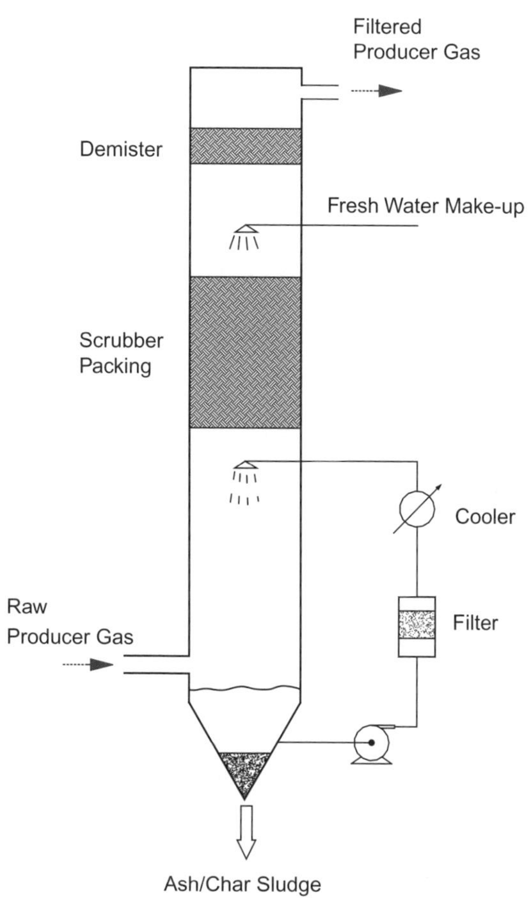

Figure 4.10 Wet scrubber particulate removal.

diameter particles from various gasifier types.[27] Raw gas is quenched with process water and collected in a wastewater basin. A portion of the oils, aerosols and tars will also be collected with the quench spray. Particulate is filtered from the collected wastewater and the water is cooled before being recycled to the tower. Moisture droplets are trapped in a demisting column prior to entering downstream equipment. A complicating factor for wet scrubbing is the capture of heavy tar and oils from the producer gas. Some of the heavier tar condensate will reach the water phase as agglomerated tar particles and foul filtration equipment and possibly the demister column.

Baghouse cloth fabric filters, shown schematically in Figure 4.11, have been used effectively for capturing particles smaller than 5 μm in hot gas environments. Baghouse collectors consist of a plurality of cloth tubes open at one end. Dirty gas enters from the outside of the cloth tube, forming a filter cake.

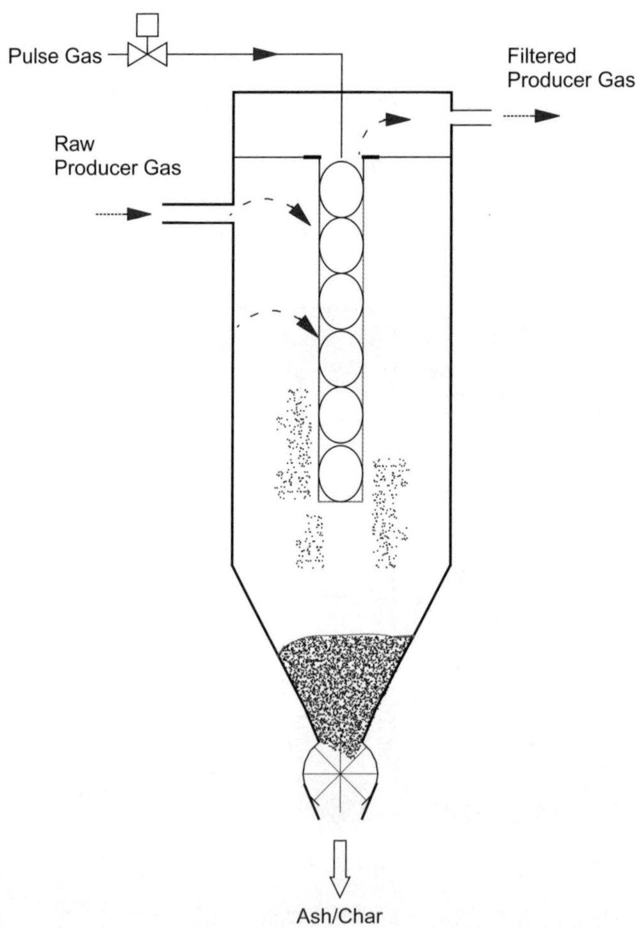

Figure 4.11 Baghouse filtration schematic.

The filter tube is structurally supported by a metal cage to prevent it from collapsing under the differential pressure. At variable time intervals, the filter cake is removed by back-pulsing an inert gas in the opposite direction to the producer-gas flow. The solids cake falls into a hopper located beneath the filter elements.

As with all filter media devices, moisture condensation combined with the fine particulate loading will result in large pressure-drop losses. Condensation of tar will tend to increase the efficiency of particle capture by developing a sticky surface that effectively traps ash and char. However, tar and moisture condensation will quickly foul most filter media, as in the case presented for cyclones. The condensation issue can be lessened by operating the filtration system at higher temperatures (*e.g.*, $>300\,^{\circ}C$). Most conventional cloth filter media tend to char and degrade over short time intervals. Consequently, this option would require selecting materials such as woven glass wool filter bags that can be used at temperatures up to 300 °C.

Taking advantage of the sticky-surface phenomenon, filter beds or other granular media have been used with limited success over time periods of a day before clogging with tars. Moving beds can prolong granular filter life by continuously adding fresh media while extracting fouled material.[28] The apparatus, shown schematically in Figure 4.12, uses a granular medium (stone pebbles) that is countercurrently contacted with raw producer gas exiting a gasifier. Particle-laden gas tangentially enters an outer chamber and contacts pebbles at the end of a bed that is flowing downward with gravity. The granular media enters at the top of the vessel *via* a gas-sealing device as the filtered producer gas exits. Along the axis of the moving bed, heat is transferred from the producer gas to the media. Consequently, a film of tar begins to condense on the incoming media developing a sticky surface area for capturing ash, char and soot particles. The tar-laden media material can be fired in a rotary kiln to recover energy and recycled back to the moving bed filter after being cooled. One major disadvantage of this system is the quantity of solids that must be continually moved.

4.5.2 Tar and Oil Removal

In addition to the major chemical components that comprise producer gas (*e.g.*, CO, H_2, CO_2, CH_4, H_2O, and N_2), a considerable quantity of organic compounds is produced during the gasification process. The quantity of tars/oils present in the producer gas will depend on the gasifier type employed[29] (see Table 4.4). These oil and tar organic compounds consist of both low molecular weight hydrocarbons and high molecular weight aromatics. One of the major technical obstacles for implementing biomass gasification technology is the contamination of tar and oil in the producer gas. For gas turbine or engine applications, the lower molecular weight hydrocarbons are easily utilized and add to the higher heating value of the producer gas. However, the higher molecular weight organics, often referred to as tar or oil, will condense in

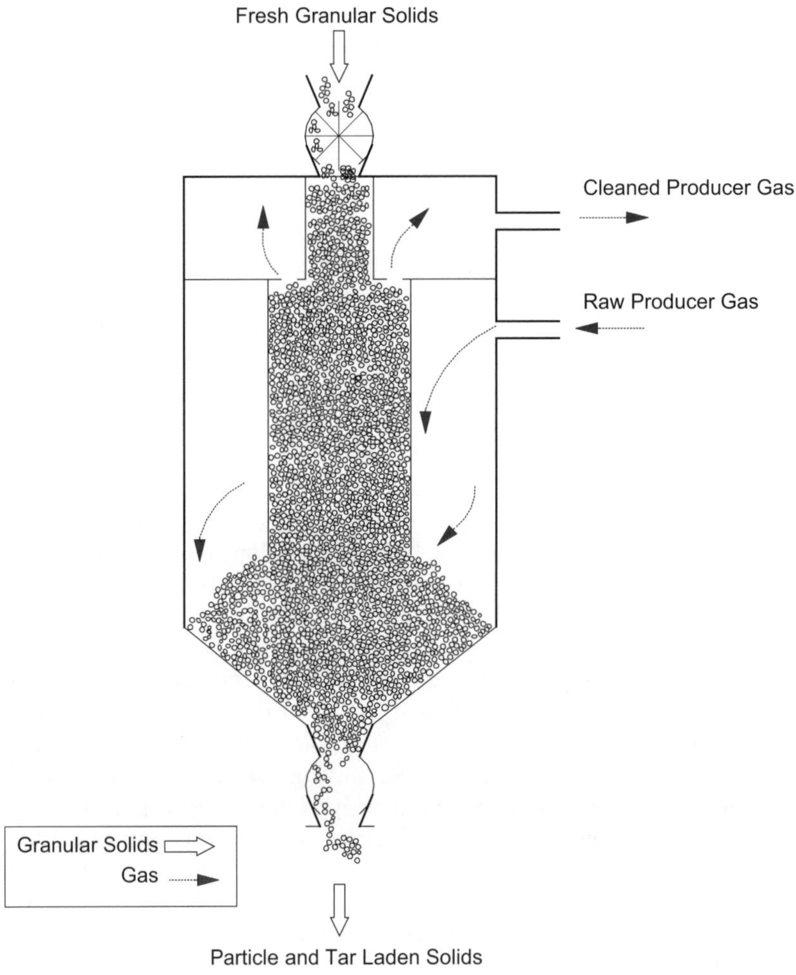

Figure 4.12 Moving-bed granular filter.

Table 4.4 Producer-gas tar concentration in various gasifier reactor designs.

	Fixed-bed updraft gasifier	*Fixed-bed downdraft gasifier*	*Fluidized-bed dense phase*	*Fluidized-bed transport reactor*
Mean tar content ($g/N\,m^3$)	50	0.5	12	8
Range of tar conc. ($g/N\,m^3$)	10–150	0.01–6	1–23	1–30

Source: Morf.[29]

process piping, thereby fouling heat exchangers and other equipment located downstream of the gasifier.

Tar compounds tend to complicate particulate removal and producer-gas drying operations. Tars will tend to condense onto particulate filters, severely increasing the pressure drop. Tars and oils condense along with water in gas condensers in which the saturated producer gas is dried, creating an oily emulsion that is difficult to treat.

A considerable research effort has recently been aimed at tar removal from producer gas.[30] Processes for tar suppression or elimination can be implemented either inside the gasifier or postgasification. Approaches inside the gasification reactor include either direct or indirect methods. The indirect approach focuses on optimizing process variables to avoid the formation of tars. Optimizing the operating parameters such as temperature, air equivalence ratio, and gas residence time can be quite effective in lowering tar formation in the producer gas for any given reactor design. The direct approach involves the use of catalysts to crack the tars into lighter hydrocarbons. Thermal catalytic gas processes crack the large molecules into smaller noncondensable (at ambient temperatures) species. These methods tend to require costly catalysts and are often difficult to control. Costly capital equipment is often required to handle the extreme process temperatures. Indeed, this area presents many challenges both from technical and economic perspectives. However, extensive coverage of this subject area is beyond the scope of this chapter.

References

1. C. Altafini, P. Wander and R. Barreto, *Energy Conv. Mgmt.*, 2003, **44**, 2763.
2. C. Blasi, *Chem. Eng. Sci.*, 2000, **55**, 2931.
3. J. Chen, "Kinetic engineering modeling of co-current moving-bed gasification reactors for carbonaceous material", Ph.D. thesis, Cornell University, New York, 1987.
4. R. Chowdhury, P. Bhattacharya and M. Chakravarty, *Int. J. Energy Res.*, 1994, **18**, 581.
5. J. Ciferno and J. Marano, "Benchmarking Biomass Gasification Technologies for Fuels, Chemicals and Hydrogen Production", US Department of Energy, National Energy Technology Laboratory Publication, June 2002.
6. B. Gøbel, U. Henriksen, T. Jensen, B. Qvale and N. Houbak, *Bioresource Technol.*, 2007, **98**, 2043.
7. T. Jayah, L. Aye, R. Fuller and D. Stewart, *Biomass Bioenergy*, 2003, **25**, 459.
8. J. Milligan, "Downdraft gasification of biomass." Ph.D. Thesis, Aston University, Birmingham, UK, 1994.
9. Z. Zainal, *Energy Conv. Mgmt.*, 2001, **42**, 1499.

10. T. Reed, "Encyclopedia of Biomass Thermal Conversion: The Principles and Technology of Pyrolysis, Gasification & Combustion", Biomass Energy Foundation, Golden, Colorado, 2002.
11. J. Kram, *Biomass Magazine*, 2008, **3**, 34.
12. A. Bridgwater, G. Evans, "An Assessment of Thermochemical Conversion Systems For Processing Biomass and Refuse", Energy Technology Support Unit (ETSU) on behalf of the Department of Trade (United States), ETSU B/T1/00207/REP, 1993.
13. Technical fact sheet from Energy Information Association (EIA) 2008, www.eia.doe.gov/cneaf/solar.renewables/page/wood/wood.html
14. Technical fact sheet from Energy Information Association (EIA) 2008, www.eia.doe.gov/cneaf/solar.renewables/page/biomass/biomass.html
15. Technical fact sheet from the United States Department of Agriculture (USDA), National Agricultural Statistics Service 2005, Pou **3-1**.
16. A. Melgar, J. Pérez, H. Laget and A. Horillo, *Energy Conv. Mgmt.*, 2007, **48**, 59.
17. J. Rezaiyan, N. P. Cheremisinoff, *Gasification Technologies: A Primer for Engineers and Scientists*, ISBN-10: 0824722477, CRC Publishing, 1st edn, 2005.
18. L. Devi, K. J. Ptasinski and F. Janssen, *Biomass Bioenergy*, 2003, **24**, 125.
19. C. Pfeifer, R. Rauch and H. Hofbauer, *Ind. Eng. Res.*, 2004, **43**, 1634.
20. K. Cummer and R. Brown, *Biomass Bioenergy*, 2002, **23**, 113.
21. C. Sagüés, *Bioresource Technol.*, 2007, **98**, 845.
22. M. Lapuerta, J. Hernández, A. Pazo and J. López, *Fuel Process. Technol.*, 2008, **89**, 828.
23. J. Neathery, US Patent Office, Patent Pending, 2009.
24. H. Daynes, *Gas Analysis by Measurement of Thermal Conductivity,* Cambridge University Press, Cambridge, 1933.
25. H. Willard, L. Merritt, J. Dean, F. Settle, "Instrumental Methods of Analysis", 7th edn, ISBN 0-534-08142-8, Wadsworth Publishing Company, Belmont, California, 1988.
26. B. Pathak, D. V. Kapatel, P. Bhoi, A. Sharma and D. Vyas, *Int. Energy J.*, 2007, **8**, 15.
27. P. Hasler and T. Nussbaumer, *Biomass Bioenergy*, 1999, **16**, 385.
28. R. Brown, "Biomass Derived Hydrogen from a Ballasted Gasifier", DoE Final Report DE-FC36-01GO119091, www.osti.gov/bridge/product.biblio.jsp?osti_id = 901792.
29. P. Morf, "Secondary Reactions of Tar During Thermochemical Biomass Conversion", Ph.D. Dissertation, Eidgenoessische Technische Hochschule Zurich (Switzerland), 2001.
30. J. Han and H. Kim, *Renew. Sustain. Energy Rev.*, 2008, **12**, 397.

CHAPTER 5

Conversion of Biomass to Liquid Fuels and Chemicals via *the Fischer–Tropsch Synthesis Route*

GARY JACOBS AND BURTRON H. DAVIS

University of Kentucky, Center for Applied Energy Research, 2540 Research Park Drive, Lexington, Kentucky, 40511, USA

5.1 Motivation

The Fischer–Tropsch (FT) synthesis reaction could play a vital role in not only bridging the gap between a petroleum-based and a hydrogen economy, but also in sustaining a hydrogen economy as an important means of producing high-density energy carriers as intermediary fuels. The list of potential benefits in switching to H_2 as an energy carrier, in conjunction with fuel-cell technology, is very attractive and includes higher efficiencies and lower emissions when considering the entire fuel cycle (*i.e.* well to wheels[1]). Since hydrogen is not a naturally occurring resource on earth, it must be generated from any of a number of energy sources, in order to deliver it to an end user. As such, the term "hydrogen economy" is somewhat of a misnomer, since it should be viewed as an energy carrier (*i.e.* in that respect, much like electricity[1]) and not an actual naturally occurring resource. There are a number of serious issues that must be overcome, not only in fuel-cell technology, but also in terms of hydrogen production and distribution, if a hydrogen economy is going to succeed. For example, hydrogen infrastructure does not currently exist. Also, although the electrochemical conversion of H_2 to H_2O in fuel cells results in

RSC Energy and Environment Series No. 1
Thermochemical Conversion of Biomass to Liquid Fuels and Chemicals
Edited by Mark Crocker

Published by the Royal Society of Chemistry, www.rsc.org

zero emissions, the production of H_2 from coal gasification and natural-gas reforming generates significant levels of CO_2, such that capture and sequestration would likely be necessary from political and environmental standpoints.[2] The major reaction pathways to generate syngas involve coal/biomass gasification, methane steam reforming, and partial oxidation of methane, the latter two reactions often being combined to offset the heat produced from partial oxidation with the heat consumed from steam reforming (*e.g.*, oxidative steam reforming or autothermal reforming).

Examples of syngas production reactions:

$$C + H_2O \rightarrow CO + H_2 \quad (119\,\text{kJ/mol}) \tag{5.1}$$

$$CH_4 + H_2O \rightarrow CO + 3H_2 \quad (206\,\text{kJ/mol}) \tag{5.2}$$

$$CH_4 + 1/2O_2 \rightarrow CO + 2H_2 \quad (-36\,\text{kJ/mol}) \tag{5.3}$$

In considering the distribution of H_2 relative to other fuels, it must be appreciated that hydrogen is difficult to compress, transport and store, and has a poor energy density, as shown in Table 5.1.

Production of liquid hydrogen is energy intensive; indeed, hydrogen liquefaction requires an energy input equivalent to one-third of its energy content, and even then, the energy density is still quite low.[1] In considering Table 5.1, one important advantage of diesel is its high energy density. In terms of energy conversion, diesel produced by the Fischer–Tropsch synthesis reaction (with or without product upgrading) is a flexible energy carrier. Not only can premium diesel be produced from syngas, but it can either be combusted or subsequently reformed at a point-of-use to release H_2 for use in fuel-cell applications. Among other liquid chemical carriers of hydrogen, methanol is another flexible energy carrier that should be strongly considered, and ethanol can be produced by fermentation processes and is a renewable energy carrier.

Fischer–Tropsch synthesis:

$$CO + 2H_2 \rightarrow -[CH_2]_n- + H_2O \quad (-165\,\text{kJ/mol}) \tag{5.4}$$

Table 5.1 Volumetric energy densities of various fuels (MJ/L).

Fuel	*Energy density (MJ/L)*
Diesel	37.3
Gasoline	34.2
Ethanol	24
Methanol	15.6
H_2 (l)[a]	10.1
H_2 (700 bar)[a]	5.6
H_2 (g)[a]	0.01079

[a]burned in air.

Methanol synthesis:

$$CO + 2H_2 \rightarrow CH_3OH \quad (-90.6\,kJ/mol) \tag{5.5}$$

$$CO_2 + 3H_2 \rightarrow CH_3OH + H_2O \quad (-49.5\,kJ/mol) \tag{5.6}$$

From an environmental perspective, the production of hydrogen from fossil fuels is justified only when higher efficiencies or lower emissions result.[1] H_2-production technologies will not significantly reduce CO_2 emissions unless capture and sequestration technology can be adequately demonstrated, and will not reduce reliance on fossil fuels.[1] However, from an environmental standpoint, the potential of hydrogen may be reached if renewable resources can be used in a sustainable manner for its production.[1]

One can envision the future role of the Fischer–Tropsch synthesis reaction both during the interim period and when (and if) a hydrogen economy is eventually in place. In the near future, an increasing amount of transportation fuels and chemicals will likely be generated from syngas derived from natural gas, coal, and biomass through the Fischer–Tropsch synthesis reaction, followed by product upgrading. The resulting premium diesel (*i.e.* virtually sulfur-free with low aromatics) would be used in diesel combustion engines. When a hydrogen infrastructure is eventually in place, Fischer–Tropsch synthesis may continue to be used to produce chemicals (*e.g.*, oxygenates, olefins, and waxes) and diesel from gasified renewables, the diesel serving as a high-density liquid energy carrier.[3] It would likely be transported from a stationary source either *via* pipeline or trucks to the point-of-use, where it could then be reformed to release hydrogen for use in either PEM fuel-cell applications[3] or used directly in combustion engines. As mentioned previously, another candidate as a flexible liquid energy carrier is methanol, but its higher toxicity may preclude its use in favor of the more benign FT product.

This chapter focuses on general aspects of the Fischer–Tropsch synthesis reaction with an emphasis on considerations that should be taken into account in converting biomass-derived syngas. The topics covered include: (1) biomass-derived syngas and potential catalyst poisons; (2) catalyst selection considerations; (3) product distribution dependencies on conditions/catalysts; and (4) catalyst structure–function properties, kinetics, and proposed mechanisms.

5.2 Syngas Derived from Biomass Gasification for Use in Fischer–Tropsch Synthesis

Before discussing further the Fischer–Tropsch synthesis reaction, it is important to briefly consider syngas derived from the gasification of biomass. Biomass gasification for transportation fuel production is in its infancy. The Energy Research Center of the Netherlands (ECN) examined many different possibilities for biomass gasification and concluded that a slagging-type O_2 blown entrained flow gasifier was likely required, due to the amount of ash that is produced.[4] Because the ash from biomass is rich in CaO, which doesn't melt under typical gasifier conditions, and because it is desirable to expel the ash as a liquid, a fluxing material (*e.g.*, silica or clay) was deemed necessary in order to obtain the right slagging properties in the temperature range of interest.[4]

Another critical point is that alkali metals melt at relatively low temperatures, and since they are volatile a significant fraction is removed in the gas phase.[4] Alkali compounds are certainly an important component in biomass ash. In fact, the word potassium comes from the word pot-ash, from the time when wood was burned in braziers, and lye was extracted from ash to make soap. ECN narrowed down the slagging-type process configurations to three.

The first process[4] was recommended when biomass is used alone, as it was determined to be the most efficient. In that case, the biomass does not need to mimic coal, and the high reactivity of biomass means that particles as large as 1 mm can probably be used. A simple flow diagram involving crushing to 1 mm, piston feeding, and screw transport was provided. According to ECN, the advantages of screw feeding include technical simplicity and lower inert-gas consumption, the latter resulting in reduced electric energy consumption for compression of the gas and higher gasifier efficiency ($\sim$84%).

If biomass is treated to mimic coal,[4] which may be useful when cofeeding biomass and coal, one must first break the particles down to around 100 micrometers. Since pulverizing is expensive, often a mild thermal treatment is used ($\sim$270 °C for 30 min) known as torrefaction, and then the particles are milled to $\sim$100 micrometers. Piston feeding or lock hoppers are used, and pneumatic transport is employed. Both lock hoppers and pneumatic transport require a lot of inert gas for compression, and the overall efficiency, including electricity generation, was suggested to be around 75%.

The third option involves flash pyrolysis of the biomass to form an oil/char slurry, which is pumpable.[4] This could be advantageous if the biomass source is remote. The overall efficiency was suggested to be about 75%. It is important to be aware that these different configurations yield syngas mixtures with varying levels of inert gas present in the syngas feed.

Since there are a number of contaminants generated by the gasifier, which are thus present in the raw syngas from biomass gasification, treatment steps are necessary. It would be helpful to gain insight into the impacts of these contaminants on the downstream process catalysts, in order to define the treatment steps and degrees of treatment that are required. ECN provided an example of the products obtained from gasification of wood. An air-blown circulating fluidized-bed gasifier operating at 850 °C and atmospheric pressure was used,[5,6] such that a significant concentration of hydrocarbons was left unconverted, requiring further processing steps. With an oxygen-blown entrained flow gasifier operating between 1200 and 1500 °C and 20–40 bar, virtually all of the hydrocarbon compounds should be converted to syngas, but the other impurities will still be present, with levels dependent on the biomass feed source used. Among the impurities listed[5,6] were NH_3 (2200 mg/m^3), HCl (130 mg/m^3), H_2S (150 mg/m^3), COS + CS_2 + HCN + HBr (<25 mg/m^3), and dust + soot + ash (2000 mg/m^3).

Table 5.2 Typical syngas specifications for Fischer–Tropsch synthesis.[6]

Contaminant	*Allowable levels*
Sulfur compounds ($H_2S + COS + CS_2$)	< 1 ppmv
Nitrogen compounds ($NH_3 + HCN$)	< 1 ppmv
Halide compounds ($HCl + HBr + HF$)	< 10 ppbv
Alkali metals (Na + K)	< 10 ppbv
Particles (soot, ash)	Almost complete removal

Syngas specifications quoted for Fischer–Tropsch synthesis processes are listed in Table 5.2. The general approach to clean syngas[6] is to quench the raw hot gas with water to cool the gas and remove solid particles (dust, soot, and ash) and the volatile alkaline metals. NH_3 is removed downstream with the halides (HCl, HBr, HF) using a water washer. H_2S is removed from the syngas (0.6% H_2S by volume for coal-derived syngas) by physical absorption in dimethyl ethers of polyethylene glycol (*e.g.*, Selexol), cold methanol (*e.g.*, Rectisol), or an amine-based solvent. The H_2S is sent to a sulfur-recovery unit comprising an air-blown Claus plant for oxidizing H_2S to elemental sulfur (with accompanying tail-gas cleanup). For example, in the Shell Claus off-gas treating process,[7] these compounds (H_2S, SO_2, COS, CS_2 and S vapor) are catalytically converted to H_2S, which is removed with an amine absorption unit and recycled back to the Claus plant. COS and HCN are hardly removed in gas cleaning. If high loads of these compounds are present, it is economically more attractive to install a hydrolysis step to convert the compounds to H_2S and NH_3. Or, if a sour shift catalyst is used (*e.g.*, a CoMo catalyst), the COS will be converted to H_2S concurrently with water-gas shift, and then both the H_2S and CO_2 can be captured by the amine solvent in one step following the shift operation. The remaining amount of H_2S is trapped in guard beds containing activated carbon.

5.3 Differences Between Co and Fe Fischer–Tropsch Synthesis Catalysts

There are several differences to be noted between cobalt and iron when considering the selection of the catalyst. One important factor is the syngas ratio, which can vary with the resource used (*e.g.*, natural-gas-derived syngas often, but not always, has higher H_2/CO than coal- or biomass-derived syngas) and the gasifier type (*e.g.*, a Lurgi gasifier converts coal but rejects enough CO_2 such that the syngas ratio is closer to 2:1,[8] whereas a GE/Texaco gasifier typically converts coal to a lower syngas ratio in the range of 0.7–1).[9]

If the syngas ratio is high (*e.g.*, 2:1), cobalt catalysts should be strongly considered, since they possess low intrinsic water-gas shift activity. On the other hand, if the syngas ratio is low (*e.g.*, 0.7–1), as with biomass-derived syngas, iron catalysts have an advantage, since they possess water-gas shift (WGS) activity, which can be used to boost the H_2/CO ratio internally (*i.e.* both FT and WGS in one reactor), rejecting CO_2 in the process. If a low syngas ratio is available, cobalt catalysts can still be used effectively, but WGS should be carried out upstream in a separate reactor using an Fe/Cr catalyst.[10]

$$\text{Water} - \text{gas shift reaction:} \quad H_2O + CO \rightarrow CO_2 + H_2 \quad (-41.4\,\text{kJ/mol}) \tag{5.7}$$

A schematic of the two scenarios is provided in Figure 5.1.

Another major consideration is the selectivity to the products. If, for example, chemicals production is desired, iron catalysts should again be considered, since they display a higher selectivity toward α-olefins, oxygenates (*e.g.*, alcohols, aldehydes, and ketones), and branched isomers.[11,12]

In surveying the literature, there is a skewed perception that cobalt catalysts generally perform better than iron catalysts. However, that view should be reconsidered, as summarized in Table 5.3. In fact, depending on the conversion level, Fe catalysts can be as or even more productive than cobalt.[14] At high conversion levels, a cobalt catalyst may be more productive than iron (0.5 to 1.0 g hydrocarbons/g_{cat} h). However, at lower conversion levels, iron catalysts exhibit high activity (*e.g.*, 5.0 g hydrocarbons/g_{cat} h). Researchers at the University of Kentucky's Center for Applied Energy Research (CAER) have tested promoted iron catalysts for long periods of time with very little deactivation,[14] so the prevailing wisdom that cobalt catalysts have longer lifetimes should also be reconsidered. It is true that cobalt catalysts are more attrition resistant than current Fe catalysts, due to the sloughing off of the rather friable iron carbide phase, which tends to blind the filters used to separate the catalyst particles

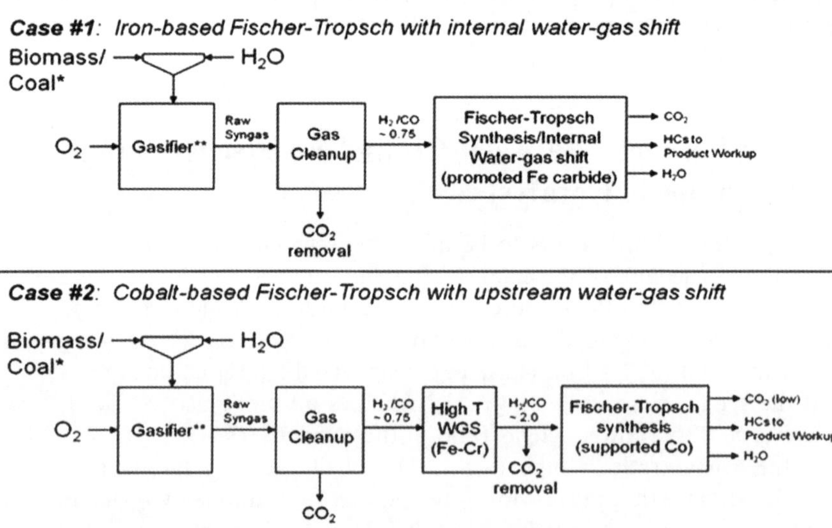

Figure 5.1 The syngas ratio influences both catalyst selection and reactor configuration.

Table 5.3 General comparison of cobalt and iron catalysts.[13]

	CATALYST COMPARISON–PERCEPTION vs. REALITY			
	Perception		*Reality*	
Concept	*Co*	*Fe*	*Co*	*Fe*
Chain Growth Probability (α)	High	Low	Intermediate	Low to High
Activity	High	Low	High	High
Olefin Selectivity	Low	High	Low	High
H_2/CO Ratio for Feed	2 +	0.5 to 2.5	1.0 to 2 +	0.5 to 2 +
Water-Gas-Shift, Activity and CO_2 Selectivity	Very Low	High	Very Low	Intermediate to High
Methane Selectivity	High	Low	Intermediate	Low to very low
Temperature	Only Low	Low to High	Only Low	Low to High
Temperature Sensitivity, Activation Energy	Very Sensitive	Slight	Slight	Slight
Temperature Sensitivity, Product Distribution	Sensitive	Slight	Very Sensitive	Very Slight
Pressure Sensitivity, Product Distribution	Very	Slight	Very	Slight
Sulfur Tolerance	Very Little	Very Little	Very Little	Intermediate
Stability, Life Time	Long	Short	Long	Long
Cost	Expensive	Cheap	Expensive	Cheap to Expensive
Separation from Wax	Easy	Difficult	Easy	Very Difficult
Robustness	Excellent	Poor	Excellent	Very Poor

from the wax.[15] Also, it is true that if high H_2/CO ratio syngas is used, Co generally outperforms Fe in terms of exhibiting lower light product selectivity. Yet, the same conclusion cannot be made for the cases of biomass or coal-derived syngas with low H_2/CO ratio. In fact, Fe possesses lower light product selectivity for the undesirable $C_1 - C_4$ products.[13]

5.4 Product Distribution Dependence on Conditions

Syngas is typically converted over a catalyst in a fixed-bed or slurry reactor to produce a distribution of hydrocarbons, which follows Anderson–Schulz–Flory (ASF) kinetics. If a plot is made of the log of the mole fraction *versus* the carbon number, a straight line will result from which alpha, the chain-growth probability factor, is calculated. Methane selectivity is typically higher than predicted by ASF kinetics, while ethylene is lower.[17] The products are fractionated into several boiling ranges of interest, including light gas, naphtha, kerosene, and wax. Process conditions for Fischer–Tropsch synthesis are chosen to favor a certain product range of interest. That is, they are selected to either increase or decrease the slope of the line of the plot, and thus, the value of α.

For example, as shown in Figure 5.2, one may be interested in producing gasoline, diesel, jet fuel, lubricants, or wax. If gasoline motor fuel is desired, one is interested in a lower boiling range fraction with octane number between 87 and 93. However, for high octane number, to prevent engine knock, one desires branched alkanes and aromatics over straight-chain paraffins. For jet fuel and diesel fuels, the boiling range is higher, and cetane number is important, as it is associated with ignition delay time, with higher numbers having a shorter delay time. In that case, straight-chain paraffins are desired.

Because of the kinetics of Fischer–Tropsch synthesis, a certain product distribution results and straight run distillation of the FT product is not sufficient to produce transportation fuels. For example, if low-temperature FT is carried out to favor diesel products, as in the commercial Sasol Arge process (now partly replaced by slurry bubble column reactors), heavier products are favored (Figure 5.3). However, there are still lighter products and heavier wax product

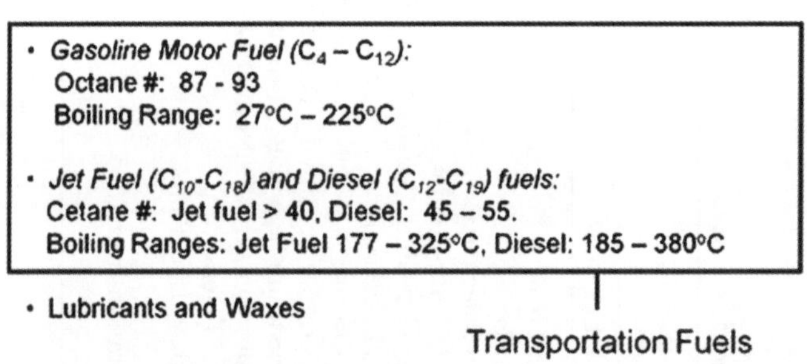

Figure 5.2 Potential products of interest.

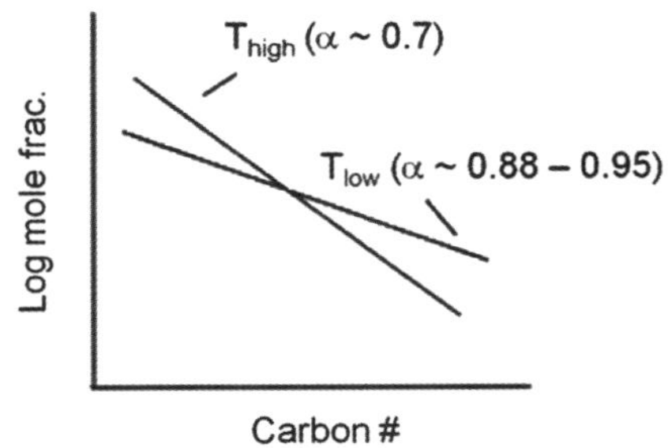

Low T	Sasol Arge	High T	Sasol Synthol
• low $C_1 - C_4$	13.3	• higher $C_1 - C_4$	43.0
• low $C_5 - C_{11}$	17.9	• higher $C_5 - C_{11}$	40.0
• low C_{12}-C_{19}	13.9	• less C_{12}-C_{19}	7.0
• 50-70% wax	51.7	• low wax	4.0
• 220-270°C		• 325 – 350°C	
• α: 0.87+		• α: ~0.7	
• gasoline/diesel: 1:2		• gasoline/diesel: 2:1	
• 80⁺ Cetane #		• 50-60 Cetane #	
• 0-20 Octane #		• 0-60 Octane #	

Figure 5.3 Dependency of Fischer–Tropsch synthesis ASF distribution on temperature. Product selectivities (in %) of the Sasol Arge (220 °C) and Sasol Synthol (325 °C) processes are on a C atom basis. Reprinted with permission from the author.[16]

that must be further processed. On the other hand, if high-temperature FT is conducted to favor gasoline boiling range product, a considerable amount of very light product must be dealt with. The overall process, depending on location, may be less benign from an environmental perspective (*e.g.*, if the light gases are flared or upgraded, the latter requiring energy intensive processes).

The reaction pressure is also important. Higher pressures tend to favor the production of heavier products.[17] At the University of Kentucky Center for Applied Energy Research, low-temperature Fischer–Tropsch synthesis is typically carried out at approximately 20 atm over cobalt catalysts in the temperature range of 200–230 °C using a H_2/CO syngas ratio of 2:1; and, at around 10–30 atm over iron catalysts at 230–270 °C using a H_2/CO syngas ratio in the neighborhood of 0.70–1.1.

5.5 Proposed Mechanisms, Kinetics, and Catalyst Structure–Function Properties

As summarized in Table 5.4, there have been a number of kinetic studies reported in the literature, including both mechanistically derived models (*e.g.*, Langmuir–Hinshelwood–Hougen–Watson expressions), as well as empirically derived mechanisms (*e.g.*, power-law kinetic models). The problem in having so many different models is that, from the standpoint of reactor design, it would

Table 5.4 Kinetic models based on mechanistic and empirical derivations.[18]

Mechanistic kinetic model of FTS	*Catalyst*	*Mechanism*	
1 $r_{CO} = -kP_{CO}P_{H2}/(1+mP_{CO})^2$	Co	Carbide	
2 $r_{CO} = -kP_{CO}^{0.5}P_{H2}/(1+m_1P_{CO}^{0.5})^3$	Co	Carbide	
3 $r_{CO} = -kP_{CO}P_{H2}^{0.5}/(1+m_1P_{CO}^{0.5}+m_2P_{H2}^{0.5})^2$	Fe	Carbide	
4 $r_{CO} = -kP_{CO}^{0.5}P_{H2}^{0.5}/(1+m_1P_{CO}^{0.5}+m_2P_{H2}^{0.5}+m_3P_{CO})^2$	Fe	Carbide	
5 $r_{CO+H2} = -kP_{H2}P_{CO}/(P_{CO}+m_1P_{CO2})$	Fe	Enol theory	
6 $r_{CO+H2} = -kP_{H2}P_{CO}/(P_{CO}+m_1P_{H2O}+m_2P_{CO2})$	Fe	Enol theory	
7 $r_{CO} = -kP_{H2}P_{CO}/(1+mP_{H2O}/P_{H2})$	Co & Fe	Carbide & CO insertion	
8 $r_{CO} = -kP_{H2}P_{CO}/(1+mP_{H2O})$	Fe	Enol theory	
		Reaction order	
Empirical kinetic model of FTS		*a*	*b*
1 $r_{CO} = -kP_{H2}$	Co & Fe	–	–
2 $r_{HC} = -kP_{H2}/P_{CO}$	Fe	–	–
3 $r_{CO} = -kP_{CO}^aP_{H2}^b$	Co & Fe	−1 to 1	0.5 to 2
4 $r_{CO} = -kP_{CO}^aP_{H2}^b/(1+mP_{CO})$	Co	0.65	0.6
5 $r_{FT} = -kP_{CO}^aP_{H2}^b/(1+mP_{CO})^2$	Fe	1	0.5
6 $r_{FT} = -k(P_{CO}^aP_{H2}^b/P_{H2O})/(1+mP_{CO}P_{H2}/P_{H2O})^2$	Fe	1	1.5
7 $r_{CO} = -kP_{CO}^aP_{H2}^b/(1+mP_{H2O}/P_{H2})$	Co	−0.8 to −0.2	0.5 to 0.6

be useful to have on hand a simple model that can fit a number of different catalysts.

The mechanistic models usually draw on any of several proposed mechanisms, two of which are provided in Figure 5.4.[19] For example, in the "enol" mechanism, popularized in the 1950s and 1960s by researchers such as Storch *et al.*[20] and Emmett *et al.*,[21] CO adsorbs linearly on the metal and is hydrogenated to an adsorbed "enol" intermediate. Chain growth proceeds *via* successive dehydration of adjacent enol intermediates. Another category is the carbene mechanism, where CO adsorbs dissociatively to make adsorbed O and carbide. The mechanism proceeds through hydrogenation of O to adsorbed H_2O, with subsequent desorption, as well as hydrogenation of the carbide to carbene. Addition of carbene leads to chain growth. A more complex version of a carbene mechanism, showing the steps for the formation of alkanes, alkenes, and even branched species, is provided.

In examining various kinetic models for cobalt, it is clear that at low contact times, most of the models display a linear trend (Figure 5.5). It is only at high contact times, where the CO conversion is high, that the line starts to bend. While slightly different degrees of bending have been observed for a variety of catalysts, it is interesting to note that the bending only occurs when the P_{H_2O}/P_{H2} in the reactor becomes high.

Before we look at the kinetic influence of H_2O, it is first important to understand how cobalt catalysts function. Iglesia,[30] in 1997, demonstrated quite convincingly that the active sites on Co catalysts are surface cobalt atoms. If one plots the CO conversion *versus* the fraction of surface atoms exposed to the surface (*i.e.* the dispersion), one essentially obtains a straight line. Said another way, if a plot is made of the turnover frequency of CO per Co metal surface active site, *versus* the dispersion, one often obtains essentially a straight flat line, showing that each site has the same activity (Figure 5.6). This may not be true at very high dispersions, however.[31]

The typical standard preparation method involves sequential aqueous impregnation of a soluble cobalt precursor salt (*e.g.*, cobalt nitrate) and promoters, with standard drying steps (*e.g.*, in a rotary evaporator), followed by a single calcination step under air flow.

By XRD, the cobalt oxide is initially present as Co_3O_4. When characterizing catalysts by hydrogen temperature-programmed reduction (TPR), reduction results in two peaks, with the second one typically consuming three times as much H_2 as the first (Figure 5.7, top left). This is why many researchers have proposed a two step reduction of cobalt oxides:

$$Co_3O_4 + H_2 \rightarrow 3CoO + H_2O, \text{ followed by } 3CoO + 3H_2 \rightarrow 3Co + 3H_2O$$

In general, for calcined catalysts of similar loading, we have found that, while the first step of reduction occurs at about the same temperature (~300–350 °C range), the second step strongly depends on the type of support.[32] While Co oxides appear to have relatively weak interaction with silica, the TPR profiles of titania and alumina-supported cobalt catalysts are much broader, some

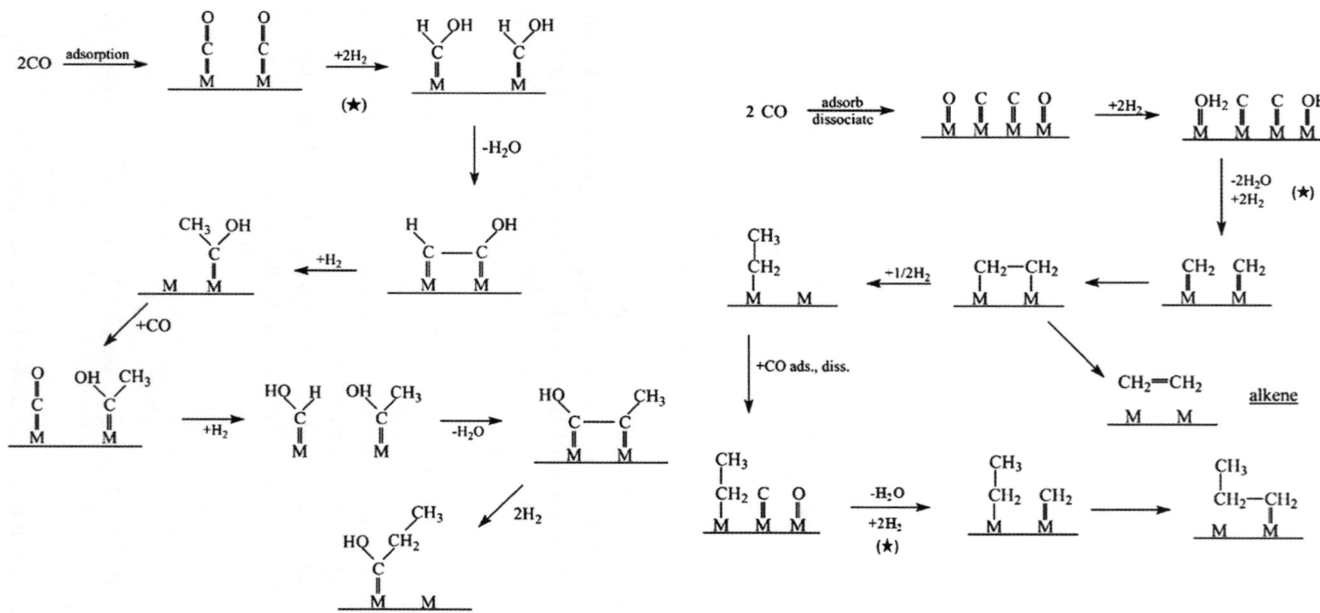

(* rate controlling)

Figure 5.4 (top left) Enol mechanism; (top right) simple carbene mechanism; (top of next page) complex carbene mechanism. Reprinted with permission from Elsevier.[19]

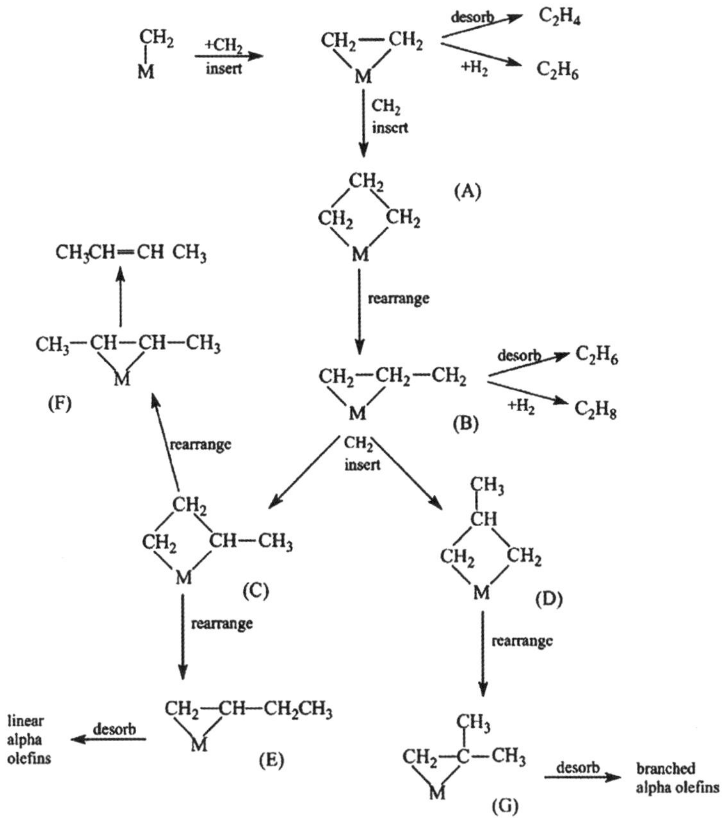

Figure 5.4 Continued.

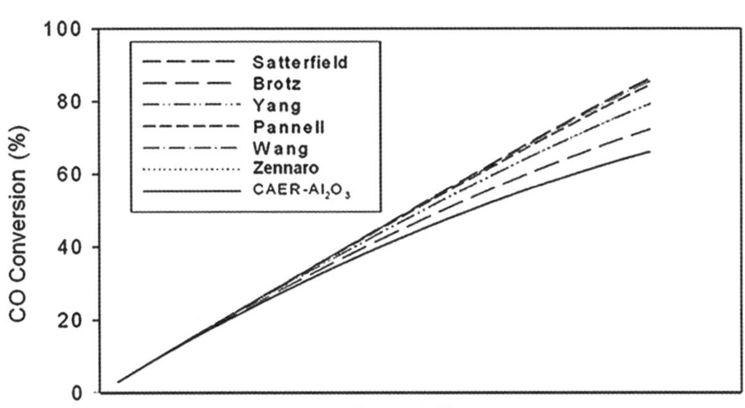

Figure 5.5 Kinetic behavior of supported cobalt catalysts as a function of contact time with H_2/CO ratio of 2:1.[22–29] Units of contact time vary by research group. Reprinted with permission from Elsevier.[29]

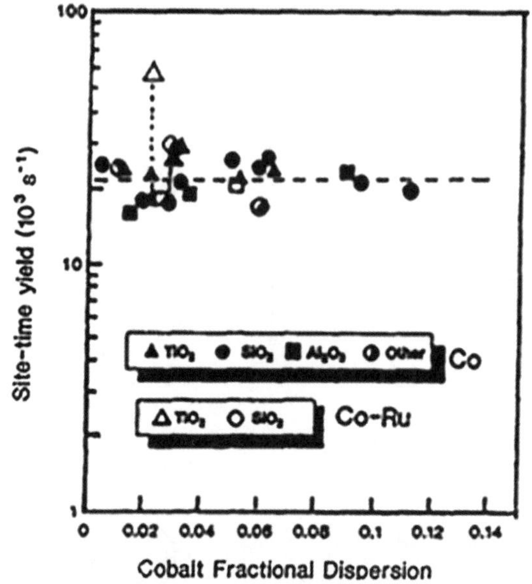

Figure 5.6 Site-time yield *versus* cobalt fractional dispersion over various catalysts from reference. Reprinted with permission from Elsevier.[30]

extending all the way beyond 700 °C. If one extends the TPR beyond 800 °C, there is typically a small peak for some cobalt that is sacrificed to the support as cobalt aluminate.

For slurry processes, the catalyst is typically reduced *ex situ* at a standard temperature (*e.g.*, 10 h reduction in 33%H_2/inert gas at 350 °C) in a tubular flow reactor, and transferred under inert gas to the slurry reactor, where it is combined with the start-up solvent and re-reduced under milder conditions. One typically adds Co at close to commercial loadings (15–30% Co) onto a γ-Al_2O_3 support (with surface area of 150–200 m^2/g). If TPR is carried out after standard reduction (*e.g.*, as for the 15%Co/Al_2O_3 catalyst in Figure 5.7 top right) a significant fraction of cobalt, likely in the form of CoO, remains unreduced. This leads to complications in estimating cobalt metal particle size by hydrogen chemisorption methods, where 100% of reduction is often, though erroneously, assumed.

X-ray absorption near edge spectroscopy (XANES) spectra of cobalt reference compounds at the Co K-edge (1s to empty 4p transition) exhibit differences in their line shapes, such that XANES can be used to probe the extent of Co reduction. Essentially, the X-rays are used to eject electrons, which at low kinetic energies will probe unoccupied electronic states above the Fermi level. The changes in the pre-edge features and white line shapes along the trajectory of a hydrogen TPR-XANES experiment (Figure 5.7, bottom) indicate that reduction does indeed proceed in two steps, where Co_3O_4 is first converted to CoO, while the reduction of CoO to Co metal is strongly influenced by the

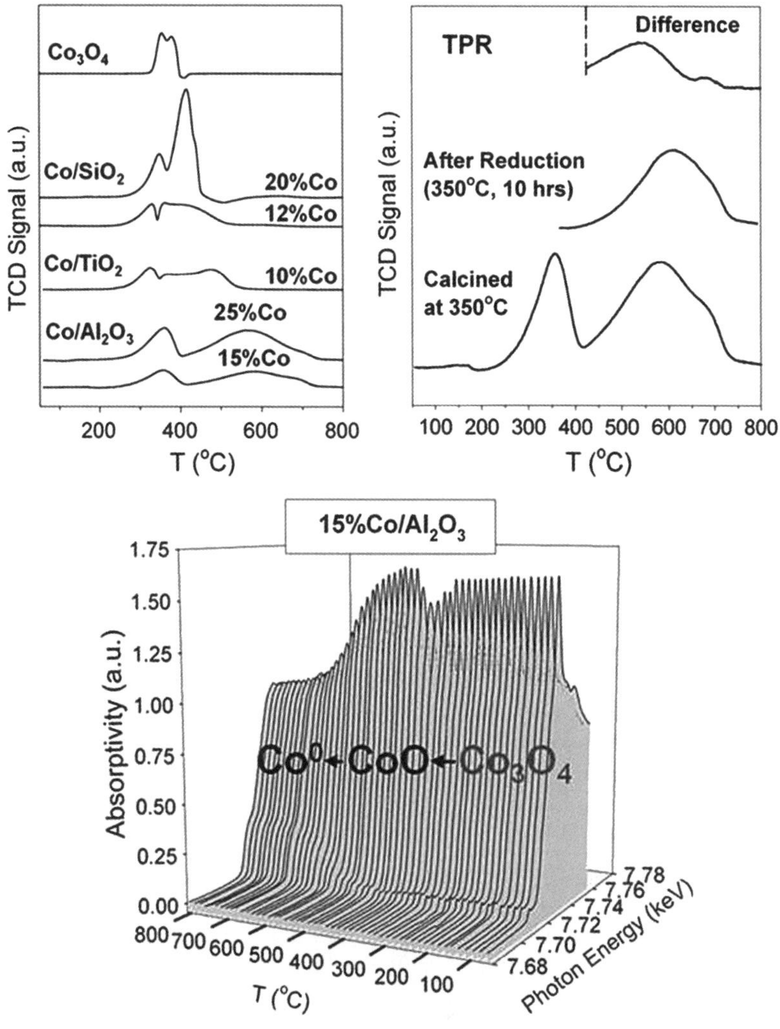

Figure 5.7 (top left) TPR profiles of unpromoted supported cobalt catalysts.[32] (top right) TPR after standard H_2-reduction of 10 h at 350 °C.[32] (bottom) TPR-XANES experiment of the reduction of 15%Co/γ-Al_2O_3.[33] Reprinted with permission from Elsevier.[32]

nature of the support and presence/absence of promoters.[33] While increasing loading to grow the cobalt cluster does somewhat facilitate breaking the γ-Al_2O_3 support interaction with cobalt oxides, adding a reduction promoter like Pt clearly shifts both reduction steps to much lower temperatures. The transitions on the weakly interacting Co/SiO_2 catalyst occur, for the most part, very rapidly, and more closely resemble reduction of bulk Co_3O_4 (*i.e.* CoO is a very short-lived intermediate). However, one should not immediately take improved cobalt reduction as a sign of a catalyst with higher surface cobalt densities. In fact, the opposite finding is often the case when comparing unpromoted cobalt catalysts with different support interactions.

While moving to a support with a stronger interaction generally results in a lower extent of reduction following a typical activation (*e.g.*, 350 °C in 33% H_2/inert), it generally leads to smaller cobalt clusters with higher densities of exposed Co surface atoms per gram of catalyst upon reduction, that can more than compensate for the loss in extent of reduction. One important tool that can be easily used to obtain information on cobalt cluster size, extent of reduction, and active-site densities, is hydrogen chemisorption with pulse reoxidation.[32] A quick hydrogen chemisorption method used to obtain surface cobalt atom site counts is hydrogen TPD. The catalyst is reduced in hydrogen at the activation temperature, and then cooled under flowing hydrogen, allowing the hydrogen to adsorb to the surface of the cobalt particles. By increasing the temperature and flowing inert gas, one can then sweep the hydrogen off the surface and, by way of stoichiometry, count the active sites using a thermal-conductivity detector (TCD). However, dispersion is defined to be the number of surface cobalt metal atoms divided by the total number of cobalt metal atoms. Thus, to obtain the true dispersion of metal, it is first necessary to estimate the fraction of the cobalt that has been reduced. This is accomplished using a pulse reoxidation after the H-TPD is completed. From the amount of O_2 consumed, one can estimate the amount of metal present by assuming that Co metal is oxidizing to Co_3O_4. It is only by combining the two measurements (*i.e.* H-TPD and O_2 titration) that one can estimate the average Co metal cluster size, as shown in Figure 5.8 for a 15%Co/γ-Al_2O_3 catalyst.

$$\%D = \frac{(\#\ Co^0 \text{ atoms on surface})}{(\#\ Co^0 \text{ atoms total})} = \frac{\#\ Co^0 \text{ atoms surface}}{(\#\ Co \text{ atoms in sample})(\%\ \text{reduced})}$$

$$\#\ Co \text{ atoms in sample} = (0.220\ g)(0.15)(AW_{Co})$$

$$\%\ \text{reduced} = \frac{(\#\ \text{moles } O_2 \text{ consumed})(3 \text{ moles } Co^0) \times 100\%}{(\#\ Co \text{ moles in sample})(2 \text{ moles } O_2)}$$

Assumption: Spherical type morphology

$$\text{Diam.} = 6\ V/A = 6/(\%D \times \text{rho} \times A) = \frac{6 \times (1000\ g/kg) / \%D}{(8.89\ g/cm^3)(654\ cm^2/kg)}$$

- TPD sweep estimates number Co surface atoms
- Pulse reoxidation estimates fraction of Co reduced.
- Both sets of data are needed to estimate Co cluster size.

Figure 5.8 Hydrogen chemisorption with pulse reoxidation.

Table 5.5 Results of hydrogen chemisorption/pulse reoxidation. Reprinted with permission from Elsevier.[32]

Catalyst	T_{red} (°C)	*μmol H_2 desorbed per g cat.*	*Uncorr. % dispersion*	*O_2 uptake (μmol/g cat.)*	*% Reduced*	*Corrected % dispersion*	*Corrected diameter (nm)*
12.4%Co/SiO_2	350	**18.2**	1.7	902	**64**	2.7	**38.2**
10.0%Co/TiO_2	300	**42.8**	5.1	593	**52**	9.7	**10.6**
15%Co/Al_2O_3	350	**66.9**	5.3	509	**30**	17.5	**5.9**

Based on this approach, it is easy to see that just because a catalyst is more easily reduced, as shown by the pulse reoxidation measurements (Table 5.5), this does not mean that the active-site density is higher. Despite half the extent of reduction for 15%Co/Al_2O_3 compared to 12.4%Co/SiO_2, the former catalyst has approximately three times the surface cobalt site density, due to the smaller average cobalt cluster size. The support interaction with cobalt oxide species, not necessarily the support surface area (SiO_2 352 m^2/g, Al_2O_3 200 m^2/g), is needed to stabilize the small clusters. Again, for unpromoted catalysts, higher extents of reduction often equate to more bulk-like Co_3O_4 character.

Recently, CAER researchers plotted the extent of reduction *versus* cobalt cluster size, and suggested that the smaller cobalt oxide species strongly interacting with the support, *i.e.* those below 10 nm, are more difficult to reduce.[32] To overcome the support interaction and improve the extent of reduction for Co/Al_2O_3 catalysts, one can either increase the loading to give a larger average Co particle size and lessen the surface contact with the support, or, add reduction promoters to facilitate the nucleation of reduced sites on cobalt oxides, probably *via* a H_2 dissociation and spillover mechanism. Both methods result in improved extent of reduction. When applying the H_2 chemisorption/pulse reoxidation methods, addition of reduction promoter does lead to greatly improved site densities, due mainly to an increase in the extent of reduction (Table 5.6). Combining the two measurements (H-TPD and O_2 titration), the resulting average Co cluster size does not change drastically.

Selected CAER testing results,[34] conducted using a continuously stirred tank reactor (CSTR), are reported in Figure 5.9, in which the catalysts are compared on an equivalent gram basis. The results are for unpromoted and promoted 15%Co catalysts. The trend in initial CO conversion rate correlates well with the trend in active-site densities from H_2 chemisorption/pulse reoxidation measurements.

It is important not only from a kinetic viewpoint, but also from a stability viewpoint, to consider the impact of H_2O on supported cobalt catalysts. Patents by Kim at Exxon Research indicate that H_2O addition can positively influence CO conversion rates and selectivity away from light products, especially methane.[35] CAER conducted testing in a CSTR and, to maintain constant space velocity and reactant partial pressures, replaced inert balancing gas with increasing amounts of steam during the run. Indeed, for Co/SiO_2, the catalyst with the weakest Co–support interaction, a modest but measurable

Table 5.6 Hydrogen chemisorption/pulse reoxidation results of unpromoted and promoted catalysts. Reprinted with permission from Elsevier.[32]

Catalyst	T_{red} (°C)	*μmol H_2 desorbed per g cat.*	*Uncorr. % dispersion*	*O_2 uptake (μmol/g cat.)*	*% Red.*	*Corr. % disp.*	*Corr. diam. (nm)*
Titania							
10.0%Co/TiO_2	300	**43**	5.1	593	**52**	9.7	**10.6**
0.2%Ru-10%Co/TiO_2	300	**67**	7.8	722	**64**	12.2	**8.5**
Alumina							
15%Co/Al_2O_3	350	**67**	5.3	509	**30**	17.5	**5.9**
0.5%Pt-15%Co/Al_2O_3	350	**141**	11.0	1024	**60**	18.4	**5.6**
1.0%Ru-15%Co/Al_2O_3	350	**116**	9.7	823	**50**	18.2	**5.7**
0.2%Re-15%Co/Al_2O_3	350	**142**	11.2	939	**55**	20.4	**5.1**
1.0%Re-15%Co/Al_2O_3	350	**168**	13.2	1187	**70**	19.4	**5.5**
25%Co/Al_2O_3	350	**78**	3.7	1174	**42**	8.7	**11.8**
0.5%Pt-25%Co/Al_2O_3	350	**149**	7.0	1994	**71**	9.4	**11.0**

increase in the CO conversion rate was identified.[36] However, at higher levels of added H_2O, accelerated deactivation rates were observed. For Co/TiO_2, a negligible to slightly negative impact of H_2O was found that was for the most part reversible, and higher deactivation rates were observed at lower space velocity.[37] For Pt promoted 15%Co/Al_2O_3, where reduction of the strongly interacting nanoscale Co clusters has occurred, a reversible negative effect of H_2O cofeeding occurred up to about 25% volume added H_2O, followed by a catastrophic decrease above that level.[38] In contrast, for a standard 25%Co/Al_2O_3 catalyst, closer to commercial loadings, the catalyst was found to be more resistant to the irreversible deactivation.[39] It appears that increasing the Co cluster size in Co/Al_2O_3 made the catalyst more robust. These effects were subsequently investigated by synchrotron methods such as EXAFS/XANES. Tests were repeated over 0.5%Pt-15%Co/Al_2O_3 and catalyst was removed at the different points of interest. For the catalyst at close to the start of the run, a combination of Co and CoO was observed in XANES spectra. Also, at low levels of added H_2O, where the kinetic effect was reversible, essentially no change was found. However, at 30% volume added H_2O, the catalyst exhibited a very sharp feature in the XANES derivative spectra very similar to bulk $CoAl_2O_4$, suggesting that Co–support compound formation occurred.[40] Over the more robust 25%Co/Al_2O_3 catalyst, for the high 25%H_2O condition the formation of CoO was detected, perhaps associated with smaller Co crystallites, or due to surface oxidation, as Co metal recovered when H_2O was switched off (*i.e.* the catalyst re-reduced).[39]

In EXAFS, similar findings were noted. For the 0.5%Pt-15%Co/Al_2O_3 catalyst, at and after 15%H_2O addition, predominantly Co–Co metal coordination was present. However, at 30% H_2O addition, a loss in Co–Co metal coordination was detected with the appearance of Co–O bonds. The catalyst

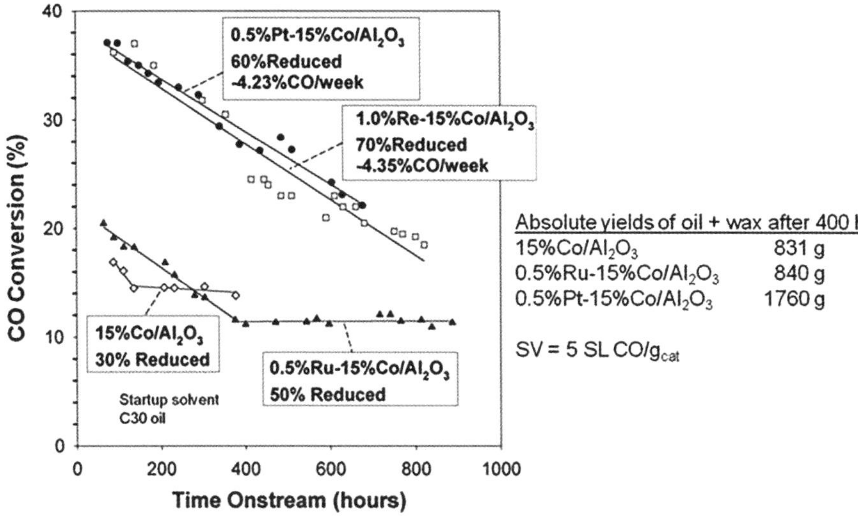

Figure 5.9 Trends in initial CO conversion rates from testing in a CSTR: 220 °C, ~20 atm, $H_2/CO = 2.0$.[34]

did not recover when H_2O was switched off, again pointing to the idea that something quite irreducible had formed, like a cobalt–support compound.[40] In contrast, for the 25%Co/Al_2O_3 catalyst, the loss of Co–Co coordination was followed by a recovery when H_2O was switched off. However, the Fourier transform magnitude was slightly larger for Co–Co coordination, suggesting that some sintering had occurred from the oxidation-reduction cycle.[39]

Another interesting finding was that there was a jump in the CO_2 selectivity observed when traversing the threshold between 25 and 28 vol% added H_2O for the 0.5%Pt-15%Co/Al_2O_3 catalyst. Co is typically used when the H_2/CO ratio is 2:1 because it exhibits very low water-gas shift activity. The jump in CO_2 selectivity suggests that cobalt changes chemical form to a higher oxidation state.[40]

To account for the kinetic influence of H_2O, CAER introduced a simple, but effective kinetic model with a water parameter "m":[29]

$$\text{Rate} = kP_{CO}^{a}P_{H2}^{b}/[1 + m(P_{H2O}/P_{H2})]$$

This was applied over a number of different supported Co/alumina[22,29] and Co/silica catalysts.[22,41] In all cases, the order in CO was found to be negative, indicating strong adsorption of CO. Likewise, in almost all cases, the order in H_2 was found to be close to "+0.5", suggesting that dissociated hydrogen is likely involved in the catalytic mechanism. Activation energies fell within a narrow range of 75–94 kJ/mol, close to the values of between 80 and 120 kJ/mol that have been reported in the open literature.[23,25,28] The calculated rate constants also fell within a rather narrow window, between about 0.13 to 0.22 mol/g-Co/h/MPa$^{(a+b)}$ on a per gram of activated-Co basis. However, it was with respect to the water-effect parameter that the clear advantages of the CAER model were observed. The model very adequately addresses the sign and sensitivity of each catalyst to the water effect. The catalysts for which the cobalt clusters were identified to be small enough to have a high fraction within the catalyst pore, displayed the greatest sensitivity to water – both negative, where the strongly interacting supports negatively influence CO conversion due to adsorption inhibition effects that can be brought on by Co oxidation; and positive, over the weakly interacting supports, where H_2O has been proposed to be able to displace heavier hydrocarbons from the pore and free up active sites for reaction[42] (Figure 5.10).

For Fe catalysts, with a low H_2/CO syngas ratio feed, it is important to consider the two competing reactions of FT and WGS (Figure 5.11). FT tends to limit CO conversion at low contact times, while WGS tends to limit the conversion at higher contact times. At higher contact times, one needs the H_2 produced from WGS in order to boost the CO conversion level, while at low contact times, little H_2O is present to carry out WGS.

To probe the mechanism of FT synthesis over Fe, isotopic tracing with radioactive carbon-14 was employed. In this case, labeled CO_2 was cofed to the reactor,[43] and ^{14}C was followed in the products. Interestingly, the amount of labeled C found in the FT product was essentially constant as a function of carbon number, if one subtracts the small amount of labeled CO that was formed from RWGS and would have propagated. The catalyst contained both K and Cu promoters (Figure 5.12).

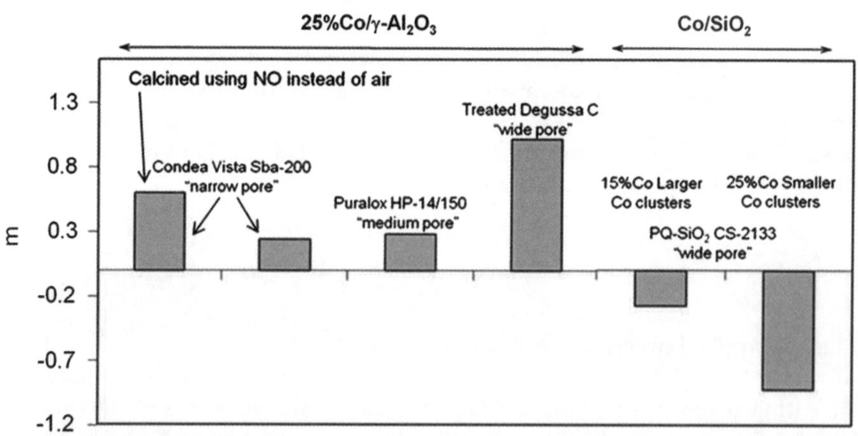

Figure 5.10 Variation of water-effect parameter "*m*" as a function of catalyst type and cluster size.[10]

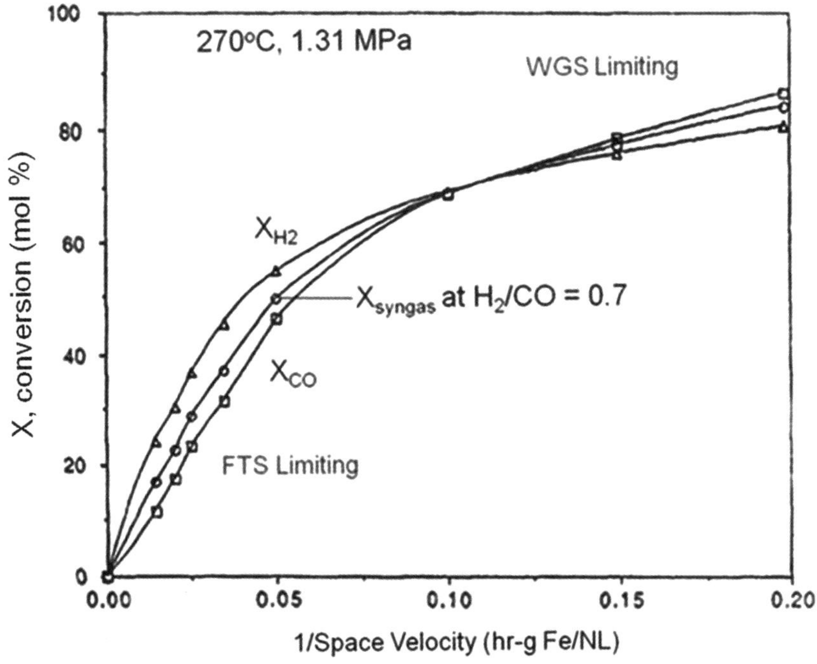

Figure 5.11 Conversion *versus* inverse space velocity over an iron catalyst with $H_2/CO = 0.7$.

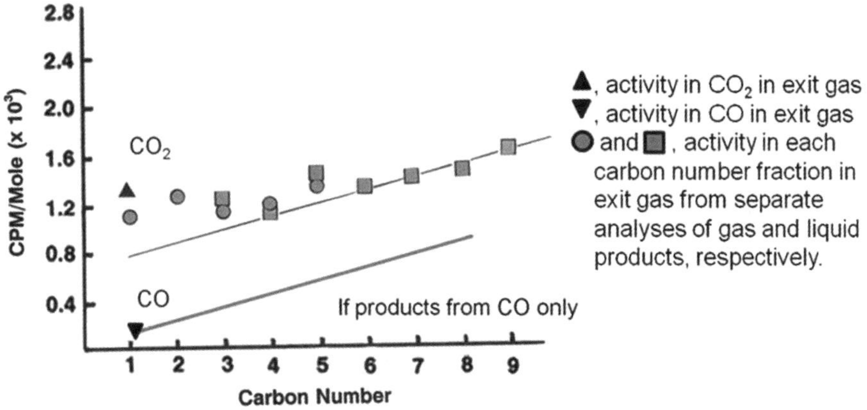

Figure 5.12 Counts per minute *versus* carbon number showing relatively constant ^{14}C content as a function of carbon number after cofeeding $^{14}CO_2$ with syngas over a Cu and K promoted Fe catalyst. Reprinted with permission from Elsevier.[43]

These data strongly suggest that, since both CO_2 and CO can initiate chain growth, the adsorbed intermediate that initiates chain growth is likely an oxygenate species (Figure 5.13).[43] One proposed species that could satisfy this hypothesis is formate (Figure 5.14), but this is still under debate. A different oxygenate species could also satisfy the findings.

The chemical composition of iron catalysts is certainly important for synthesis. Figure 5.15 (top) displays results for a potassium-promoted Fe catalyst stabilized with Si under a low H_2/CO ratio feed. If the catalyst is activated in CO, followed by synthesis over Fe, the conversion starts out very high. However, if activated in synthesis gas, the conversion starts out very low. However, if the catalyst with low conversion is then activated for 24 h in CO by

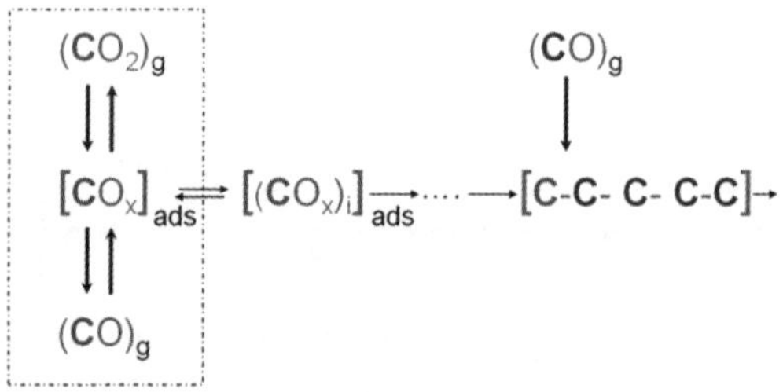

Figure 5.13 Since both CO and CO_2 initiate chains, an oxygenate species is a likely intermediate for chain initiation, prior to propagation. Reprinted with permission from Elsevier.[43]

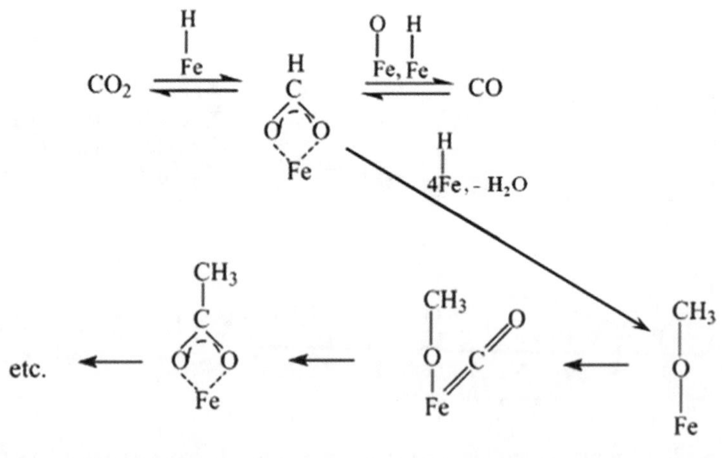

Figure 5.14 One possible candidate for the proposed oxygenate intermediate that can be initiated by CO_2 and CO is formate. Reprinted with permission from Elsevier.[43]

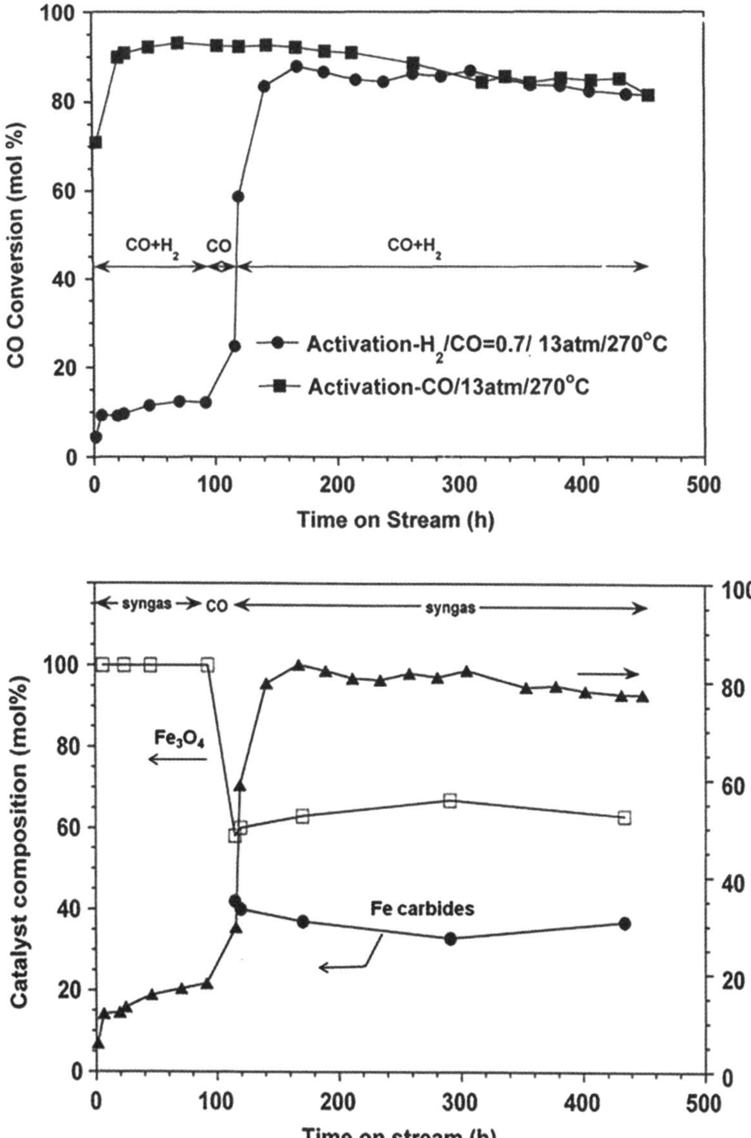

Figure 5.15 Effect of different pretreatments on a 100 Fe: 3.7 Si: 0.7 K catalyst. Reprinted with permission from Elsevier.[43]

stopping the H_2 flow, and the syngas is restarted, the catalyst quickly achieves the same high conversion rate as the CO pretreated sample.

Figure 5.15 (bottom) shows the same catalyst activated in syngas, where small amounts of catalyst samples were retrieved from the reactor during synthesis and solidified in the waxy matrix. Mössbauer spectroscopy data were used to follow the changes in the catalyst composition. Initially, exposure of the catalyst to syngas alone led to a high concentration of Fe_3O_4. Yet, after it was treated with CO alone for 24 h, a much higher concentration of Fe carbides was present and the syngas conversion increased remarkably.[43] Fe_3O_4 was still present, but in lower concentration.

Thus, it was proposed that the working catalyst exists in the phase-boundary zone between Fe carbides and Fe_3O_4 oxide, in good agreement with the ternary phase diagram provided in Figure 5.16.[43] At low H_2/CO ratio syngas and at low conversion, FT kinetics is controlling the product distribution, relatively more H_2O is present, since at low conversion CO is being consumed by the H_2 present, and H_2O is formed as a product. Thus, a higher fraction of Fe_3O_4 is present. At higher conversion levels, though, the FT reaction needs the H_2 generated from WGS, so the H_2O level goes down, and the composition moves to the left, where more Fe_3C is present.

According to HR-TEM studies, the iron catalyst typically exists as an Fe_3O_4 oxide core surrounded by an iron-carbide matrix. However, there are numerous types of iron carbides, the most common in the literature being cementite, epsilon carbide, and the Hägg carbide, also known as Chi-carbide. The molecular structures and corresponding EXAFS spectra generated by theoretical software are displayed in Figure 5.17.

Figure 5.18 demonstrates results from a collaboration between researchers at Argonne National Laboratory and CAER to demonstrate with EXAFS (*i.e.* using synchrotron radiation) that promoters like K and Cu assist in promoting the carburization rate during Fe catalyst activation. The final spectrum suggests a mixture of Hägg and epsilon carbides has formed. Not only are K and Cu promoters important from the standpoint of activation, they also help

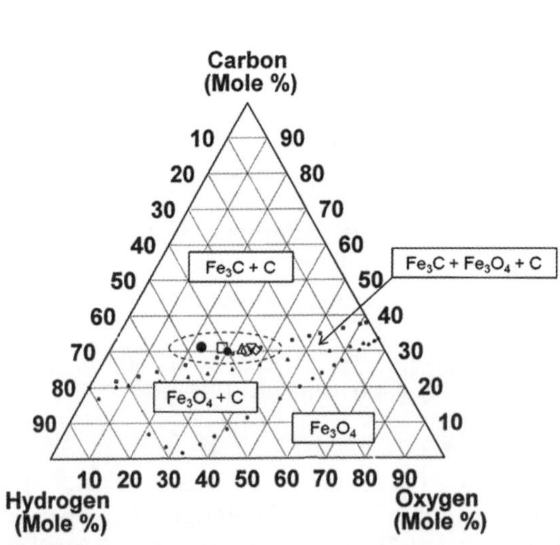

Figure 5.16 Ternary phase diagram – working catalyst exists in zone between Fe carbide and Fe_3O_4. Reprinted with permission from Elsevier.[43]

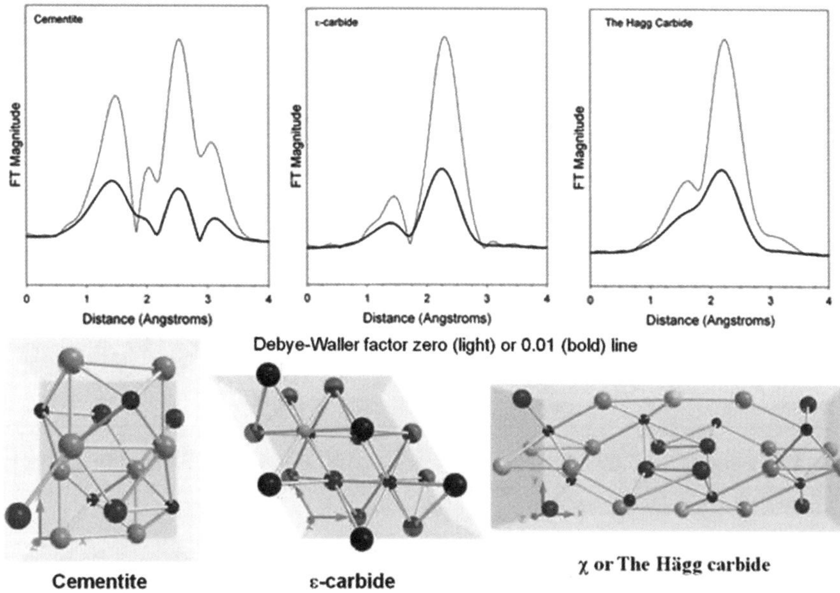

Figure 5.17 (top) EXAFS calculated spectra and (bottom) corresponding molecular arrangements of Fe carbides. Reprinted with permission from Taylor & Francis.[44]

to maintain the presence of Fe carbide as a function of time on-stream. Without the presence of promoters, the catalysts exhibit a faster deactivation rate *via* oxidation by H_2O generated by the FT reaction.[44]

5.6 Conclusions

Once treated, biomass-derived syngas will either have a high H_2/CO ratio (*e.g.*, close to 2:1) or a low ratio (*e.g.*, 0.7–1.0), depending on how much CO_2 the gasifier rejects. If high H_2/CO ratio syngas is available, a cobalt catalyst may be used, since cobalt possesses low intrinsic water-gas shift activity. If low H_2/CO ratio syngas is produced (*e.g.*, from an oxygen-blown entrained-flow Texaco/GE gasifier), then either a cobalt or an iron catalyst can be used. If cobalt is selected, then an upstream water-gas shift reaction is needed to adjust the H_2/CO ratio upward, while iron has the advantage of possessing intrinsic shift activity, such that both FT and WGS may be carried out in a single reactor. The sensitivity of poisons (*e.g.*, alkali halides, sulfur compounds, nitrogen compounds, *etc.*) produced from biomass gasification on downstream WGS and FT processes is a concern and is currently being investigated.

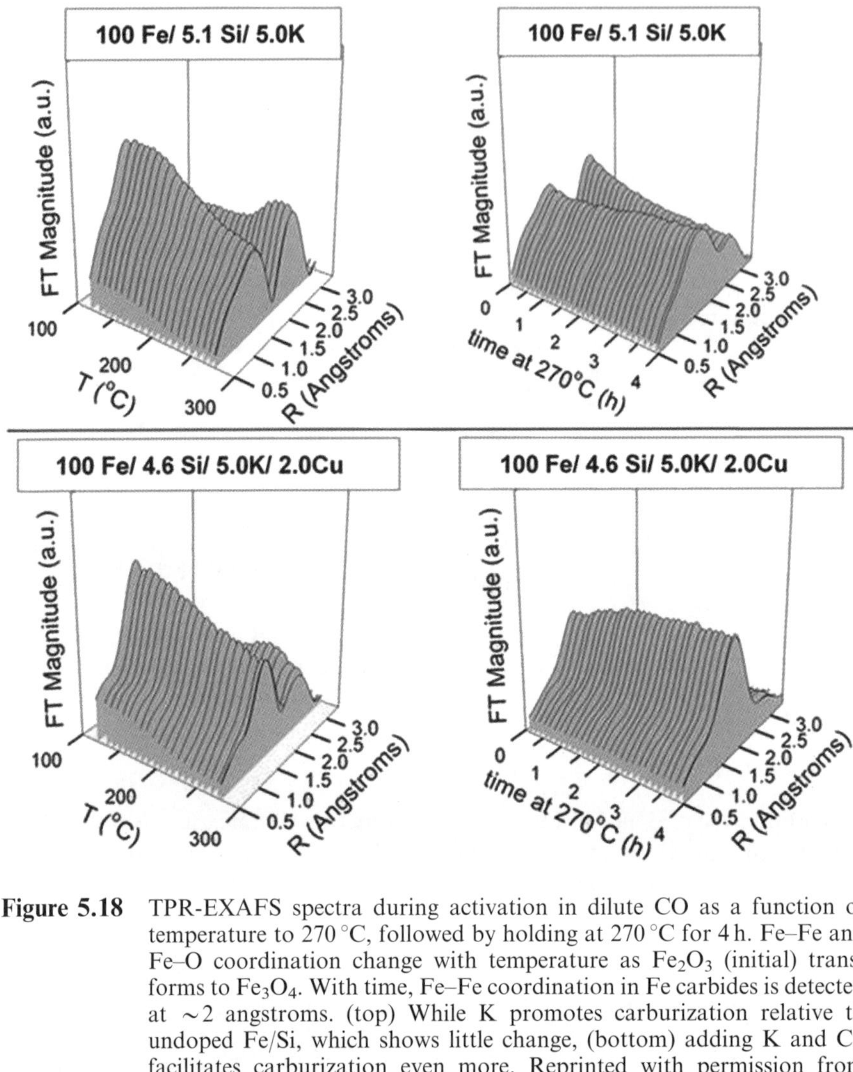

Figure 5.18 TPR-EXAFS spectra during activation in dilute CO as a function of temperature to 270 °C, followed by holding at 270 °C for 4 h. Fe–Fe and Fe–O coordination change with temperature as Fe_2O_3 (initial) transforms to Fe_3O_4. With time, Fe–Fe coordination in Fe carbides is detected at ~2 angstroms. (top) While K promotes carburization relative to undoped Fe/Si, which shows little change, (bottom) adding K and Cu facilitates carburization even more. Reprinted with permission from Taylor & Francis.[44]

The active sites on cobalt catalysts are surface cobalt metal atoms. For standard catalysts prepared by aqueous impregnation methods and air calcination, it is known that the support type, *via* the interaction with cobalt oxides, governs the extent of reduction and the average cobalt cluster size, both of which influence active-site densities. The commercial support, γ-Al_2O_3, exhibits a strong interaction and generates small cobalt clusters with relatively lower reducibility. Reduction promoters (*e.g.*, Ru, Pt, Re, *etc.*)

can be used to facilitate reduction and boost active-site densities considerably. The support type influences the kinetic behavior of H_2O, such that the effect can be positive (weakly interacting support) or negative (strongly interacting support). The CAER kinetic model includes a water-effect parameter "*m*" that is useful in describing catalytic behavior. Support type also has been demonstrated to impact catalyst stability in H_2O cofeeding studies. In some cases (*e.g.*, low loading, strongly interacting support, high partial pressure of H_2O), cobalt support compound formation may occur.

Iron catalysts should be strongly considered for converting syngas derived from high carbon content material (*e.g.*, coal, biomass, or coal/biomass blends), not only because of the WGS advantage, but also because iron catalysts can be designed to have a higher selectivity to chemicals such as α-olefins, oxygenates, and branched hydrocarbons. Iron catalysts can be designed to have high stability, and can be operated to produce low selectivities to light compounds (*e.g.*, at lower H_2/CO ratios) and high hydrocarbon productivities (*e.g.*, at lower CO conversion rates). Iron catalysts should be promoted with elements like alkali, copper, and manganese to achieve the product selectivities of interest, as well as stability. The active working catalyst is a combination of Fe carbides and Fe_3O_4. Promoters such as alkali and Cu, among other factors, facilitate carburization and prevent deactivation of the catalyst by H_2O-induced reoxidation. Isotope tracer studies (*e.g.*, $^{14}CO_2$ cofeeding) point to a reaction mechanism that involves an oxygenate species as a chain initiator.

References

1. F. Barbir, *PEM Fuel Cells: Theory and Practice,* Elsevier Academic Press, Amsterdam, The Netherlands, 2005.
2. A. B. Rao and E. S. Rubin, *Environ. Sci. Technol.*, 2002, **36**, 4467.
3. "Hydrogen, Fuel Cells, & Infrastructure Technologies Program: Multi-year Research, Development, and Demonstration Plan: Planned program activities for 2003–2010," US Dept. of Energy, Energy Efficiency and Renewable Energy, 2003.
4. A. van der Drift, H. Boerrigter, B. Coda, M. K. Cieplik, and K. Hemmes, "Entrained Flow Gasification of Biomass: Ash behavior, feeding issues, and system analyses," Energy Research Center of the Netherlands, Technical document ECN-C-04-039, April 2004.
5. H. Boerrigter and A. van der Drift, "Large-scale Production of Fischer–Tropsch Diesel From Biomass: Optimal Gasification and Gas Cleaning Systesm," Energy Research Center of the Netherlands, Technical document ECN-RX-04-119, November 2004.
6. H. Boerrigter, H. P. Calis, D. J. Slort, H. Bodenstaff, A. J. Kaandorp, H. den Uil and L. P. L. M. Rabou, "Gas Cleaning for Integrated Biomass

Gasification (BG) and Fischer–Tropsch (FT) Systems: Experimental demonstration of two BG-FT systems ("Proof-of-Principle")," Energy Research Center of the Netherlands, Technical document ECN-C--04-056, December 2004.
7. Shell technical brochure, "Gas/Liquid Treating and Sulfur Processes: Helping refiners meet product specifications and emission limits," www-static.shell.com.
8. H. R. Linden, W. W. Bodle, B. S. Lee and K. C. Vyas, *Ann. Rev. Energy*, 1976, **1**, 65.
9. M. Jazbec, K. Sendt and B. S. Haynes, *Fuel*, 2004, **83**, 2133.
10. R. O'Brien, X. D. Hu, R. Tuell and Y. Cai, "High temperature shift catalyst prepared with a high purity iron precursor", Süd-Chemie, Inc., PCT Int. Appl., 2005, 25 pp.
11. L. Xu, S. Bao, R. J. O'Brien, A. Raje and B. H. Davis, *CHEMTECH*, 1998, **28**(1), 47.
12. A. Sarkar, R. A. Keogh, S. Bao and B. H. Davis, *Appl. Catal. A: General*, 2008, **341**(1–2), 146.
13. M. Luo, S. Bao, R. S. Keogh, A. Sarkar, G. Jacobs and B. H. Davis, "Fischer–Tropsch synthesis: a comparison of iron and cobalt catalysts," AIChE Annual Meeting, Conference Proceedings, San Francisco, CA, United States, Nov. 12–17, 2006.
14. B. H. Davis, *Ind. Eng. Chem. Res.*, 2007, **46**, 8938.
15. A. Sarkar, J. K. Neathery, R. L. Spicer and B. H. Davis, "Continuous process for separation of Fischer–Trospch wax from ultrafine iron nanocatalyst particles by using a cross-flow filtration module," Abstracts of Papers, 236th ACS National Meeting, Philadelphia, PA, United States, August 17–21, 2008 (2008).
16. M. E. Dry, "The Sasol Route to Fuels," *CHEMTECH*, Dec. 1982, pp. 744–750.
17. G. P. van der Laan and A. A. C. M. Beenackers, *Catal. Rev.–Sci. Eng.*, 1999, **41**(3&4), 255.
18. W. Ma, G. Jacobs, D. E. Sparks, M. K. Gnanamani, R. R. Pendyala, B. H. Davis, C. H. Yen and T. M. Tomsik, "Fischer–Tropsch synthesis: Support and cobalt cluster size effects on kinetics and hydrocarbon selectivity over Co/Al_2O_3 and Co/SiO_2 catalysts," *in progress*.
19. B. H. Davis, *Fuel Proc. Tech.*, 2001, **71**(1–3), 157.
20. H. H. Storch, N. Golumbic and R. B. Anderson, *The Fischer–Tropsch and Related Synthesis,* Wiley, New York, 1951.
21. L. C. Browning and P. H. Emmett, *J. Am. Chem. Soc.*, 1952, **74**, 1680.
22. W. Ma, G. Jacobs, D. E. Sparks, R. L. Spicer, U. M. Graham and B. H. Davis, "Comparison of the kinetics of the Fischer–Tropsch synthesis reaction between structured alumina supported cobalt catalysts with different pore size," Abstracts of Papers, 235th ACS National Meeting, New Orleans, LA, United States, April 6–10, 2008.

23. I. C. Yates and C. N. Satterfield, *Energy & Fuels*, 1991, **5**(1), 168.
24. W. Brotz, *Z. Eletrochem.*, 1949, **53**, 301.
25. C.-H. Yang, F. E. Massoth and A. G. Oblad, *Adv. in Chem. Ser.*, 1979, **178** (Hydrocarbon Synth. Carbon Monoxide Hydrogen), 35.
26. R. B. Pannell, C. L. Kibby and T. P. Kobylinski, *Stud. Surf. Sci. and Catal.*, 1981, **7**(1) (New Horiz. Catal.), 447.
27. J. Wang, "Physical, chemical, and catalytic properties of borided cobalt Fischer–Tropsch catalysts," Ph.D. Thesis, Brigham Young University, Provo, Utah, 1987, 171 pp.
28. R. Zennaro, M. Tagliabue and C. H. Bartholomew, *Catal. Today*, 2000, **58**(2000), 309.
29. T. K. Das, X. Zhan, J. Li, G. Jacobs, M. E. Dry and B. H. Davis, *Stud. Surf. Sci. and Catal.*, 2007, **163** (Fischer–Tropsch Synthesis, Catalysts and Catalysis), 289.
30. E. Iglesia, *Appl. Catal. A: General*, 1997, **161**(1–2), 59.
31. J. P. den Breejen, P. B. Radstake, G. L. Bezemer, J. H. Bitter, V. Froeseth, A. Holmen and K. P. de Jong, *J. Am. Chem. Soc.*, 2009, **131**(20), 7197.
32. G. Jacobs, T. K. Das, Y. Zhang, J. Li, G. Racoillet and B. H. Davis, *Appl. Catal. A: General*, 2002, **233**(1–2), 263.
33. G. Jacobs, Y. Ji, B. H. Davis, D. Cronauer, A. J. Kropf and C. L. Marshall, *Appl. Catal. A: General*, 2007, **233**(2), 177.
34. G. Jacobs, A. Sarkar, Y. Ji, P. M. Patterson, T. K. Das, M. Luo and B. H. Davis, "Fischer–Tropsch synthesis: characterization of interactions between reduction promoters and Co for Co/Al_2O_3-based GTL catalysts," AIChE Annual Meeting, Conference Proceedings, San Francisco, CA, United States, Nov. 12–17, 2006 (2006).
35. C.J. Kim, "Water addition for increased carbon monoxide/hydrogen hydrocarbon synthesis activity over catalysts comprising cobalt, ruthenium and mixtures thereof which may include a promoter metal," (Exxon Research and Engineering Co., USA), US 5,227,407 (1993), 9 pp.
36. J. Li, G. Jacobs, T. K. Das, Y. Zhang and B. H. Davis, *Appl. Catal. A: General*, 2002, **236**(1–2), 67.
37. J. Li, G. Jacobs, T. K. Das and B. H. Davis, *Appl. Catal. A: General*, 2002, **233**(1-2), 255.
38. J. Li, X. Zhan, Y. Zhang, G. Jacobs, T. K. Das and B. H. Davis, *Appl. Catal. A: General*, 2002, **228**(1–2), 203.
39. G. Jacobs, P. M. Patterson, T. K. Das, M. Luo and B. H. Davis, *Appl. Catal. A: General*, 2004, **270**(1–2), 65.
40. G. Jacobs, T. K. Das, P. M. Patterson, J. Li, L. Sanchez and B. H. Davis, *Appl. Catal. A: General*, 2003, **247**(2), 335.
41. T. K. Das, W. A. Conner, J. Li, G. Jacobs, M. E. Dry and B. H. Davis, *Energy & Fuels*, 2005, **19**(4), 1430.

42. A. K. Dalai, T. K. Das, K. V. Chaudhari, G. Jacobs and B. H. Davis, *Appl. Catal. A: General*, 2005, **289**(2), 135.
43. B. H. Davis, *Catalysis Today*, 2009, **141**, 25.
44. G. Jacobs, A. Sarkar, Y. Ji, B. H. Davis, D. C. Cronauer, A. J. Kropf and C. L. Marshall, "Application of a novel TPR EXAFS/XANES method using a multisample holder to characterize promoted iron and cobalt Fischer–Tropsch synthesis catalysts," Abstracts of Papers, 235th ACS National Meeting, New Orleans, LA, United States, April 6-10, 2008.

CHAPTER 6

Bioderived Syngas to Alcohols

ADEFEMI EGBEBI AND JAMES SPIVEY

Louisiana State University, Cain Department of Chemical Engineering, Baton Rouge, LA 70803, USA

6.1 Introduction

Ethanol is the most widely used liquid biofuel in the transportation sector worldwide.[1] The United States, Brazil and several EU countries have programs directed at promoting the use of biofuels. In the United States, it is blended at up to 10% with gasoline (E10) and can be used in most vehicles manufactured since the 1970s. A limited number of cars can also use a blend of gasoline with up to 85% ethanol by volume (E85). In Brazil, 80% of automobile production has flexible-fuel capability with choices of either hydrous (~100%) ethanol or 20–25% ethanol/gasoline blend.[1] The use of ethanol blends as transportation fuels can potentially reduce both the consumption of crude oil and emissions from combustion processes.[2–4] Figure 6.1 shows that the consumption and production of ethanol in the US has been increasing over the years as ethanol has gradually replaced MTBE as the oxygenated gasoline blend of choice because of concerns about the possibility of groundwater contamination by MTBE.

This trend is expected to continue as demand for gasoline and the percent of ethanol in the gasoline pool are projected to rise.

Most ethanol is obtained *via* biochemical processing of biomass such as sugarcane (Brazil), sugar beets (Europe) and corn (US). These raw crops can be processed to provide the sugars that can be fermented into alcohols/ethanol. In the US, the cost of ethanol production is likely to increase as a result of the

RSC Energy and Environment Series No. 1
Thermochemical Conversion of Biomass to Liquid Fuels and Chemicals
Edited by Mark Crocker

Published by the Royal Society of Chemistry, www.rsc.org

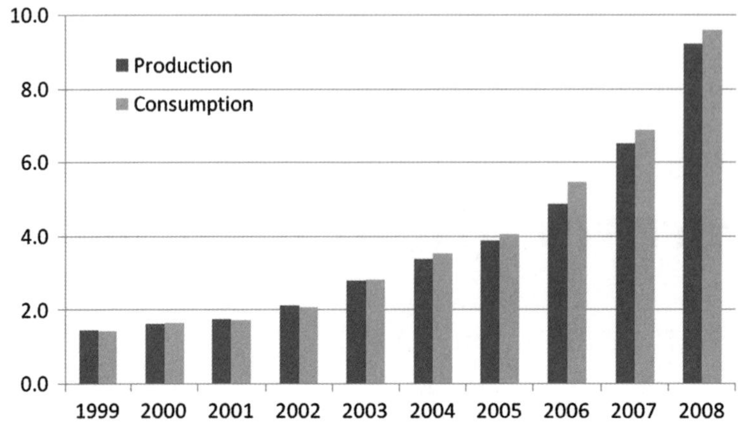

Figure 6.1 US ethanol production and production capacity, 1999–2007 (billion gallons).[5]

interplay among the price of corn, competition for corn supplies between fuel and food markets, and competition between corn and other crops for limited arable-land use.[6]

Thermochemical conversion of biomass to syngas *via* gasfication, with the subsequent catalytic conversion of syngas to alcohols, provides an alternative production route for fuel ethanol. Bioderived syngas can also be converted to other alcohol products that can be used as fuel, *e.g.*, methanol, butanol and mixed alcohols. This broadens the feedstock base of biomass ethanol because thermochemical processes can convert lignocellulosic biomass like wood and straw that are not readily converted biochemically. Although biochemical processes have the advantages of high selectivity to specific endproducts (including ethanol) and lower operating temperatures, the reaction rates of thermochemical processes are orders of magnitude higher. A hybrid route that combines biochemical and thermochemical processes can also be used to take advantage of both processes. This involves first thermochemically gasifying biomass materials followed by fermentation of the resultant syngas using micro-organisms under controlled conditions.[1]

Alcohols can also be used in fuel-cell applications as hydrogen carriers. In one scenario, ethanol would be produced at a central site, transported as a liquid to the point of use (*e.g.*, a fuel cell), where it is either used directly in a fuel cell[7–10] or reformed to produce a hydrogen-rich gas *via* steam reforming[11–20] or partial oxidation.[21,22] Analysis of studies on ethanol reforming is beyond the scope of this chapter, and the reader is referred to papers cited above for further information.

Recent reports have suggested that there are advantages to higher alcohols such as 1-butanol as a blending agent with gasoline and as a neat transportation fuel.[23] These advantages include lower solubility in water than ethanol (and therefore more easily separated from groundwater in case of a leak), higher energy content, less corrosion of existing infrastructure, and greater compatibility with gasoline. Like methanol, 1-butanol can also be used as an esterifying reagent for the synthesis of biodiesel from fatty acids.[24–26] In addition, research shows that if 1-butanol is the desired product, it can be produced by the self-condensation of ethanol on catalysts such as alkali zeolites[27] and $Ni/\gamma\text{-}Al_2O_3$.[28]

Challenges that remain for the conversion of biomass-derived syngas to ethanol include novel catalytic reactor designs tailored to the typically smaller scale of biomass-conversion processes,[29] catalysts for downstream adjustment of the H_2/CO ratio for specific endproducts,[30] and catalysts for the conversion of biomass-derived syngas to various alcohol products.

This chapter gives an overview of the conversion of biomass-derived syngas to alcohols with emphasis on the current developments in the catalysts, challenges facing the process, and future prospects.

6.2 Syngas to Alcohols: Thermodynamics

Most literature on the conversion of synthesis gas to alcohols has focused on methanol as an endproduct, with far less work directed at ethanol. However, because the synthesis of methanol from syngas is a mature technology, we shall focus on the syntheses of higher alcohol products, which are still at the research stage.

Biomass-derived syngas consists mostly of CO, CO_2, H_2 and H_2O, making this process more than just a CO hydrogenation reaction. In addition, the overall conversion of syngas involves CO_2 hydrogenation, methanation and water-gas shift reactions. A summary of thermodynamic analyses of these individual reactions is presented here to help understand the equilibrium limits imposed on the overall reaction.

6.2.1 Hydrogenation of CO

The CO hydrogenation reactions leading to alcohols can be summarized as:

$$n\mathrm{CO} + 2n\mathrm{H_2} \rightarrow \mathrm{C}_n\mathrm{H}_{2n+1}\mathrm{OH} + (n-1)\mathrm{H_2O} \quad n = 1,2,3,4 \qquad (6.1)$$

For the alcohols of interest here ($n = 1$ to 4), the thermodynamic values are listed below:

n	Reaction	ΔH_r°	ΔG_r°	
1	$CO + 2H_2 \rightarrow CH_3OH$	−21.62	−6.00	(6.2)
2	$2CO + 4H_2 \rightarrow C_2H_5OH + H_2O$	−61.08	−29.18	(6.3)
3	$3CO + 6H_2 \rightarrow C_3H_7OH + 2H_2O$	−97.31	−49.08	(6.4)
4	$4CO + 8H_2 \rightarrow C_4H_9OH + 3H_2O$	−133.44	−68.76	(6.5)

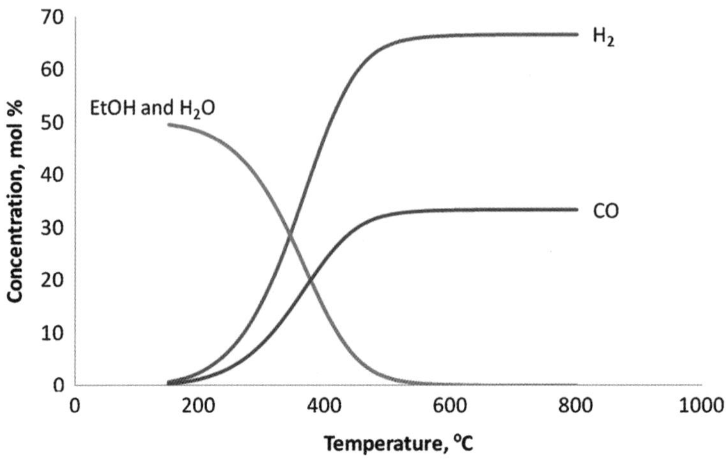

Figure 6.2 Equilibrium composition for the hydrogenation of CO to ethanol (H_2/CO = 2.0, 30 bar, calculated using HSC Chemistry® software[32]).

These reactions are favorable and highly exothermic regardless of the carbon number. The stoichiometry of all the alcohol-forming reactions requires a H_2/CO ratio of 2. When treated individually, the equilibrium yield of the alcohol is substantial at lower temperatures, and decreases with increasing temperature, *e.g.*, as shown in Figure 6.2 for ethanol at 30 bar.[31]

However, in a typical CO hydrogenation reaction, the alcohol-forming reactions 6.2–6.5 occur simultaneously with side reactions like methanation and water-gas shift. The equilibrium yield of the alcohol-forming reactions increases with number of carbon atoms in the desired alcohol. Therefore under equilibrium conditions, for C_1–C_4 alcohols, the yield of the alcohol products would decrease in the following order: butanol > propanol > ethanol > methanol as reported in the thermodynamic studies of Mawson and Roberts.[33] However, many experimental studies on the synthesis of higher alcohols have resulted in higher yields of methanol or methane than higher alcohols. This suggests that the synthesis of higher alcohols is kinetically controlled. Ultimately, the relative degree of completion of the individual reactions, and hence the selectivity to various products, depends on a variety of factors like the catalyst type, reactor type, syngas composition, and reaction conditions.

6.2.2 Hydrogenation of CO_2

The hydrogenation of CO_2 to alcohols can be summarized as:

$$nCO_2 + 3nH_2 \rightarrow C_nH_{2n+1}OH + (2n-1)H_2O \quad n = 1,2,3,4 \qquad (6.6)$$

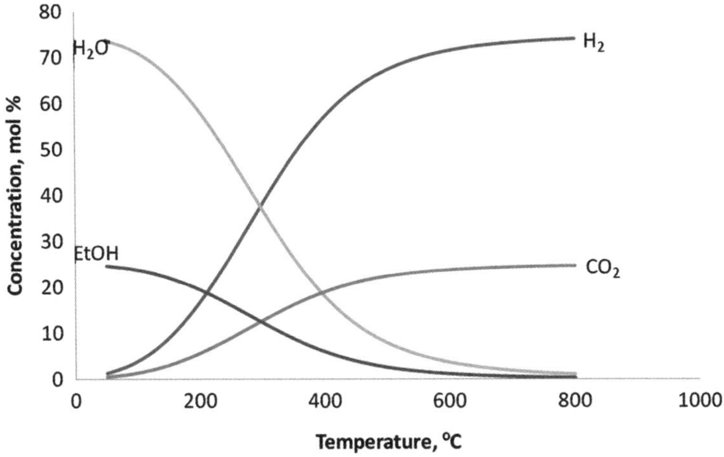

Figure 6.3 Equilibrium composition for the hydrogenation of CO_2 to ethanol ($H_2/CO = 3.0$, at 30 bar, calculated using HSC Chemistry® software[32]).

e.g. for ethanol,

$$2CO_2 + 6H_2 \rightarrow C_2H_5OH + 3H_2O \tag{6.7}$$

$$\Delta H_r^\circ = -41.54\,\text{kcal/mol}, \quad \Delta G_r^\circ = -15.70\,\text{kcal/mol}$$

This reaction is also exothermic and thermodynamically favorable although it is less exothermic and less thermodynamically favorable than the corresponding CO hydrogenation reactions. The equilibrium concentration trends follow those of CO hydrogenation provided the reaction products are restricted to only alcohol and water, *e.g.*, ethanol and water concentrations decrease while those of CO_2 and H_2 increase with temperature (Figure 6.3).

6.2.3 Side Reactions

The water-gas shift (WGS) reaction:

$$CO + H_2O \leftrightarrow CO_2 + H_2 \tag{6.8}$$

is a very important side reaction that affects the equilibrium of both CO and CO_2 hydrogenation reactions. In the hydrogenation of CO to ethanol, the H_2O formed can react with CO to produce CO_2 and H_2, while in CO_2 hydrogenation the reverse WGS reaction may occur. The reverse WGS is essentially a partial reduction of CO_2 to CO, which has been identified as an elementary step involved in the synthesis of ethanol from CO_2 hydrogenation.[34,35] This suggests that the hydrogenation of both CO and CO_2 can proceed through a common intermediate.

When methanation of CO and/or CO_2:

$$CO_2 + 4H_2 \rightarrow CH_4 + 2H_2O \quad (6.9)$$

$$CO + 3H_2 \rightarrow CH_4 + H_2O \quad (6.10)$$

is allowed to occur along with the hydrogenation reactions, methane is the most thermodynamically significant product.[33] Thermodynamic simulations of equilibrium concentrations of an initial mixture composition corresponding to syngas produced by biomass gasification showed ethanol concentrations dropping drastically from more than 20% at ~400 °C and 30 bar when methane is not allowed as a product of reaction, to virtually zero at all temperatures when methane is allowed as a product.[31] Therefore, it is necessary that the thermodynamically favored formation of methane is kinetically limited for alcohols to be produced from syngas.

6.2.4 Effect of Pressure

All alcohol-forming reactions are volume contracting and thus the effect of increasing total pressure is to increase the equilibrium concentration of the reaction products, *i.e.* the alcohols. However side reactions like the methanation reactions are also volume contracting and might be more pressure sensitive under certain conditions. Therefore monotonic increase in reaction pressure does not necessarily increase alcohol selectivity in the product streams.

6.3 Catalyst Types

A variety of catalyst types have been studied for the alcohol-forming reactions. However, most of these studies are focused on the hydrogenation of CO, and do not include concentrations of CO_2 and H_2O representative of those in biomass-derived syngas. We shall group these catalysts in terms of the desired alcohol products: methanol, ethanol and mixed-alcohol synthesis.

6.3.1 Methanol Synthesis

Methanol synthesis from conventional syngas is a well-established industrial process using mostly syngas from the steam reforming of natural gas, with a typical H_2/CO ratio of ~2, which is similar to the stoichiometric requirement for methanol formation from CO and H_2. However, biomass-derived syngas is generally CO_2-rich and H_2-deficient (*i.e.* low H_2/C) and would require additional processing downstream of the gasifier,[36] *e.g.*, water-gas shift reaction, methane reforming, CO_2 removal or H_2 addition, to adjust the H_2/CO ratio to that required for methanol synthesis.

Cu-based catalysts are traditionally used for the synthesis of methanol from CO hydrogenation. However, industrial methanol synthesis typically uses syngas feedstocks containing about 5% CO_2,[37,38] which makes the synthesis of methanol from biomass-derived syngas closely related to current industrial methanol synthesis. Numerous papers have been published on the synthesis of methanol from $CO/CO_2/H_2$ mixtures using similar Cu-based (Cu/ZnO) catalysts[37–42] although there are fewer reports on the conversion of "biomass-derived syngas" to methanol.[36,43–45]

Yin and Leung[36] studied commercial $Cu/ZnO/Al_2O_3$ catalysts for synthesis of methanol using biomass-derived syngas with varying $H_2/CO+CO_2$ (H/C) and CO_2/CO ratios that are representative of the composition from different gasification processes. At reaction conditions of $P = 3.6$ MPa, $T = 240$–$260\,°C$ and $SV = 7000\,h^{-1}$ with four different syngas compositions [Syngas 1 ($H_2/CO+CO_2 = 0.7$, $CO_2/CO = 0.56$); Syngas 2 (1.1, 0.8); Syngas 3 (2.2, 0.1); Syngas 4 (1.1, 1.1)], they reported that while the yield of methanol decreases with decreasing H_2/C ($H_2/CO+CO_2$) ratio, the selectivity to methanol is even more sensitive to the CO_2/CO ratio – decreasing with increasing CO_2/CO ratio (Figures 6.4 and 6.5). This study suggests that high CO_2 content inhibits methanol formation at the conditions tested, and that a combination of low CO_2/CO ratio and high H_2/C ratio can be used to maximize methanol yield from biomass-derived syngas on $Cu/ZnO/Al_2O_3$ catalysts.

However, it appears that some amount of CO_2 is necessary for optimal methanol synthesis from syngas. For example, Klier *et al.*[46] reported a maximum methanol synthesis rate from a syngas containing 2% CO_2 (plus 28%

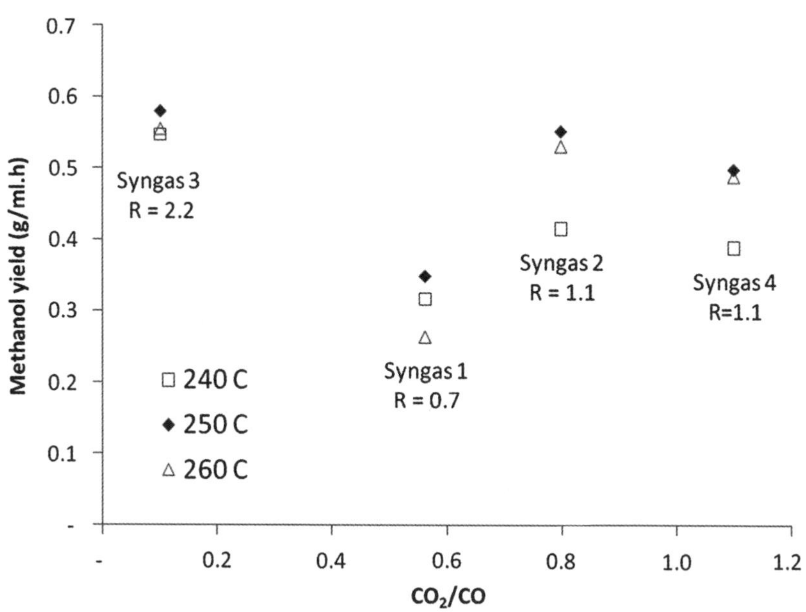

Figure 6.4 Relationships between methanol yield and the CO_2/CO and $H_2/(CO+CO_2)$ ratios at $T = 240$, 250, 260 °C, $P = 3.6$ MPa. *R* represents $H_2/(CO+CO_2)$ ratio. Reprinted with permission from Yin *et al.*[36] Copyright 2005 American Chemical Society.

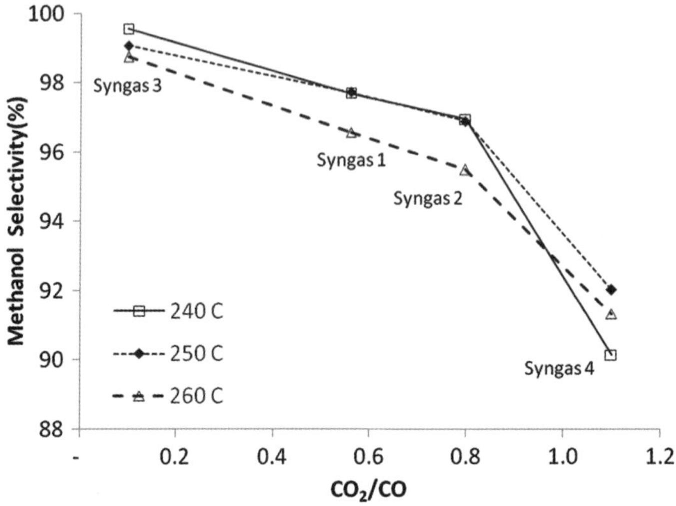

Figure 6.5 Relationships between methanol selectivity and the CO_2/CO ratio at T = 240, 250, 260 °C, *P* = 3.6 MPa. Reprinted with permission from Yin *et al.*[36] Copyright 2005 American Chemical Society.

CO, 70% H_2), claiming that the effect of CO_2 is mainly promotional because CO hydrogenation is the primary source of methanol in a $CO/CO_2/H_2$ feed mix, and that CO_2 only becomes a significant source of carbon when the syngas feed is CO_2-rich.[46]

Conversely, Chinchen *et al.*[42] showed from isotopic-labeling tests that, under industrial conditions (50 bar and 250 °C), CO_2 (and not CO) is the predominant source of methanol from a CO_2-rich syngas on $Cu/ZnO/Al_2O_3$ and that CO only becomes a major reactant to methanol at very low concentrations of CO_2. Skrzypek *et al.*[47] agreed that methanol synthesis is mainly from CO_2 hydrogenation. Their work shows that methanol is readily formed from CO_2 and H_2 (without CO present), while there is no methanol formation when the feed contains only CO and H_2 (with no CO_2 or steam). In the latter case, a small amount of water is necessary to initiate methanol synthesis by first forming CO_2 *via* the water-gas shift reaction.[47] Interestingly, while water has been shown to have some promotional effect on methanol synthesis from CO/H_2 syngas, its effects on predominantly CO_2/H_2 feeds has been shown to be inhibitive.[37]

The exact effect of CO_2 in syngas feeds on methanol synthesis is still a subject of debate. It seems to be dependent on its concentration in the syngas feed – at higher concentrations, especially when the CO_2/CO ratio is greater than unity, the effect of the water formation from reverse WGS reduces selectivity to methanol and decreases catalytic activity.[39] The adsorption of CO_2 on the Cu catalyst is more intense than CO[46] and at high CO_2 content, strong CO_2 adsorption inhibits methanol synthesis. At intermediate concentrations, it is beneficial to methanol synthesis because there is enough CO_2 to keep the Cu in an intermediate oxidized state necessary for methanol synthesis, while at very low or zero concentrations of CO_2, the catalyst is said to be "overreduced".[37,46]

The effect could be dependent on the age of the catalyst, where it has been reported that deactivated Cu/ZnO catalysts yield a maximum methanol synthesis rate from 30% CO_2, 70% H_2 (no CO) syngas, while on fresh catalysts, the maximum rate is achieved when only 2% CO_2 is included in syngas of 70% H_2, balance 28% CO.[38]

Reaction conditions could also determine how CO_2 affects methanol synthesis from syngas. Below industrial temperatures, methanol synthesis rate increases with CO_2 feed concentration but at temperatures higher than industrial, the rate of methanol formation peaks.[37]

The key to the differences highlighted above might be the oxidation state of Cu catalysts. It appears that CO_2 hydrogenation and CO hydrogenation to methanol require different catalytic sites[41] – CO hydrogenation requiring oxidized Cu sites while metallic Cu is necessary for CO_2 hydrogenation.[48] An active catalyst for the synthesis of methanol from a mixture of CO, CO_2 and H_2 as contained in a biomass-derived syngas would need to have the right balance of oxidized and metallic Cu sites. Despite this controversy, it is generally accepted that methanol can be synthesized from both CO and/or CO_2 hydrogenation over Cu-ZnO catalysts. As a result, methanol may be the best target product for the conversion of CO_2-rich syngas, such as that produced from biomass gasification.

6.3.2 Ethanol Synthesis

The conversion of syngas containing mainly CO and hydrogen to ethanol has been well studied.[31,49] However, there are relatively few studies based on hydrogenation of CO_2 and CO/CO_2 mixtures. Very few of these studies contain concentrations of H_2, CO, CO_2 typical of those found in biomass-derived syngas. The majority of the catalysts that have been tested for this reaction are modified versions of commercial methanol synthesis catalysts (Cu-based ones), or those used for Fischer–Tropsch (FT) synthesis. However, it has been difficult to modify these two types of catalysts to shift the product selectivity from hydrocarbons (for FT catalysts) and methanol (for Cu-based catalysts) despite a number of attempts at catalyst modification.

Catalysts for ethanol synthesis from the hydrogenation of CO or CO_2 can be broadly grouped into four categories: Rh-based catalysts, modified methanol synthesis catalysts (based on Cu), modified Fischer–Tropsch type catalysts and modified Mo-based catalysts.[31,49–51]

6.3.2.1 Rh-based Catalysts

Rh-based catalysts have been widely studied for the conversion of syngas to ethanol because of their ability to produce C_2^+ oxygenates such as ethanol, acetaldehyde and acetic acid selectively from syngas.[52] Rh occupies an interesting position in the periodic table among its other Group VIII counterparts, lying between metals that easily dissociate CO to form higher hydrocarbons (*e.g.*, Fe and Co) and those that do not dissociate CO and produce methanol (*e.g.*, Pd, Pt and Ir).[51,53] Its catalytic properties can be easily modified to form methane, hydrocarbons, alcohols, or other oxygenates from CO hydrogenation, depending on support, promoter, and reaction conditions.[51,54–60]

The general reaction sequence for the formation of ethanol from CO and CO_2 hydrogenation can be represented by the steps shown in Figure 6.6.[61–64] The first step involves the adsorption of CO, CO_2 and H_2, with both CO and CO_2 adsorbing to form either surface CO species or dissociating into surface C and O species. It is interesting to note that the adsorption of CO_2 results in the formation of linearly and bridge-bonded CO, which has been identified by IR spectroscopy on Rh-Mo/ZrO_2,[65] Rh/Al_2O_3,[66] and Rh-Li/Y.[67] In one of these studies,[66] the presence of hydrogen strongly enhanced the formation of CO, possibly by reacting with the surface O atom formed in the initial adsorption of CO_2 and driving the adsorption process forward. This suggests that CO_2 hydrogenation proceeds *via* the dissociative adsorption of CO_2 to form CO and O atoms on the surface.[63,64]

The nondissociatively adsorbed CO can either be hydrogenated to form methanol[62,68] or dissociated and hydrogenated to form a surface hydrocarbon $(CH_x)_{ad}$. These $(CH_x)_{ad}$ species can be hydrogenated to form methane (or higher hydrocarbons *via* a chain-growth mechanism) or an undissociated CO can be inserted into the metal–carbon bond in $(CH_x)_{ad}$[61] resulting in an oxygenated C_2 intermediate that could be further hydrogenated to ethanol.

However, studies comparing $CO + H_2$ *versus* $CO_2 + H_2$ over Rh-based catalysts show some differences in these two reactions.[63,69–71] Specifically, the rate of CO_2 hydrogenation to methane is greater than CO hydrogenation for identical levels of Rh (2.3%) supported on ZrO_2, Al_2O_3, SiO_2, and MgO, and the activation energy for CO_2 hydrogenation is always less than for CO hydrogenation.[63] Stronger adsorption and surface coverage of CO compared to

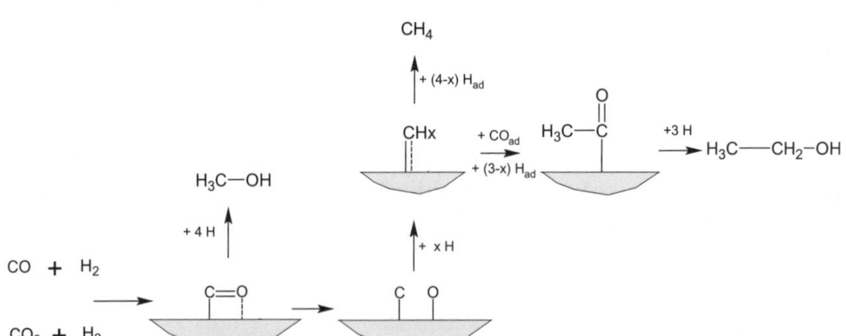

Figure 6.6 A simplified sequence for ethanol formation by CO hydrogenation on Rh-based catalysts.

CO_2 is said to be the reason for the difference in hydrogenation activities of CO and CO_2.[72]

Previous work on the hydrogenation of CO/CO_2 mixtures focuses on the effect of including small concentrations of CO_2 in CO hydrogenation experiments. Yields of methanol and ethanol have been reported to increase when a portion of CO in the feed was replaced with low levels of CO_2 during CO hydrogenation[73] and when small amounts of CO are added to the feed during CO_2 hydrogenation.[67] Increasing concentrations of CO_2 resulted in a maximum increase of alcohol yield at about 5–10% CO_2 on 1% Rh-Mo/ZrO_2.[73] The authors attribute this to the reverse WGS reaction, which presumably produces additional CO that is converted to the alcohols. However, methane yield increased continuously over the range of CO_2 concentrations studied. The decline in alcohol yield at higher levels of CO_2 is attributed to strong adsorption of CO_2 on sites that form the alcohols, with the reaction then shifted toward methanation at higher concentrations of CO_2.[73] An alternative explanation is that CO_2 reacts more readily to form methane than CO over the entire range of CO_2 concentrations, causing the monotonic increase in methane yield with CO_2 content.

On virtually all catalysts on which high ethanol selectivities have been reported, CO conversions are low because hydrocarbon formation, which typically accompanies high catalyst activity, is suppressed. The observed trend is that selectivity to C_2 oxygenates decreases with increasing CO conversion. Therefore, there has to be a balance between the catalyst activity and selectivity to obtain a high yield of ethanol.

6.3.2.2 Modified Methanol Synthesis Catalysts (Based on Cu)

The use of alkali promoters on traditional Cu-based methanol synthesis catalysts for higher alcohol synthesis dates back to the early 1920s when results showed that the yield of higher alcohols increases during methanol synthesis on catalysts precipitated with alkali (as a result of the traces of alkali left on the catalyst during preparation).[74] This has led to extensive study of alkali promoters[75–83] as well as some nonalkaline additives[84–86] for the formation of higher alcohols, including ethanol.

Alkali compounds including Li, Na, K, Rb and Cs generally shift the selectivity of Cu-based catalysts to higher alcohols (from methanol) by increasing the chain-growth activity, but this almost always comes at the expense of CO conversion.[85] K and Cs have been extensively used on these catalysts and their functions have been suggested to be dual in nature:

(a) suppression of surface acidity by the titration of acid sites that lead to dimethylether (DME).[76] Reducing DME selectivity effectively leads to higher alcohol selectivity because DME is formed by the condensation reaction of methanol.
(b) providing basic sites necessary for the various C–C and C–O bond-forming reactions.[79]

The optimum degree of promotion is dependent on the promoter type, concentration, and catalyst support, among other factors. Cu/Zn/Cr has been shown to require more Cs (or alkali) doping than Cu/Zn/Al and binary Cu/Zn for optimal higher alcohol synthesis because the chromia content introduces acidity to the catalyst which leads to DME.[79] The yield of higher alcohols has been shown to go through a maximum as the promoter concentration is increased.[79,83] This is because as the promoter concentration increases, more alkali sites are created, increasing the yield of higher alcohols. However, eventually the promoter blocks the Cu/Zn sites that are required for methanol synthesis. When this occurs, methanol formation is hindered, thereby limiting higher alcohol synthesis.[83]

Even though alkalis promote the formation higher alcohols on Cu catalysts, the effect of alkali doping on ethanol yield is not as significant. Table 6.1 shows the product yield of CO hydrogenation over Cu/ZnO doped with various levels of Cs. The addition of 0.25 mol% Cs to unpromoted Cu/ZnO leads to an insignificant increase in ethanol yield, but 1-propanol and isobutanol yields increase considerably. Ethanol yield decreases, while other higher alcohols yields increase at higher levels of Cs doping. This is attributed to the unique carbon chain-growth mechanism of higher alcohol formation on modified methanol catalysts, where the rate-determining C_1 to C_2 step is followed by a relatively faster C_2 to C_3 step and C_3 to isobutanol step (terminal product of higher alcohol synthesis on modified Cu catalysts[78]) leading to a decrease in ethanol yield and an increase in propanol and isobutanol yields.

Synthesis of higher alcohols is favored over Cu-based catalysts by CO-rich feed mixtures because the rate of chain growth increases with increasing partial pressure of CO, while the termination rate increases with increasing H_2 partial pressure.[74,75] High H_2 partial pressures have the effect of inhibiting the rate of C_1–C_2 chain growth step by enhancing the conversion of C_1 intermediates to methanol.[81]

The suitability of Cu-based catalysts for ethanol synthesis from biomass-derived syngas, however, remains to be seen. When 6% CO_2 is cofed with CO and H_2, Elliot reported an increase in both methanol as well as higher alcohols synthesis.[87] Conversely, over alkali-promoted catalysts, Hilmen *et al.*[76] reported inhibition effects of CO_2 on both methanol and higher alcohols synthesis by increasing oxygen coverages on Cu surfaces and titrating the basic sites

Table 6.1 Product yields of CO hydrogenation over Cs-doped Cu/ZnO.[78]

	Product yield (g/kg cat/h)						
Catalyst	*CO_2*	*water*	*alkanes*	*methanol*	*ethanol*	*1-propanol*	*isobutanol*
Undoped Cu/ZnO	367	1.3	16.8	204	22.6	10.1	20.7
0.25 mol% Cs/Cu/ZnO	412	1.3	16.2	181	22.7	29.6	28.9
0.34 mol% Cs/Cu/ZnO	403	1.7	13.4	157	17.0	38.1	48.6
0.43 mol% Cs/Cu/ZnO	430	1.3	14.0	162	18.2	24.1	33.6
1.5 mol% Cs/Cu/ZnO	403	0.04	4.3	213	8.1	18.0	4.8

583 K, 7.6 MPa, H_2/CO = 0.45, GHSV = 3260 L/kg cat/h.

necessary for condensation reactions. The degree of inhibition depended on the concentration of Cu sites on the catalyst – the inhibition for methanol synthesis is weaker on catalysts with high Cu site density while those with lower Cu site densities are more affected.[76]

In summary, while lower concentrations of CO_2 seem to have a promoting effect for methanol synthesis on Cu-based catalysts, high concentrations of CO_2 generally inhibit higher alcohol formation on modified methanol catalysts.

6.3.2.3 Modified Fischer–Tropsch-Type Catalysts

Evidence that alcohols, with ethanol present in the largest proportion, are precursors to the formation of hydrocarbons on Fischer–Tropsch type catalysts has been reported since 1952.[88] These types of catalysts, based on Co, Ru and Fe, have been reported to give higher alcohols when suitably modified with additives.[89–93] Some authors reported that the synthesis of higher alcohols on Ir/Ru-SiO_2[94] and Ir/Co-SiO_2[95] might be promoted by a synergistic interaction between metals that readily dissociate CO (Ru and Co) and Ir, which does not dissociate CO. A combination of two such metals might produce a catalyst that has the proper combination of CO dissociation and CO insertion, which are necessary for higher alcohol formation. The mechanism for alcohol formation on modified Fisher–Tropsch catalysts is essentially the same as described above for Rh catalysts – starting with CO dissociation and hydrogenation of the adsorbed carbon into CH_x surface species, followed by CO insertion into the CH_x species as shown in Figure 6.6.[96]

Kintaichi *et al.*[97] tested a series of bimetallic catalysts containing a pair of group VIII metals, one of which dissociates CO and one that does not. They reported that Ir-Ru/SiO_2 gives the highest CO conversion, least methanol selectivity and the highest selectivity for higher alcohols. The addition of alkali improved C_2-oxygenate selectivity. The properties of these catalysts are said to be significantly affected by the preparation procedure, especially factors like impregnation sequence,[97] precursors,[92,93,98] metal and promoter loading.[97] A close interaction between the catalyst and promoter is important for higher alcohol yield on Ir-Ru[97] and Pd-Co/CeO_2.[99] A coimpregnated Ir-Ru/SiO_2 catalyst showed higher alcohol selectivity than those in which Ir and Ru were sequentially impregnated.

Different promoters have been shown to have different effects on Fischer–Tropsch-type catalysts. On a Co/SiO_2 catalyst promoted with Re-Sr[93]and Sr,[100] Takeuchi *et al.* reported deviations from the Schulz–Flory distribution of C_2 hydrocarbon and C_2 oxygenates. They saw a deficit in C_2 hydrocarbon and an excess in C_2 oxygenates and suggested a mechanism in which oxygenates and hydrocarbons are formed through the same intermediates. The effect of the promoters therefore would be the preferential conversion of the intermediate to ethanol at the expense of C_2 hydrocarbons. Alkali dopants promote activity and selectivity to C_2^+ oxygenates by depressing hydrocarbon formation.[97] Transition metals like Ir, Re, Pt and Os help to reduce inactive Co (II) acetate species to the active metallic state by H_2 activation (during pretreatment) while keeping the Co highly dispersed. High Co dispersion is absolutely necessary for oxygenate synthesis; agglomeration of Co particles tends to catalyze hydrocarbon formation.[98] Although the promoted Co catalysts showed enhanced selectivity towards ethanol, hydrocarbon selectivities remain high (above 60%) in virtually all reported studies.[89,92–94,101]

The incorporation of Cu into Fischer–Tropsch catalysts has also been explored – Inui and coworkers report the synthesis of ethanol *via* CO_2 hydrogenation using such multifunctional catalysts.[34,102–105] These catalysts are a mixture of Rh, Fe and Cu designed to partially reduce CO_2 to CO, propagate chain growth (C–C bond formation), and insert an –OH group. An Fe-Cu-Al-K Fischer–Tropsch catalyst gave 8% ethanol selectivity from a CO_2/H_2 (25:75) mixture. The selectivity increased to 11% when 3% CO was substituted for CO_2. While the increase in CO concentration increased the ethanol yield, a CO-rich gas reduced ethanol selectivity because CO_2 was formed (rather than ethanol) *via* the shift reaction. The performance of this catalyst is said to depend on the oxidation-reduction state of the Fe catalyst during reaction – the active phase for CO_2 hydrogenation to ethanol is Fe_3O_4 and is a function of the reduction temperature. Reduction at about 450 °C gives Fe_3O_4, insufficient reduction leaves Fe in the inactive Fe_2O_3 phase and overreduction leads to metallic Fe. Combining Fe with other catalysts and suitable promotion from metals like Pd and Ga (which exhibit H_2 spillover and reverse-spillover, respectively) maintains the oxidation state of the catalysts during reaction conditions.[34]

6.3.2.4 *Modified (Sulfide and Unsulfided) Mo-based Catalysts*

When alkali metals are added to Mo-based catalysts, the selectivity for CO hydrogenation has been shown to shift from hydrocarbons to alcohols.[106] The promoting effect of alkalis (on MoS_2) for alcohol formation was found to increase in the order $Li<Na<Cs<Rb<K$, suggesting that moderate basic promotion is desired.[107] Muramatsu *et al.* claimed that the role of K on Mo/SiO_2 is to preserve the surface MoO_2 species, which is active for alcohols, by retarding the reduction of Mo to metal.[108] Selectivity to alcohols on alkali-promoted Mo catalysts normally follows the Schulz–Flory distribution, which limits higher alcohol formation. However, further promotion with transition metals like Co and Ni has been shown to improve C_2^+ alcohol selectivity.[109–111] When K/MoS_2 catalyst is comodified with Ni and Mn, the synergistic effect of both promoters is said to enhance the catalytic activity and the formation of C_2–C_3 alcohols. Table 6.2 shows the catalytic performance of K/MoS_2 catalyst in which the addition of Ni improved catalyst activity and selectivity to alcohols. Ni is thought to enhance the $C_1 \rightarrow C_2$ homologation step, which might explain the high ethanol selectivity. The further addition of Mn inhibits the enrichment of Ni, leading to the suppression of methanation on Ni while improving the dispersion of the

Table 6.2 Catalytic performance of K/MoS_2 catalysts in CO hydrogenation.[109]

		Selectivity (C-atom%)		*Alcohol distribution (wt%)*				
Catalyst	*CO conv. (%)*	*alcohols*	CH_x^a	C_1	C_2	C_3	C_4	C_5
K/MoS_2	11.4	55.20	44.80	71.6	18.8	7.4	2.0	0.2
$Ni/K/MoS_2$	23.3	59.98	40.02	56.3	26.9	11.6	4.5	0.7
$Ni/Mn/K/MoS_2$	17.8	81.72	18.82	45.8	32.9	15.4	5.0	0.9

$T = 588$ K; $P = 9.5$ MPa; $H_2/CO = 2.0$; GHSV = 6000 h^{-1}; $^a CH_x$ = hydrocarbons.

catalyst.[109] The main mechanism for ethanol formation on alkali-promoted Mo-based catalyst is *via* the insertion of CO into the metal–CH_x bond as depicted in Figure 6.6.[112,113]

Preparation techniques have been reported to affect the selectivity and activity of Mo catalysts. KCl promoted Mo/SiO_2, which was prepared by a successive impregnation method in which K was added to silica gel followed by impregnation with Mo solution, was found to give higher activity and selectivity for alcohol formation than when Mo was added first. The sequence was found to greatly affect the activity and selectivity because certain interactions between Mo and SiO_2, which inhibit higher-alcohol formation, are said to be less pronounced when K was added first.[114] A rapid drying procedure was found to improve alcohol selectivity and activity on K-Mo/C catalyst.[115] A modified Mo/SiO_2 catalyst prepared using metal oxide vapor synthesis (MOVS) exhibited much higher activity and selectivity to higher alcohols than a nominally similar catalyst prepared by the conventional impregnation method.[110] These improved activities and selectivities result from higher dispersion of active species.

Significant amounts of CO_2 are formed on MoS_2 catalyst when the feed is CO_2-free because of its high activity for the water-gas shift (WGS) reaction.[115,116] However, the inclusion of CO_2 in a syngas feed shifts the WGS reaction equilibrium toward H_2O formation, causing large amounts of water to be formed instead of CO_2. CO_2 in the feed also reduces the formation of higher alcohols, possibly due to the inhibition of the chain growth process by CO_2 or by H_2O produced by the reverse WGS reaction.[115]

6.3.3 Mixed-Alcohol Synthesis

Mixed alcohols generally refers to a mixture of C_1–C_6 alcohols,[50] although with some preference towards higher (C_2–C_6) alcohols.[117] Most ethanol synthesis catalyst systems also promote the formation of other higher alcohols besides ethanol from syngas. Therefore, the four broad catalyst groups discussed for ethanol synthesis are applicable to mixed-alcohols synthesis. In fact, because of the nature of the chain-growth mechanism of higher-alcohol formation on modified methanol catalysts, the reaction mixture of CO hydrogenation always

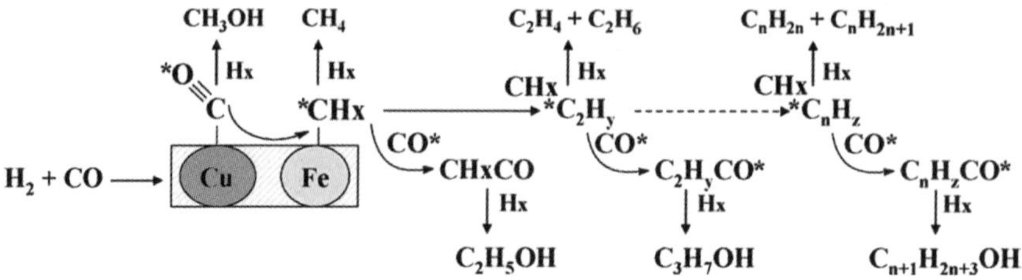

Figure 6.7 Reaction scheme for CO hydrogenation over Fe-Cu catalysts. Reprinted with permission from Elsevier.[120]

contains higher concentrations of C_3 and C_4 alcohols than ethanol.[78] Catalysts containing both Fischer–Tropsch elements (Co, Fe or Ni) and Cu have also been investigated with the ratios of these two major components adjusted to change the methanol and higher alcohol yields.[51] In these catalysts, the Cu component is believed to be responsible for methanol, while the FT portion is the site for chain growth as depicted in Figure 6.7. Co–Cu-based catalysts, patented by Institut Francais du Petrole (IFP),[118,119] Fe-Cu[89] and Ni-Cu[120] have been shown to be successful for this synthesis. Alkali promotion of all these catalysts remains necessary for increased chain-growth and alcohol formation and reduced hydrocarbon formation.[117]

MoS_2[96,109] and Mo_2C[121–125] catalysts copromoted with alkali and either Fe or Cu or Ni have also been demonstrated to be active for the mixed-alcohols synthesis from syngas feeds.

For CO_2-containing syngas, IFP's CuCo-based catalysts have shown relatively high yields of higher alcohols even with syngas containing up to 13% CO_2[49] although it was reported by Boz that up to 10% CO_2 content in syngas feed reduced the selectivity to higher alcohols.[126]

6.4 Conclusions

The conversion of biomass-derived syngas to alcohols is identical in principle to the conversion of syngas from any other source, once differences in the composition of the syngas are taken into account. Though not addressed here, the removal of contaminants and adjustment of the concentrations of CO, CO_2, H_2, and H_2O in the syngas from biomass sources can, in practice, lead to a syngas that is indistinguishable from one produced from other feedstocks.

The synthesis of methanol from syngas is a mature technology, despite some remaining questions about the exact mechanism. Interest in the formation of higher alcohols appears to be increasing as these compounds are finding uses as fuel additives and hydrogen carriers. Formation of these alcohols from syngas is fundamentally different from methanol synthesis, since both C–H bonds and C–OH bonds must be formed from the CO and CO_2 in the syngas.

The wide range of catalysts investigated for the formation of higher alcohols from syngas can be divided into four general classes:

(a) Rh-based catalysts;
(b) modified methanol synthesis catalysts (based on Cu);
(c) modified Fischer–Tropsch type catalysts;
(d) modified Mo-based catalysts.

Typically, alkali promoters are used to modify these catalysts to promote chain growth and increase selectivity to higher alcohols.

In addition to the effect of promoters, the effect of the presence of CO_2 in biomass-derived syngas can greatly affect the selectivity to alcohols. While CO_2 seems to have some promotional effect for methanol synthesis, its effect is largely negative for the synthesis of other alcohols, suggesting that some CO_2 removal step might be necessary after gasification, prior to catalytic conversion of the syngas.

References

1. M. Balat and H. Balat, *Appl. Energy*, 2009, **86**, 2273.
2. M. Fatih Demirbas, *Appl. Energy*, 2009, **86**, S151.
3. S. K. Hoekman, *Renewable Energy*, 2009, **34**, 14.
4. K. R. Jegannathan, E.-S. Chan and P. Ravindra, *Renew. Sustain. Energy Rev.*, 2009, **13**, 2163.
5. EIA, Annual Energy Review 2008, (US Department of Energy, Energy Information Administration), 2009.
6. D. J. Hayes, *Catal. Today*, 2009, **145**, 138.
7. P. Ferreira-Aparicio, M. J. Benito and J. L. Sanz, *Catal. Rev.-Sci. Eng.*, 2005, **47**, 491.
8. G. Apanel and J. Eric, *Fuel Cells Bull.*, 2004, **11**, 12.
9. G. Apanel and J. Eric, *Fuel Cells Bull.*, 2004, **5**, 3.
10. L. Hernández and V. Kafarov, *Int. J. Hydrogen Energy*, 2009, **34**, 7041.
11. J. R. Rostrup-Nielsen, *Phys. Chem. Chem. Phys.*, 2001, **3**, 283.
12. F. Joensen and J. R. Rostrup-Nielsen, *J. Power Sources*, 2002, **105**, 195.
13. E. Y. Garcia and M. A. Laborde, *Int. J. Hydrogen Energy*, 1991, **16**, 307.
14. I. Fishtik, A. Alexander and R. Datta, 16th Meeting of the North American Catalysis Society, Boston, MA, 1999.
15. J. Llorca, P. R. de la Piscina, J. A. Dalmon, J. Sales and N. Homs, *Appl. Catal. B*, 2003, **43**, 355.
16. G. A. Deluga, J. R. Salge, L. D. Schmidt and X. E. Verykios, *Science*, 2004, **303**, 993.
17. F. Aupretre, C. Descorme, D. Duprez, D. Casanave and D. Uzio, *J. Catal.*, 2005, **233**, 464.
18. V. Fierro, V. Klouz, O. Akdim and C. Mirodatos, *Catal. Today*, 2002, **75**, 141.

19. V. Fierro, O. Akdim, H. Provendier and C. Mirodatos, *J. Power Sources*, 2005, **145**, 659.
20. O. Akdim, W. J. Cai, V. Fierro, H. Provendier, A. Van Veen, W. J. Shen and C. Mirodatos, *Top. Catal.*, 2008, **51**, 22.
21. T. Ioannides and S. Neophytides, *J. Power Sources*, 2000, **91**, 150.
22. T. Ioannides, *J. Power Sources*, 2001, **92**, 17.
23. R. Chase, *USA Today*, 2006, June 23.
24. M. Canakci and J. Van Gerpen, *Trans. ASAE*, 1999, **42**, 1203.
25. J. U. Obibuzor, R. D. Abigor and D. A. Okiy, *J. Am. Oil Chem. Soc.*, 2003, **80**, 77.
26. Y. Warabi, D. Kusdiana and S. Saka, *Appl. Biochem. Biotechnol.*, 2004, **113–116**, 793.
27. C. Yang and Z. Y. Meng, *J. Catal.*, 1993, **142**, 37.
28. K. W. Yang, X. Z. Jiang and W. C. Zhang, *Chin. Chem. Lett.*, 2004, **15**, 1497.
29. C. Cao, Y. Wang, D. C. Elliott, J. Hu and D. Stevens, Abstracts of Papers, 226th ACS National Meeting, New York, NY, United States, 2003.
30. T. J. Wang, J. Chang, J. X. Zhu and Y. Fu, *Ranliao Huaxue Xuebao*, 2004, **32**, 297.
31. J. J. Spivey and A. A. Egbebi, *Chem. Soc. Rev.*, 2007, **36**, 1514.
32. A. Roine, Outokumpu Research Oy., Pori, Finland, 5.11 edn, 2003.
33. S. Mawson, M. S. Mccutchen, P. K. Lim and G. W. Roberts, *Energy & Fuels*, 1993, **7**, 257.
34. T. Inui, T. Yamamoto, M. Inoue, H. Hara, T. Takeguchi and J. B. Kim, *Appl. Catal. A*, 1999, **186**, 395.
35. P. Gronchi, E. Tempesti and C. Mazzocchia, *Appl. Catal. A*, 1994, **120**, 115.
36. X. L. Yin, D. Y. C. Leung, J. Chang, J. F. Wang, Y. Fu and C. Z. Wu, *Energy Fuels*, 2005, **19**, 305–310.
37. J. S. Lee, K. H. Lee, S. Y. Lee and Y. G. Kim, *J. Catal.*, 1993, **144**, 414–424.
38. C. Kuechen and U. Hoffmann, *Chem. Eng. Sci.*, 1993, **48**, 3767.
39. J. Wu, M. Saito, M. Takeuchi and T. Watanabe, *Appl. Catal. A*, 2001, **218**, 235.
40. C. Yang, Z. Ma, N. Zhao, W. Wei, T. Hu and Y. Sun, *Catal. Today*, 2006, **115**, 222.
41. Y. Choi, K. Futagami, T. Fujitani and J. Nakamura, *Catal. Lett.*, 2001, **73**, 27.
42. G. C. Chinchen, P. J. Denny, D. G. Parker, M. S. Spencer and D. A. Whan, *Appl. Catal.*, 1987, **30**, 333.
43. K. Kumabe, S. Fujimoto, T. Yanagida, M. Ogata, T. Fukuda, A. Yabe and T. Minowa, *Fuel*, 2008, **87**, 1422.
44. J.-W. Bae, H. S. Potdar, S.-H. Kang and K.-W. Jun, *Energy & Fuels*, 2007, **22**, 223.
45. Y. Zhang, J. Xiao and L. Shen, *Ind. Eng. Chem. Res.*, 2009, **48**, 5351.

46. K. Klier, V. Chatikavanij, R. G. Herman and G. W. Simmons, *J. Catal.*, 1982, **74**, 343.
47. J. Skrzypek, M. Lachowska and H. Moroz, *Chem. Eng. Sci.*, 1991, **46**, 2809.
48. J. Nakamura, Y. Choi and T. Fujitani, *Top. Catal.*, 2003, **22**, 277.
49. V. Subramani and S. K. Gangwal, *Energy & Fuels*, 2008, **22**, 814.
50. L. Nowicki and T. Olewski, *Chemical Papers-Chemicke Zvesti*, 2003, **57**, 21.
51. P. Forzatti, E. Tronconi and I. Pasquon, *Catal. Rev.-Sci. Eng.*, 1991, **33**, 109.
52. M. M. Bhasin, W. J. Bartley, P. C. Ellgen and T. P. Wilson, *J. Catal.*, 1978, **54**, 120.
53. M. L. Poutsma, L. F. Elek, P. A. Ibarbia, A. P. Risch and J. A. Rabo, *J. Catal.*, 1978, **52**, 157.
54. M. Ichikawa, T. Fukushima and K. Shikakura, Proc. 8th Int. Cong. Catal., Berlin, Germany, 1984.
55. M. Ichikawa, T. Fukushima, T. Yokoyama, N. Kosugi and H. Kuroda, *J. Phys. Chem.*, 1986, **90**, 1222.
56. M. Ichikawa, *Chemtech*, 1982, **12**, 674.
57. M. Ichikawa, *Bull. Chem. Soc. Jpn.*, 1978, **51**, 2268.
58. M. Bowker, *Catal. Today*, 1992, **15**, 77.
59. X. L. Pan, Z. L. Fan, W. Chen, Y. J. Ding, H. Y. Luo and X. H. Bao, *Nature Mater.*, 2007, **6**, 507.
60. J. Hu, Y. Wang, C. Cao, D. C. Elliott, D. J. Stevens and J. F. White, *Catal. Today*, 2007, **120**, 90.
61. M. Ichikawa and T. Fukushima, *J. Chem. Soc., Chem. Commun.*, 1985, 321.
62. A. Takeuchi and J. R. Katzer, *J. Phys. Chem.*, 1982, **86**, 2438.
63. T. Iizuka, Y. Tanaka and K. Tanabe, *J. Catal.*, 1982, **76**, 1.
64. M. F. H. Vantol, A. Gielbert and B. E. Nieuwenhuys, *Appl. Surf. Sci.*, 1993, **67**, 166.
65. E. Guglielminotti, E. Giamello, F. Pinna, G. Strukul, S. Martinengo and L. Zanderighi, *J. Catal.*, 1994, **146**, 422.
66. T. Iizuka and Y. Tanaka, *J. Catal.*, 1981, **70**, 449.
67. K. K. Bando, K. Soga, K. Kunimori and H. Arakawa, *Appl. Catal. A*, 1998, **175**, 67.
68. A. Takeuchi and J. R. Katzer, *J. Phys. Chem.*, 1981, **85**, 937.
69. F. Solymosi, A. Erdohelyi and T. Bansagi, *J. Catal.*, 1981, **68**, 371.
70. B. A. Sexton and G. A. Somorjai, *J. Catal.*, 1977, **46**, 167.
71. F. Solymosi and A. Erdohelyi, *J. Mol. Catal.*, 1980, **8**, 471.
72. S. Ichikawa, *J. Mol. Catal.*, 1989, **53**, 53.
73. S. Marengo, S. Martinengo and L. Zanderighi, *Chem. Eng. Sci.*, 1992, **47**, 2793.
74. K. J. Smith and R. B. Anderson, *Can. J. Chem. Eng.*, 1983, **61**, 40.
75. K. J. Smith and R. B. Anderson, *J. Catal.*, 1984, **85**, 428.
76. A. M. Hilmen, M. T. Xu, M. J. L. Gines and E. Iglesia, *Appl. Catal. A*, 1998, **169**, 355.

77. J. G. Nunan, C. E. Bogdan, K. Klier, K. J. Smith, C. W. Young and R. G. Herman, *J. Catal.*, 1988, **113**, 410.
78. J. G. Nunan, C. E. Bogdan, K. Klier, K. J. Smith, C. W. Young and R. G. Herman, *J. Catal.*, 1989, **116**, 195.
79. J. G. Nunan, R. G. Herman and K. Klier, *J. Catal.*, 1989, **116**, 222.
80. M. T. Xu and E. Iglesia, *Catal. Lett.*, 1998, **51**, 47.
81. L. Majocchi, L. Lietti, A. Beretta, P. Forzatti, E. Micheli and L. Tagliabue, *Appl. Catal. A*, 1998, **166**, 393.
82. J. M. Campos-Martín, J. L. G. Fierro, A. Guerrero-Ruiz, R. G. Herman and K. Klier, *J. Catal.*, 1996, **163**, 418.
83. E. M. Calverley and K. J. Smith, *J. Catal.*, 1991, **130**, 616.
84. J. E. Baker, R. Burch and S. E. Golunski, *Appl. Catal.*, 1989, **53**, 279.
85. J. C. Slaa, J. G. Vanommen and J. R. H. Ross, *Catal. Today*, 1992, **15**, 129.
86. N. Zhao, R. Xu, W. Wei and Y. H. Sun, *React. Kinet. Catal. Lett.*, 2002, **75**, 297.
87. D. J. Elliott, *J. Catal.*, 1988, **111**, 445.
88. D. Gall, E. J. Gibson and C. C. Hall, *J. Appl. Chem.*, 1952, **2**, 371.
89. A. Razzaghi, J. P. Hindermann and A. Kiennemann, *Appl. Catal.*, 1984, **13**, 193.
90. M. Pijolat and V. Perrichon, *Appl. Catal.*, 1985, **13**, 321.
91. K. Fujimoto and T. Oba, *Appl. Catal.*, 1985, **13**, 289.
92. M. Inoue, T. Miyake, Y. Takegami and T. Inui, *Appl. Catal.*, 1984, **11**, 103.
93. K. Takeuchi, T. Matsuzaki, H. Arakawa and Y. Sugi, *Appl. Catal.*, 1985, **18**, 325.
94. H. Hamada, Y. Kuwahara, Y. Kintaichi, T. Ito, K. Wakabayashi, H. Iijima and K. Sano, *Chem. Lett.*, 1984, 1611.
95. Y. Kintaichi, Y. Kuwahara, H. Hamada, T. Ito and K. Wakabayashi, *Chem. Lett.*, 1985, 1305.
96. S. A. Hedrick, S. S. C. Chuang, A. Pant and A. G. Dastidar, *Catal. Today*, 2000, **55**, 247.
97. Y. Kintaichi, T. Ito, H. Hamada, H. Nagata and K. Wakabayashi, *Sekiyu Gakkaishi-J. Jpn. Petrol. Inst.*, 1998, **41**, 66.
98. T. Matsuzaki, K. Takeuchi, T. Hanaoka, H. Arawaka and Y. Sugi, *Appl. Catal. A*, 1993, **105**, 159.
99. H. Idriss, C. Diagne, J. P. Hindermann, A. Kinnemann and M. A. Barteau, *Stud. Surf. Sci. Catal.*, 1993, **75**, 2119.
100. K. Takeuchi, T. Matsuzaki, T. A. Hanaoka, H. Arakawa, Y. Sugi and K. Wei, *J. Mol. Catal.*, 1989, **55**, 361.
101. T. Matsuzaki and T. A. Hanaoka, *Abstracts of Papers of the American Chemical Society*, 1996, **211,** 14-Petr.
102. T. Inui, *Catal. Today*, 1996, **29**, 329.
103. T. Inui and T. Yamamoto, *Catal. Today*, 1998, **45**, 209.
104. T. Yamamoto and T. Inui, *Advances in Chemical Conversions for Mitigating Carbon Dioxide*, 1998, **114**, 513.

105. T. Inui, *Abstracts of Papers of the American Chemical Society*, 2000, **219**, U250.
106. T. Tatsumi, A. Muramatsu and H. O. Tominaga, *Chem. Lett.*, 1984, 685.
107. H. C. Woo, T. Y. Park, Y. G. Kim, I. S. Nam, J. S. Lee and J. S. Chung, *Stud. Surf. Sci. Catal.*, 1993, **75**, 2749.
108. A. Muramatsu, T. Tatsumi and H. Tominaga, *Bull. Chem. Soc. Jpn.*, 1987, **60**, 3157.
109. H. J. Qi, D. B. Li, C. Yang, Y. G. Ma, W. H. Li, Y. H. Sun and B. Zhong, *Catal. Commun.*, 2003, **4**, 339.
110. E. C. Alyea, D. He and J. Wang, *Appl. Catal A*, 1993, **104**, 77.
111. J. G. Santiesteban, C. E. Bogdan, R. G. Herman and K. Klier, 9th Int. Congr. Catal, 1988.
112. K. J. Smith, R. G. Herman and K. Klier, *Chem. Eng. Sci.*, 1990, **45**, 2639.
113. D. B. Li, C. Yang, W. H. Li, Y. H. Sun and B. Zhong, *Top. Catal.*, 2005, **32**, 233.
114. T. Tatsumi, A. Muramatsu and H. Tominaga, *J. Catal.*, 1986, **101**, 553.
115. G. Lu, C. F. Zhang, Y. Q. Gang, Z. B. Zhu, Y. H. Ni, L. J. Cheng and F. Yu, *Appl. Catal. A*, 1997, **150**, 243.
116. Y. M. Li, R. J. Wang and L. Chang, *Catal. Today*, 1999, **51**, 25.
117. S. Phillips, A. Aden, J. Jechura, D. Dayton and T. Eggeman, Thermochemical Ethanol *via* Indirect Gasification and Mixed Alcohol Synthesis of Lignocellulosic Biomass, NREL/TP-510-41168, National Renewable Energy Laboratory, Golden, CO, 2007.
118. P. Chaumette, P. Courty, A. Kiennemann, R. Kieffer, S. Boujana, G. A. Martin, J. A. Dalmon, P. Meriaudeau and C. Mirodatos, *Ind. Eng. Chem. Res.*, 1994, **33**, 1460.
119. P. Courty, P. Chaumette, C. Raimbault and P. Travers, *Oil & Gas Science and Technology-Rev. IFP*, 1990, **45**, 561.
120. K. Fang, D. Li, M. Lin, M. Xiang, W. Wei and Y. Sun, *Catal. Today*, 2009, **147**, 133.
121. H. C. Woo, K. Y. Park, Y. G. Kim, I.-S. Namau, Jong ShikChung and J. S. Lee, *Appl. Catal.*, 1991, **75**, 267.
122. M. Xiang, D. Li, W. Li, B. Zhong and Y. Sun, *Catal. Commun.*, 2007, **8**, 88.
123. M. Xiang, D. Li, W. Li, B. Zhong and Y. Sun, *Catal. Commun.*, 2007, **8**, 503.
124. M. Xiang, D. Li, H. Qi, W. Li, B. Zhong and Y. Sun, *Fuel*, 2007, **86**, 1298.
125. M. Xiang, D. Li, H. Xiao, J. Zhang, W. Li, B. Zhong and Y. Sun, *Catal. Today*, 2008, **131**, 489.
126. I. Boz, *Catal. Lett.*, 2003, **87**, 187.

CHAPTER 7

Fast Pyrolysis of Biomass for Energy and Fuels

A. V. BRIDGWATER

Bioenergy Research Group, Aston University, Birmingham B4 7ET, UK

7.1 Introduction

Renewable energy is of growing importance in responding to concerns over the environment and security of energy supply. Biomass is unique in providing the only renewable source of fixed carbon, which is an essential ingredient in meeting many of our fuel and consumer-goods requirements. Wood and annual crops and agricultural and forestry residues are some of the main renewable-energy resources available. The biodegradable components of MSW (municipal solid waste) and commercial and industrial wastes are also significant bioenergy resources, although particularly in the case of MSW, they may require extensive processing before conversion. Biomass is considered the renewable-energy source with the highest potential to contribute to the energy needs of modern society for both the developed and developing economies worldwide. In addition, energy from biomass based on short rotation forestry and other energy crops can contribute significantly towards the objectives of the Kyoto Protocol in reducing greenhouse-gas emissions and alleviating problems related to climate change.

Biomass fuels and residues can be converted to energy *via* thermal, biological, and mechanical or physical processes as summarized in Figure 7.1. The key difference between thermal and biological conversion is that biological conversion usually gives single or specific products such as ethanol or biogas

RSC Energy and Environment Series No. 1
Thermochemical Conversion of Biomass to Liquid Fuels and Chemicals
Edited by Mark Crocker

Published by the Royal Society of Chemistry, www.rsc.org

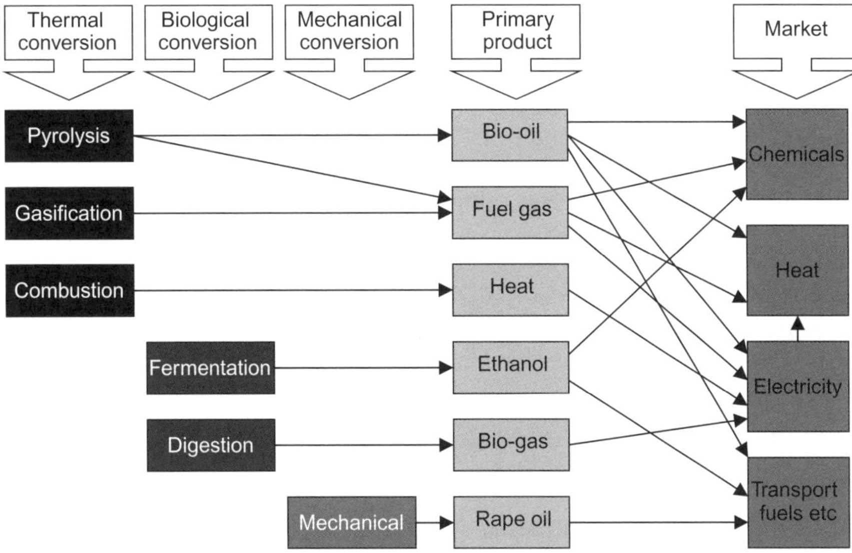

Figure 7.1 Conversion processes, products and applications.

(which contains up to 60% methane) and is a relatively slow process, typically taking hours, days, weeks (anaerobic fermentation and farm digestion) or years (landfill gas by digestion) for reactions to be completed. Thermal conversion gives multiple and often complex products, with catalysts often used to improve the product quality or spectrum, and takes place in very short reaction times of typically seconds or minutes. Figure 7.1 summarizes some of the main products that can be obtained by processing biomass. Of particular current interest is processing byproducts, residues and wastes from biological and mechanical processes such as lignin as a byproduct from bioethanol production and glycerol from biodiesel production.

7.1.1 Biomass Resources

Biomass is a diffuse resource, arising over very large areas, and thus requiring large land areas with substantial logistical problems in collection and transport as well as high costs. Typically, a sustainable crop of 10 dry t/ha/y of woody biomass can currently be produced in Northern Europe, rising to perhaps 15 or maybe 20 dry t/ha/y for energy crops in Southern Europe. Crop development through breeding and modification will increase these yields eventually. Thus, an area of 1 km^2 or 100 ha will currently produce around 1000 dry t/y, enough for a power output of around 150 kW_e at low conversion efficiencies or up to 300 kW_e at high conversion efficiencies. A 50 MW_e biomass-fired power station will thus require a planted area of around 200 km^2. This is one reason for the increased interest in internationally traded biomass, using dock-side conversion facilities.

A further consideration with almost all forms of biomass is their seasonality: coppiced crops can usually only be harvested during the winter months, and energy crops and agricultural residues are even more seasonal, typically being available for only a few months a year. Either extensive provision for storage thus has to be made, or a multifuel system is required. Increasing efforts are underway to develop processes that can handle a number of different fuels, either mixed or separately. This option also addresses the increased difficulties of processing high-ash materials in fast pyrolysis when blending can provide a sufficiently low ash fuel to produce a good-quality oil.

An alternative solution to the dispersed nature of biomass is to "concentrate" it locally through conversion to a higher-density liquid by fast pyrolysis. This is being actively examined as an alternative way of improving the logistics and economics of moving solid biomass from field or forest to the conversion or utilization facility in order to benefit from economies of scale.

7.2 Thermal Conversion Processes

There are three main thermal processes available for converting biomass to a more useful energy form – pyrolysis, gasification and combustion. Figure 7.2 summarizes their products and applications. Combustion is already a well-established commercial technology with applications in most industrialized and developing countries and development is concentrated on resolving environmental problems. Gasification has been practiced for many years and there are many examples of demonstration and precommercial activities, which have been widely reviewed.[1]

This review focuses on the emerging advanced technology of fast pyrolysis both as an integrated process for production of a liquid fuel that can be used

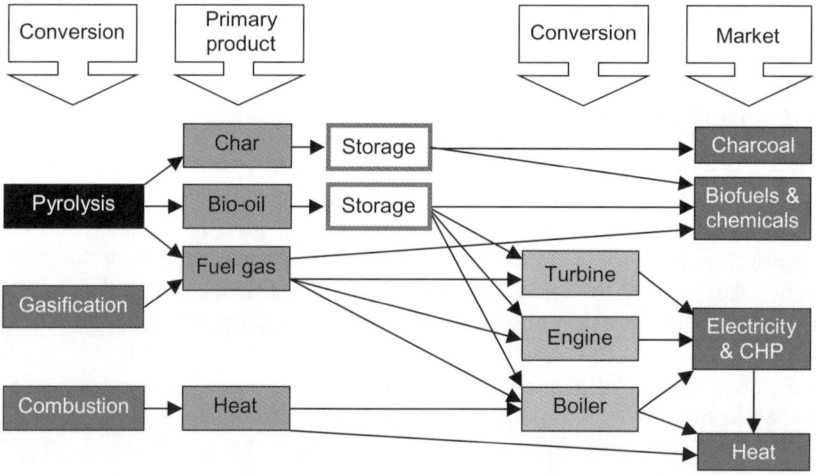

Figure 7.2 Products from thermal biomass conversion.

directly and as an intermediate pretreatment step to convert solid biomass into a higher energy content transportable liquid for subsequent processing for heat, power, biofuels, and chemicals. This technology is expected to offer the greatest contribution in the short term in terms of versatility, improved efficiency and environmental acceptability.

In all cases, a commercial process comprises three main stages from feed reception to delivery of one or more useful products:

- Feed reception, storage, handling, preparation and pretreatment;
- Conversion of solid biomass by fast pyrolysis to a more usable form of energy in liquid form which is known as bio-oil;
- Conversion of this primary liquid product by processing, refining or clean-up to a marketable end product such as electricity, heat, biofuels and/or chemicals.

7.3 Fast Pyrolysis

Pyrolysis is thermal decomposition occurring in the absence of oxygen. It is always also the first step in combustion and gasification, but in these processes it is followed by total or partial oxidation of the primary products. Lower process temperatures and longer vapor residence times favor the production of charcoal. High temperatures and longer residence times increase biomass conversion to gas, and moderate temperatures and short vapor residence time are optimum for producing liquids. Three products are always produced, but the proportions can be varied over a wide range by adjustment of the process parameters. Table 7.1 and Figure 7.3 indicate the product distribution obtained from different modes of pyrolysis, showing the considerable flexibility achievable by changing process conditions. Fast pyrolysis for liquids production is currently of particular interest as the liquid can be stored and transported, and used for energy, chemicals or as an energy carrier.

Table 7.1 Typical product yields (dry wood basis) obtained by different modes of pyrolysis of wood.

Mode	*Conditions*	*Liquid*	*Solid*	*Gas*
Fast	~500 °C, short hot vapor residence time ~1 s	75%	12% char	13%
Intermediate	~500 °C, hot vapor residence time ~10–30 s	50% in 2 phases	25% char	25%
Slow – Torrefaction	~290 °C, solids residence time ~30 min	0% if vapours are burned	80% solid	20%
Slow – Carbonization	~400 °C, long vapor residence hours→days	30%	35% char	35%
Gasification	~750–900 °C	5%	10% char	85%

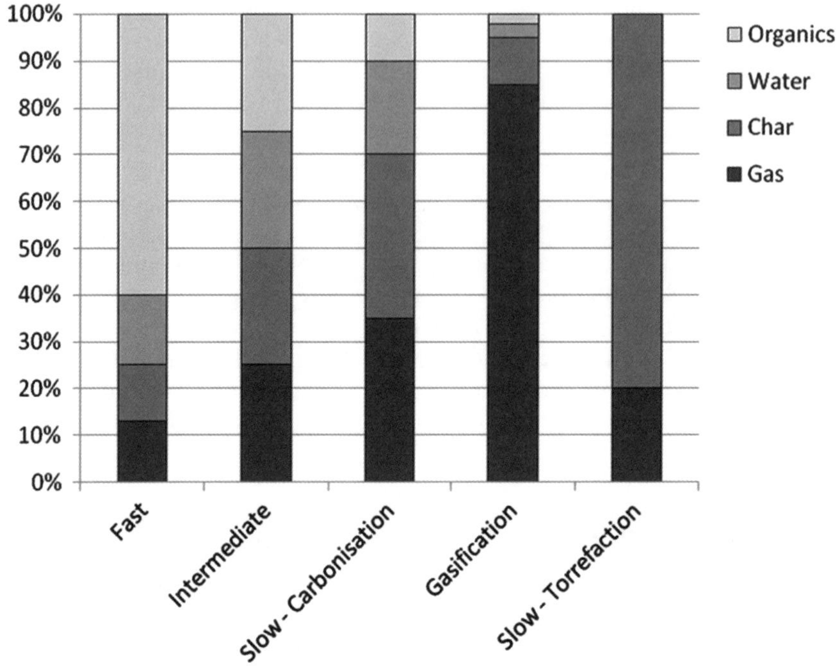

Figure 7.3 Product spectrum from pyrolysis.

7.3.1 Principles

In fast pyrolysis, biomass decomposes very quickly to generate mostly vapors and aerosols and some charcoal. After cooling and condensation, a dark brown homogenous mobile liquid is formed that has a heating value about half that of conventional fuel oil. While it is related to the traditional pyrolysis processes for making charcoal, fast pyrolysis is an advanced process, with carefully controlled parameters to give high yields of liquid. The essential features of a fast-pyrolysis process for producing liquids are:

- very high heating and heat-transfer rates at the reaction interface, which usually requires a finely ground biomass feed of typically less than 3 mm;
- carefully controlled pyrolysis reaction temperature of around 500 °C and vapor phase temperature of 400–450 °C;
- short hot vapor residence times of typically less than 2 s;
- rapid cooling of the pyrolysis vapors to give the bio-oil product.

As fast pyrolysis for liquids occurs in a few seconds or less, heat- and mass-transfer processes and phase transition phenomena, as well as chemical reaction kinetics, play important roles. The critical issue is to bring the reacting biomass particles to the optimum process temperature and minimize their exposure to the intermediate (lower) temperatures that favor formation of charcoal. One way this objective can be achieved is by using small particles, for example in the fluidized-bed processes that are described later. Another possibility is to transfer heat very fast only to the particle surface that contacts the heat source. This second method is applied in ablative processes that are described later.

The main product, bio-oil, is obtained in yields of up to 75 wt% on a dry-feed basis, together with byproduct char and gas, which can be used within the process to provide the process heat requirements so there are no waste streams other than flue gas and ash. Liquid yield depends on temperature, hot vapor residence time, char separation and biomass ash content, the last two having a catalytic effect on vapor cracking.

A fast-pyrolysis process includes drying the feed to typically less than 10% water in order to minimize the water in the product liquid oil (although up to 15% can be acceptable in some process configurations), grinding the feed (to around 2–3 mm particle size in the case of fluid-bed reactors) to give sufficiently small particles to ensure rapid reaction, pyrolysis reaction, separation of solids (char), and quenching and collection of the liquid product (often referred to as bio-oil).

Virtually any form of biomass can be considered for fast pyrolysis. While most work has been carried out on wood because of its consistency and comparability between tests, over 100 different biomass types have been tested by many laboratories, ranging from agricultural wastes such as straw, olive pits and nut shells to energy crops such as miscanthus and sorghum, forestry wastes such as bark and solid wastes such as sewage sludge and leather wastes.

7.3.2 Reactors

At the heart of a fast-pyrolysis process is the reactor. Although it probably represents at most only about 10–15% of the total capital cost of an integrated system, most research and development has focused on the reactor, although increasing attention is now being paid to control and improvement of liquid quality and improvement of collection systems. The rest of the process consists of biomass reception, storage and handling, biomass drying and grinding, product collection, storage and, when relevant, upgrading. Several comprehensive reviews of fast-pyrolysis processes for liquids production are available.[1–8]

7.3.2.1 Bubbling Fluid Beds

Bubbling fluid beds–usually referred to as just fluid beds as opposed to circulating fluid beds–have the advantages of a well-understood technology that is simple in construction and operation, good temperature control and very efficient heat transfer to biomass particles arising from the high solids density. Figure 7.4 shows a typical configuration using electrostatic precipitators for coalescence and collection of aerosols, which seem to be the separation process of choice for aerosols in most commercial processes. Heating can be achieved in

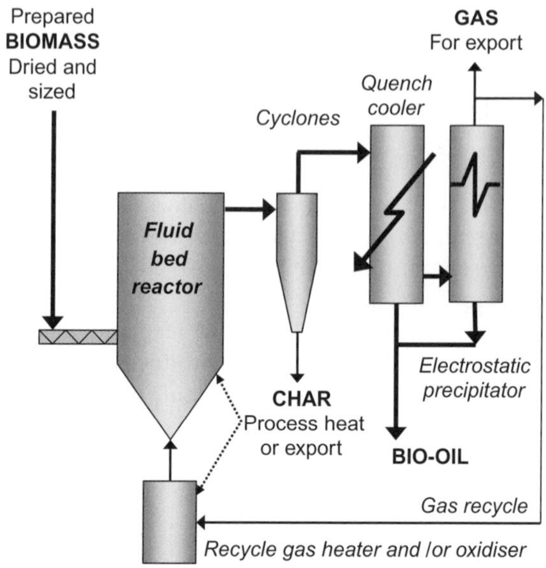

Figure 7.4 Bubbling fluid-bed reactor with electrostatic precipitator.

a variety of ways and scaling is well understood. However, heat transfer to bed at large scales of operation has to be considered carefully because of the scale-up limitations of different methods of heat transfer, as discussed later. Fluid-bed pyrolyzers give good and consistent performance with high liquid yields of typically 70–75 wt% from wood on a dry-feed basis. Small biomass particle sizes of less than 2–3 mm are needed to achieve high biomass heating rates, and the rate of particle heating is usually the rate-limiting step.

Residence time of solids and vapors is controlled by the fluidizing gas flowrate and is higher for char than for vapors. As char acts as an effective vapor cracking catalyst at fast-pyrolysis reaction temperatures, rapid and effective char separation/elutriation is important. This is usually achieved by ejection and entrainment followed by separation in one or more cyclones, so careful design of sand and biomass/char hydrodynamics is important. The high level of inert gases arising from the high permanent gas flows required for fluidization result in very low partial pressures for the condensable vapors and thus care is needed to design and operate efficient heat-exchange and liquid-collection systems. In addition, the large inert gas flowrates result in relatively large equipment thus increasing cost.

All the early work on fluid beds was carried out at the University of Waterloo in Canada, which pioneered the science of fast pyrolysis and established a clear lead in this area for many years.[9–11] Bubbling fluid beds have been selected for further development by several companies, including Union Fenosa,[12] who built and operated a 200-kg/h pilot unit in Spain based on the University of Waterloo process, which was dismantled some years ago; Dynamotive, who have built and tested a 75-kg/h and 400-kg/h pilot unit[13] in Canada based on an RTI design, and have subsequently built a 100-t/d and a 200 t/d plant now in operation in Canada; Wellman, who built a 250-kg/h unit[14] in the UK, which has not operated; Biomass Engineering Ltd in the UK who are finalising construction of a 250 kg/h pilot unit and Fortum who built and extensively tested a 500-kg/h plant in Finland which has now been dismantled. Many research units have also been built at universities and research institutions as they are relatively easy to construct and operate and give good results.

7.3.2.2 Circulating Fluid Beds and Transported Bed

Circulating fluid bed (CFB) and transported-bed reactor systems have many of the features of bubbling beds described above, except that the residence time of the char is almost the same as for vapors and gas, and the char is more attrited due to the higher gas velocities. This can lead to higher char contents in the collected bio-oil unless more extensive char removal is included. This char is recovered with the sand and burned in a close coupled combustor to reheat the sand which is then recirculated. A typical layout is shown in Figure 7.5. An added advantage is that CFBs are potentially suitable for larger throughputs even though the hydrodynamics are more complex, as this technology is widely used at very high throughputs in the petroleum and petrochemical industry. Heat supply is usually from recirculation of heated sand from a secondary char combustor, which can be either a bubbling or circulating fluid bed. In this respect the process is similar to a twin fluid-bed gasifier except that the reactor

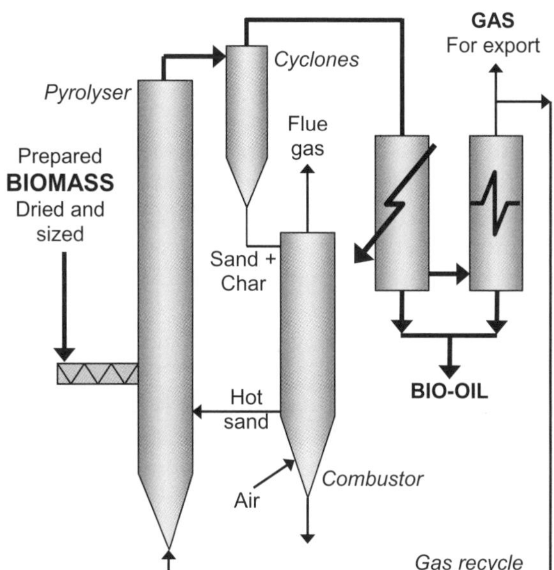

Figure 7.5 Circulating fluid-bed reactor.

(pyrolyzer) temperature is much lower and the closely integrated char combustion in a second reactor requires careful control to ensure that the temperature, heat flux and solids flow match the process and feed requirements. One of the unproven areas is heat transfer at large scales of throughput.

Examples include the 650-kg/h ENEL plant in Italy built by Ensyn,[15,16] the several Ensyn units in the USA at Red Arrow in Wisconsin for production of food flavorings up to 1700 kg/h, the research unit at VTT in Finland built by Ensyn, the CRES pilot plant in Greece[17] that is awaiting rebuilding and a laboratory unit at the University of Leuven in Belgium.[18]

7.3.2.3 Ablative Pyrolysis

Ablative pyrolysis is substantially different in concept compared with other methods of fast pyrolysis. In all the other methods, the rate of reaction is limited by the rate of heat transfer through the biomass particles, which is why small particles are required. The mode of reaction in ablative pyrolysis is like melting butter in a frying pan–the rate of melting can be significantly enhanced by pressing the butter down and moving it over the heated pan surface. In ablative pyrolysis, heat is transferred from the hot reactor wall to "melt" wood that is in contact with it under pressure. The molten wood then vaporizes to a product very similar to that derived from fluid-bed systems.

The pyrolysis front thus moves unidirectionally through the biomass particle. As the wood is mechanically moved away, the residual oil film both provides lubrication for successive biomass particles and also rapidly evaporates to give pyrolysis vapors for collection in the same way as other processes. The rate of reaction is strongly influenced by pressure, the relative velocity of the wood and the heat-exchange surface and the reactor surface temperature. The key features of ablative pyrolysis are therefore as follows:

- high pressure of particle on hot reactor wall, achieved by centrifugal force in the NREL, USA, concept (no longer operational[19]) or mechanically at Aston University, UK (Figure 7.6) and at PyTec in Germany;
- high relative motion between particle and reactor wall;
- reactor wall temperature less than 600 °C.

As reaction rates are not limited by heat transfer through the biomass particles, larger particles can be used and in principle there is no upper limit to the size that can be processed. The process, in fact, is limited by the rate of heat supply to the reactor rather than the rate of heat absorption by the pyrolyzing biomass, as in other reactors. There is no requirement for inert gas, so the processing equipment is smaller and the reaction system is thus more intensive. In addition, the absence of fluidizing gas substantially increases the partial pressure of the condensable vapors, leading to more efficient collection and smaller equipment. However, the process is surface-area-controlled so scaling is more costly and the reactor is mechanically driven, and is thus more complex.

Much of the pioneering fundamental work on ablative pyrolysis reactors was performed by the CNRS laboratories in Nancy, France where extensive basic research has been carried out into the relationships between pressure, motion and temperature.[20] The National Renewable Energy Laboratory (NREL) in Boulder, Colorado developed the ablative vortex reactor, in which the biomass is accelerated to supersonic velocities to derive high tangential pressures inside a heated cylinder.[19] Unreacted particles are recycled and the vapors and char fines leave the reactor axially for collection. Liquid yields of 60–65 wt% on dry-feed basis are typically obtained.

Aston University has developed the ablative plate reactor shown in Figure 7.6. This has a prototype rotating blade reactor in which pressure and motion is derived mechanically, obviating the need for a carrier gas.[21] Liquid yields of 70–75 wt% on dry-feed basis are typically obtained. A second-generation reactor has recently been built and commissioned and has been patented (Figure 7.7).[22]

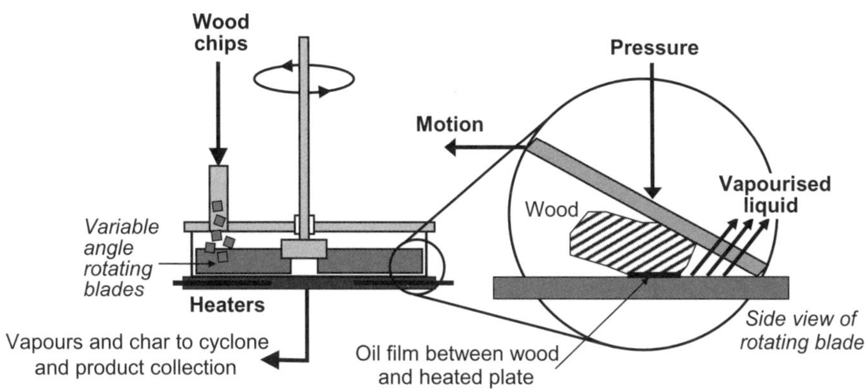

Figure 7.6 Aston University original ablative plate fast-pyrolysis reactor.

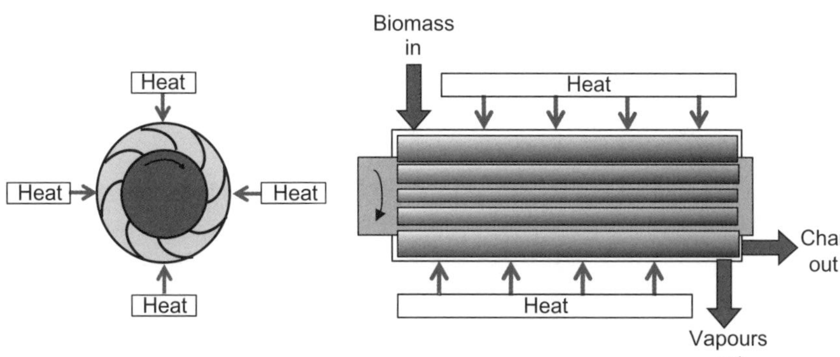

Figure 7.7 Aston University Mark 2 ablative fast-pyrolysis reactor.

Another configuration is the mechanically driven PyTec process in Germany.[23] The company has built and tested a laboratory unit based on hydraulically feeding wood rods onto a rotating electrically heated cone. The liquid collection system is analogous to the other systems described above.[23] A 6-t/d unit has been built in north Germany in 2006 that is currently undergoing testing and designs are in progress for a 50 t/d unit. The liquid is used in an engine for power generation. Other units include the now abandoned coiled tube at Castle Capital,[24] and cyclonic reactors.[25]

7.3.2.4 Entrained Flow

Entrained flow fast pyrolysis is in principle a simple technology, but most developments have not been as successful as had been hoped, mostly because of the poor heat transfer between a hot gas and a solid particle. High gas flows are required to effect sufficient heat transfer, which requires large plant sizes and entails difficult liquid collection from the low vapor partial pressure. Liquid yields have usually been lower than fluid-bed and CFB systems.

This configuration has been developed at Georgia Tech Research Institute[26] and scaled up by Egemin.[27] Bridgwater and Peacocke[21] give further details. Neither the GTRI nor Egemin process is now operational nor are there any known plans for further development, probably because of the difficulties encountered in achieving good heat transfer from a gaseous heat carrier to solid biomass. Liquid yields of up to 50–60 wt% on dry feed have been reported.

7.3.2.5 Rotating Cone

The rotating-cone reactor, invented at the University of Twente[28] and under development by BTG,[29] is a relatively recent development and effectively operates as a transported-bed reactor, but with transport effected by centrifugal forces in a rotating cone rather than gas. A 250-kg/h unit is now operational, and a scaled up version of 50 t/d was commissioned in Malaysia in mid-2005. Figure 7.8 shows an early prototype on the left and its role in an integrated fast-pyrolysis process on the right. The key features are:

- centrifugation (at $\sim$10 Hz) drives hot sand and biomass up a rotating heated cone;
- vapors are collected and processed conventionally;
- char and sand drop into a fluid bed surrounding the cone, whence they are lifted to a separate fluid-bed combustor where char is burned to heat the sand, which is then dropped back into the rotating cone;
- char is burned in a secondary bubbling fluid-bed combustor. The hot sand is recirculated to the pyrolyzer;
- carrier gas requirements in the pyrolysis reactor are much less than for fluid-bed and transported-bed systems; however, gas is needed for char burn-off and sand transport;

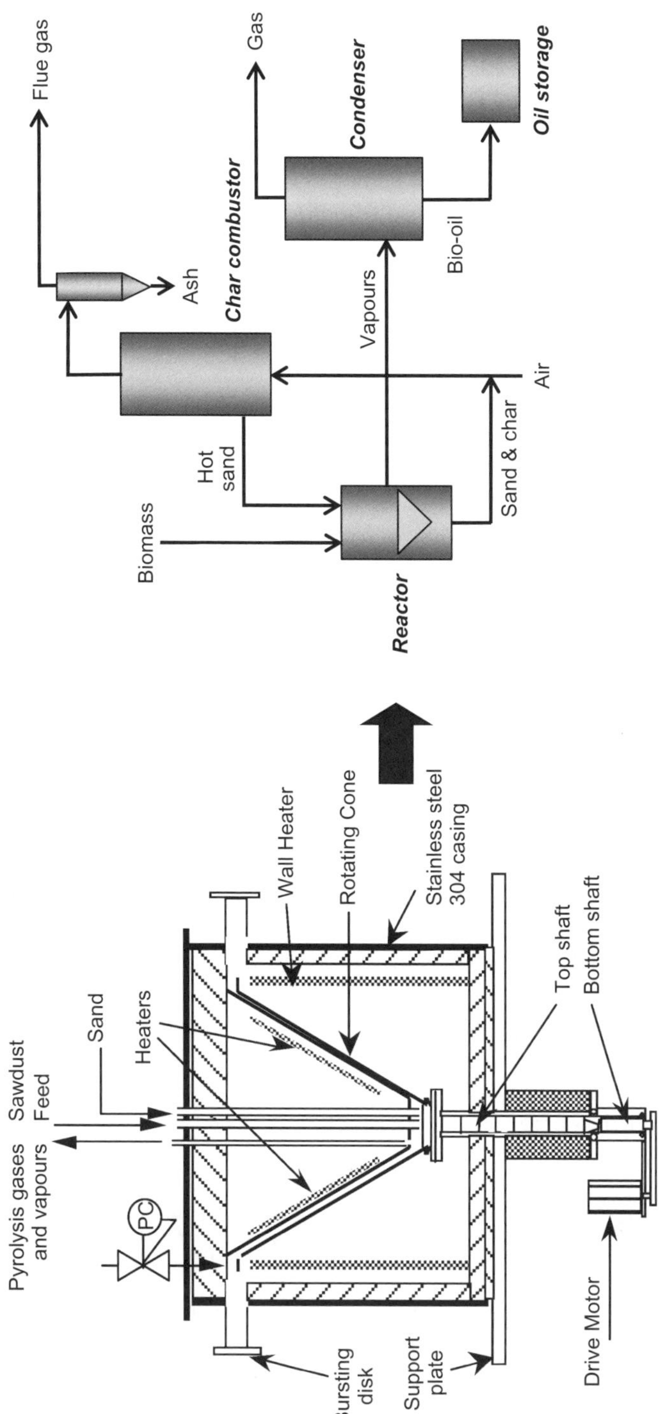

Figure 7.8 Rotating-cone pyrolysis reactor and integrated process.

- complex integrated operation of three subsystems is required: rotating-cone pyrolyzer, riser for sand recycling, and bubbling-bed char combustor;
- liquid yields of 60–70% on dry feed are typically obtained.

7.3.2.6 Vacuum Pyrolysis

Vacuum pyrolysis is arguably not a true fast pyrolysis as the heat-transfer rate to and through the solid biomass is much slower than in the previously described reactors. However, the vapor residence time is comparable and the liquid product has some similar characteristics.

The basic technology was developed at the University of Laval using a multiple hearth furnace but is now based on a purpose-designed horizontal moving bed. A 50-kg/h unit is available for research and the technology was scaled up by Pyrovac to a 3.5-t/h unit, which was operating at Jonquiére in Canada up to 2001.[30] The process operated at 450 °C and 100 kPa. Liquid yields of 35–50% on dry feed were typically obtained with higher char yields than fast-pyrolysis systems. Conversely, the liquid yields are higher than in slow pyrolysis technologies because the vapors are removed quickly from the reaction zone, thus minimizing secondary reactions. The process is complex and costly because the high vacuum necessitates the use of very large vessels and piping. The advantages of the process are that it can process larger particles than most fast-pyrolysis reactors, there is less char in the liquid product because of the lower gas velocities, and no carrier gas is needed.

7.3.3 Heat Transfer in Fast Pyrolysis

There are a number of technical challenges facing the development of fast pyrolysis, of which the most significant is heat transfer to the reactor. Pyrolysis is an endothermic process, requiring a substantial heat input to raise the biomass to reaction temperature, although the heat of reaction is relatively insignificant. Heat transfer in commercial reactors is a significant design feature and the energy in the byproduct charcoal would typically be used in a commercial process by combustion of the char in air. The char typically contains about 25% of the energy of the feedstock, and about 50–75% of this energy is required to drive the process. The byproduct gas only contains about 5% of the energy in the feed and this is not sufficient for pyrolysis. Figure 7.9 summarizes the main methods of providing the necessary heat, which can be added in several ways:

- through heat-transfer surfaces located in suitable positions in the reactor (1 and 2 in the figure);
- by heating the fluidization gas in the case of a fluid bed or circulating fluid-bed reactor (3 in the figure), although excessive gas temperatures may be

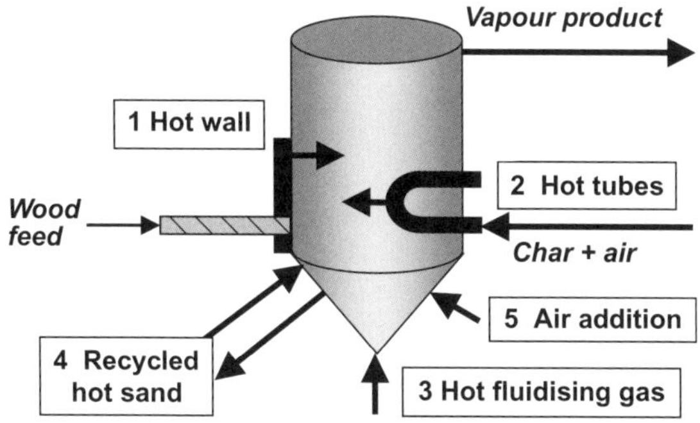

Figure 7.9 Methods of providing process heat for fast pyrolysis.

needed to input the necessary heat resulting in local overheating and reduced liquid yield, or alternatively very high gas flows are needed, resulting in unstable hydrodynamics. Partial heating is usually satisfactory and desirable to optimize energy efficiency;
- by removing and reheating the bed material in a separate reactor (4 in the figure) as used in most CFB and transported-bed reactors;
- by the addition of some air, although this can create hotspots and increase cracking of the liquids to tars (5 in the figure).

There are a variety of ways of providing the process heat from byproduct char or gas; or from fresh biomass. This facet of pyrolysis reactor design and optimization is most important for commercial units and will attract increasing attention as plants become bigger. Examples of options include:

- combustion of fresh biomass instead of char, particularly where there is a lucrative market for the char;
- gasification of the byproduct char and combustion of the resultant producer gas to provide greater control and avoid alkali metal problems in the char combustor;
- use of byproduct gas with similar advantages as above, although there is unlikely to be sufficient energy available in this gas without some supplementation;
- use of bio-oil product;
- use of fossil fuels where these are available at low cost, do not affect any interventions allowable on the process or product, and the byproducts have a sufficiently high value.

7.3.4 Summary and Status of Reactors

Table 7.2 summarizes known operational fast and slow pyrolysis systems for liquid production as of 2009.

Table 7.2 Summary of fast-pyrolysis reaction systems for liquids recently and currently operational.

Fast pyrolysis	*Industrial*	*Units built*	*Max size kg/h*	*Research*	*Max size kg/h*
Fluid bed	Agritherm, Canada	2	200	Adelaide U., Australia	1
	Biomass Engineering Ltd, UK	1	200	Aston U., UK	5
	Dynamotive, Canada	4	8,000	Cirad, France	2
	RTI, Canada	5	20	Curtin U., Australia	2
				ECN, NL	1
				East China U. Science and Technology, Shanghai, China	nk
				Gent U., Belgium	0.3
				Guangzhou Inst. Energy Conversion, China	10
				Harbin Institute of Technology	nk
				Iowa State U., USA	6
				Monash U. Australia	1
				NREL, USA	10
				PNNL, USA	1
				Shandong U. Technology	nk
				Shanghai JiaoTong U.,	1
				Shenyang U., China	1
				South East U., China	1
				Texas A&M U., USA	42
				TNO, Netherlands	10
				U. Basque Country, Spain	nk
				U. Campinas, Brazil	100
				U. Maine, USA	0.1
				U. Melbourne, Australia	0.1
				U. Naples, Italy	1
				U. Science and Technology of China	650
				U. Seoul, Korea	nk
				U. Twente, Netherlands	1

				U. Western Ontario, Canada	nk
				U. Zaragoza, Spain	nk
				USDA, ARS, ERRC, USA	1
				Virginia Tech. U., USA	0.1
				VTT, Finland	1
				vTI, Germany	6
				Zhejiang U., China	3
				Zhengzhou U., China	2
Spouted fluid bed	Ikerlan, Spain	1	10	Anhui U. of Science & Technology, China	5
				U. Basque Country, Spain	nk
Transported bed & CFB	Ensyn, Canada	8	4,000	CPERI, Greece	1
	Metso/UPM, Finland	1	400	Guangzhou Inst. Energy Conversion, China	nk
				U. Birmingham, UK	nk
				U. Nottingham, UK	nk
				VTT, Finland	20
Rotating cone	BTG, Netherlands	4	2,000	BTG, Netherlands	10
Integral catalytic pyrolysis	BioEcon, Netherlands + Kior USA	nk	nk	Battelle Columbus, USA	1
				PNNL, USA	1
				Technical U. of Munich	nk
				U. Massachusetts-Amhurst, USA	nk
				Virginia Tech. U., USA	3?
Vortex				TNO, Netherlands	30
Centrifuge reactor				Technical U. Denmark	nk
Ablative	PyTec, Germany	2	250	Aston U., UK	20
				Institute of Engineering Thermophysics, Ukraine	15
				Latvian State Institute, Latvia	0.15
				Technical U. Denmark	1.5
Augur or Screw	Abritech, Canada	4	2083	Auburn U. USA	1.0
	Lurgi LR, Germany	1	500	KIT (FZK), Germany	500
	Renewable Oil Intl, USA	4	200	Mississippi State U., USA	2
				Michigan State U. USA	0.5
				Texas A&M U., USA	30

Table 7.2 *(continued)*

Fast pyrolysis	*Industrial*	*Units built*	*Max size kg/h*	*Research*	*Max size kg/h*
Radiative-Convective				CNRS – Nancy U., France	nk
Entrained flow				Dalian U. of Technology, China	nk
				Institute for Wood Chemistry, Latvia	nk
				Shandong U. of Technology, China	0.05
Microwave	Carbonscape New Zealand & UK	nk	nk	Chinese Academy of Sciences, Dalian, China	nk
	Bioenergy 2020+ gmbh, Austria	1	nk	National Inst. Advanced Industrial Sci. & Technol., Japan	<0.1
				Shandong U. China	<0.1
				Technical U. Vienna, Austria	nk
				U. Malaysia Sarawak	<0.1
				U. Minnesota, USA	10
				U. Mississippi	nk
				U. Nottingham, UK and China	nk
				U. York, UK	nk
				Washington State U.-Tricities, USA	<1
Moving bed and fixed bed	Anhui Yineng Bio-energy Ltd., China	3	600	Anadolu University, Turkey	nk
				U. Autònoma de Barcelona, Spain	nk
				U. Science & Technology of China	~0.5
Ceramic ball downflow				Shandong University of Technology, China	110
Unspecified				U. Kentucky, USA	nk
				U. Texas, USA	nk
				Technical U. Compiegne, France	nk
Vacuum	Pyrovac, Canada	1	3500	None known	

nk = not known

7.3.5 Char Removal

Char acts as a vapor-cracking catalyst, so rapid and effective separation from the pyrolysis product vapors is essential. Cyclones are the usual method of char removal, however, some fines always pass through the cyclones and collect in the liquid product where they accelerate aging and exacerbate the instability problem that is described below.

Pressure filtration of the liquid for substantial removal of particulates (down to $<5\,\mu m$) is reportedly very difficult because of the complex interaction of the char and pyrolytic lignin, which appears to form a gel-like phase that rapidly blocks the filter. Modification of the liquid microstructure by addition of solvents such as methanol or ethanol that solubilize the less-soluble constituents can improve this problem and contribute to improvements in liquid stability, as described below.

7.3.6 Liquids Collection

The gaseous products from fast pyrolysis consist of aerosols, true vapors and noncondensable gases. These require rapid cooling to minimize secondary reactions and condense the true vapors, while the aerosols require additional coalescence or agglomeration. Simple indirect heat exchange can cause preferential deposition of lignin-derived components leading to liquid fractionation and eventually blockage in pipelines and heat exchanges. Quenching in product bio-oil or in an immiscible hydrocarbon solvent is widely practised.

Orthodox aerosol capture devices such as demisters and other commonly used impingement devices are not very effective and electrostatic precipitation is currently the preferred method at both laboratory- and commercial-scale units. The vapor product from fluid-bed and transported-bed reactors has a low partial pressure of collectible products due to the large volumes of fluidizing gas, and this is an important design consideration in liquid collection. This disadvantage is reduced in the rotating-cone and ablative reaction systems, both of which exclude inert gas that leads to more compact equipment and lower costs.[31]

7.3.7 Byproducts

Char and gas are byproducts, typically containing about 25 and 5% of the energy in the feed material, respectively. The pyrolysis process itself requires about 15% of the energy in the feed, and of the byproducts, only the char has sufficient energy to provide this heat. The heat can be derived by burning char in orthodox reaction system designs, which makes the process energy self-sufficient. More advanced configurations could gasify the char to a LCV gas and then burn the resultant gas more effectively to provide process heat with the advantage that the alkali metals in the char can be much better controlled and avoid potential slagging problems from direct char combustion.

The waste heat from char combustion and any heat from surplus gas or byproduct gas can be used for feed drying and in large installations could be used for export or power generation. An important principle of fast pyrolysis is that there are no waste products from a well-designed and well-run process other than clean flue gas and ash.

7.4 Pyrolysis Liquid – Bio-Oil

Crude pyrolysis liquid or bio-oil is dark brown and approximates to biomass in elemental composition. It is composed of a very complex mixture of oxygenated hydrocarbons with an appreciable proportion of water from both the original moisture and reaction product. Solid char may also be present. Typical organics yields from different feedstocks and their variation with temperature is shown in Figure 7.10 and Figure 7.11 shows the temperature dependence of the four main products from a variety of feedstocks.[32] Similar results are obtained for most biomass feedstocks, although the maximum yield can occur between 480 and 520 °C depending on feedstock. Grasses, for example, tend to give maximum liquid yields of around 55–60 wt% on a dry feed basis at the lower end of this temperature range, depending on the ash content of the grass.

The liquid is formed by rapidly quenching and thus "freezing" the intermediate products of flash degradation of hemicellulose, cellulose and lignin. The liquid thus contains many reactive species, which contribute to its unusual attributes. Bio-oil can be considered a microemulsion in which the continuous phase is an aqueous solution of holocellulose decomposition products, that stabilizes the discontinuous phase of pyrolytic lignin macro-molecules through mechanisms such as hydrogen bonding. Aging or instability is believed to result

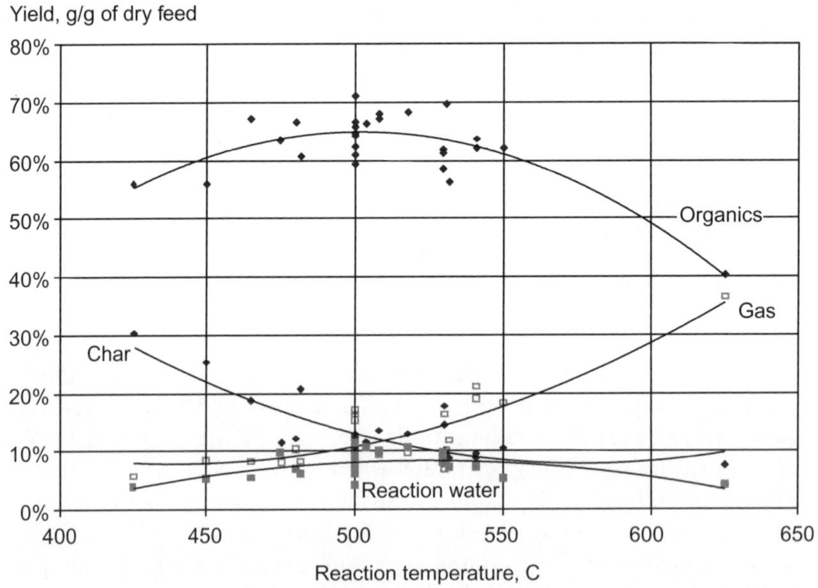

Figure 7.10 Variation of products from Aspen Poplar with temperature.[32]

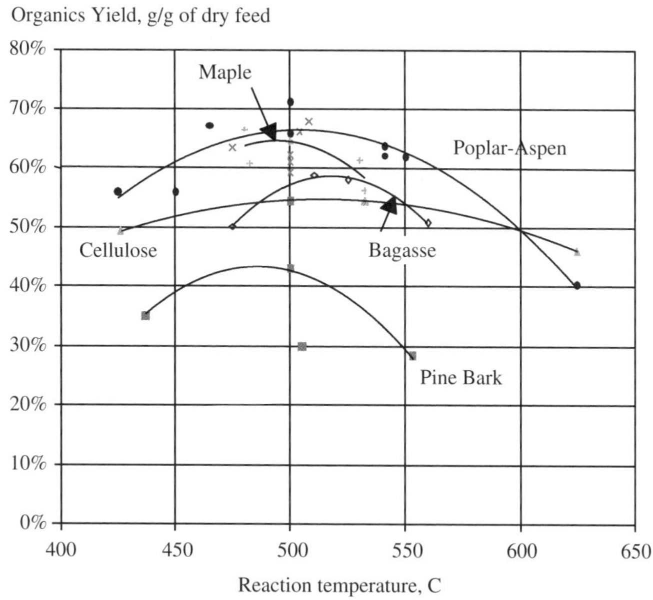

Figure 7.11 Organics yield from different feedstocks.[32]

from a breakdown in this emulsion. In some ways it is analogous to asphaltenes found in petroleum.

7.4.1 Bio-Oil Characteristics

Fast pyrolysis liquid has a higher heating value of about 17 MJ/kg as produced with about 25 wt% water that cannot readily be separated. While the liquid is widely referred to as "bio-oil", it will not mix with any hydrocarbon liquids nor can it be distilled. It is composed of a complex mixture of oxygenated compounds that provide both the potential and challenge for utilization. There are some important characteristics of this liquid that are summarized in Table 7.3 and discussed briefly below.

Pyrolysis oil typically is a dark brown, free-flowing liquid. Depending on the initial feedstock and the mode of fast pyrolysis, the color can be almost black through dark red-brown to dark green, being influenced by the presence of microcarbon in the liquid and chemical composition. Hot vapor filtration gives a more translucent red-brown appearance owing to the absence of char. High nitrogen content can impart a dark green tinge to the liquid.

There are many particular characteristics of bio-oil that require consideration for any heat, power or biofuel application. These are summarized in Table 7.4 with effects, solutions and relevance to heat, power and biofuel applications.

The liquid has a distinctive odor, an acrid smoky smell due to the low molecular weight aldehydes and acids, which can irritate the eyes on prolonged exposure. The liquid contains several hundred different chemicals in widely

Table 7.3 Typical properties of wood-derived crude bio-oil.

Physical property		*Typical value*
Moisture content		25%
pH		2.5
Specific gravity		1.20
Elemental analysis	C	56%
	H	6%
	O	38%
	N	0–0.1%
HHV as produced		17 MJ/kg
Viscosity (40 °C and 25% water)		40–100 cp
Solids (char)		0.1%
Vacuum-distillation residue		up to 50%

Characteristics
Liquid fuel
Ready substitution for conventional fuels in many stationary applications such as boilers, engines, turbines
Heating value of 17 MJ/kg at 25 wt% water, is about 40% that of fuel oil/diesel
Does not mix with hydrocarbon fuels
Not as stable as fossil fuels
Quality needs definition for each application

varying proportions, ranging from formaldehyde and acetic acid to complex high molecular weight phenols, anhydrosugars and other oligosaccharides.

The liquid contains varying quantities of water, which forms a stable single-phase mixture, ranging from about 15 wt% to an upper limit of about 30–50 wt% water, depending on the feed material, how it was produced and subsequently collected. A typical feed material specification is a maximum of 10% moisture in the dried feed material, as both this feed moisture and the water of reaction from pyrolysis, typically about 12% based on dry feed, both report to the liquid product. Pyrolysis liquids can tolerate the addition of some water, but there is a limit to the amount of water that can be added to the liquid before phase separation occurs, in other words the liquid cannot be dissolved in water. Water addition reduces viscosity, which is useful; reduces the heating value, which means that more liquid is required to meet a given duty; and can improve stability. The effect of water is therefore complex and important. It is miscible with polar solvents such as methanol, acetone, *etc.* but totally immiscible with petroleum-derived fuels.

The density of the liquid is very high at around 1.2 kg/l, compared with light fuel oil at around 0.85 kg/l. This means that the liquid has about 42% of the energy content of fuel oil on a weight basis, but 61% on a volumetric basis. This has implications for the design and specification of equipment such as pumps and atomizers in boilers and engines.

Viscosity is important in many fuel applications.[34] The viscosity of the bio-oil as produced can vary from as low as 25 cSt to as high as 1000 cSt (measured at 40 °C) or more depending on the feedstock, the water content of the oil, the amount of light ends collected and the extent to which the oil has aged.

Pyrolysis liquids cannot be completely vaporized once they have been recovered from the vapor phase. If the liquid is heated to 100 °C or more to try to remove water or distil off lighter fractions, it rapidly reacts and eventually produces a solid residue of around 50 wt% of the original liquid and some distillate containing volatile organic compounds and water. While bio-oil has been successfully stored for several years in normal storage conditions in steel and plastic drums without any deterioration that would prevent its use in any of the applications tested to date, it does change slowly with time, most noticeably there is a gradual increase in viscosity. More recent samples that have been distributed for testing have shown substantial improvements in consistency and stability, demonstrating the improvement in process design and control as the technology develops.

7.4.2 Environment, Health and Safety

As bio-oil becomes more widely available, attention will be increasingly placed on environment, health and safety aspects. A study was completed in 2005 to assess the ecotoxicity and toxicity of 21 bio-oils from most commercial producers of bio-oil around the world in a screening study, with a complete assessment of a representative bio-oil. The study includes a comprehensive evaluation of transportation requirements as an update of an earlier study[35] and an assessment of the biodegradability.[36] The results are complex and require more comprehensive analysis but the overall conclusion is that bio-oil offers no significant health, environment or safety risks.

7.5 Bio-Oil Upgrading

Bio-oil can be upgraded in a number of ways – physically, chemically and catalytically. This has been extensively reviewed[37–41] and only the more significant features and recent developments are reported here.

7.5.1 Physical Upgrading of Bio-Oil

The most important properties that may adversely affect bio-oil fuel quality are incompatibility with conventional fuels from the high oxygen content of the bio-oil, high solids content, high viscosity, and chemical instability.

7.5.1.1 Filtration

Hot-vapor filtration can reduce the ash content of the oil to less than 0.01% and the alkali content to less than 10 ppm, much lower than reported for biomass oils produced in systems using only cyclones. This gives a higher quality product with lower char,[41] however the liquid yield is reduced by about 10–20% due to the char accumulating on the filter surface that cracks the vapors. There is limited information available on the performance or operation of hot vapor filters, but they can be specified and perform similar to hot gas filters in gasification processes.

Table 7.4 Characteristics of bio-oil.[33]

Characteristic	*Effect*	*Solution*	*Importance for heat and power*	*Importance for biofuels*
Suspended char	Erosion; Equipment blockage, High CO emissions in combustion	Filter vapor Filter liquid Modify char Modify application	High	High – catalyst blockage
Alkali metals	Catalyst poisoning Deposition of solids in combustion Erosion and corrosion Slag formation Damage to turbines	Pretreat feed to remove ash Hot vapor filtration Process oil Modify application	High – causes deposits, but most alkali metal reports to the char so need not be a problem	High – catalyst poisoning
Inhomogeneity	Phase separation, Partial phase separation, Layering; Poor mixing, Inconsistency in handling, storage and processing	Modify or change process Modify or change feedstock Change feedstock to low lignin Add cosolvents Control water content	Very high	Very high
Trace contaminants *e.g.* S, Cl	Catalyst poisoning	Include suitable cleaning processes	Low	High
Low pH	Corrosion of vessels and pipework	Careful materials selection such as polyolefins or stainless steel	Low	Low

Temperature sensitivity	Irreversible decomposition of liquid into two phases $>100\,^{\circ}C$ Potential phase separation $>60\,^{\circ}C$	Store liquids at room temperature preferably in absence of air	Low if controlled	Medium
Materials incompatibility	Destruction of seals and gaskets	Careful materials selection	Low – choose materials	Low– choose materials
High viscosity	Gives high pressure drop increasing equipment cost High pumping cost Poor atomization	Careful heating up to $50\,^{\circ}C$, rapid in-line heating to $80\,^{\circ}C$ is also possible Add water Add cosolvents	Medium for atomization	Low
Aging	Slow increase in viscosity from secondary reactions such as condensation Potential phase separation	Do not store for long periods Avoid exposure to air Add water Add cosolvents	Low	Low
Water content	Complex effect on viscosity and stability: Increased water lowers heating value, density, stability, raises pH	Optimize at 25% for consistency and miscibility Optimize for application NB. Water removal is very difficult	Low, as long as it is controlled	Depends on catalyst

Diesel-engine tests performed on crude and on hot-filtered oil showed a substantial increase in burning rate and a lower ignition delay for the latter, due to the lower average molecular weight for the filtered oil.[42] Hot gas filtration has not yet been demonstrated over a long-term process operation. A consequence of hot vapor filtration to remove char is the catalytic effect of the accumulated char on the filter surface, which potentially cracks the vapors, reduces yield by up to 20%, reduces viscosity and lowers the average molecular weight of the liquid product. Little work has been done in this area by NREL,[41] VTT and Aston University,[43] and very little has been published.

Liquid filtration to very low particle sizes of below around 5 μm is very difficult due to the physicochemical nature of the liquid and usually requires very high pressure drops and self-cleaning filters.

7.5.1.2 Emulsions

Pyrolysis oils are not miscible with hydrocarbon fuels but they can be emulsified with diesel oil with the aid of surfactants. A process for producing stable microemulsions with 5–30% of bio-oil in diesel has been developed at CANMET.[44] The University of Florence, Italy, has been working on emulsions of 5 to 95% bio-oil in diesel[45,46] to make either a transport fuel or a fuel for power generation in engines that does not require engine modification to dual-fuel operation. There is limited experience of using such fuels in engines or burners, but significantly higher levels of corrosion/erosion were observed in engine applications compared to bio-oil or diesel alone. A further drawback of this approach is the cost of surfactants and the high energy required for emulsification.

Polar solvents have been used for many years to homogenize and reduce the viscosity of biomass oils. The addition of solvents, especially methanol, showed a significant effect on the oil stability. Diebold and Czernik[47] found that the rate of viscosity increase ("aging") for the oil with 10 wt% of methanol was almost twenty times less than for the oil without additives.

7.5.2 Chemical and Catalytic Upgrading of Bio-Oil

Upgrading bio-oil to a conventional transport fuel such as diesel, gasoline and kerosene requires full deoxygenation, which can be accomplished by three main routes, as summarized below. There is also interest in partial upgrading to a product that is compatible with refinery streams in order to take advantage of the economy of scale and experience in a conventional refinery.

- Hydrotreating;
- catalytic vapor cracking;
- gasification to syngas followed by synthesis to hydrocarbons or alcohols.

Chemical and catalytic methods for upgrading bio-oil by hydrotreating and zeolite cracking have been reviewed.[5,39,40] There are three main routes to biofuels, as summarized in Figure 7.12: Gasification followed by hydrocarbon or

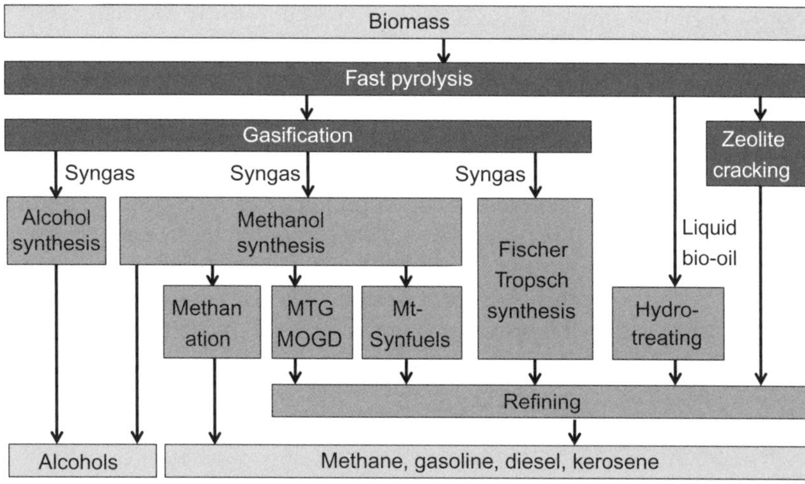

Figure 7.12 Upgrading of bio-oil to biofuels and chemicals.

alcohol synthesis; zeolite cracking of pyrolysis vapors; and hydrotreating or hydroprocessing of liquid bio-oil.

7.5.2.1 Hydrotreating

Hydroprocessing rejects oxygen as water by catalytic reaction with hydrogen. This is a separate process to fast pyrolysis that can therefore be carried out remotely. The process is high pressure (up to 200 bars) and moderate temperature (up to 400 °C) and requires a hydrogen supply or source. The projected yield is around 25% weight of a naphtha-like product on a dry biomass basis or 55% in energy terms,[40] which does not consider the hydrogen requirement. If hydrogen is produced from biomass, the overall yield drops to around 15%. Full hydrotreating gives a naphtha-like product that requires orthodox refining to derive conventional transport fuels, probably in a conventional refinery to take advantage of know-how and existing processes. The process can be depicted by the following conceptual reaction:

$$C_1H_{1.33}O_{0.43} + 0.77H_2 \rightarrow CH_2 + 0.43\mathbf{H_2O} \qquad (7.1)$$

The process is carried out at high temperature, high hydrogen partial pressure, and in the presence of catalysts.[48] This results in elimination of oxygen as water and in hydrogenation-hydrocracking of larger molecules. The catalysts originally tested were based on sulfided CoMo or NiMo supported on alumina and the process conditions are similar to those used in the desulfurization of petroleum fractions. However, the catalyst supports of typically aluminosilicates were found to be unstable in the high water content environment of bio-oil and some progress has been made on the use of more-resistant catalyst supports. This area has been thoroughly reviewed.[49] More recently UOP has been working with Pacific Northwest National Laboratory in the USA to address the scientific and technical challenges and develop a cost-effective process.[50]

There is a substantial hydrogen requirement to both hydrogenate the organic constituents of bio-oil and remove the oxygen as water. The hydrogen requirement can be represented by a requirement to process an additional amount of biomass equal to about 80% of that required to produce the bio-oil. The process is thus less inefficient than simple yield figures presented. If only the organic fraction of bio-oil is hydroprocessed by phase separation, the hydrogen required can be produced by steam reforming the aqueous phase. There is also a high cost from the high-pressure requirement.[51]

7.5.2.2 Zeolite Cracking

Zeolite cracking rejects oxygen as CO_2, as summarized in the conceptual reaction below:

$$C_1H_{1.33}O_{0.43} + 0.13O_2 \rightarrow 0.66CH_2 + 0.34\mathbf{CO_2} \qquad (7.2)$$

This is a close-coupled process carried out on the primary vapors from fast pyrolysis and can thus be integrated into the fast-pyrolysis process, thereby obviating the need for a separate upgrading process and need for high pressure.

Catalytic vapor cracking over acidic zeolite catalysts provides deoxygenation by simultaneous dehydration-decarboxylation producing mostly aromatics[52] at 450 °C and atmospheric pressure. Oxygen is ultimately rejected as CO_2 and CO from a secondary oxidizing reactor to burn off the coke deposited on the catalyst. This would operate much as an FCC unit in a refinery. The low H/C ratio in the bio-oils imposes a relatively low limit on the hydrocarbon yield and, in addition, the technical feasibility is not yet proven. A projected typical yield of aromatics suitable for gasoline blending from biomass is about 20% by weight or 45% in energy terms.[53] The crude aromatic product would be sent for refining in a conventional refinery.

A key feature of this route is the absence of a hydrogen requirement and operation at atmospheric pressure. Some of the earlier work was carried out at NREL in a close coupled reactor with their ablative fast-pyrolysis process.[54] The process was modelled technically and economically by an IEA Bioenergy Task.[53] Extensive upgrading work on zeolite cracking of pyrolysis vapors formed by revaporization of bio-oil has been performed by Bahkshi *et al.*[55]

Catalyst deactivation remains a concern for both routes, although the coking problem with zeolites can, in principle, be overcome by a conventional FCC arrangement with continuous catalyst regeneration by oxidation of the coke. Some concern has been expressed over the poor control of molecular size and shape with orthodox zeolites and the propensity for formation of more noxious hydrocarbons.[56] The processing costs are high and the products are not competitive with fossil fuels.[57] This approach has only been studied at the basic research level and considerably more development is necessary.

7.5.2.3 Gasification for Synfuels

A recent concept that has attracted much interest in Europe is the decentralized production of bio-oil and bio-oil-char slurries for transportation to a central process plant for gasification and synthesis of hydrocarbon transport fuels, by, for example, Fischer–Tropsch synthesis. This is depicted in Figure 7.12. Although there is a small energy penalty from the lower pyrolysis energy efficiency, transportation energy and additional bio-oil gasification stage, this is more than compensated by the economies of scale achievable on a commercial-sized gasification and transport fuel synthesis plant.[58]

Although the concept of very large gasification plants of 5 GWe or more has been promoted[59] based on importation of biomass on a massive scale to an integrated plant, based for example at Rotterdam, there are significant obstacles to be overcome. Decentralized fast pyrolysis plants of 100 000 t/y or 12 t/h are currently feasible and close to being commercially realized. Bio-oil gasification in an entrained-flow oxygen blown pressurized gasifier is also feasible based on a Texaco or Shell system, with the added advantage that feeding a liquid at pressure is easier than solid biomass, and the gas quality under such conditions is likely to be higher than from solid biomass. Finally, transport fuel synthesis at 50 000 to 200 000 t/d is also commercially realizable from the extensive gas-to-liquid plants currently operational around the world.

Synthetic hydrocarbons include diesel, gasoline, kerosene and methane (synthetic natural gas – SNG). These liquid fuels are entirely compatible with conventional fuels in all proportions, but are much cleaner. At least in the medium term, these are likely to be the biofuel of choice due to the ease of assimilation into the market place. It is also important that consumers have confidence in transport fuels, and conventional but synthetic hydrocarbon fuels are likely to cause the least consumer resistance, particularly since there are well-established and well-defined quality requirements that would normally be applied in existing refineries if these are used for finishing products prior to marketing.

7.5.2.4 Hydrogen

As referred to above, the water-soluble (carbohydrate-derived) fraction of bio-oil can be processed to hydrogen by steam reforming.[60,61] This has been accomplished in a fluidized-bed process by several researchers using commercial, nickel-based catalysts under conditions similar to those for reforming natural gas. The process depends on a viable use for the organic lignin-derived fraction of bio-oil, for example, use as a phenol replacement in phenol–formaldehyde resins[62] or for upgrading this organic fraction into, for example, hydrocarbons or phenolic products.

7.6 Applications of Bio-Oil

Bio-oil can substitute for fuel oil or diesel in many static applications including boilers, furnaces, engines and turbines for electricity generation. Figure 7.13 summarizes the possibilities.

7.6.1 Energy Carrier

Biomass is a widely dispersed resource that has to be harvested, collected and transported to the conversion facility. The low bulk density of biomass, which can be as low as 150 kg/m^3, means that transport costs are high and the number of vehicle movements for transportation to a large-scale processing facility are also very high, with consequent substantial environmental impacts.

Conversion of biomass to a liquid by fast pyrolysis at or near the biomass source will reduce transport costs and reduce environmental concerns as the liquid has a density of 1.2 kg/m^3 – eight times higher than lower-density crops and residues. This not only reduces the number of vehicle movements and costs by up to 87%, it also reduces costs of handling and transportation by virtue of it being a liquid that can be pumped. It is also possible to consider mixing the by-poduct char with the bio-oil to make a slurry to improve the energy content of the product, but the pyrolysis process will then require that its process energy needs are met from another source. This leads to the concept shown in Figure 7.14, with small decentralized fast pyrolysis plants of 100 000 to 300 000 t/y for production of liquids to be transported to a central processing

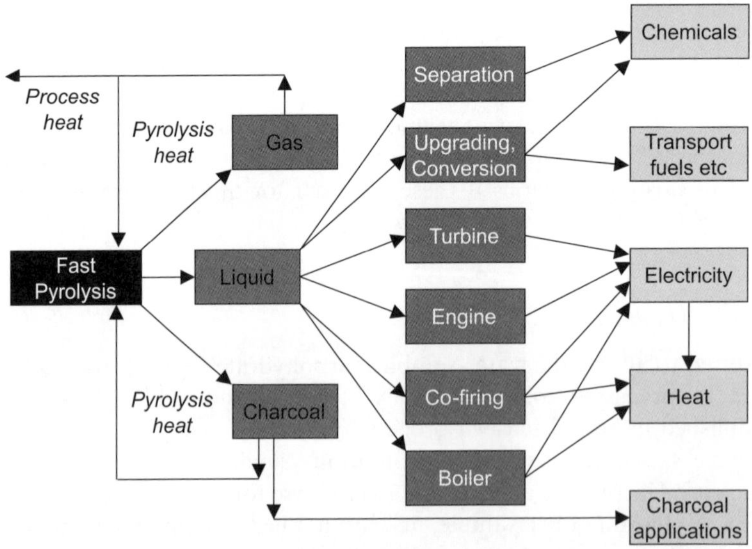

Figure 7.13 Applications for fast-pyrolysis products.

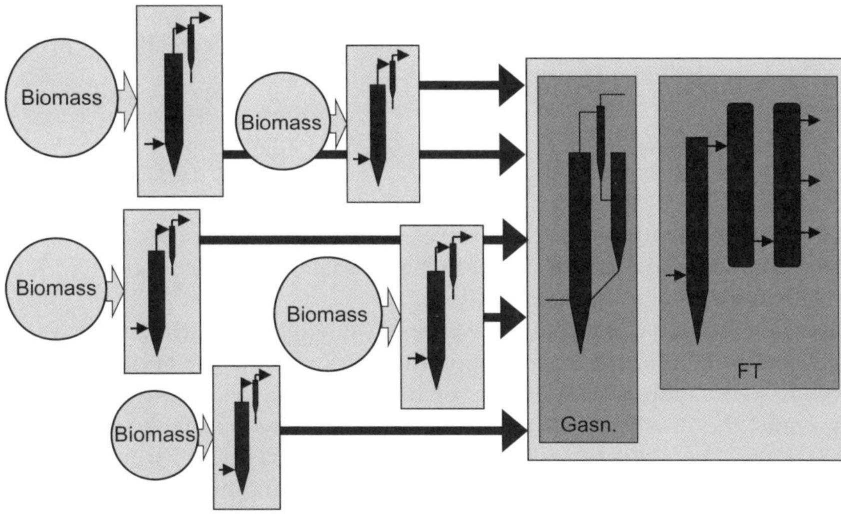

Figure 7.14 Decentralized fast pyrolysis with centralized gasification hydrocarbon synthesis.

Table 7.5 Comparison of solid biomass gasification vs pyrolysis liquid gasification.

Advantages	*Disadvantages*
Capital cost reduction of ~10% due to lower raw material handling costs	Capital cost increase of ~10% due to diseconomies of scale in small pyrolysis plants
Capital cost reduction of ~10% due to lower gasification costs in feeding a liquid at pressure compared to solid biomass	Efficiency loss of ~6–7% due to additional processing step
Product gas requires less cleaning giving capital cost reduction of ~5%	

plant – depicted here as a gasification + Fischer–Tropsch or methanol synthesis plant.

Adoption of decentralized fast pyrolysis with transportation of the resultant liquid to a central gasification and fuel synthesis plant has both advantages and disadvantages as summarized in Table 7.5.

7.6.2 Combustion

Bio-oils are combustible but not flammable. Bio-oil requires significant energy for ignition due to the high level of nonvolatile and nonflammable constituents, but once ignited, it burns with a stable self-sustaining flame. An extensive study on the fundamentals of bio-oil combustion of single droplets was carried out at Sandia National Laboratory.[63,64] This showed a unique multistep process consisting of the following stages:

- ignition;
- quiescent burning;
- droplet microexplosion;
- disruptive sooty burning of droplet fragments (bright yellow);
- formation and burnout of cenosphere particles.

Under the same conditions, conventional fossil-fuel oil demonstrated only quiescent, sooty burning from ignition through burnout. The burning times of bio-oils were comparable to those of No. 2 fuel oil under the same conditions and the adiabatic flame temperature for bio-oil is relatively high at 1400–1700 °C compared to 1900–2000 °C for orthodox fuels. This is due to the significantly lower stoichiometric air-to-fuel ratio for bio-oil (about 7 compared to 14 for orthodox fuels).

Most research on bio-oil combustion in boilers has been carried out in Finland, particularly by Neste Oy,[65] in a 2.5-MW Danstoker boiler supplied with a dual fuel burner and by VTT Energy in collaboration with Oilon Oy[66] in an 8-MWth nominal capacity furnace operated at 4 MWth output. The main findings of these tests were:

- Some modifications of the burner and boiler sections were required.
- The flame from bio-oil combustion was longer than that from burning standard fuel oil.
- A support fuel was necessary during start up and usually also with lower quality bio-oils.

There were clear differences in combustion behavior and emissions for different bio-oils tested; lower-quality oils with higher viscosity, water and solids content showed significantly worse performance. Emissions were generally lower than from burning heavy fuel oil except for particulates. They strongly depended on the proper handling of bio-oil (optimum was preheating to 50 °C and compressed air atomization) and its quality, for example, higher water content leads to lower NO_x but higher particulates in flue gases. Other test work in North America and Europe, summarized in ref. 38, confirmed the similarity in combustion behavior with conventional fuel oils.

The only commercial system that regularly uses bio-oil to generate heat is at the Red Arrow Products pyrolysis plant in Wisconsin,[67] which has been operated for over ten years. The 5-MWth swirl burner used different mixtures of byproducts, water-insoluble fraction of bio-oil (pyrolytic lignin), char and gas from the food-flavorings plant.

In conclusion, a constant and better quality bio-oil available at attractive price is necessary for commercial, large-scale applications. Storage, pumping, filtration, atomization and optimization of the burner/boiler design to improve performances and reduce emissions only require relatively minor modifications to the existing equipment. Bio-oil is a suitable boiler fuel as long as it has consistent characteristics, provides acceptable emissions level, and is economically feasible.

7.6.3 Cofiring

Coprocessing of biomass with conventional fuels is potentially a very attractive option that enables full economies of scale to be realized as well as reducing the problems of product quality and clean up.

Most current cofiring applications are those where the biomass fuels are added to the coal feed and this is widely practised at up to 5% on the energy demand of the power station. A few applications involve conversion to a fuel gas *via* gasification followed by close coupled firing to the power station boiler. There are a few examples of cofiring fast pyrolysis liquids including a coal-fired power station at Manitowac in the USA using the byproduct from liquid-smoke production[67] at the Red Arrow plant.[68] The combustion of bio-oil was clean and efficient with no adverse changes on the boiler operation or on the emission levels. There is also some limited experience of successfully firing bio-oil in a natural-gas-fired power station in the Netherlands.

At present, cofiring offers the best opportunities for market penetration of biomass as the overall costs are relatively low because the power cycle in the coal-fired power plant is already there. In addition, cofiring has the advantage over cocombustion (where the biomass fuels are mixed with coal before or during the combustion process), in that the biomass residual ash is not mixed with the coal ash, which has an existing market as a construction material and can thus be recycled back to the forest or field. Also, the technical risks are low as the gas is utilized hot and therefore there is no tar problem. In reburning applications, (where the fuel gas is introduced almost at the top of the coal boiler), the environmental performance of the power station is significantly improved, in addition to the replacement of fossil fuels by renewable biomass fuels.[69]

7.6.4 Engines and Turbines

Diesel engines offer a high efficiency (up to 45%) in power generation and can also be adapted to combined heat and power systems (CHP). Medium- and slow-speed engines are known for fuel flexibility and can operate on low-grade fuels. The main concerns for operating diesel engines on bio-oils are some specific properties of these liquids, such as difficult ignition (resulting from low heating value and high water content), corrosiveness (acids), and coking (thermally unstable compounds). Injection with a pilot fuel has been found advantageous in lower compression diesels. Turbines are more efficient at higher capacities but modern microturbines offer improved performance. The potential advantages of using bio-oils for power generation have led to a range of research activities that have been reported in ref. 5.

7.6.5 Chemicals

For many centuries, wood pyrolysis liquids were a major source of chemicals such as methanol, acetic acid, turpentine, tars. At present, most of these compounds can be produced at a lower cost from fossil-fuel feedstocks. Although over 300 compounds have been identified in wood fast-pyrolysis oil, their concentrations are usually too low to consider separation and recovery. Up to now therefore, technologies for producing products from the whole bio-oil or from its major, relatively easy separable fractions are the most developed. A more detailed review on this subject, including consideration of higher-value products, was published by Radlein[70] and a thorough review of the literature on production of chemicals utilizing whole oil, fractions of bio-oil and specific chemicals was published in 2004.[38] Some of the chemical products that have been reported are shown in Table 7.6.

Chemicals are always attractive commercial possibilities owing to their much higher added value compared to fuels and energy products. This suggests a biorefinery concept in which the optimum combinations of fuels and chemicals are produced and this is discussed below.

7.6.5.1 Chemicals in Bio-Oil

The chemicals in bio-oil are derived from random thermal decomposition of hemicellulose, cellulose and lignin. Over 400 individual chemicals have been identified in bio-oil and the numbers of chemicals by group regularly identified in the literature are shown in Table 7.7.

A more comprehensive list was also published by Diebold,[37] of which the most abundant are listed in Table 7.8.

Strategies for separation or recovery of any of these chemicals need to consider market, value, costs and process. The natural first step is to evaluate components with the highest concentration since the processing will be easier and the costs lower. However, this may not prove the best strategy, and methodologies need to carefully consider capital and operating costs, product values and residue or waste utilization or disposal. The opportunities for optimization are considerable and challenging, and are likely to involve a range of process generation, evaluation and optimization tools including process synthesis and linear programming.

Table 7.6 Some chemicals produced from fast pyrolysis.

Acetic acid	Adhesives	Calcium-enriched bio-oil	Deicer
Food flavorings	Hydrogen	Hydroxyacetaldehyde	Levoglucosan
Levoglucosenone	Maltol	Phenol	Polyphenols
Preservatives	Resins	Slow-release fertilizers	Sugars

Table 7.7 Numbers of chemicals by group regularly identified in the literature.[37]

Acids	12
Alcohols	4
Aldehydes	5
Esters	1
Hydrocarbons	6
Ketones	32
Oxygenates	16
Phenolics	56
Steroids	15
Sugars	8
TOTAL	**162**

7.6.5.2 Other Potential Products

There are a wide range of potential products from a biorefinery. Just as the petroleum and petrochemical industry produce a vast array of industrial and consumer products, so a biorefinery could eventually meet all our needs, since technology is not the limiting factor but economics. A few examples of potential products are given below to illustrate the diversity of opportunities:

- furfural from xylose for tetrahydrofuran and butanediol for engineering elastomers;
- 5-hydroxymethylfurfural from glucose for furan-di-carboxylic acid for engineering polymers;
- levulinic acid for levulinic-acid-based esters as "green" solvents and fuel additives; levulinic-acid-based lactones for polymers and nylons;
- phenol from lignin for phenol esters and ethers for surfactants and performance fluids; substituted phenols as antioxidants and fire retardant agents for plastics;
- thermoplastic polymers such as tetrahydrofuran-based high-performance resins synthesis and processing; polylacton-based segmented block copolymer resins synthesis and processing; furan-di-carboxylic-acid-based thermoplastic copolyesters and copolyamides;
- thermoset polymers including phenolic resins and epoxy resins synthesis and processing.

7.6.6 Biorefinery

The large majority of chemicals are manufactured from petroleum feedstocks. Only a small proportion of the total oil production, around 5%, is used in chemical manufacture but the value of these chemicals is high and contributes a comparable revenue as fuel and energy products. There is a clear economic

Table 7.8 Chemicals ranked by maximum reported yield, wt%.[37]

Bio-oils wt%	*Compound*	*Group*
13.0	Hydroxyacetaldehyde	Misc oxygenates
12.0	Acetic (ethanoic)	Acid
9.1	Formic (methanoic)	Acid
8.5	Acetaldehyde	Aldehyde
7.4	Acetol (hydroxyacetone)	Misc oxygenates
7.2	Isoeugenol	Guaiacol
5.2	Furfural alcohol	Furan
4.8	2,6-DiOMe phenol	Syringol
4.6	Ethanedial	Aldehyde
3.8	Phenol	Phenol
3.3	Formaldehyde	Aldehyde
3.2	Cellobiosan	Sugar
3.1	1,6-Anhydroglucofuranose	Sugar
2.9	Fructose	Sugar
2.8	Acetone	Ketone
2.4	Methanol	Alcohol
2.3	Eugenol	Guaiacol
2.2	5-OH-methyl-2-furfural	Furan
2.0	Ethylene glycol	Alcohol
1.9	1,4-DiOH benzene	Phenol
1.9	4-Methyl guaiacol	Guaiacol
1.9	Methyl cyclopentenolone	Misc oxygenates
1.8	Propanoic	Acid
1.5	Propyl syringol	Syringol
1.5	Syringaldehyde	Syringol
1.4	Ethanol	Alcohol
1.4	Levoglucosan	Sugar
1.4	D-xylose	Sugar
1.3	2-Ethylphenol	Phenol
1.3	Glucose	Sugar
1.2	Angelicalactone	Ester
1.1	2-Methoxy phenol	Guaiacol
1.1	2-Furanone	Furan
1.1	Furfural	Furan
1.1	4-OH-3-methoxybenzaldehyde	Misc. oxygenates

advantage in building a similar flexibility into the biofuels market by devoting part of the biomass production to the manufacture of chemicals. In fact, this concept makes even more sense in the context of biomass because it is chemically more heterogeneous than crude oil and conversion to fuels, particularly hydrocarbons, is not so cost effective. Figure 7.15 shows fast pyrolysis at the heart of a biorefinery.

As the prices of oil and coal rise, there will be an increasing incentive to find alternative sources of energy and chemicals. However, there is an important difference between energy and chemicals. Biomass is just one possible energy source that must compete with renewables (wind, solar, tidal) or nuclear but, for chemicals, there are no viable alternative sources of carbon apart from

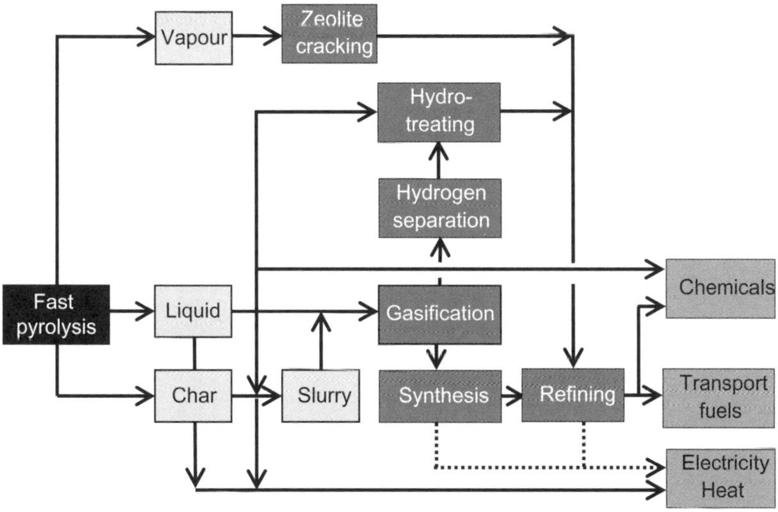

Figure 7.15 Fast-pyrolysis-based biorefinery.

biomass. Therefore, it will become increasingly attractive to produce chemicals, including hydrocarbon fuels, from biomass and, unlike energy, there is likely to be a long-term and sustainable market for technologies for deriving chemicals from biomass.

Since the empirical chemical composition of biomass, approximately $(CH_2O)_n$, is quite different from that of oil $(CH_2)_n$, the range of primary chemicals that can be easily derived from biomass and oil are quite different. Hence, any biomass-based chemical industry will necessarily be constructed on quite a different selection of simple "platform" chemicals than those currently used in the petrochemical industry. Given the chemical complexity of biomass, there is some choice of which platform chemicals to produce since, within limits, different processing strategies of the same biomass can lead to different breakdown products. However, once a set of platform chemicals has been chosen for biobased chemical production and the appropriate network of production plants is established, it will become increasingly difficult to change that choice without disrupting the whole manufacturing infrastructure. Since the available biomass will inevitably show major regional differences, it is quite possible that the choice of platform chemicals derived from biomass will show much more geographical variation than in petrochemical production. Figure 7.16 shows schematically how an integrated pyrolysis-based biorefinery might operate by integrating several of the different conversion technologies discussed above to create a single processing facility.

A key feature of the biorefinery concept is the coproduction of fuels, chemicals and energy. As explained earlier, there is also the possibility of gasifying biomass to make syngas, a mixture of hydrogen and carbon monoxide for subsequent synthesis of hydrocarbons, alcohols and other chemicals. However,

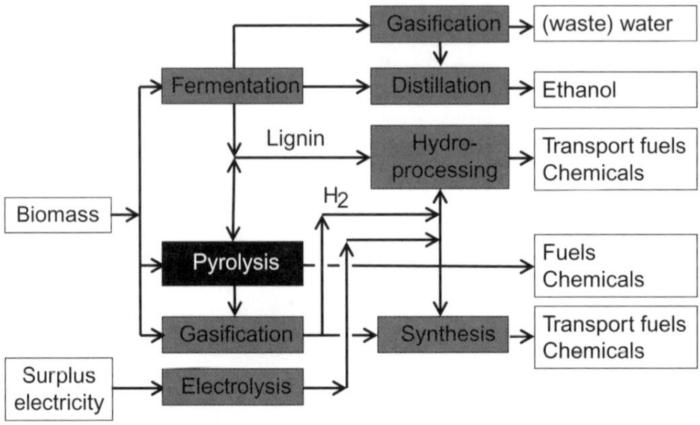

Figure 7.16 A biorefinery concept based on integrated biological and thermal processing for transport fuels and chemicals.

this route is quite energy intensive so much of the energy content of the biomass is lost in the processing. Electricity generation may be the most efficient use of forestry products.[71]

It is currently envisaged that biorefineries should focus on a limited number of platform chemicals for earlier implementation. Evolution will enable nonedible feedstock such as cellulose or lignin into well-defined, useful carbon sources at high purity but at a lower cost than with starch or sugar feed-stocks. For example, vegetable oils can be transformed into a variety of high-value chemicals, particularly surface-active chemicals, therefore it is wasteful merely to burn them as biodiesel, which should be derived from less valuable biofeedstocks.[72]

While biorefineries are not new, the recognition of their strategic and economic potential is recent. A biorefinery can be defined as the optimized performance of the use of biomass for materials, chemicals, fuels and energy applications, where performance relates to costs, economics, markets, yield, environment, impact, carbon balance and social aspects. In other words, there is optimized use of resources, maximized profitability, maximized benefits and minimized wastes.

These are not necessarily compatible or consistent in defining the best process system, so optimization usually requires their careful consideration. A biorefinery needs to optimize the biomass production, pretreatment and preparation subsystem, the pyrolysis subsystem and the applications subsystem dependently and independently. Economics will usually dominate where commercial considerations are involved.

7.6.6.1 Biorefinery Examples

Although biorefineries have been explained, defined and promoted for many years, such as by Radlein,[70] and there are many examples of biorefineries such as in the pulp and paper industry, there are few examples based on or derived from fast pyrolysis. These include:

- NREL in the USA who have developed processes to separate phenolics for resins[73] leaving an aqueous fraction, which is then reformed for hydrogen[61] as a product or for hydrogenation or hydrotreating;
- Red Arrow in the USA who produce liquid smoke and other specialities and burn the organic residue as fuel both within the factory and in a local power station.[67]

7.6.6.2 Markets

Finding and securing markets for renewable products is a key feature of development of a biorefinery. Existing markets will substitute fossil-fuel-derived products with the "bonus" of a renewable product. This will aid marketing and may command a small premium in the early years. For new products the market needs to be developed and established. In this regard, large companies will contribute considerably more and will be key to establishment of a green chemical industry. In both cases, inducements will accelerate progress

7.6.6.3 Summary

- Economically viable processes are essential for industry to be seriously involved. Inducements are likely to be necessary for some time.
- Economy of scale is important.
- Markets need to be identified and possibly developed for new products.
- Biomass cost is important for low-value products.
- Fast pyrolysis offers the advantage of a storable and transportable product that permits economies of scale.
- Early developments will be more successful with simple processes and fewer products.
- Processes, products and markets will need to be developed.
- Optimization of processes, products & systems is essential.
- Process synthesis offers a family of useful design and evaluation heuristics.

7.7 Overall Fast-Pyrolysis System

7.7.1 System Components and Integration

A fast-pyrolysis system consists of an integrated series of operations starting with delivery of a roughly prepared feedstock, such as whole tree chips from short-rotation coppice, wood waste from furniture manufacture, energy crops such as miscanthus or sorghum, or agricultural residues such as straw.

Table 7.9 Main process steps in a fast-pyrolysis system.

Reception and storage	Low-capacity systems of up to around 3 t/h feed typically consist of a concrete pad for tipping delivered feed and a front-end loader to move it between reception, storage and handling steps. Larger-size plants will increasingly require more sophisticated systems like those employed in pulp and paper mills. These will include weighbridge, tipping units, conveyors and bunker storage.
Feed drying	This is usually essential unless a naturally dry material such as straw is available. Sources of heat for drying include waste heat from combustion of byproduct char for pyrolysis process heat, combustion of byproduct gas, combustion of biomass, waste heat from engine or turbine if power generation is on site, combustion of wastes from chemicals recover or bio-oil refining. Equipment and processes for drying have been reviewed.[74]
Comminution	As explained earlier, particles have to be very small to allow rapid heating and achieve high liquid yields. This is costly and reactors, such as ablative pyrolyzers, that can use larger particles have an advantage.
Reactor configuration	As described above, a wide variety of configurations that show considerable diversity and innovation in meeting the basic requirements of fast pyrolysis have been tested. However, the "best" method is not yet established.
Char + ash separation	Some char is inevitably carried over from cyclones and collects in the liquid. Almost all of the ash in the biomass is retained in the char, so successful char removal gives successful ash removal. The char may be separated and exported if there is a viable market; otherwise it would be used to provide process heat either directly as in circulating fluid-bed reactors or indirectly as in fluid-bed systems.
Liquids collection	To collect the bioliquids, quenching with an immiscible liquid such as a hydrocarbon or a cooled liquid product would usually be used in larger-scale processes. Although collection of aerosols is difficult, there has been considerable success with electrostatic precipitators. This technology is widely used on very large scales, so at least in principle this should not be a problem. Careful design is needed to avoid blockage from differential condensation of heavy ends. Light-ends collection is important in reducing liquid viscosity.
Storage and transport	The bio-oil will require some local storage before local or remote use. A tank farm will provide storage and blending facilities. Both storage and transport are features unique to fast pyrolysis and permit economies of scale to be realized from building as large a conversion plant as possible as well as offering economic supplies of bio-oil for distributed or decentralized small-scale power and heat applications.

A complete integrated fast-pyrolysis process for power generation consists of three major stages: feedstock reception, storage and pretreatment; fast pyrolysis; and power generation. The main steps and their characteristics are summarized in Table 7.9.

7.8 Barriers

Thermal biomass-conversion technologies have to overcome a number of technical and nontechnical barriers before industry will undertake their commercialization. The technical barriers have been described above, and some of the more important nontechnical barriers are now explored.

For industry to implement renewable energy technologies to meet greenhouse-gas-mitigation targets, investment has to be attractive. Without some fiscal incentives (or disincentives against fossil fuels), companies will only invest in those projects that are sufficiently profitable, and most of these will be in niche markets and special opportunities. Only governments can create the necessary instruments. So economics is a critical barrier, as all biofuels have to compete with fossil fuels. Industry will only invest in technology that has an acceptable return at an acceptable risk.[75] Acceptable returns will come from incentives for capital expenditure of product purchase, or to disincentives to orthodox options by, for example, taxation on fossil fuels or legislation, *etc.* Investors are generally risk-averse and always prefer low-risk investments, but if risks have to be accepted, a higher return is expected. Technology developers can do much to minimize technical risk and this topic has been thoroughly described and discussed.[75] This is one of the major roles that governments have to play.

NIMBY attitudes present another barrier that has a major impact on permitting. There is widespread approval of the interest in, and move towards, renewable energy and bioenergy–as long as it is not near me! Increasing attention will need to be placed on "selling" the idea to the populace where the plant is to be built, and this problem could be exacerbated by the need to build smaller plants, and thus many more plants, than conventional power stations and refineries. Early plants will have the curiosity factor and may enjoy popularity as an attraction by themselves or as part of a "green" site. However, the increasing acceptance of biofuels such as pellets in many parts of Europe should lead to greater acceptance of the value and virtues of these technologies.

Economies of scale are a vital feature of the development of industry and technology: the larger a process can be built, the cheaper it becomes. This is particularly important in the energy and process industries that will shoulder responsibility for technology development and implementation. However, biomass is a diffuse resource that has to be harvested from large areas. A modest 10 MW_e power station operating at a modest efficiency of 35% will require about 40 000 t/year of wood on a dry basis, which will require about 4000 ha of land or 40 square km. This could reduce to 2000 ha or 20 square km if the promise of high-yielding short-rotation forestry is realized. Neither figure makes any allowance for nonproductive land. The cost and logistics of transporting biomass to a processing plant impose an upper limit on plant size. The maximum size that has been suggested in Europe ranges from 30 to 80 MW_e in the short to medium term, and 100 to 150 MW_e in North America. This places a practical upper limit on the benefits of scale.

The final nontechnical barrier to biomass that needs to be acknowledged is that created by the established energy suppliers and providers, who have considerable investments in conventional energy systems and who will naturally seek to maximize the returns and maintain the competitive edge of these. Most major energy companies have their own programmes of supporting renewable energy, but there have always been concerns over the extent to which they will seek to protect their other interests in energy supply.

7.9 Conclusions

The liquid bio-oil produced by fast pyrolysis has the considerable advantage of being storable and transportable, as well as having the potential to supply a number of valuable chemicals. In this respect it offers a unique advantage and its production should be considered as complementary to other biomass-conversion processes. Although fast-pyrolysis technologies for the production of liquid fuel have been successfully demonstrated on a small scale, and several large pilot plants or demonstration projects are in operation or at an advanced stage of construction, they are still relatively expensive compared with fossil-based energy, and thus face economic and other nontechnical barriers when trying to penetrate the energy markets.[76,77]

As with biomass gasification, fast pyrolysis will only be able to penetrate energy markets if completely integrated into a biomass system. Thus, the innovation in practically all demonstration projects under implementation lies not only in the technical aspects of the various processes but also in the integration of pyrolysis technologies in existing or newly developed systems, where it can be demonstrated that the overall system offers better prospects for economic development. Niche markets, such as utilization of bio-oil in a marine environment (where bio-oil is much less damaging than fuel oil in case of accidents) are likely to be the most attractive in the short term. The overall challenges include small-scale efficient and cost-effective power generation (electricity and mechanical power) and heat-production systems, while also taking advantage of the more economically attractive chemicals markets.

There are also specific challenges facing fast pyrolysis that relate to technology, product and applications, all of which need to be addressed. These include:

- The cost of the bio-oil, which is 10% to 100% more than fossil fuel.
- Availability: there are limited supplies of bio-oil for testing and development of applications.
- A lack of standards for use and distribution of bio-oil. Inconsistent quality inhibits wider usage, and considerable work is required to characterize and standardize these liquids and develop a wider range of energy applications.
- Bio-oil is incompatible with conventional fuels.
- Users are unfamiliar with this material.

- Dedicated fuel handling systems are needed.
- Pyrolysis as a technology does not enjoy a good image.

The most important issues that need to be addressed are:

- scale-up;
- cost reduction;
- improving product quality including setting norms and standards for producers and users;
- environment health and safety issues in handling, transport and usage;
- encouragement for developers to implement processes; and users to implement applications;
- information dissemination.

For pretreatment, fast pyrolysis offers the benefits of a liquid fuel with concomitant advantages of easy storage and transport as well as comparable higher power generation efficiencies at the smaller scales of operation that are likely to be realized from bioenergy systems compared to fossil-fuelled systems.

Finally, it is important to note that a complete bioenergy process starts with reception of biomass fuel and ends with delivery of an end product such as electricity or bio-oil to the customer. At each stage of the overall process there are potential impacts on the workers at the plant, on the surrounding population and on the environment generally. All of these must be properly considered by reference to international, national and local legislation.

References

1. A. V. Bridgwater and K. Maniatis, in *Molecular to Global Photosynthesis*, M. D. Archer and J. Barber, (ed.), IC Press, 2004, pp. 521–612.
2. D. Mohan, C. U. Pittman and P. Steele, *Energy & Fuels*, 2006, **20**, 848.
3. A. V. Bridgwater, *Thermal Sci.*, 2004, **8**(2), 21.
4. S. R. A. Kersten, X. Wang, W. Prins and W. P. M. van Swaaij, *Ind. Eng. Chem. Res.*, 2005, **44**, 8773.
5. S. Czernik and A. V. Bridgwater, *Energy & Fuels*, 2004, **18**, 590.
6. A. V. Bridgwater, *Chem. Eng. J.*, 2003, **91**, 87.
7. A. V. Bridgwater, S. Czernik and J. Piskorz, in "Fast Pyrolysis of Biomass a Handbook Vol. 2", A. V. Bridgwater, (ed.), CPL Press, Newbury, UK, 2002, pp. 1–22.
8. A. V. Bridgwater, in "Thermal Biomass Conversion", A.V. Bridgwater, H. Hofbauer, S. van Loo, (ed.), CPL Press, Newbury, UK, 2009.
9. D. S. Scott, J. Piskorz and D. Radlein, *Ind. Eng. Chem. Process Des. Dev.*, 1985, **24**, 581.
10. D. S. Scott and J. Piskorz, *Can. J. Chem. Eng.*, 1982, **60**, 666.
11. D. S. Scott, R. L. Legge, J. Piskorz, P. Majerski and D. Radlein, in "Developments in Thermochemical Biomass Conversion", A. V. Bridgwater and D. G. G. Boocock, (ed.), 1997, pp. 523–535.

12. A. Cuevas, C. Reinoso and D. S. Scott, in "Proc. Power Production from Biomass II", VTT Espoo, Espoo, March 1995.
13. A. Robson, *PyNe newsletter No. 11*, June 2001, ISSN 1470–3521, Aston University, Birmingham, UK, pp. 1–2.
14. R. McLellan, *PyNe Newsletter No. 10*, December 2000, ISSN 1470–3521, Aston University, Birmingham, UK, p. 12.
15. G. Trebbi, C. Rossi and G. Pedrelli, in *Developments in Thermochemical Biomass Conversion*, A. V. Bridgwater and D. G. B. Boocock, (ed.), 1997, pp. 378–387.
16. C. Rossi and R. G. Graham, in *Biomass Gasification and Pyrolysis*, M. K. Kaltschmitt and A. V. Bridgwater, (ed.), CPL Scientific Ltd, UK, 1997, pp. 300–306.
17. I. Boukis, M. E. Gyftopoulou and I. Papamichael, in *Progress in Thermochemical Biomasss Conversion*, A.V. Bridgwater, (ed.), 2001, pp. 1259–1267.
18. L. Braeckman, W. Prins and J. G. Pieters, *Commun. Agric. Appl. Biol. Sci.*, 2008, **73**(1), 3.
19. J. P. Diebold and J. Scahill, in *Production, Analysis and Upgrading of Oils from Biomass*, C. T. Ratcliffe, E. M. Suuberg and K. S. Vorres, (ed.), American Chemical Society, 1987, **32(2)**, 21.
20. J. Lédé, J. Panagopoulos, H. Z. Li and J. Villermaux, *Fuel*, 1985, **64**, 1514.
21. G. V. C. Peacocke and A. V. Bridgwater, *Biomass Bioenergy*, 1995, **7**, 147.
22. A.V. Bridgwater, G.V.C. Peacocke and N.M. Robinson, US patent 7,625,532, 2007.
23. D. Meier, *PyNe newsletter 17*, April 2005, Aston University, Birmingham, UK, pp. 1–3.
24. J. W. Black and D. B. Brown, in *Biomass Thermal Processing, Proc. 1st. Canada/ European Community R&D Contractors' Meeting*, E. Hogan, J. Robert, G. Grassi and A.V. Bridgwater, (ed.), 1990, pp. 123–125.
25. J. Lédé, *Ind. Eng. Chem. Res.*, 2000, **39**, 893.
26. R. J. Kovac and D. J. O'Neil, in *Pyrolysis and Gasification*, G. L. Ferrero, K. Maniatis, A. Buekens and A. V. Bridgwater, (ed.), Elsevier Applied Science, 1989, pp. 169–179.
27. K. Maniatis, J. Baeyens, H. Peeters and G. Roggeman, in *Developments in Thermochemical Biomass Conversion*, A. V. Bridgwater and D. G. B. Boocock, (ed.), 1997, pp. 1257–1264.
28. W. Prins and B. M. Wagenaar, in *Biomass Gasification and Pyrolysis*, M. K. Kaltschmitt and A. V. Bridgwater, (ed.), CPL Scientific Ltd., UK, 1997, pp. 316–326.
29. B. M. Wagenaar, R. H. Venderbosch, J. Carrasco, R. Strenziok and B. J. van der Aa, in *Progress in Thermochemical Biomasss Conversion*, A.V. Bridgwater, (ed.), 2001, 1268–1280.
30. J. Yang, D. Blanchette, B. de Caumia and C. Roy, in *Progress in Thermochemical Biomass Conversion*, A. V. Bridgwater, (ed.), 2001, pp. 1296–1311.
31. G. V. C. Peacocke, A. V. Bridgwater and J. G. Brammer, in *Science in Thermal and Chemical Biomass Conversion*, A.V. Bridgwater and D.G.B. Boocock, (ed.), CPL Press, Newbury, UK, 2006, pp. 1785–1802.

32. A.J. Toft, Ph.D. Thesis, Aston University, Birmingham, UK, 1998.
33. A. Fivga, Aston University, Birmingham, UK, private communication, 2008.
34. J. P. Diebold, T. A. Milne, S. Czernik, A. Oasmaa, A. V. Bridgwater, A. Cuevas, S. Gust, D. Huffman and J. Piskorz, in *Developments in Thermochemical Biomass Conversion*, A.V. Bridgwater and D.G.B. Boocock, (ed.), 1997, pp. 433–447.
35. G. V. C. Peacocke, in *Fast Pyrolysis of Biomass: A Handbook, Volume 2*, CPL Press, Newbury, UK, 2002, pp. 293–338.
36. J. Blin, G. Volle, P. Girard, A. V. Bridgwater and D. Meier, *Fuel*, 2007, **86**, 2679.
37. J. P. Diebold, in *Fast Pyrolysis of Biomass: A Handbook, Volume 2*, A.V. Bridgwater, (ed.), CPL Press, Newbury, UK, 2002, pp. 243–292.
38. S. Czernik and A. V. Bridgwater, *Energy & Fuels*, 2004, **18**, 590.
39. A. V. Bridgwater, *Catal. Today*, 1996, **29**, 285.
40. A. V. Bridgwater, *Appl. Catal. A*, 1994, **116**, 5.
41. J. P. Diebold, S. Czernik, J. W. Scahill, S. D. Philips and C. J. Feik, in *Biomass Pyrolysis Oil Properties and Combustion Meeting*, T. A. Milne, (ed.), National Renewable Energy Laboratory, Boulder, CO, 1994, pp. 90–108.
42. A.L. Shihadeh, Ph.D. Thesis, Massachusetts Institute of Technology, Boston, USA, 1998.
43. J. Sitzmann and A.V. Bridgwater, 15th European Energy from Biomass Conference, Berlin, 7–11 May, 2007.
44. M. Ikura, M. Slamak and H. Sawatzky, US Patent 5,820,640 (1998).
45. P. Baglioni, D. Chiaramonti, M. Bonini, I. Soldaini and G. Tondi, in *Progress in Thermochemical Biomass Conversion*, A. V. Bridgwater, (ed.), 2001, 1525–1539.
46. P. Baglioni, D. Chiaramonti, K. Gartner, H. P. Grimm, I. Soldaini, G. Tondi, A. V. Bridgwater and A. Webster, *Biomass Bioenergy*, 2003, **25**, 85, *ibid.*, 2003, **25**, 101.
47. J. P. Diebold and S. Czernik, *Energy & Fuels*, 1997, **11**, 1081.
48. D. C. Elliott and E. Baker, in *Energy from Biomass and Wastes X*, D. Klass, (ed.), 1983, pp. 765–782.
49. R. Maggi and D. Elliott, in *Developments in Thermochemical Biomass Conversion*, A. V. Bridgwater and D. G. B. Boocock, (ed.), Blackie Academic and Professional, 1997, pp. 575–588.
50. UOP press release at www.uop.com/renewables/UOP_Ensyn_Final.pdf.
51. M.-L. Cottam and A. V. Bridgwater, *Biomass Bioenergy*, 1994, **7**, 267.
52. C. Chang and A. Silvestri, *J. Catal.*, 1977, **47**, 249.
53. J. P. Diebold, D. Beckman, A. V. Bridgwater, D. C. Elliott and Y. Solantausta, in *Advances in Thermochemical Biomass Conversion*, A. V. Bridgwater, (ed.), Blackie, 1994, p. 1325–1342.
54. J. P. Diebold and J. W. Scahill, *Energy Prog.*, 1988, **8**(1), 59.
55. (a) J. D. Adjaye and N. N. Bakhshi, *Fuel Proc. Technol.*, 1995, **45**, 161; (b) J. D. Adjaye and N. N. Bakhshi, *Fuel Proc. Technol.*, 1995, **45**, 185.

56. P. T. Williams and P. A. Horne, *Biomass Bioenergy*, 1994, **7**, 223.
57. A. V. Bridgwater and M.-L. Cottam, *Energy & Fuels*, 1992, **6**, 113.
58. A. V. Bridgwater, *Technical and Economic Assessment of Thermal Processes for Biofuels, Report to NNFCC, June 2009,* Cope Ltd UK.
59. H. P. A. Calis, J. P. Haan, H. Boerrigter, A. van der Drift, G. Peppink, R. van den Broek, A. P. C. Faaij and R. H. Venderbosch, in *Proceedings of Pyrolysis and Gasification of Biomass and Waste,* A.V. Bridgwater, (ed.), Strasbourg, September 2002, CPL Scientific Press, Newbury, UK, 2003.
60. D. Wang, S. Czernik, D. Montane, M. Mann and E. Chornet, *Ind. Eng. Chem. Res.*, 1997, **36**, 1507.
61. S. Czernik, R. French, C. Feik and E. Chornet, *Ind. Eng. Chem. Res.*, 2002, **41**, 4209.
62. M. K. Mann, P. L. Spath, K. Kadam, in *Proceedings of the 11th World Hydrogen Energy Conference*, Stuttgart, Germany, June 23–28, 1996.
63. M. Wornat, G. Bradley and N. Yang, *Energy & Fuels*, 1994, **8**, 1131.
64. R. Shaddix and S. Huey, in *Developments in Thermochemical Biomass Conversion*, A. V. Bridgwater and D. G. B. Boocock, (ed.), Blackie Academic & Professional, London, 1997, pp. 465–480.
65. S. Gust, in *Developments in Thermochemical Biomass Conversion*, A. V. Bridgwater and D. G. B. Boocock, (ed.), Blackie Academic & Professional, London, 1997, pp. 481–488.
66. A. Oasmaa, M. Kytö and K. Sipilä, in *Progress in Thermochemical Biomass Conversion*, A. V. Bridgwater, (ed.), Blackwell Science, Oxford, 2001, pp. 1468–1481.
67. B. A. Freel, R. G. Graham and D. R. Huffman, in *Bio-oil Production and Utilization*, A. V. Bridgwater and E. Hogan, (ed.), CPL Press, Newbury, UK, 1996, pp. 86–95.
68. R. Sturzl, The commercial cofiring of RTP^{TM} bio-oil at the Manitowoc Public Utilities power generation station, available at http://www.ensyn.com
69. CRE Group Ltd, Technical review on opportunities and markets for co-utilization of biomass and waste with fossil fuels for power generation, report prepared for the European Commission, Brussels, 2000.
70. D. Radlein, in *Fast Pyrolysis of Biomass: A Handbook*, A. Bridgwater, S. Czernik, J. Diebold, D. Meier, A. Oasmaa, C. Peacocke, J. Piskorz, D. Radlein, (ed.), CPL Press, Newbury, UK, 1999, pp. 164–188.
71. *Biorefineries – Industrial Processes & Products Vols 1 & 2*, B. Kamm, P. R. Gruber, M. Kamm, ed., Wiley-VCH, Weinheim, Germany, 2006, Vol. **1**, p. 45.
72. *Biorefineries – Industrial Processes & Products Vols 1 & 2*, B. Kamm, P. R. Gruber, M. Kamm, (ed.), Wiley-VCH, Weinheim, Germany, 2006, Vol. **2**, p. 312.
73. S. Kelly, X. Wang, M. Myers, D. Johnson and J. Scahill, in *Developments in Thermochemical Biomass Conversion*, A. V. Bridgwater and D. G. B. Boocock, (ed.), Blackie Academic & Professional, London, 1997, pp. 557–572.

74. J. Brammer and A. V. Bridgwater, *Renew. Sustain. Energy Rev1999,* **3***(4), 243.*
75. P. Thornley and E. Wright, *Evaluation of bioenergy projects, in PyNe Final Report to the EC, October 2001.*
76. A. V. Bridgwater, S. Czernik and J. Piskorz, in *Progress in Thermochemical Biomass Conversion*, A. V. Bridgwater, (ed.), Blackwell, 2001, pp. 977–997.
77. A. V. Bridgwater, A. J. Toft and J. G. Brammer, *Sustainable Renewable Energy Rev.*, 2002, **6**, 181.

CHAPTER 8
Hydrothermal Processing of Biomass

P. E. SAVAGE, R. B. LEVINE AND C. M. HUELSMAN

University of Michigan, Department of Chemical Engineering,
2300 Hayward Dr., Ann Arbor, MI, USA

8.1 Introduction

Water is essential for life and a major component of plant biomass. The conversion of solar energy stored in photosynthetic organisms into useful fuels for society must therefore contend with moisture contents that range from 10–60 wt% for lignocellulosic biomass and even higher for aquatic biomass such as microalgae. Some biomass valorization processes, such as conventional gasification and fast pyrolysis, require dry feedstocks or else suffer tremendous penalties in thermal efficiency. Water removal can be accomplished through solar drying where climate permits, or else requires heat inputs that typically diminish process net energy production. In an effort to obviate the capital, energy, and time required to dry biomass, many investigators have studied and developed hydrothermal processes for biomass valorization. Hydrothermal processing, which involves the application of heat and pressure to biomass in an aqueous medium, mimics the conversion of ancient plant material into the crude oil and natural gas reservoirs we rely upon today. This chapter focuses on hydrothermal processes that generate liquid and gaseous fuels from biomass. Though it is also possible to generate high energy density solids from biomass *via* hydrothermal chemistry,[1,2] the production of solid biochar is outside the scope of this chapter.

RSC Energy and Environment Series No. 1
Thermochemical Conversion of Biomass to Liquid Fuels and Chemicals
Edited by Mark Crocker

Published by the Royal Society of Chemistry, www.rsc.org

8.1.1 Background on Hydrothermal Processing

A hydrothermal process is one that involves water at elevated temperatures and pressures. High-temperature water (HTW) refers to water in its liquid state below its critical temperature and pressure (374 °C, 221 bar), whereas it becomes a highly compressible fluid called supercritical water (SCW) above this point. An advantage of hydrothermal processing for biomass is that hot water can serve as a solvent, a reactant, and even a catalyst or catalyst precursor. While many biomass compounds (*e.g.*, lignin, cellulose) are not water soluble at ambient conditions, most are readily solubilized in HTW or SCW.[3,4] These soluble components can then be subject to hydrolytic attack, engendering fragmentation of biomacromolecules. Water, both in its dissociated and native form, can help catalyze hydrolysis and other reactions.[5,6]

Hydrothermal processing can occur over a continuum of temperature and pressure regimes to yield desired solid, liquid and gaseous products. Under mild conditions (250–350 °C, 40–165 bar), biomacromolecules hydrolyze and react in a process called hydrothermal liquefaction, yielding a viscous biocrude oil. If a suitable catalyst is employed, it is possible to convert these primary fragments to gases (mainly CH_4) through low-temperature hydrothermal gasification. At higher supercritical temperatures, primary fragments decompose further, even without added catalyst, to form permanent gases. In this regime, the process is referred to as supercritical water gasification (SCWG). Understanding the fundamental reaction pathways occurring in each of these hydrothermal processes is important to ensure their rational design and optimization.

As outlined in recent reviews,[7–9] hydrothermal processing has unique advantages for wet biomass related to both the chemistry (*e.g.*, rapid hydrolysis, enhanced deoxygenation) and engineering considerations (*e.g.*, higher energy efficiency). Hydrothermal processing obviates the need for feedstock dewatering and drying. For example, SCWG of biomass with at least 30% moisture requires less energy than drying,[10] mainly because hydrothermal processing avoids the energy penalty associated with the phase change from liquid to vapor. For instance, the enthalpy change required to take liquid water at 25 °C to a liquid at 300 °C is about half of that required to vaporize the water (*i.e.* dry the biomass).[7,11,12] In addition, hydrothermal processing allows for integrated energy recovery because the hot reactor effluent stream can be used to preheat the ambient-temperature feed stream.[7] In comparison, conventional biomass-conversion processes that employ drying use lower temperatures, making energy recovery from process streams more difficult.

Energy efficiency is typically high for hydrothermal processes. Hydrothermal upgrading (HTU), a commercial-scale process for hydrothermal liquefaction, reportedly achieves 75% thermal efficiency and requires just 2% of the input material's energy content to meet its heat demands.[13] For hydrothermal gasification, process energy efficiencies are typically reported to be in the 45% to 70% range.[14,15] Ro *et al.* (2007) concluded that hydrothermal gasification of manure is net-energy positive when the feedstock is above 2% solids and appropriate heat recovery is in place.[16] In addition, analysis of corn starch

SCWG (745 °C and 280 bar) indicated an overall energy efficiency for the process of 76%, and the product gas contained about 92% of the chemical energy of the biomass.[17] This process yielded 4.5 J of energy in the product gas for every 1.0 J of unrecovered heat or work input when proper heat integration was employed. In related work on hydrothermal gasification of chicken manure, process energy efficiencies as high as 70% were obtained with a 15% solids feedstock.[18] Process energy efficiency was found to depend heavily upon the energy content of the feedstock, the feedstock loading, and the heat-recovery efficiency. The most effective way to improve energy efficiency was to increase the feedstock loading, which means less water must be heated per unit of treated biomass. However, overall gas yields from SCWG can be lower at higher biomass loadings, drawing attention to the detailed modeling required for process optimization (Figure 8.1).[19] It is generally thought that a solids loading of 15–20% or more is needed to achieve practical economies.

The operating cost associated with high-pressure processing is not prohibitive because the feed stream is a liquid or slurry, which can be transported and pressurized with pumps. Also, recall that high-temperature, high-pressure processes are used routinely in the chemical processing industry, even for commodity chemicals with a modest selling price (*e.g.*, NH_3 synthesis). Of course, there are important solids-handling issues that need to be addressed in any biomass-conversion process involving water and elevated pressures.

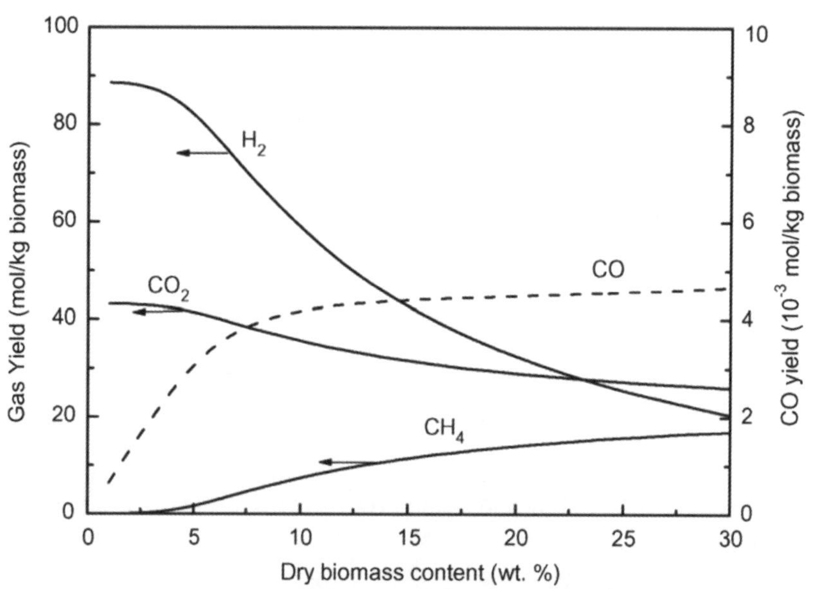

Figure 8.1 Calculated equilibrium gas yields for hydrothermal gasification of wood sawdust ($CH_{1.35}O_{0.617}$) as a function of dry biomass concentration. Temperature is 600 °C and pressure is 250 bar. Reprinted with permission from Elsevier.[14]

Moving a pressurized slurry through process equipment will require careful design of pumps and extruders. In addition, residual solids will need to be recovered from the process, and inorganic components in biomass can precipitate in SCW, causing reactor plugging and/or catalyst deactivation.

8.1.2 Properties of Water

Hydrothermal processing relies on the unique properties of water at high temperature and pressure. Most importantly, elevated temperatures diminish hydrogen bonding and reduce water's dielectric constant (Figure 8.2).[20] Thus, many organic compounds become completely miscible in HTW. From a solubility perspective, HTW behaves much like a polar organic solvent. Furthermore, the ion product ($K_W = [OH^-][H^+]$) for liquid water increases with temperature, being nearly three orders of magnitude higher at 250 °C than it is at ambient temperatures. Accordingly, HTW boasts higher H^+ and OH^- concentrations than ambient liquid water, increasing the effectiveness of hydrolysis and dehydration reactions, which are known to be acid or base catalyzed. The properties and applications of SCW, which is more gas-like, are discussed elsewhere in detail.[21–24]

The balance of this chapter focuses on two topics: hydrothermal liquefaction and hydrothermal gasification. For each topic, we provide an introduction to the technology and some background material. We then discuss the chemistry involved and highlight some recently published studies to show the progress being made. Each section closes with a discussion about the outlook for the process.

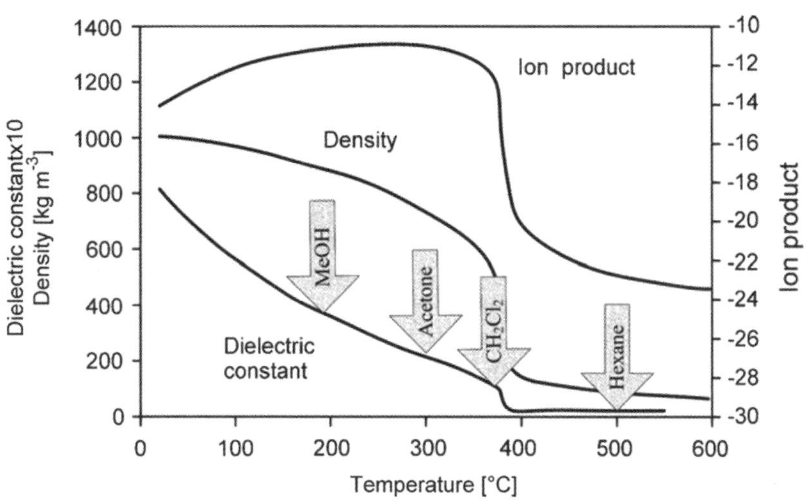

Figure 8.2 Temperature dependence of physical properties of water at 240 bar (dielectric constants of typical organic solvents at room temperature are indicated). Reprinted with permission from Elsevier.[20]

8.2 Hydrothermal Liquefaction

Hydrothermal liquefaction produces a high energy density organic liquid *via* the conversion of biomass in liquid water at elevated temperatures. The influence of catalysts (*e.g.*, alkali salts, metals), reducing gases (*e.g.*, H_2 and CO), and cosolvents (*e.g.*, acetone, glycerol, recycled bio-oil) on product yield and composition have been investigated. In general, feedstocks containing ~80% water are subjected to subcritical temperatures (250–350 °C) to create a hydrophobic bio-oil with reduced oxygen content (10–18%) compared to the parent material (ca. 40%). Bio-oils have been found to contain over 400 compounds, depending on the feedstock and processing conditions.[25] This bio-oil can be used directly as a heavy petroleum oil replacement, for cofiring with coal, and is a candidate for upgrading to high-quality distillate fuels (*e.g.*, diesel and gasoline). Oxygen removal through decarboxylation and hydrodeoxygenation is critical to ensure fuel quality because oxygenated hydrocarbons have lower energy contents and higher melting points, boiling points, and viscosities compared to hydrocarbons of a similar molecular weight.[26] Compared to pyrolysis, hydrothermal liquefaction occurs at less severe temperatures and produces bio-oil with a lower oxygen content, less moisture, and a higher heating value.[8] In addition to traditional lignocellulosic feedstocks such as wood,[27–31] other renewable feedstocks have recently been liquefied: switchgrass,[32] high-diversity grassland perennials,[33,34] straw,[35] corn cobs,[36] secondary pulp/paper sludge,[37] and animal manures.[38]

8.2.1 Process-Variable Effects

Several process variables influence the yields and composition of the bio-oils produced. These variables include the heating rate, reaction time, temperature, biomass particle size,[39] biomass loading (% solids), catalysts, and the liquefaction atmosphere (inert (N_2 or Ar) or reducing (CO, H_2)). The heating rate is a relevant variable because the majority of work in liquefaction has been done in batch reactors or simple continuous-flow systems with reactor heating times often on the order of several minutes.[40] Faster heating rates (140 °C/min), however, have yielded substantially more oil.[41] The impact of heating rate on product formation and process economics deserves further attention.

8.2.1.1 Feedstock Composition

Feedstock composition directly affects liquefaction results. Early work on petroleum formation from marine kerogen explored how model biomass compounds decomposed in the presence of sedimentary constituents (*e.g.*, $CaCO_3$, clay) to form *n*-alkanes.[42] Although $CaCO_3$ in the presence of water significantly increased conversion to all organic classes, fatty-acid decarboxylation alone could not account for the distribution of *n*-alkanes typically found in petroleum. In later work, the liquefaction of terrestrial carbohydrates (cellulose, lignin, sawdust) yielded few *n*-alkanes, while soy protein and soy oil resulted in high *n*-alkane yields.[43] In particular, proteins yielded *n*-alkanes with a distribution similar to petroleum, while soy oil did not, indicating the importance of feedstock composition and reactions between different classes of biopolymers during liquefaction.

This work was extended to the liquefaction of the marine microalga *Nannochloropsis*, which was grown in a low-N medium to induce the accumulation of lipids.[44] Harvested algae (90% moisture; 34% total extracted lipids) with different amounts of triacylglycerides (TAG) in the lipid fraction (*cf.* 7.7% and 36.6%) were liquefied at 300 °C for 10 min. The cells with a higher TAG content yielded about 10% more oil. In the same study, standards composed of triolein and lecithin showed a similar trend. Similar findings were recently reported for two different microalgae.[45] This pattern was also apparent in sewage sludges with similar elemental compositions but different amounts of protein, fiber, and oil: higher feedstock oil content (determined as the ether-soluble portion) led to increased bio-oil yields.[46]

8.2.1.2 Nitrogen Partitioning

N partitioning during liquefaction has been examined to reduce the potential for NO_x emissions from bio-oil combustion. In a study with cellulose and ammonia, bio-oil yield and N content increased with the N/C ratio of the initial material, while the amount of solid residue formed declined.[47] Ammonia was suspected of acting as a base catalyst (much like alkali salts) and a reactant with cellulose. In a separate study, albumin (13.9% N) liquefaction yielded at most 10% oil with about 7–9% N (or 5% of the feedstock N) while liquefaction of the other feedstocks (*e.g.*, stillage and sewage sludge; typically 4–7% N) under identical conditions yielded between 32–43% oil containing 4–6% N (equivalent to 28–46% of the original feedstock N).[48] The remaining N was found as ammonia in the aqueous phase. Sodium carbonate reduced the N content of the bio-oil at 250 °C. Likewise, oil yield, N content of the bio-oil, and aqueous concentrations of NH_4^+ increased with reaction temperature for the liquefaction (100–350 °C for 1 h) of sewage sludge (84% moisture; 8.2% N).[49] Further work investigated the liquefaction of 19 amino acids (0.4 g/L in H_2O, 300 °C, ~1 h) and revealed that most had oil yields between 2–10%, except for tryptophan, which resulted in 50% oil yield.[50] The majority of N was found in the aqueous phase as ammonia and organic N, with only 2–7% of N distributed to the oil for most amino acids. However, for tryptophan, an indole-containing amino acid, approximately 50% of the N was distributed to the oil. It was hypothesized that the N in the stable heterocyclic group would not be decomposed to aqueous N compounds and therefore resided in the oil. Previous work has suggested that N in this form may be difficult to remove through traditional upgrading processes.[51] Overall, it was concluded that the hydrolysis of proteins to amino acids prior to liquefaction may help reduce the N content of bio-oil.

Under supercritical reaction conditions (500 °C, 30 MPa), feedstocks containing protein produced more oil and less gas than did solely lignocellulosic feedstocks.[52] The authors concluded that products of carbohydrate and protein degradation most likely react with each other to form free-radical scavengers *via* the Maillard reaction. These scavengers decrease the rate of free-radical chain reactions that play a large role in gas formation, thereby increasing bio-oil yields at the expense of gaseous products. This result is significant because it shows that the hydrothermal behavior of an individual compound may differ when present in a mixture.

8.2.1.3 Reaction Atmosphere

Reductive gases (*e.g.*, CO and H_2) generally improve liquefaction oil yields and increase product H/C ratios.[53,54] Early work in hydrothermal processing by Appell *et al.* (1971) was an extension of efforts to liquefy coal into a low-sulfur oil without using H_2.[53] Biomass conversion to a benzene-soluble oil at 350–400 °C was more rapid in $H_2O + CO$ than with H_2 alone for a variety of cellulosic feedstocks (*e.g.*, wood waste, urban waste, sewage sludge). Sodium carbonate increased the oil yield. It was hypothesized that CO could react with alkaline salts to form alkali formate, which could transfer H as a hydride ion to the substrate. Or, CO may react with hydroxyl and/or carboxyl groups to form CO_2, thereby removing O. It was also hypothesized that CO blocks the polymerization of unsaturated substances formed from the dehydration of cellulose, thereby preventing solid-char formation. In general, most recent laboratory studies on liquefaction have been carried out without a reducing atmosphere.[9]

8.2.2 Liquefaction Chemistry

This section focuses on the chemistry of biological compounds under subcritical hydrothermal conditions. The aqueous-phase chemistry of several biomass model compounds (*e.g.*, glucose, cellulose, triacylglycerols, amino acids) has been extensively studied and recently reviewed,[7] so here we briefly recount the most recent advancements in this field, while focusing mostly on the more complex chemistry of whole biomass and aquatic feedstocks. Understanding the fundamental reaction pathways in hydrothermal liquefaction is crucial to controlling product formation, designing optimized reactors, and improving the overall performance of the process. Regrettably, the mechanisms of hydrothermal liquefaction of biomass remain incompletely understood. The major chemical reactions in HTW involve hydrolysis, pyrolysis, decomposition of monomers and smaller molecules, and inter- and intramolecular rearrangements through bond rupture, dehydration, and decarboxylation. Biomacromolecules are hydrolyzed to water-soluble oligomers and monomers, which can undergo further degradation, repolymerize, or have functional groups reduced. Bio-oil process conditions can be used to optimize conversion and product selectivity to obtain intermediates of interest (*e.g.*, glucose for bioethanol production) and must be feedstock specific to account for variable rates of decomposition of different biopolymers. Some excellent reviews have covered the hydrothermal chemistry of biopolymers.[55–57]

Considering their abundance and relatively low cost, lignocellulosic feedstocks have been of great interest to produce liquid fuels (*i.e.* bio-oil) directly and as a source of fermentable sugars for bioethanol production.[58,59] Generally, these feedstocks contain 60–80% water when harvested and 30–50% cellulose, 15–30% hemicellulose, and 20–35% lignin on a dry weight basis.[9,55] Some additional amount of organics (*e.g.*, terpenes, tannins, waxes, fatty acids and proteins) and inorganic minerals (*i.e.* ash) are also present and may play a role in liquefaction.

8.2.2.1 Cellulose

Cellulose is a homopolymer typically containing 100–3000 glucose residues joined by ß(1,4) linkages. Hydrogen bonding creates a low-surface area crystalline form that is not soluble in water at standard conditions but partially dissolves in subcritical water at 302 °C and completely dissolves at 330 °C.[3] Dissolution and hydrolysis is a rapid stepwise process.[4] First, cellulose depolymerizes into oligosaccharides in a heterogeneous manner with water diffusing to the cellulose surface. Next, oligosaccharides hydrolyze and decompose to monosaccharides and pyrolysis products, respectively. Finally, additional decomposition of the monosaccharides may occur. The major products of cellulose decomposition (2% solids, 25 MPa, 320–400 °C, and 0.05–10.0 s) include aqueous oligomers, monomers (glucose and fructose), aqueous decomposition products of glucose, and lower molecular weight acids (Figure 8.3).[40,60] About 65% of cellulose converts to the oligomers and monomers in 4.8 s at 335 °C.[3]

Bio-oil production from cellulose has been studied in the presence and absence of several catalysts, most commonly alkali carbonates such as Na_2CO_3. Oil yields increased and char production decreased in the presence of Na_2CO_3, which most likely reacts with H_2O and CO to form sodium formate, a source of H that can transfer to the oil phase.[40,53,61] Reaction conditions that favored homogeneous conditions (faster heating rate, temperature closer to critical point) also increased oil yields. The catalytic effects of heterocyclic nitrogenous bases, like isoquinoline, have also been recognized. These may be effective H donors to the oil phase by forming salts (*e.g.*, formate) like alkali metals do or by receiving and then donating H directly.[53]

8.2.2.2 Lignin

Lignin is an amorphic, phenolic biopolymer that provides strength and rigidity in the cell walls of vascular plants. It consists of an irregular array of hydroxyl- and methoxy-substituted phenylpropane units connected mainly through

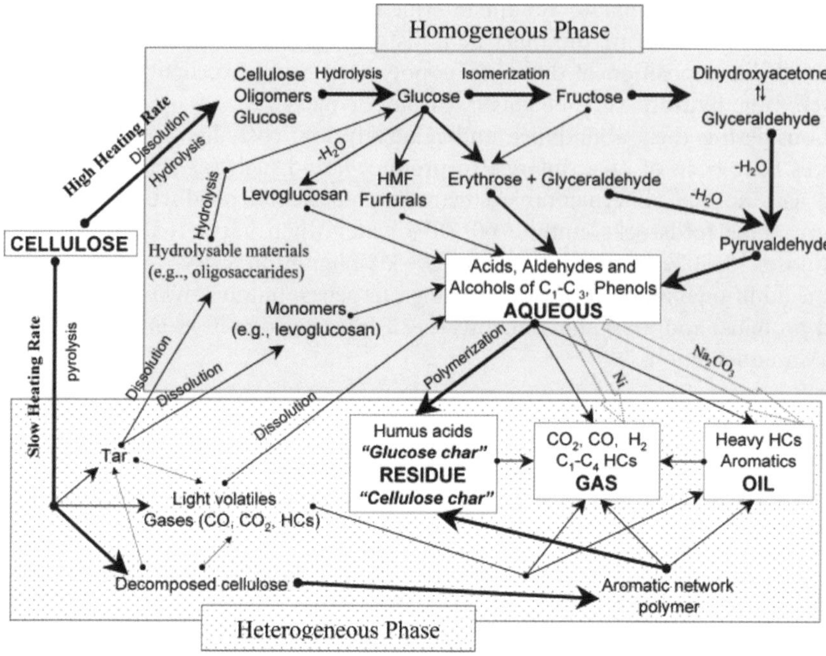

Figure 8.3 Reaction network for cellulose decomposition in homogeneous and heterogeneous environments in subcritical water. Reprinted with permission.[40] Copyright 2004 American Chemical Society.

ether bonds.[56,62,63] Liquid yields of 32–79% were achieved when liquefying various lignin sources (5–6% moisture) in water at 374 °C and 22 MPa for 10 min.[64] Lignin structure and pretreatment has a significant impact on yields and products. Major products identified included: guaiacol, methyl dehydroabietate, and ethyl- and methyl-substituted phenols. Guaicol further hydrolyzes to catechol, which has also been investigated recently as a model for lignin liquefaction.[63] Several authors have proposed that lignin degradation begins with a fast reaction yielding water-soluble fragments and then a slow reaction in which these compounds repolymerize and condense into solid residue (*i.e.* phenolic char) (Figure 8.4).[62,64] Recent work using diamond anvil cells (DACs) to observe *in situ* phase behavior of water/phenol and lignin systems suggests this first step is mostly a heterogeneous reaction, much like cellulose hydrolysis.[62,65] Work with lignin from high-diversity grassland perennials found that higher heating rates (140 °C/min) reduced the formation of solid residue, possibly by reducing side reactions in the aqueous phase, achieving a maximum liquid yield of 82.1% at 374 °C for 1 min residence time (4 min total time).[33] In addition, various catalysts were examined (0.1% sulfuric acid, 1% sulfuric acid, 1% NaOH, solid superacid $SO_4^{2-}/ZrO_2–Al_2O_3$, and solid alkali $CaO–ZrO_2$), though none demonstrated significantly improved liquid yields.

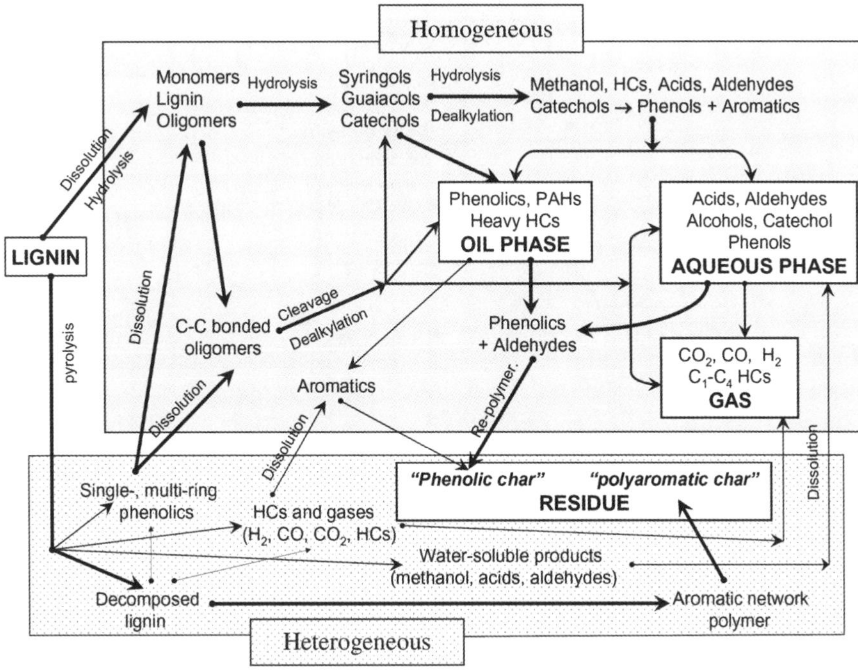

Figure 8.4 Reaction network for lignin decomposition in homogeneous and heterogeneous environments in supercritical water. Reprinted with permission from Elsevier.[62]

8.2.2.3 Lipids

Hydrolysis of fats and oils has been commonplace in the oleochemical industry for decades. The hydrolysis of vegetable oils occurs quickly (15–20 min) in water at 270–280 °C with no degradation of saturated fatty acids (FA) but a significant reduction in the amount of polyunsaturated FA due to isomeratization (cis to trans) and degradation.[66,67] As mentioned previously, liquefaction bio-oil yields tend to increase with feedstock lipid content. Current work in our lab indicates that bio-oils from liquefied (350 °C; 1 h) lipid-rich marine feedstocks (~30% lipid) contain fatty acids from hydrolyzed triglycerides.

Ideally, if long-chain FA can be stabilized in water at high temperatures and decarboxylated, biomass-derived sources of oils (*e.g.*, soy, rapeseed, jatropha, algae) could be used to make long-chain hydrocarbons resembling petroleum. When stearic acid ($C_{17}H_{35}COOH$) was reacted in SCW (0.17 g/cm^3 water) at 400 °C for 30 min, conversion was only 2% with C_{16} alkene being the major product (Figure 8.5).[68] Alkali hydroxides increased conversion substantially (to 13% with NaOH and 32% with KOH), with the major liquid product being C_{17} alkane. In the presence of metal oxides, the conversion increased to 30%, 62% and 68% with CeO_2, Y_2O_3 and ZrO_2, respectively, producing mostly C_{16} alkene.

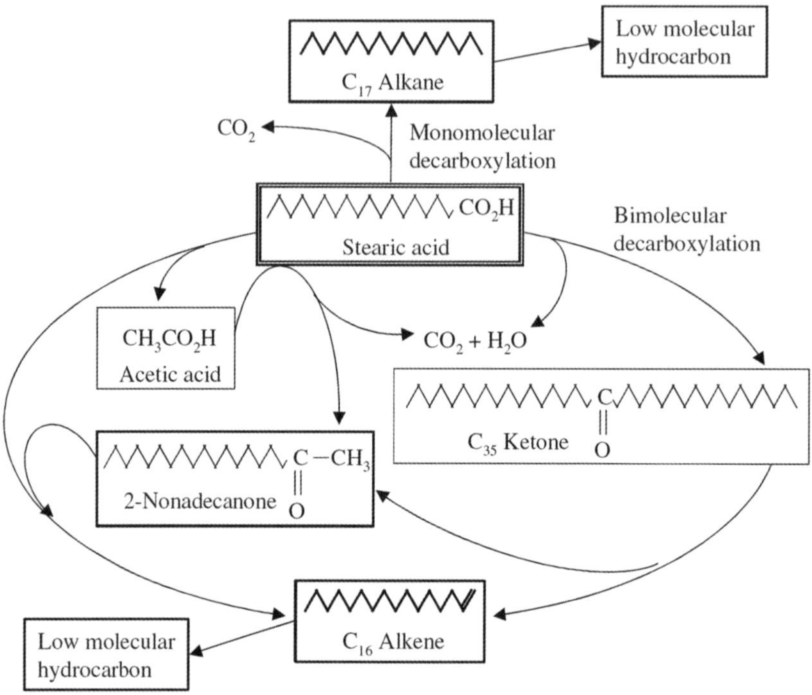

Figure 8.5 Conversion of stearic acid in supercritical water with and without additives. Reprinted with permission from Elsevier.[68]

8.2.2.4 *Wastes*

Animal and human wastes have long been targeted for conversion into fuels *via* liquefaction in order to reduce the environmental and capital costs associated with waste disposal.[69] For example, swine manure liquefaction has been extensively studied at the University of Illinois.[38,54,70,71] Oil yields from hydrothermal liquefaction (285 °C, $P_{CO} = 690$ kPa) increased from 10% to 60% as reaction time increased from 30 to 120 min. Recently, work at lower temperatures (240–280 °C, 10–120 min, 0.7 MPa initial N_2) resulted in ~33% oil yields, largely as the result of decarboxylation of noncellulose components in the manure (*e.g.*, lipids and proteins).[38]

8.2.2.5 *Aquatic Feedstocks*

Aquatic biomass, such as duckweed, macroalgae, and microalgae, has been investigated as a suitable feedstock for hydrothermal processing. Interest in converting aquatic biomass to transportation fuels has grown considerably, largely on the promise of high areal biomass and oil yields (15–100 Mg dry wt. ha^{-1} y^{-1} and 1000–100 000 L ha^{-1} y^{-1}), the opportunity to recycle CO_2

from flue gases, and the ability to cultivate it on abandoned or unproductive land using brackish, salt, or wastewaters instead of freshwater.[72–75] From a technoeconomic perspective, harvesting, dewatering, and extracting triglycerides are the most difficult steps in conventional biofuel production from microalgae. For example, oil extraction alone can account for up to 90% of the energy needed to synthesize biodiesel from algae oil,[76] and harvesting is energy and cost intensive.[77] Among the various conversion processes that have been reviewed,[78–81] liquefaction has the distinct advantage of obviating biomass dewatering and accomplishing bio-oil production with high energy efficiency. However, high feedstock N concentrations pose a challenge for bio-oil upgrading.[82] The prospect of making biodiesel directly *via* low-temperature liquefaction, as proposed by Patil *et al.* (2008), remains unknown.[83]

When duckweed (90% moisture; 35% protein, 12% cellulose, 14% hemicellulose, 3% lignin, and 16.5% ash) was liquefied, oil first appeared at 260 °C, and its yield increased with temperature to 30% at 340 °C, after which yields declined due to increasing decomposition.[84] Components in crude oil are typically classified as saturated compounds, aromatic compounds, resin, and asphalt. The bio-oil produced at 340 °C was primarily asphalt (54%) and resin (44%) with small amounts of saturate (1%) and aromatic compounds (0.8%), making it most suitable for use as a heavy oil or a candidate for upgrading.

Macroalgae, which are easy to harvest given their large size, are available from two major sources: waterbodies experiencing eutrophication and as a product of algae-based wastewater-treatment systems.[85,86] Most macroalgae lack cellulose and lignin and have a lower heating value and higher ash content than terrestrial biomass.[87] Metals present in macroalgae exhibit catalytic effects during pyrolysis and require further study.[87] When a freshly harvested macroalga (*Chaetomorpha linum*) was liquefied (250 °C to 395 °C; 1 h), bio-oil yields increased from 2% to 8% over this temperature range.[88] A life-cycle assessment revealed that hydrothermal liquefaction was one of the best processing options for this renewable feedstock.[89]

Microalgae are unicellular, photosynthetic organisms that grow in both freshwater and marine environments. When cultivated for fuel production, harvesting is a significant challenge because algal densities rarely reach above 0.5–2 g dry wt./L. After dewatering, aqueous slurries with 80–90% moisture are typical, suggesting this feedstock may be uniquely suited to hydrothermal processing. When heated to 250–350 °C, biomass in this aqueous slurry decomposes to form various amino acids, sugars, and fatty acids. Under these conditions, a bio-oil is produced. Previous reports on the hydrothermal processing of microalgae[81,83,90–97] indicate that 30–60% of the algal biomass can be converted to bio-oils. The oils have a high heteroatom content (about 10–20% O, 6% N) and an energy content of about 30–50 MJ/kg. The molecular components of the oil include fatty acids, alkanes, and aromatics. Unlike conventional biodiesel synthesis, hydrothermal processing does not require algae with a high lipid content, though lipid-rich feedstocks may improve bio-oil yields, as discussed previously. Indeed, as nonlipid portions are converted into bio-oil, the amount of oil generated can be greater than the original hydrocarbon content in the algae.[91–94,97]

Early work in the 1990s on microalgae liquefaction focused almost exclusively on two green algae: halotolerant *Dunaliella terciolecta* and hydrocarbon-rich *Botryococcus braunii*.[92–94,96] Liquefaction at 250–350 °C with 5 wt% Na_2CO_3 typically yielded between 30–40% and 55–65% bio-oil for *D. terciolecta* and *B. braunii*, respectively. Net energy calculations indicated that *B. braunii* production and subsequent liquefaction would be the most net energy positive given its lower nutrient requirements and easier harvestability.[93] However, these net energy assessments have been called into question and much work remains to be done to determine the overall environmental impacts of algae cultivation and subsequent liquefaction.[98]

More recently, the liquefaction of harvested *Microcystis viridis* (95% moisture; containing 46% C, 7.3% H, 9.5% N) was examined. The maximum oil yield (33%) occurred at 340 °C, 5 wt% Na_2CO_3, and 30 min.[90] Longer times (60 min) decreased oil yields, likely due to degradation. The bio-oil was classified as a heavy oil, consisting mainly of C_{17}–C_{18} *n*-alkanes and aromatic compounds typically found in heavy oils (*e.g.*, naphthalene, dibenzothiophene, phenanthrene). Its elemental composition was 62% C, 8% H, 8% N, and 2% S, and it had a heating value of 31 MJ/kg. The remaining N was found in the aqueous phase (ca. 1000 mg N/L).

Jena and Das extended previous efforts at microalgae liquefaction by considering both a single (*Spirulina*) and mixed algae (from open ponds with wastewater) under conditions of different solids concentrations (10% and 20%), catalyst type (NiO, Na_2CO_3), catalyst loading (0 and 5%), and residence time (30 or 60 min).[45] Reaction temperature was 350 °C and ~20 MPa. Mixed algae contained higher ash content than *Spirulina* (*cf.* 22.5% and 6.6; largely due to harvesting with alum) but less protein (*cf.* 54.5% and 65.4%) and fewer lipids (*cf.* 6.7% and 13.1%). In both catalyzed and uncatalyzed reactions, bio-oil production was significantly less for mixed algae, likely because of its reduced protein and lipid fractions. Na_2CO_3 increased bio-oil yields, while NiO decreased oil yields but led to a higher fraction of C in gaseous products. Higher organic loading resulted in lower bio-oil yields at 60 min in uncatalyzed conditions (*cf.* 48% to 37%), though the reverse was true for catalyzed reactions. Bio-oil was alkaline (pH 9.3), contained 5–9% water and 6.5% N, and had an energy content of 30–36 MJ/kg. Interestingly, HPLC analysis of the aqueous phase found that ethanol (25 g/kg algae) was the major product, with formate (15 g/kg algae) and succinate (trace) present as well.

Yu *et al.* also liquefied algae at 20% solids and studied the impact of reaction conditions.[99] Bio-oil yield from commercially available *Chlorella pyrenoidosa* (74% protein, <0.1% crude fat) increased from 24% to 35% when the reaction temperature increased from 200 to 300 °C. Higher temperatures also significantly reduced the toluene insolubility value of the bio-oil, indicating higher-quality oil with lower viscosity. Initial N_2 pressure was found to have no effect on yield at lower temperatures.

To summarize this section on hydrothermal liquefaction, we recapitulate some of the main points:

1) The major feedstocks for liquefaction have been lignocellulosics, aquatic biomass (micro- and macroalgae), and wastes (animal manure and human sewage).
2) Bio-oil yields depend on feedstock composition; in particular, higher amounts of lipid and protein increase bio-oil yields.
3) Bio-oil yields tend to increase with temperature up to about 350 °C, then decline as more gaseous products are formed at higher temperatures.
4) Lower heating rates and longer processing times (beyond that necessary to achieve a nearly maximum conversion) often increase side reactions (*e.g.*, repolymerization) that increase residue formation and decrease bio-oil yields.
5) Alkali salts (e.g, Na_2CO_3) enhance bio-oil production and reduce residue formation at subcritical temperatures.
6) N content in the bio-oil is dependent on feedstock N composition and may pose a challenge for bio-oil upgrading and subsequent NO_x emissions.
7) Reducing gases (H_2 and CO) can improve bio-oil yield and composition, but are not required.

8.2.3 Outlook

While biomass liquefaction appears promising, its commercialization has been limited.[9] One of the most successful processes operating on a large scale is hydrothermal upgrading (HTU). The HTU process was developed largely in the Netherlands with support from Shell and other partners. Biomass is liquefied (300–350 °C, 120–180 bar, 5–20 min) to create a bio-oil (45% of input material), gas (25%; mostly CO_2), water (20%), and dissolved organics (10%; *e.g.*, acetic acid and ethanol). The heavy bio-oil typically has a heating value around 30–35 MJ/kg, H/C atomic ratio around 1:1, and an oxygen content of 10–18%. As mentioned previously, the process achieves about 75% thermal efficiency with proper heat integration.[13]

Hydrothermal liquefaction is a promising method for converting biomass into liquid fuels, but care should be taken to align scientific achievements with broader expectations beyond the laboratory. The majority of work to date has been carried out with model compounds that have allowed us to elucidate basic reaction pathways but provide limited information on the behavior of real biomass feedstocks. Future work should consider the effects of feedstock components that may provide *in situ* catalysis or inhibit desired reactions. There is also a need to more closely analyze all product streams, not just fuel components. Plus, the catalytic effects of metal-walled reactors have not been sufficiently explored and must be considered in any scale-up process. Though methods of bio-oil analysis have improved since the early work in this field,[100] standardization is lacking.[9] It is difficult to compare data across laboratories because definitions of yield and system phase behavior vary, pressures are often unspecified, and product analysis varies considerably. Finally, few if any studies have reported bio-oil production without the use of solvents (*e.g.*, acetone, dichloromethane) to separate this product from its surroundings. Recovered bio-oil N content may differ depending on the recovery solvent employed, as additional organic compounds may also be included in the extraction. Attempts to limit the heteroatom content of bio-oil should be made both before and after liquefaction, potentially as part of the feedstock-production scheme.

Liquefaction is most likely to develop as part of an integrated critical fluid biorefinery concept[101] or by integration with existing oil refineries for bio-oil upgrading.[102] Ultimately, small-scale bio-oil production at the site of biomass generation and subsequent transport to a central facility for bio-oil processing may be an efficient and cost-effective model for large-scale biomass conversion to liquid fuels.[57] While technoeconomic analyses have been done for several processes based on lignocellulosic feedstocks, more work with aquatic biomass and waste streams should help elucidate how liquefaction can serve as a waste-to-energy platform. In any form, this technology certainly faces challenges on scale-up: highly variable biomass compositions; feedstock slurries that pose reactor feeding and plugging difficulties; inorganic contaminants or components that impact product formation and may precipitate; and maintenance issues all must be considered in practical applications. Moreover, bio-oils suffer from high viscosity, corrosiveness, cold-flow problems, and degradation upon storage that has limited their commercial relevance.[8] Finally, the utility of liquefaction must be weighed against other alternatives; for example, hydrothermal gasification as discussed in the next section. Detailed life-cycle assessments should be carried out to compare these processes so the most effective use of renewable resources can be assured.

8.3 Hydrothermal Gasification

Wet biomass can be gasified at low temperatures ($\sim$300–400 °C) to make primarily CH_4 or at higher temperatures ($\sim$600–700 °C) to make primarily H_2. As previously mentioned, hydrothermal gasification has distinct advantages over conventional gasification of wet biomass and more quickly converts wet organic materials to fuel gases than biomethanation processes (*i.e.* anaerobic digestion).[15] Moreover, hydrothermal gasification is a highly efficient and economically attractive biomass and waste valorization technology.[103] Recently, an LCA of low-temperature hydrothermal gasification of waste biomass (20% solids manure or wood) to synthetic natural gas (SNG) demonstrated that 60–70% of the feedstock energy content was converted to SNG.[104] The fertilizer coproduct reduced greenhouse-gas (GHG) emissions by 0.6 $kg_{eq.}$ CO_2/MJ SNG, accounting for 97% of the beneficial impact. Not surprisingly, the major environmental impact for hydrothermal gasification is the CO_2 that is separated from CH_4 during treatment of the product gas. A large number of excellent review articles pertaining to hydrothermal gasification have appeared recently.[7,11,24,105–111]

Hydrothermal gasification was first described by Modell in 1985, who processed sawdust in supercritical water (SCW).[112] Since then, a substantial amount of work has been done to develop this technology by studying the effects of different feedstocks, reaction conditions, and catalysts on product yield and composition. Despite such attention, SCWG remains a chemically complex process that is not fully understood. This is due, in part, to the structural complexity of biomass, which makes it difficult to reduce the process to a network of reaction pathways with well-defined steps. Furthermore, research in this area tends to take a pragmatic approach that focuses on maximizing gas yields rather than elucidating and characterizing the mechanisms that constitute the underlying chemistry. In this portion of the chapter, we discuss the chemistry of hydrothermal gasification, including known reaction pathways, catalysis, and the influence of different process variables on product yields and composition.

8.3.1 Gas-Phase Reactions and Kinetics

The overall process that takes place during SCWG to produce H_2 can be summarized by the following endothermic reaction:

$$C_xH_yO_z + (2x - z)H_2O \rightarrow xCO_2 + (2x - z + y/2)H_2 \quad (8.1)$$

The elemental composition of the biomass (x, y and z) governs the composition of the gas mixture that forms at equilibrium. Typically, a feedstock with higher H content will result in increased gasification yields. Importantly, water is not only the reaction medium, but also a reactant and source of hydrogen. The presence of water in SCWG leads to hydrolysis reactions that help decompose the biomass feedstock into intermediate compounds.[24]

Recently, a kinetics model of cellulose and lignin conversion in uncatalyzed SCWG provided insight into the dominant intermediate reactions.[113] The model incorporates 11 reactions, including hydrolysis, steam-reforming, thermal decomposition, water-gas shift (WGS), and methanation. Due to the complexity of the feedstock, reaction intermediate species were not tracked individually; rather, they were considered together as a single pseudo-component. Nevertheless, the model predicted gas yields at biomass loadings and water densities different from those used in its construction, and simulations at long times produced equilibrium predictions that agreed with thermodynamic calculations.

In this model, biomass-derived intermediate compounds participate in steam reforming to generate syngas, a mixture of CO and H_2 (eqn 8.2).

$$C_xH_yO_z + (x - z)H_2O \rightarrow xCO + (x - z + y/2)H_2 \quad (8.2)$$

This endothermic reaction is the workhorse for generating H_2, and is particularly dominant at short times, when little gasification has taken place.[113] Syngas participates in two important reversible, exothermic gas-phase reactions: water-gas shift (eqn 8.3) and methanation (eqn 8.4).

$$CO + H_2O \leftrightarrow CO_2 + H_2 \quad (8.3)$$

$$CO + 3H_2 \leftrightarrow CH_4 + H_2O \quad (8.4)$$

These gas-interconversion reactions, particularly WGS, are most dominant at long reaction times when a majority of product gases have already been released.[113]

Pyrolytic reactions also contribute to hydrothermal gasification, particularly at high temperatures. Pyrolysis cleaves bonds in biomass intermediate compounds directly, leading to the formation of H_2, CH_4, CO, and CO_2 gases as well as char and tar. Solids production represents an effective loss of gas and is undesirable. One of the advantages of gasification in water is that char formation is suppressed,[114] except at slow heating rates.[115] The Resende model showed that H_2 is primarily produced *via* steam-reforming (eqn 8.2) at short times and WGS (eqn 8.3) at long times, whereas CO_2, CO, and CH_4 form predominantly from intermediate species *via* hydrothermal pyrolysis reactions.[113] This type of modeling analysis can serve as a foundation for future studies with other feedstocks and at other reaction conditions. Other than our recent work in quartz reactors[116,117] and that of Buhler *et al.* in a metal reactor[118] there are no published articles in which SCWG kinetics have been reported.

In an effort to eliminate confounding catalytic effects due to metal reactor walls, quartz capillary reactors have been used to study SCWG of several biomass model compounds: phenol and guaiacol (lignin model compounds),[117] glycine,[119] and methanol.[116] In these works, the effects of temperature, water density, reactant loading, and the presence of catalytic Ni wire on the yields and kinetics of liquid- and gas-phase reaction products were examined. Experiments with phenol led to rapid gasification at temperatures of 600 and 700 °C, and the presence of Ni increased the WGS reaction rate, resulting in higher H_2 yields than reactions without catalyst. Glycine experiments resulted in char formation, and although the presence of Ni catalyst did not promote gasification to completion, it did increase H_2 and CO_2 formation *via* the WGS reaction. Notably, gasification of methanol did not approach completion without the addition of Ni catalyst, and the major gas product was H_2. Currently, we are investigating SCWG of phenol in more detail to obtain a better fundamental understanding of the underlying chemistry. To date, we have identified more than 30 different reaction intermediates that form when phenol is gasified in SCW (Figure 8.6).

8.3.2 Regimes in Hydrothermal Gasification

In general, lower temperatures (350–500 °C) favor CH_4 production, while higher temperatures (400–600 °C) result in greater H_2 production (Figure 8.7).[19] Except at very high temperatures or long reaction times, uncatalyzed hydrothermal

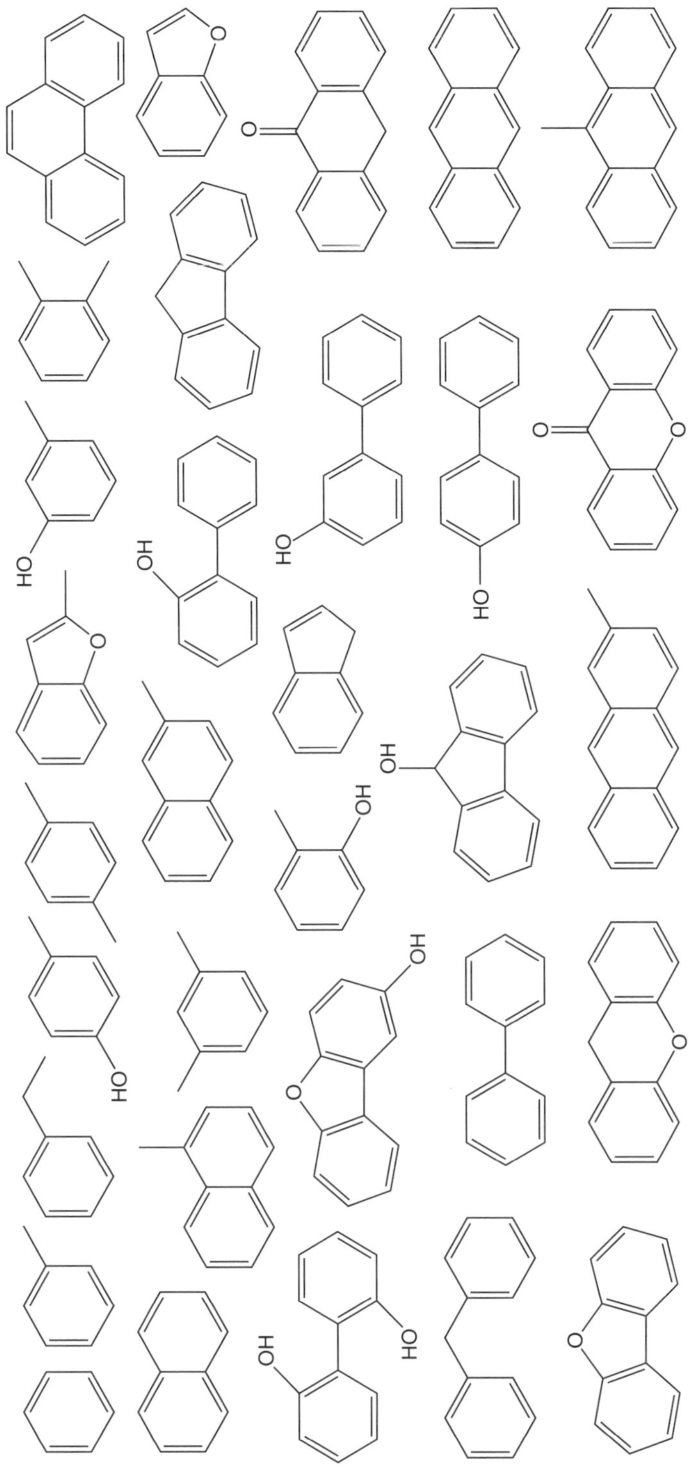

Figure 8.6 Liquid-phase products identified from phenol SCWG at 600 °C after ~ 1 h.

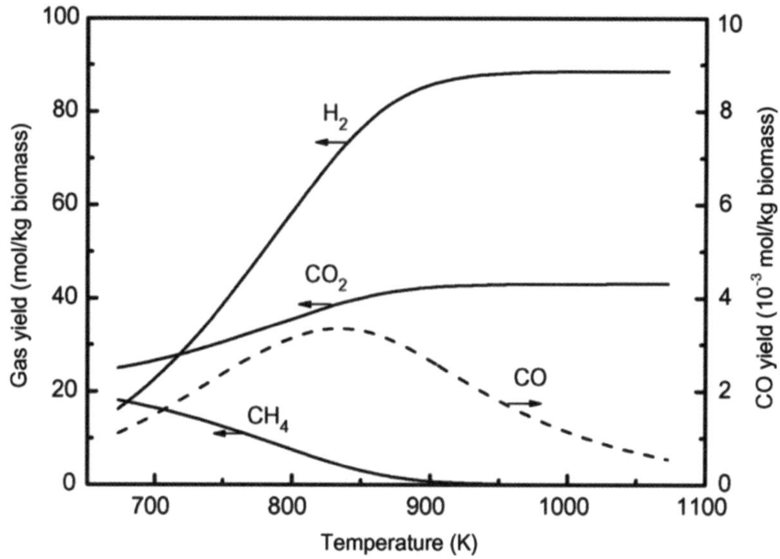

Figure 8.7 Calculated equilibrium gas yields for hydrothermal gasification of wood sawdust ($CH_{1.35}O_{0.617}$) as a function of temperature. Dry biomass concentration is 5 wt% and pressure is 250 bar. Reprinted with permission from Elsevier.[14]

gasification of biomass does not approach complete conversion, and measured gas yields tend to be far from their equilibrium values.[113] As a result, catalysis has been a major focus of research efforts to improve gas yields. Due to the different product yields and catalytic demands of each process, a distinction has emerged between low-temperature (near or supercritical water) catalytic gasification to obtain CH_4-rich gas and high-temperature SCWG to obtain H_2-rich gas.

8.3.2.1 Low-Temperature Catalytic Gasification

Low-temperature hydrothermal gasification generally ranges from approximately 350 to 500 °C and always involves catalysis.[12] Consequently, catalyst activity and stability is a major research focus. Early work demonstrated the effectiveness of Ni metal at promoting hydrothermal gasification of biomass at 350–450 °C and 20 MPa.[120] Consistent with the overall thermodynamics, CH_4 is predominant at these temperatures. These initial findings suggested that employing metal catalysts in a high-pressure aqueous environment could compensate for operation at lower temperatures.

Additional work showed that many commercial base- and noble-metal catalysts are inactive or present stability issues stemming from carbon deposition, oxidation, and sintering in a hot pressurized water environment.[121] Only Ni,

Ru, and Rh were reported to have a significant effect on hydrothermal gasification at 350 °C and 20 MPa. Although Ru showed long-term stability, Ni suffered from rapid sintering but could be stabilized by another metal.[122] The stability of common support materials in hot water was also investigated; some reacted with water or lost their physical integrity (*e.g.*, alumina), while carbon, monoclinic zirconia, and rutile titania were stable.[122] Results have been promising in subsequent studies with nickel and ruthenium catalyst formulations.[105,123,124] Alkali salts have also been studied as homogeneous catalysts for low-temperature hydrothermal gasification.[61,123,125–131] These catalysts, which are naturally present in real biomass, lower the onset temperature for cellulose degradation, depress char formation, and increase H_2 and CO_2 production by promoting the WGS reaction.[132,133] The activity of alkalis has been postulated to arise from their ability to suppress dehydration reactions that form furfural and 5-hydroxymethylfurfural (5-HMF), which are tar and char precursors, and from their ability to capture CO_2 as a metal carbonate in order to enhance the WGS reaction and promote H_2 formation.[131]

8.3.2.2 High-Temperature SCWG

In high-temperature gasification (500–800 °C), free-radical mechanisms dominate.[118] Pyrolysis reactions that directly release gases and the WGS reaction (eqn 8.3) are heavily promoted, leading to high yields of H_2-rich gaseous product. Lee *et al.* gasified glucose in SCW at 480–750 °C and 28 MPa.[134] The H_2 yield increased sharply with increasing temperature above 660 °C, while the CO yield peaked at this temperature and decreased at higher temperatures. This effect can be attributed to the very high rate of the WGS reaction at these high temperatures (~700 °C). Although the WGS reaction is exothermic and thermodynamically less favorable with increasing temperature, the overall H_2 yield at equilibrium remains high in this range due to high water concentrations (Figure 8.7).

Heterogeneous metal catalysts are often avoided at SCWG conditions due to degradation and deactivation,[135] but activated carbon and alkali have been employed as catalysts to aid high-temperature gasification. Carbon-gasification efficiencies near 100% were achieved for glucose gasification in supercritical water at 600 °C and 34.5 MPa with activated carbon.[136] Similar experiments with more complex biomass feedstocks resulted in reduced reaction rates with higher temperatures (650–715 °C) required to achieve high conversions.[135] Deactivation of the carbon was observed after several hours in both cases. Significantly, activated carbon catalyst can be recovered and separated from the ash produced by SCWG of biomass.[137] Recovery of catalyst and removal of ash will be necessary in large-scale continuous processes. Homogeneous alkali catalysts have also been employed to improve high temperature gasification. In experiments at 600–700 °C and 20–40 MPa, KOH as a catalyst increased H_2 and CO_2 yields and decreased CO yield by accelerating the WGS reaction.[130] Moreover, carbohydrates, aromatics, glycine, and real biomass were completely gasified to a H_2-rich product gas at 550–600 °C and 25 MPa with the addition of KOH and K_2CO_3.[114]

Recently, glucose was gasified in SCW with Ni and activated carbon as heterogeneous catalysts.[138] SEM was used to characterize the catalysts, demonstrating that although they were effective, combined Ni/activated carbon catalysts suffered from coke deposition at low temperatures and Ni crystallization, both of which inhibit catalytic activity by increasing pore size and reducing surface area. Catalysts with mesosize pores (2–50 nm) may be preferred for minimizing mass transport limitations when gasifying glucose in SCW.

Resende *et al.* (2009) recently reported on the noncatalytic gasification of lignin and cellulose in SCW.[139] Batch reactions were carried out in quartz reactors to avoid catalytic wall effects. The influence of different process variables on gasification yield and composition was systematically studied. The apparent activation energies suggested that H_2 and CO_2 form *via* different pathways, but these pathways are similar for both lignin and cellulose feedstocks.

8.3.3 Hydrothermal Gasification Chemistry

The chemical compositions of different biomass feedstocks are complex and vary depending on the source. For this reason, model compounds that are structurally similar to the components of biomass are often selected for study to obtain insight about their chemical behavior when undergoing hydrothermal gasification. As in studies on liquefaction, cellulose and lignin are commonly selected as biomass model compounds. Derivatives of biomass are also commonly used; glucose, xylose, and other monosaccharides form when cellulose and hemicellulose undergo hydrolysis. Phenols and aromatics result when lignin breaks down under hydrothermal conditions. Much of the previous work with simple model compounds has been exploratory, with little attention paid to the identities and yields of intermediate products in the liquid phase and along reaction pathways.

8.3.3.1 Cellulose

As noted in the section on liquefaction, the glycosidic bonds in cellulose and hemicellulose hydrolyze very quickly in HTW. Biomacromolecule monomers can form a bio-oil (at low temperatures) or be gasified (at higher temperatures). Reaction schemes for different temperature regimes have been proposed. At temperatures near 300 °C, glucose and oligosaccharides react much faster than cellulose hydrolyzes, resulting in a low yield of these hydrolysis products.[140] Mostly char is formed, unless catalysts are used to promote low-temperature gasification. Above the critical point, however, the cellulose hydrolysis rate jumps more than an order of magnitude and it becomes faster than the glucose decomposition rate, leading to an accumulation of hydrolysis products in the liquid phase. This phenomenon has been studied in a DAC,[140–142] which led to a mapping of the main reaction pathways (Figure 8.3). Furthermore, gas formation is favored at higher temperatures as the main reaction mechanism shifts from ionic to free radical, as mentioned earlier.

8.3.3.2 Lignin

Lignin decomposes in a hydrothermal environment to form low molecular weight phenolic compounds such as syringols, guaiacols, and catechols. Elimination reactions simultaneously produce formaldehyde, which immediately instigates condensation reactions with these alkylphenols.[143,144] As a result, the alkylphenols polymerize to form higher molecular weight compounds with formaldehyde acting as a crosslinking agent.[145] In short, without the addition of catalysts, hydrothermal gasification of lignin leads to substantial char formation and, therefore, low gas yields.[146] Ru and Rh with γ-alumina and carbon supports have been shown to be very effective catalysts for gasifying alkylphenols in water at 500 °C and 0.3 g/cm^3, whereas Pt and Pd were not effective.[147] However, char formation was suppressed by the presence of many of these noble metal catalysts. In experiments with lignin at 500 °C and 0.35 g/cm^3, zirconia and NaOH increased the H_2 yield by a factor of two and four, respectively, by inhibiting char formation.[148] Some char formation was observed with NaOH and Ni catalysts when gasifying lignin in water at 400 °C and 0.33 g/cm^3 water density, but the H_2 yield was still greater than without a catalyst.[149] No char and a high CH_4 yield were obtained with Ru.

8.3.3.3 Protein

Studies of protein-containing biomass resulted in lower than expected gasification yields.[52] Experiments with glucose, alanine, and $KHCO_3$ as models for cellulose, protein, and natural salts, respectively, confirmed that the presence of the amino acid decreases gas yields significantly and increases the dissolved organic carbon content.[129] As discussed previously, heterocyclic N-compounds formed from protein and carbohydrate degradation act as free-radical scavengers to limit the free-radical chain reactions important for gas formation. The presence of alkali salts has been found to counteract this inhibitory effect.

8.3.3.4 Interactions Between Model Compounds

In an effort to understand the mechanisms of hydrothermal gasification, we must move beyond the reaction pathways of individual feedstock components and investigate their interactions. Yoshida *et al.* (2001) studied the hydrothermal gasification of mixtures of cellulose, xylan, and lignin to represent the three main components of terrestrial biomass and determined that they do not collectively behave as a sum of the degradation reactions for the individual components.[150] Cellulose and hemicellulose mixtures produced gas distributions and yields similar to a weighted average of the expectation for each individual component, indicating the behavior of these two components is additive and no significant interaction exists. However, introducing lignin into a mixture with either cellulose or hemicellulose greatly suppressed H_2 production.[133] It was concluded that the degradation intermediates from cellulose and xylan react with lignin, perhaps donating H atoms *via* hydrogenation, resulting in decreased H_2 formation. The chemistry behind this behavior is still unclear.

8.3.3.5 Aquatic Feedstocks

In addition to lignocellulosics, catalytic hydrothermal gasification at subcritical temperatures for the production of CH_4-rich gaseous fuels has been explored for macroalgae[12] and microalgae[81,151] and is now being tested at a pilot scale.[152] Its usefulness in on-farm waste management has also been explored.[153] Minowa and Sawayama (1999) used a 50% Ni on silica-alumina catalyst and reacted harvested *Chlorella vulgaris* (87% moisture, 7% N, 6% ash, 49% C) at 350 °C and 18 MPa for ~1 h.[151] Higher catalyst loading led to increased conversion to gas (max 70% at 0.5 g catalyst/g wet algae) and increased the proportion of CH_4. Notably, nutrients (*e.g.*, N) recovered in the aqueous phase can be recycled for continued algal growth, though questions of its suitability remain due to catalyst (Ni) contamination.[151,154] A similar type of nutrient cycling may also be possible with the aqueous phase of liquefaction, though it has yet to be investigated.

SCWG has been explored with microalgae, with a focus on decreasing catalyst loss by precipitating N, S, and P as salts under SCW conditions in the appropriate pH range prior to the catalytic reactor.[155] In the conceptual process design, a dewatered algae slurry (~20% solids) would be pumped to ~30 MPa and preheated to 300–350 °C. Salts would then be precipitated and removed continuously from the fluid stream, which then flows to a hydrothermal methanation reactor at ~400 °C. Salt-based complications in SCW processing, which may be particularly acute in marine biomass, have been explored in continuously stirred tank reactors[127] and were recently studied *in situ* with neutron radiography to visualize salt buildup.[156] The continuous-flow process was designed to be thermally self-sufficient, with heat demands met by methane combustion or heat recovery. Ruthenium on activated coconut carbon (Ru/C) and ruthenium on zirconia (Ru/ZrO_2) were used for batch experiments without salt precipitation and showed promising results. Future work must validate this approach. Unlike liquefaction, the product of which is feedstock-specific, there is little evidence to date that CH_4 yields would rely on the specific composition of algae (*e.g.*, high lipids). Rather, one could exploit algae with the highest photosynthetic efficiency for maximal biomass production.

8.3.4 Outlook

Hydrothermal gasification of biomass shows promise as a clean and renewable energy conversion technology. Though SCWG is still not fully understood, many studies have revealed important aspects of the underlying chemistry. The main pathways for the hydrothermal gasification reactions that convert relatively low-value natural biomass to high-value synthetic fuel gas have been elucidated. Specifically, numerous model compound studies have clarified the role played by individual biomass components, such as cellulose, lignin, proteins, and salts. The influence of process variables such as temperature, pressure, water density, and biomass concentration are well understood. Work in the area of catalysis has revealed some of the reactions that are catalyzed by different metals, activated carbon, and alkali salts and under which conditions these catalysts are stable.

Nevertheless, there are still many areas where research is needed before commercial plants that gasify biomass in water can become economical. Biomass feedstocks are very complex, and so too is the chemistry needed to convert them to gases. A more fundamental mechanistic understanding of this chemistry is needed if detailed kinetic models are to be developed for process optimization and if catalytic opportunities are to be fully realized. However, while model compounds will continue to be useful for evincing reaction pathways of individual biomass components, the interactions between these pathways are not currently well understood and must be investigated. Eventually, the detailed reaction network must be built into a model that explains the hydrothermal gasification of whole biomass. Understanding how feedstocks other than lignocellulosics behave under hydrothermal gasification conditions will be important too. Recent work with aquatic biomass[155] and wastes[103,137] has demonstrated the importance of extending this technology beyond traditional renewable feedstocks. In particular, these studies have begun addressing some of the challenges this technology faces regarding feedstock ash content, catalyst deactivation, and product separation and recovery. A host of other technological issues must be considered as well: feedstock transport, solids handling, heat exchange efficiencies, and the need for robust, specialized equipment to maintain high pressures and temperatures. Many of these problems are currently being investigated at the pilot scale[157,158] and recent work in computational fluid dynamics is improving reactor design to enhance gasification and improve process efficiency.[159]

References

1. M. M. Titirici, A. Thomas, S.-H. Yu, J.-O. Muller and M. Antonietti, *Chem. Mater.*, 2007, **19**(17), 4205.
2. M. Sevilla and A. Fuertes, *Chemistry*, 2009, **15**(16), 4195.
3. S. Kumar and R. Gupta, *Ind. Eng. Chem. Res.*, 2008, **47**(23), 9321.
4. E. Kamio, H. Sato, S. Takahashi, H. Noda, C. Fukuhara and T. Okamura, *J. Mater. Sci.*, 2008, **43**(7), 2179.
5. S. Hunter and P.E. Savage, *Chem. Eng. Sci.*, 2004, **59**, 4903.
6. C. Comisar, S. Hunter, A. Walton and P.E. Savage, *Ind. Eng. Chem. Res.*, 2008, **47**, 577.

7. A. Peterson, F. Vogel, R. Lachance, M. Froelling, M. Antal and J. Tester, *Energy Environ. Sci.*, 2008, **1**(1), 32.
8. G. Huber, S. Iborra and A. Corma, *Cheminform*, 2006, **37**(52).
9. F. Behrendt, Y. Neubauer, M. Oevermann, B. Wilmes and N. Zobel, *Chem. Eng. Technol.*, 2008, **31**(5), 667.
10. Y. Yoshida, K. Dowaki, Y. Matsumura, R. Matsuhashi, D. Li, H. Ishitani and H. Komiyama, *Biomass Bioeng.*, 2003, **25**(3), 257.
11. A. Kruse, *Biofuel Bioprod. Bior.*, 2008, **2**(5), 415.
12. D.C. Elliot, *Biofuel Bioprod. Bior.*, 2008, **2**, 254.
13. F. Goudriaan, B. Van De Beld, F.R. Boerefijn, G.M. Bos, J.E. Naber, S. Van Der Wal and J.E. Zeevalkink, *Proceedings of Progress in Thermochemical Biomass Conversion*, 2000, 1312.
14. Y. Lu, L. Guo, X. Zhang and Q. Yan, *Chem. Eng. J.*, 2007, **131**(1–3), 233.
15. Y. Matsumura, *Energy Convers. Manage.*, 2002, **43**(9–12), 1301.
16. K. Ro, K. Cantrell, D. Elliott and P. Hunt, *Ind. Eng. Chem. Res.*, 2007, **46**(26), 8839.
17. Y. Calzavara, C. Joussot-Dubien, G. Boissonnet and S. Sarrade, *Energy Convers. Manage.*, 2005, **46**(4), 615.
18. A. Nakamura, E. Kiyonaga, Y. Yamamura, Y. Shimizu, T. Minowa, Y. Noda and Y. Matsumura, *J. Chem. Eng. Jpn.*, 2008, **41**(8), 817.
19. L.J. Guo, Y.J. Lu, X.M. Zhang, C.M. Ji, Y. Guan and A.X. Pei, *Catal. Today*, 2007, **129**(3–4), 275.
20. P. Kritzer and E. Dinjus, *Chem. Eng. J.*, 2001, **83**, 207.
21. A. Kruse and E. Dinjus, *J. Supercrit. Fluid*, 2007, **39**(3), 362.
22. A. Kruse and E. Dinjus, *J. Supercrit. Fluid*, 2007, **41**(3), 361.
23. A. Plugatyr and I. Svishchev, *Fluid Phase Equilib.*, 2009, **277**(2), 145.
24. P.E. Savage, *J. Supercrit. Fluid*, 2009, **47**(3), 407.
25. T. A. Milne, F. Agblevor, M. Davis, S. Deutch and D. Johnson, in *Developments in Thermal Biomass Conversion*, A. V. Bidgwater, D. G. B. Boocock, (ed.), Blackie Academic and Professional, London, UK, 1997.
26. D. Beckman and D. Elliott, *Can. J. Chem. Eng.*, 1985, **63**(1), 99.
27. Y. Yang, A. Gilbert and C. Xu, *AIChE J.*, 2009, **55**(3), 807.
28. Y. Qian, C. Zuo, J. Tan and J. He, *Energy*, 2007, **32**(3), 196.
29. T. Bhaskar, A. Sera, A. Muto and Y. Sakata, *Fuel*, 2008, **87**(10–11), 2236.
30. A. Demirbas, *Energy Sources*, 2005, **27**(13), 1235.
31. C. B. Xu and N. Lad, *Energy Fuel*, 2008, **22**(1), 635.
32. L. Cheng, X. P. Ye, R. He and S. Liu, *Fuel Process. Technol.*, 2009, **90**(2), 301.
33. B. Zhang, K. Von and M. K. Valentas, *J. Anal. Appl. Pyrol.*, 2009, **84**(1), 18.
34. B. Zhang, Z. Zhang, K. Von, Marc and K. Valentas, *Appl. Biochem. Biotechnol.*, 2009, **154**(1–3), 238.
35. X. Z. Yuan, J. Y. Tong, G. M. Zeng, H. Li and W. Xie, *Energy Fuels*, 2009, **23**, 3262.
36. Y. Zheng, L. Liang, Z. Mao, *2009 ASABE Annual International Meeting*, 2009, Paper Number: 096102.

37. C. Xu and J. Lancaster, *Water Res.*, 2008, **42**(6–7), 1571.
38. R. Dong, Y. Zhang, L. L. Christianson, T. L. Funk, X. Wang, Z. Wang, M. Minarick, G. Yu and Y. Email, *T. ASABE*, 2009, **52**(4), 1239.
39. N. Kobayashi, P. Guilin, J. Kobayashi, S. Hatano, Y. Itaya and S. Mori, *Powder Technol.*, 2008, **180**(3), 272.
40. Z. Fang, T. Minowa, R. Smith, T. Ogi and J. Kozinski, *Ind. Eng. Chem. Res.*, 2004, **43**(10), 2454.
41. B. Zhang, M. Von Keitz and K. Valentas, *Appl. Biochem. Biotechnol.*, 2008, **147**(1–3), 143.
42. R. J. Evans and J. Felbeck, G. T., *Org. Geochem.*, 1983, **4**(3–4), 145.
43. R. J. Evans and J. Felbeck, G. T., *Org. Geochem.*, 1983, **4**(3–4), 153.
44. K. Yamaberi, M. Takagi and T. Yoshida, *J. Mar. Biotechnol.*, 1998, **6**(1), 44.
45. U. Jena, K. C. Das, *2009 Bioenergy Engineering Conference*, 2009, BIO-098024.
46. A. Suzuki, *J. Chem. Eng. Jpn.*, 1988, **21**, 288.
47. S. Inoue, K. Okigawa, T. Minowa and T. Ogi, *Biomass Bioeng.*, 1999, **16**(5), 377.
48. Y. Dote, S. Inoue, T. Ogi and S. Yokoyama, *Biomass Bioeng.*, 1996, **11**(6), 491.
49. S. Inoue, S. Sawayama, Y. Dote and T. Ogi, *Biomass Bioeng.*, 1997, **12**(6), 473.
50. Y. Dote, S. Inoue, T. Ogi and S. Yokoyama, *Bioresour. Technol.*, 1998, **64**(2), 157.
51. Y. Dote, S. Yokoyama, T. Ogi, T. Minowa and M. Murakami, *Biomass Bioeng.*, 1991, **1**(1), 55.
52. A. Kruse, A. Krupka, V. Schwarzkopf, C. Gamard and T. Henningsen, *Ind. Eng. Chem. Res.*, 2005, **44**(9), 3013.
53. H. R. Appell, Y. C. Fu, S. Friedman, P. M. Yavorsky, I. Wender, *US Dept. of the Interior, Bureau of Mines*, 1971, Report 7560.
54. B. He, Y. Zhang, Y. Yin, T. Funk and G. Riskowski, *T. ASABE*, 2001, **44**(3), 697.
55. O. Bobleter, *Prog. Polym. Sci.*, 1994, **19**(5), 797.
56. Y. Yu, X. Lou and H.W. Wu, *Energy Fuels*, 2008, **22**(1), 46.
57. G. Brunner, *J. Supercrit. Fluid*, 2009, **47**, 373.
58. T. Nakata, H. Miyafuji and S. Saka, *J. Wood Sci.*, 2009, **55**(4), 289.
59. L. Suryawati, M. Wilkins, D. Bellmer, R. Huhnke, N. Maness and I. Banat, *Process Biochem.*, 2009, **44**(5), 540.
60. M. Sasaki, Z. Fang, Y. Fukushima, T. Adschiri and K. Arai, *Ind. Eng. Chem. Res.*, 2000, **39**(8), 2883.
61. T. Minowa, F. Zhen and T. Ogi, *J. Supercrit. Fluid*, 1998, **13**(1–3), 253.
62. Z. Fang, T. Sato, R. L. Smith Jr., H. Inomata, K. Arai and J. Kozinski, *Bioresour. Technol.*, 2008, **99**(9), 3424.
63. Wahyudiono, M. Sasaki and M. Goto, *Fuel*, 2009, **88**(9), 1656.
64. B. Zhang, H.-J. Huang and S. Ramaswamy, *Appl. Biochem. Biotechnol.*, 2008, **147**(1–3), 119.

65. R. Hashaikeh, Z. Fang, I. Butler, J. Hawari and J. Kozinski, *Fuel*, 2007, **86**(10–11), 1614.
66. R. Holliday, J. King and G. List, *Ind. Eng. Chem. Res.*, 1997, **36**(3), 932.
67. J. King, R. Holliday and G. List, *Green Chemistry*, 1999, **1**, 261.
68. M. Watanabe, T. Iida and H. Inomata, *Energy Convers. Manage.*, 2006, **47**(18–19), 3344.
69. J. M. Appleford, Y. Zhang, L. Christianson, T. L. Funk, K. Ocfemia, *Proceedings of the Seventh International Symposium*, 2005, ASAE Publication Number 701P0205.
70. B. He, Y. Zhang, T. Funk, G. Riskowski and Y. Yin, *T. ASABE*, 2000, **43**(6), 1827.
71. B.J. He, Y. Zhang, Y. Yin, T. Funk and G. Riskowski, *T. ASABE*, 2001, **44**(6), 1865.
72. G. Dismukes, D. Carrieri, N. Bennette, G. Ananyev and M. Posewitz, *Curr. Opin. Biotechnol.*, 2008, **19**(3), 235.
73. Q. Hu, M. Sommerfeld, E. Jarvis, M. Ghirardi, M. Posewitz, M. Seibert and A. Darzins, *The Plant Journal*, 2008, **54**(4), 621.
74. P. Schenk, S. Thomas-Hall, E. Stephens, U. Marx, J. Mussgnug, C. Posten, O. Kruse and B. Hankamer, *Bioenerg. Res.*, 2008, **1**(1), 20.
75. Y. Chisti, *Biotechnol. Adv.*, 2007, **25**(3), 294.
76. L. Lardon, A. Hélias, B. Sialve, J. Steyer and O. Bernard, *Environ. Sci. Technol.*, 2009, **43**(17), 6475.
77. E. Grima, E. Belarbia, F. Fernández, A. Medina and Y. Chisti, *Biotechnol. Adv.*, 2003, **20**, 491.
78. S. Amin, *Energy Convers. Manage.*, 2009, **50**(7), 1834.
79. B. Wang, Y. Li, N. Wu and C. Lan, *Appl. Microbiol. Biotechnol.*, 2008, **79**(5), 707.
80. C. Ryan, "Cultivating Clean Energy: The Promise of Algae Biofuels", NRDC, 2009.
81. K. Tsukahara and S. Sawayama, *J. Jpn. Petrol. Inst.*, 2005, **48**(5), 251.
82. N. H. Tran, J. R. Bartlett, G. S. K. Kannangara, A. S. Milev, H. Volk, M.A. Wilson, *Fuel*, 2009, **89**, 265.
83. V. Patil, K. Tran and H.R. Giselrød, *Int. J. Mol. Sci.*, 2008, **9**(7), 1188.
84. S. Xiu, A.C. Shahbazi, J., L. Wang, *2008 ASABE Annual International Meeting*, 2008, Paper Number: 083588.
85. W. Mulbry, S. Kondrad, C. Pizarro and E. Kebedewesthead, *Bioresour. Technol.*, 2008, **99**(17), 8137.
86. HydroMentia, Inc., http://www.hydromentia.com (retrieved 11/20/09).
87. A. Ross, J. Jones, M. Kubacki and T. Bridgeman, *Bioresour. Technol.*, 2008, **99**(14), 6494.
88. M. Aresta, A. Dibenedetto, M. Carone, T. Colonna and C. Fragale, *Environ. Chem. Lett.*, 2005, **3**(3), 136.
89. M. Aresta, A. Dibenedetto and G. Barberio, *Fuel Process. Technol.*, 2005, **86**(14–15), 1679.
90. Y. Yang, C. Feng, Y. Inamori and T. Maekawa, *Resour. Conserv. Recycl.*, 2004, **43**(1), 21.

91. S. Sawayama, S. Inoue, Y. Dote and S. Uokoyama, *Energy Convers. Manage.*, 1995, **36**(6–9), 729.
92. T. Minowa, S. Yokoyama, M. Kishimoto and T. Okakura, *Fuel*, 1995, **74**(12), 1735.
93. S. Sawayama, T. Minowa and S. Yokoyama, *Biomass Bioeng.*, 1999, **1733**.
94. Y. Dote, S. Sawayama, S. Inoue, T. Minowa and S. Yokoyama, *Fuel*, 1994, **73**(12), 1855.
95. S. Inoue, Y. Dote, S. Sawayama, T. Minowa, T. Ogi and S. Yokoyama, *Biomass Bioeng.*, 1994, **6**(4), 269.
96. M. Kishimoto, T. Okakura, H. Nagashima, T. Minowa, S. Yokoyama and K. Yamaberi, *J. Ferment. Bioeng.*, 1994, **78**(6), 479.
97. T. Matsui, A. Nishihara, C. Ueda, M. Ohtsuki, N. Ikenaga and T. Suzuki, *Fuel*, 1997, **76**(11), 1043.
98. L. Reijnders and M. A. J. Huibregts, *Biofuels for Road Transport: A Seed to Wheel Perspective*, Springer, London, 2009.
99. G. Yu, Y. Zhang, L. Schideman, T. L. Funk, Z. Wang, *2009 Bioenergy Engineering Conference*, 2009, BIO-097898.
100. D. Knezevic, D. Schmiedl, D. Meier, S. Kersten and S. Van, *W, Ind. Eng. Chem. Res.*, 2007, **46**(6), 1810.
101. J. King and K. Srinivas, *J. Supercrit. Fluid*, 2009, **47**(3), 598.
102. G. Huber and A. Corma, *Angew. Chem. Int. ed.*, 2007, **46**(38), 7184.
103. E. Gasafi, M.-Y. Reinecke, A. Kruse and L. Schebek, *Biomass Bioeng.*, 2008, **32**(12), 1085.
104. J. S. Luterbacher, M. Froling, F. Vogel, F. Marechal and J. W. Tester, *Environ. Sci. Technol.*, 2009, **43**(5), 1578.
105. D. C. Elliott, *Biofuel Bioprod. Bior.*, 2008, **2**(3), 254.
106. A. Kruse, *J. Supercrit. Fluid*, 2009, **47**(3), 391.
107. G. Brunner, *J. Supercrit. Fluid*, 2009, **47**, 373.
108. G. Van Rossum, B. Potic, S. Kersten and W. Vanswaaij, *Catal. Today*, 2009, **145**(1–2), 10.
109. P. Basu and V. Mettanant, *Int. J. Chem. Reactor Eng.*, 2009, **7**, R3.
110. Y. Guo, S. Wang, D.H. Xu, Y. Gong, H. H. Ma and X. Tang, *Renew. Sustain. Energy Rev.*, 2010, **14**(1), 334.
111. B. Yan and C. Wei, *Prog. Chem.*, 2008, **20**(10), 1553.
112. M. Modell, in *Fundamentals of Thermochemical Biomass Conversion*, R. P. Overend, T. A. Milne, L. K. Mudge, (ed.), Elsevier, 1985, p. 95.
113. F. Resende, P.E. Savage, *AIChE J.*, 2010, DOI: 10.1002/aic.12165.
114. H. Schmieder, J. Abeln, N. Boukis, E. Dinjus, a. Kruse, M. Kluth, G. Petrich, E. Sadri and M. Schacht, *J. Supercrit. Fluid*, 2000, **17**(2), 145.
115. A. Kruse, T. Henningsen, A. Sinag and J. Pfeiffer, *Ind. Eng. Chem. Res.*, 2003, **42**(16), 3711.
116. G. J. Dileo and P. E. Savage, *J. Supercrit. Fluid*, 2006, **39**(2), 228.
117. G. J. Dileo, M. E. Neff and P. E. Savage, *Energy Fuels*, 2007, **21**(4), 2340.
118. W. Buhler, E. Dinjus, H. Ederer, a. Kruse and C. Mas, *J. Supercrit. Fluid*, 2002, **22**(1), 37.

119. G. J. Dileo, M. E. Neff, S. Kim and P. E. Savage, *Energy Fuels*, 2008, **22**(2), 871.
120. D. Elliott and J. Sealock, in *Fundamentals of Thermochemical Biomass Conversion*, R. P. Overend, T. A. Milne, L. K. Mudge, (ed.), Elsevier Appl. Sci. Publ., 1985, p. 937.
121. D. Elliott, L. Sealock and E. Baker, *Ind. Eng. Chem. Res.*, 1993, **32**(8), 1542.
122. D. Elliott, T. Hart and G. Neuenschwander, *Ind. Eng. Chem. Res.*, 2006, **45**(11), 3776.
123. A. Sinag, A. Kruse and J. Rathert, *Ind. Eng. Chem. Res.*, 2004, **43**(2), 502.
124. T. Yoshida, Y. Oshima and Y. Matsumura, *Biomass Bioeng.*, 2004, **26**(1), 71.
125. A. Sinag, A. Kruse and V. Schwarzkopf, *Ind. Eng. Chem. Res.*, 2003, **42**(15), 3516.
126. A. Kruse and E. Dinjus, *Int. J. Res. Phys. Chem. & Chem. Phys.*, 2005, **219**(3), 341.
127. A. Kruse, H. Ederer, D. Ernst and B. Seine, *Chem. Eng. Technol.*, 2009, **32**(6), 971.
128. A. Sinag, A. Kruse and V. Schwarzkopf, *Chemie Ingenieur Technik*, 2003, **75**(9), 1351.
129. A. Kruse, P. Maniam and F. Spieler, *Ind. Eng. Chem. Res.*, 2007, **46**(1), 87.
130. K. Andrea, M. Danny, R. Pia and S. Michael, *Ind. Eng. Chem. Res.*, 2000, **39**(12), 4842.
131. J. A. Onwudili and P. T. Williams, *Int. J. Hydrogen Energy*, 2009, **34**(14), 5645.
132. D. Elliott and L. Sealock, *Ind. Eng. Chem. Prod. Res. Dev.*, 1983, **22**(3), 426.
133. Y. Matsumura, T. Minowa, B. Potic, S. Kersten, W. Prins, W. Van Swaaij, B. Van De Beld, D. Elliott, G. Neuenschwander, A. Kruse and M. Antal, *Biomass Bioeng.*, 2005, **29**(4), 269.
134. I. Lee, M. Kim and S. Ihm, *Ind. Eng. Chem. Res.*, 2002, **41**(5), 1182.
135. M. Antal, S. Allen, D. Schulman, X. Xu and R. Divilio, *Ind. Eng. Chem. Res.*, 2000, **39**(11), 4040.
136. X. Xu, Y. Matsumura, J. Stenberg and M. Antal, *Ind. Eng. Chem. Res.*, 1996, **35**(8), 2522.
137. T. Yanagida, T. Minowa, Y. Shimizu, Y. Matsumura and Y. Noda, *Bioresour. Technol.*, 2009, **100**(20), 4884.
138. I.- Lee and S.- Ihm, *Ind. Eng. Chem. Res.*, 2009, **48**(3), 1435.
139. F. Resende, P.E. Savage, *Energy Fuels*, 2009, **23**, 6213.
140. M. Sasaki, B. Kabyemela, R. Malaluan, S. Hirose, N. Takeda, T. Adschiri and K. Arai, *J. Supercrit. Fluid*, 1998, **13**(1–3), 261.
141. M. Sasaki, Z. Fang, Y. Fukushima, T. Adschiri and K. Arai, *Ind. Eng. Chem. Res.*, 2000, **39**(8), 2883.
142. M. Sasaki, T. Adschiri and K. Arai, *AIChE J.*, 2004, **50**(1), 192.

143. M. Saisu, T. Sato, M. Watanabe, T. Adschiri and K. Arai, *Energy Fuels*, 2003, **17**(4), 922.
144. K. Okuda, M. Umetsu, S. Takami and T. Adschiri, *Fuel Process. Technol.*, 2004, **85**(8–10), 803.
145. M. Osada, T. Sato, M. Watanabe, M. Shirai and K. Arai, *Combust. Sci. Technol.*, 2006, **178**(1–3), 537.
146. Y. Matsumura, M. Sasaki, K. Okuda, S. Takami, S. Ohara, M. Umetsu and T. Adschiri, *Combust. Sci. Technol.*, 2006, **178**(1–3), 509.
147. T. Sato, M. Osada, M. Watanabe, M. Shirai and K. Arai, *Ind. Eng. Chem. Res.*, 2003, **42**(19), 4277.
148. M. Watanabe, H. Inomata, M. Osada, T. Sato, T. Adschiri and K. Arai, *Fuel*, 2003, **82**(5), 545.
149. M. Osada, T. Sato, M. Watanabe, T. Adschiri and K. Arai, *Energy Fuels*, 2004, **18**(2), 327.
150. T. Yoshida and Y. Matsumura, *Ind. Eng. Chem. Res.*, 2001, **40**(23), 5469.
151. T. Minowa and S. Sawayama, *Fuel*, 1999, **1213**.
152. Genifuel Corporation, http://www.genifuel.com (retrieved on 11/20/09).
153. K. Cantrell, T. Ducey, K. Ro and P. Hunt, *Bioresour. Technol.*, 2008, **99**(17), 7941.
154. K. Tsukahara, T. Kimura, T. Minowa, S. Sawayama, T. Yagishita, S. Inoue, T. Hanaoka, Y. Usui and T. Ogi, *J. Biosci. Bioeng.*, 2001, **91**(3), 311.
155. S. Stucki, F. Vogel, C. Ludwig, A. Haiduc and M. Brandenberger, *Energy Environ. Sci.*, 2009, **2**(5), 535.
156. A.A. Peterson, P. Vontobel, F. Vogel and J.W. Tester, *J. Supercrit. Fluid*, 2009, **49**(1), 71.
157. A. Nakamura, E. Kiyonaga, Y. Yamamura, Y. Shimizu, T. Minowa, Y. Noda and Y. Matsumura, *J. Chem. Eng. Jpn.*, 2008, **41**(5), 433.
158. N. Boukis, U. Galla, H. Muller, E. Dinjus, 2007, 15th European Biomass Conference & Exhibition, Berlin.
159. T. Yoshida and Y. Matsumura, *Ind. Eng. Chem. Res.*, 2009, **48**(18), 8381.

CHAPTER 9

Lignin Utilization

GUNNAR HENRIKSSON, JIEBING LI, LIMING ZHANG AND MIKAEL E. LINDSTRÖM

Department of Fiber and Polymer Technology, Royal Institute of Technology, KTH, 100 44 Stockholm, Sweden

9.1 Lignin – a Large and Incompletely Utilized Natural Resource

If one opens an average biochemistry textbook and searches for the word "lignin", only a short notice or nothing at all will be found in most cases. This low interest is in large contrast to the sections on other plant polymers, such as starch and cellulose, which normally are thoroughly discussed both functionally and structurally. This small attention to lignin is neither motivated by the rarity of lignin – in fact, as one of the main components in wood, it actually belongs to the three biopolymers that occur in the largest amount by weight on Earth (the others being cellulose and chitin) – nor that lignin does not carry out any biologically important function; on the contrary, the development of lignin was a prerequisite for the development of higher land plants (see below) and thereby also for the development of higher animals on land.[1,2] Furthermore, the main organic compound in soil, *humus*, as well as the fossil resources low-rank coal, brown coal, stone coal and petroleum, originate to a large extent from microbiologically modified lignin.[3,4] Thereby, lignin indirectly has central importance both for agriculture and technology.

The relatively low interest in lignin in traditional biochemical research might have two very different explanations; first, lignin is in many ways an oddity among biopolymers – in contrast to biopolymers such as proteins, DNA, and

RSC Energy and Environment Series No. 1
Thermochemical Conversion of Biomass to Liquid Fuels and Chemicals
Edited by Mark Crocker

Published by the Royal Society of Chemistry, www.rsc.org

polysaccharides, lignin is optically inactive (meaning all carbons are racemic), its monomers are connected by several different types of linkages, the structure is most likely web-shaped rather than linear, and it lacks a well-defined secondary structure. These unusual properties are mainly due to the fact that lignin is biopolymerized according to different principles[5] than other biopolymers, as will be discussed below, and they make the polymer difficult to study using normal biochemical methods. A second reason is that the technical uses of lignin are relatively few, and do not generate high profits, compared with, for instance, cellulose. In fact, the large majority of lignins produced today are just burned for heat generation in the process industry. Since lignin is produced in large amounts as a byproduct from the chemical pulping of wood, it is one of the world's largest under-utilized renewable raw materials. One shall, however, keep in mind that wood and many wood-based materials such as medium density fiberboards, contain up to 30% lignin, albeit that the lignin is not used in a pure form, but as an integrated part in the fiber cell wall.

The complex structure of lignin and its close association with polysaccharides in wood make it very difficult to extract lignin from intact wood. During chemical pulping and certain other treatments of wood, both the lignin and the polysaccharide undergo chemical modifications, such that it is possible to extract the lignin in part. Such *technical lignins* differ, however, tremendously from the native lignin both in structure and properties. This chapter will discuss the structure and properties of natural lignin, the reactions it undergoes during different technical processes, and the possible applications of technical lignin with a focus on energy-related applications.

9.2 Structure, Properties and Biological Function of Lignin

The possibilities to use lignin in technical applications are highly dependent on its properties and thereby its structure. The structure and properties of lignin are intimately associated with the biological function of the polymer, and this is in turn associated with the evolution of plants. Thus, a deeper understanding of the structure and function of lignin requires a thorough discussion that includes aspects of science from botany to organic chemistry.

9.2.1 Occurrence and Biological Function of Lignin

The first land plants that developed were *bryophytes*, meaning that humidity is taken up from the surroundings by the entire body of the plant. They lacked roots and efficient tissues for transporting water, and hence they could not develop into large plants. Today, the phylum of mosses is a representative of this group. The more advanced land plants, *the vascular plants*, have developed a cell type, *the tracheid*, which can efficiently transport water over long distances, allowing vascular plants to develop into trees over 100 meters high.[6,7] The tracheid is in its mature stage dead, with only the cell wall left, and can be described as an empty tube that transports the water in the lumen. For a cell of this function it is, of course, very important that the water cannot penetrate the cell wall. The tracheid is also a strong cell type; inside the primary cell wall it contains a thick secondary cell wall that provides rigidity, and that plays an important mechanical role for the development of trees and other woody plants.[8,9]

The most primitive vascular plants reproduce with spores, and are called *vascular cryptogams* or *fern plants*. From these *the gymnosperms*, that reproduce with seed developed, of which the most important group today is *the conifers*. From the gymnosperms flowering plants, *the angiosperms,* developed, which are divided into the classes of *monocotyledons* (grasses including bamboo, palms, lilies, orchids and banana plants), and eudicotyledones (most other leaf-carrying trees and herbs).[9,10] There are also some more primitive angiosperms that are neither monocotyledons nor eudicotyledons, such as the water lilies. The angiosperms are more specialized and advanced than gymnosperms and vascular cryptogams; among others, tracheids were replaced by more specialized cells, *vessel elements* responsible for the water transport, and *librioform fibers*, taking care of the mechanical support.[8,9] The molecular structure of lignin and other cell-wall polymers also underwent modification during the development from gymnosperms to angiosperms. The evolutionary relationship between different groups of plant thus corresponds rather well with differences in the structure of lignin. Therefore, anyone interested in lignin is recommended to familiarize themselves with the simplified phylogenetic tree in Figure 9.1. Woody eudicotyledons are often called *hardwoods*, whereas conifers are called *softwoods*. These are the technically most important groups of woods, used as sawed timber and for pulping.

Bryophytes lack lignin, whereas all vascular plants contain the polymer in different amounts as part of the cell wall in tracheids and other cell types. This suggests that *lignin is involved in the "waterproof" properties of tracheid (and vessel element) cell walls*. Other functions that are attributed to lignin are, to *make the cell wall stiff*, to *work as a gluing substance in the middle lamella between the cells*, and to *make cell walls more resistant to biological degradation.*[11] Lignin is believed to carry out these functions by working as a cementing component in the cell wall. Theories for how lignin works on a molecular level will be discussed below, but first the chemical structure of lignin will be discussed.

9.2.2 Monomers and Intermonolignol Bonds

Lignin consists of monomers called *monolignols*, which chemically are *phenyl propanoids*, consisting of a phenolic unit with an aliphatic side chain located in the *para*-position. A monolignol consists of 9 carbons, and lignin can therefore be regarded as a polymer of *C9-units*. The three most common monolignols, *p-coumaryl alcohol, coniferyl alcohol* and *sinapyl alcohol*, are shown in Figure 9.2, together with some rarer monolignol lignin structures. The three

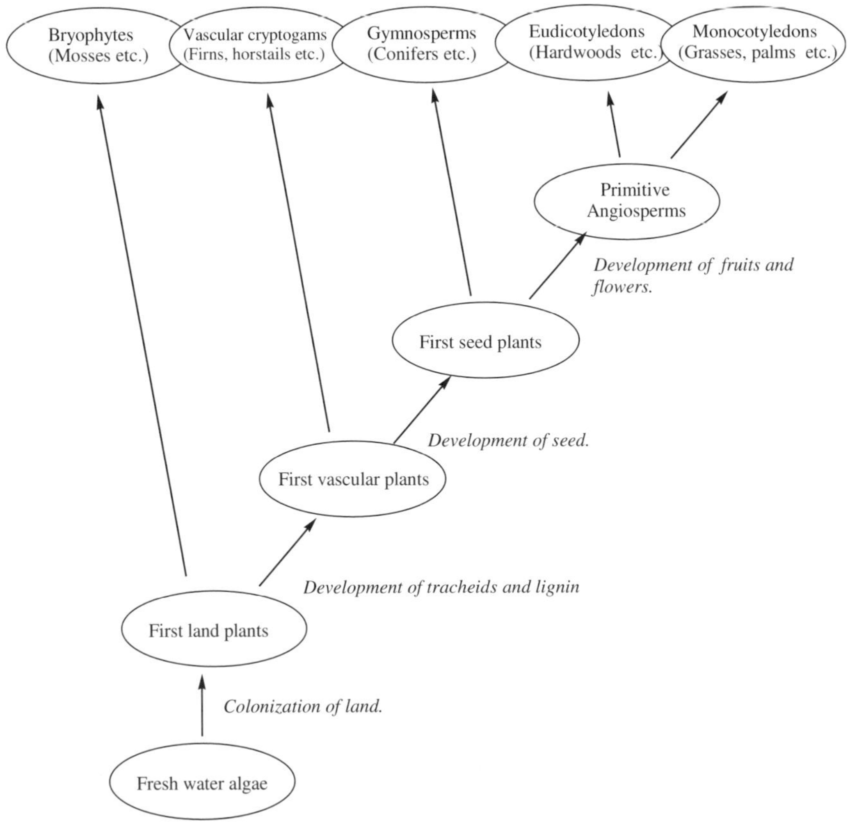

Figure 9.1 Simplified phylogenetic tree of the true plants. The figure shows the development of different groups of land plants from freshwater algae, and the development of important features associated with the plants are marked. The different phylogenic groups correlate well with chemical properties of lignin produced. Only the most important plant groups are shown in this figure.

most important monolignol structures differ in the number of methoxy structures attached to the phenol ring in the *ortho* position, where *p*-coumaryl alcohol has *no* methoxy group, conifeyl alcohol has *one*, and sinapyl alcohol has *two*. These structural differences between the common monolignols influence structural differences in the lignin polymer, which will be discussed below.[12–14] Lignin in conifers consists almost exclusively of coniferyl alcohol (with the exception of compression wood, where up to 30% of the monolignols are *p*-coumaryl alcohol). Lignin in hardwood and other eudicotyledons consists mainly of mixtures of sinapyl and coniferyl alcohol, although some species may contain low amounts of *p*-coumaryl alcohol. Grasses and other monocotyledons contain a mixture of all three main monolignols, and have significantly more *p*-coumaryl alcohol than other types of lignins. Other

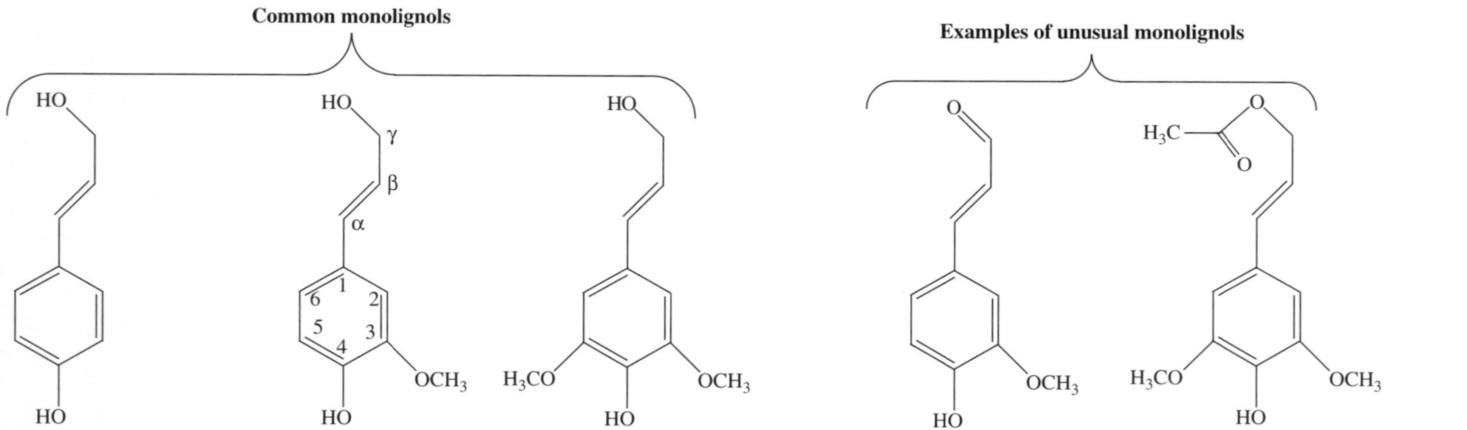

p-coumaryl alcohol mainly in compression wood in softwoods, and lignin of monocotyledons and vascular cryptogames.

Coniferyl alcohol in all kinds of lignin. In softwood it is the dominant monolignol.

Sinapyl alcohol in all kinds of lignin except softwood lignin. In the secondary cell wall of hardwood it is the dominat monomer.

Conifer aldehyde in small amounts in the primary cell wall of softwoods.

γ-acetylated sinapyl alcohol occurs in certain eudicotyledonic herbs, such as kenaf and jute.

Figure 9.2 The monomers of lignin, monolignols. Lignin consists mainly of the three common monolignols to the left, of which coniferyl alcohol is the most widespread monolignol within the plant kingdom. Lignins based on this monolignol, *i.e.* softwood lignin, have become the "standard" lignins in research. Some plants also use other monolignols in addition to the "standard set", of which two examples are shown here. They occur normally in relatively small amounts.

monolignols are rare and occur normally in small amounts, such as coniferaldehyde in certain softwoods, or as acetylated sinapyl alcohol in the eudicotyledonic herb kenaf. Special types of monolignols are especially common among monocotyledons. Lignin in vascular cryptogams seems to have similarities with the lignin in monocotyledons, and lignin in gymnosperms other than conifers is similar to softwood lignin with the exception of gnetophyte lignin that is similar to hardwood lignin.[15–17]

One of the unusual properties of lignin is that its monomers are connected by several different types of covalent bonds, which most likely are randomly distributed along the lignin chain. Monolignols are connected by different ethers and carbon–carbon bonds, while the aliphatic double bond is often lost.[5,12] The most widely used system for notation of the different monolignol bonds is based on notation of the carbons of the monolignols as shown in Figure 9.2, where the carbons in the aromatic ring are numbered with Arabic figures starting with the one connected to the propyl unit, and the aliphatic carbons are denoted with Greek letters, starting with the one closest to the aromatic ring.[18] The most important intermonolignol bond is an ether between the β-carbon and the 4-carbon, a β–O–4′ bond. This is by far the most common bond in all lignins, corresponding to half or more of intermonolignol bonds, and it is also the main bond that is broken in chemical pulping processes. There is also an ether that connects aromatic rings, *the 4–O–5′ bond*, and several types of carbon–carbon bonds, β–β, 5–5, and β–5. These bonds are more chemically stable than the β–O–4′ bond and therefore represent an obstacle in chemical pulping.[18,19] They are often termed, somewhat inappropriately, *condensed bonds*. In Table 9.1, the structures and occurrence of intermonolignol bonds are shown. Note that the quantification of lignin structures is an intricate problem – not only are the detection methods complex, it is also difficult to prepare lignin in pure form without damaging it, and therefore the exact figures concerning the frequency with which the bonds occur are uncertain. However, the importance of the β–O–4′ bond is without doubt. It is also clear that hardwoods generally have a higher content of β–O–4′ bonds and lower content of condensed bonds than softwood. These differences are due to the different monolignol structures – due to its methoxy group in the 5-position, sinapyl alcohol has difficulty in forming condensed bonds – and are technically very important; due to these differences hardwoods generally delignify faster than softwoods.

If the oxygen in a monolignol residue is involved in an ether bond, it is called a *nonphenolic residue*, whereas if the hydroxyl is free it is called a *phenolic residue* (Figure 9.3). The latter is much easier to oxidize, which is important in technical processes such as bleaching, and biological processes such as biopolymerization and biodegradation.[19,20] In natural lignin the nonphenolic structures dominate the phenolic ones, by a factor of approximately 10. It should be stressed that the lignin structure is racemic – something very unusual among biomolecules.[5,21]

Aside from the structural differences between lignin of different sources described above, there are also differences in lignin structures within the same plant; as mentioned above, lignin in compression wood from softwoods

Table 9.1 Intermonolignol bonds in natural lignin. The frequency with which the different monolignol bonds occur is uncertain, since the figures vary depending on the analytical method and how the lignin is prepared. The figures presented here represent our assumptions based on recent analytical data.[18,86–88]

Name	*Structure*	*Frequency in softwood (Bonds/100 C9 units)*	*Frequency in hardwood (Bonds/100 C9 units)*	*Comment*
β–O–4 bond (β-aryl ether)	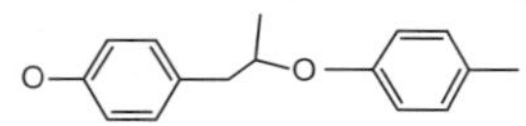	40–50	50–60	By far the most common bond in all lignin. It is the easiest intermonolignol bond to be broken in technical processes.
β–5 bond	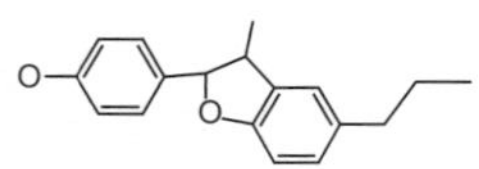	10–12	3	The most common carbon–carbon bond in lignin.
5–5	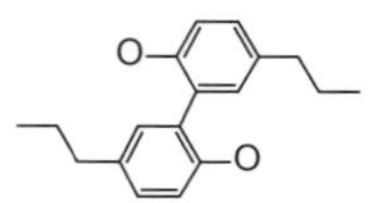	13	3	This structure is believed to be mainly formed by endgroup radicals. It can be involved in branching points in lignin.
4–O–5	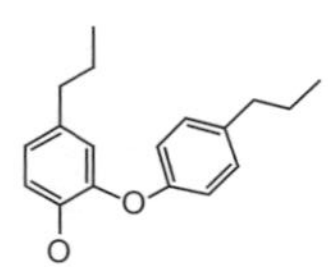	3	3	This structure is believed to be mainly formed by endgroup radicals. It can be involved in branching points in lignin.
β–β	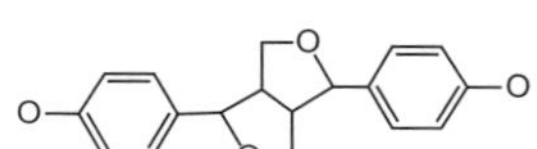	3	3	Theoretically interesting bond, since it appears to be non-randomly distributed.
Bonds to 1-position	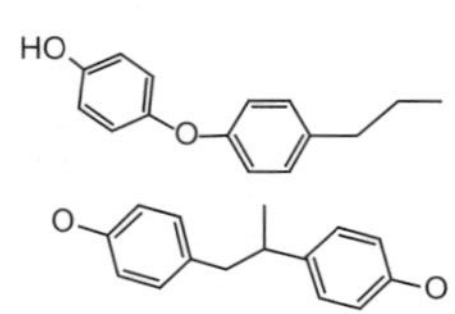	1–3	3	This type of bond "splits" a monolignol unit.

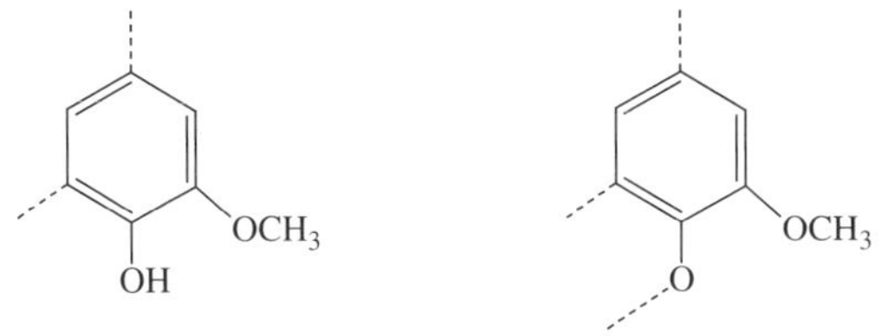

Figure 9.3 Phenolic and nonphenolic endgroups in lignin.

contains much more *p*-coumaryl alcohol than "normal" softwood lignin, and thereby has a higher content of condensed bonds (since branches consist, to a large extent, of compression wood with its condensed lignin, it is well known that branches are less well defibrillated than normal wood, and residues of incompletely defibrillated branches are often left after pulping).[22] The lignin in the middle lamella and primary cell wall of hardwoods has a higher content of coniferyl alcohol than the secondary cell wall, and thereby also a higher content of condensed bonds. Also in softwood, primary cell wall and middle lamella lignin seems to be somewhat more condensed than the secondary cell wall, although the differences are smaller, and no difference in monolignol composition is obvious.[23–26] This leads to the result that hardwoods need to be delignified to a lower lignin content than softwoods before defibrillation can occur. Furthermore, it is believed that (in hardwoods) vessel cells have a lignin richer in condensed bonds than librioform fibers.[27] This might be due to the fact that the secondary cell walls are thinner than in vessels.

In the literature, there are also a number of other types of bonds between monolignols reported. These are of lesser importance than the ones described in Table 9.1, and some structures may even have been formed during the lignin preparation process. More important are probably the covalent bonds between lignin and polysaccharides that are the next subject to be discussed.

9.2.3 Covalent Bonds Between Lignin and Polysaccharides

Several types of covalent bonds have been reported to occur between lignin and the polysaccharides of wood – in most cases hemicelluloses, but also cellulose and pectin. It seems clear that phenyl glycoside bonds and ethers to the α-carbon occur, but also carbon–carbon bonds and esters to the α-carbon have been suggested (Figure 9.4).[28–30]

Knowledge of these types of bonds, often called *lignin–carbohydrate complexes (LCC)*, was for a long time limited by a lack of suitable methods for their quantification. In recent years a number of methods have been developed, partly based on enzymatic treatments, which suggest that LCC are very common.[31] In wood, lignin covalently crosslinks different types of polysaccharides, *i.e.* the hemicelluloses xylan and glucomannan, and cellulose.[32] Figure 9.5 shows this schematically. This covalent crosslinking of wood polysaccharides

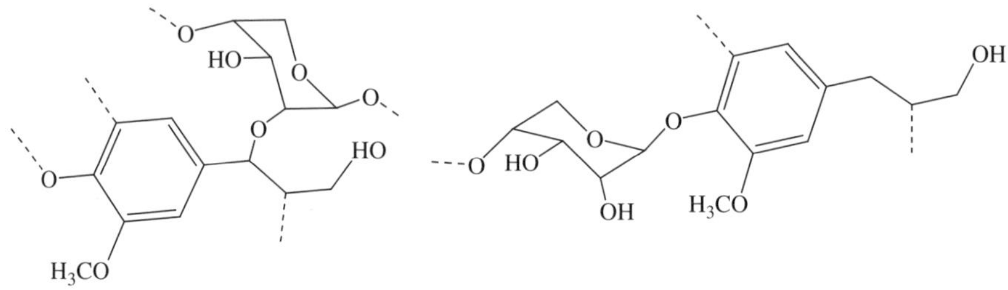

α-ether Phenyl glucoside

Figure 9.4 LCC, covalent bonds between polysaccharides and lignin. In the figure the structure of the two best-documented forms of LCC, α-ethers and phenyl glucosides, are shown. Here, the monosaccharide residue involved in the bond is a xylose residue, but similar bonds do occur to other monosaccharide residues in wood polysaccharides. Also, other types of LCC bonds have been suggested – α-esters and carbon–carbon bonds.

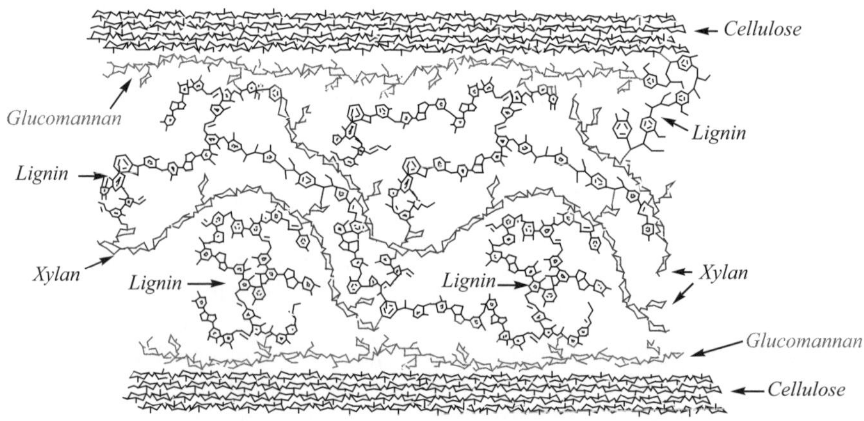

Figure 9.5 LCC networks. The figure shows a hypothetical organization of lignin and polysaccharides in wood. Cellulose fibrils are surrounded by a matrix of the hemicelluloses xylan and glucomannan, and lignin, where the latter covalently crosslinks the three polysaccharides. In this way the lignin may "lock" the polysaccharides to each other, thereby preventing swelling in a "curing" reaction.

might be very important both biologically and technically; biologically, since the function of lignin in part is to prevent the plant wall swelling in water and to make the cell wall stiff and rigid. Covalent crosslinking of the different cell wall polysaccharides obviously adds the functionalities of resistance to swelling and stiffness to the polysaccharides. Lignin can thus be described as a *curing agent in wood*. Indeed, the LCC seem to be one of the main obstacles for removal of lignin from plant cell walls.[33]

9.2.4 Physical and Chemical Properties

Lignin has, in many ways, different properties from cellulose and other polysaccharides, which are important from a technical perspective.

- *Lignin is amorphous.* Due to its branched, racemic and partly random structure, lignin is very amorphous, and no good lignin crystal structures have been reported (in contrast to cellulose). This property might be important for the biological function of lignin, *i.e.* to fill up cavities in the polysaccharide network. It is also important for certain technical applications. For instance, it is more difficult to obtain homogeneous low molecular weight fractions from lignin than from cellulose.
- *Lignin is aromatic.* In contrast to polysaccharides, lignin contains aromatic structures. This affords possibilities for charge-transfer interactions (π-binding, aromatic interaction) to polysaccharides as a complement to the covalent bonds discussed above. The aromatic structure is important for some technical applications, such as the use of lignin as an adhesive, and for the manufacture of chemicals from lignin.
- *Lignin is moderately hydrophobic.* Due to the aromatic structures lignin is more hydrophobic than polysaccharides. However, lignin is not nearly as hydrophobic as lipids. Therefore, the relatively low degree of water adsorption by the ligninfied cell wall can probably not be explained by the hydrophobicity of the lignin as such; the LCC probably play an important role here as discussed above. Lignin possessing a high degree of polymerization has poor solubility in water at neutral pH, but can be dissolved at high pH, or in organic solvents, such as chloroform and acetone. Mixtures of water and dioxane are also good solvents for lignin.
- *Lignin is more energy rich than polysaccharides.* Due to the fact that the carbons of lignin are more reduced than carbohydrates – the average redox number of a carbon in lignin is around –0.4, whereas it is 0 in carbohydrates (it is however considerably lower than fatty acids and saturated hydrocarbons, that often have redox numbers around –II) it has a higher energy content than the wood polysaccharides. The energy content can be utilized in combustion. However, the complex structure of lignin makes it difficult to use the energy for biological purposes, such as an energy/carbon source for fermentations, compared with cellulose and other polysaccharides that are relatively easy to use for ethanol fermentation. Furthermore, during different technical processes for lignin preparation, such as pulping, the lignin structure is modified, which may result in a lower energy content.

9.2.5 Biosynthesis

Most biopolymers are formed by biologically controlled condensations thermodynamically driven by phosphorylations.[21] These reactions are carried out inside enzyme complexes such as cellulosomes and ribosomes, and give optically active polymers often with perfectly determined primary structure.[21] Lignin, on the other hand, is racemic as described above, and lacks well-defined primary structure. Consequently, lignin is also synthesized according to different principles.[34] The monolignols (Figure 9.2), are oxidized, probably by peroxidases, to resonance stabilized radicals[35] (Figure 9.6). As seen in the figure, the unpaired electron is delocalized to the β-, 4-O-, 5-, 1-, and 3-positions. Covalent bonds are formed when two unpaired electrons meet and form an electron pair.[36] In principle, all kinds of combinations of unpaired electrons are possible, but in practice several of them are impossible; the product can be chemically unstable, as when two 4-O radicals couple and a peroxide is formed, or it can be sterically hindered; for example, the methoxy group in coniferyl alcohol efficiently prevents coupling at the 3-position, and the methoxy groups in sinapyl alcohol prevent the possibility for coupling at both the 3- and 5-positions. Thus, the monolignol composition determines the frequency of different intermonolignol bonds in lignin, *i.e.* lignin made of coniferyl alcohol (as for soft wood lignin), has fewer β–O–4 bonds and more condensed bonds than lignin made of mixtures of coniferyl alcohol and sinapyl alcohol (as for hardwood lignin).[37] This is important for technical reasons, since hardwood lignin is degraded faster by chemical pulping than softwood lignin due to the higher content of β–O–4 bonds in the former. Coupling at the 1-position is, however, sterically possible, although these types of bonds are relatively rare in lignin (Table 9.1).

However, most likely it is only a minor fraction of the covalent bonds in lignin that are formed by coupling of two monolignol radicals; phenolic endgroups (phenolic endgroups are created if the monolignol couples at the β- or 5-position) on the lignin polymer/oligomer can also be oxidized to radicals (Figure 9.6). The unpaired electrons on phenolic endgroups are normally located at the 4-O-, 5- and 1-positions. Endgroups can form covalent bonds by coupling with monolignol radicals, or other endgroup radicals. It is believed that 5-5 and 4-O-5 bonds are mainly created by the coupling of endgroups, and β–O–4-and β-5 bonds are believed to mainly be created by the coupling of monolignol radicals to endgroup radicals.[38]

The mechanisms for the formation of LCC are incompletely understood; α-ethers and α-esters can be created if a sugar alcohol or carboxylic acid replaces the water in the formation of a β–O–4 bond (Figure 9.7), but it is not known how phenyl glucosides are created.[39]

In principal, the lignin polymer can be created according to two different processes:

1. By *batchwise polymerization*. This means that monolignol-radicals form dimers, those in turn get oxidized and form tetramers, which in turn couple to larger structures, *etc*.
2. By *endwise polymerization*. This means that monolignol radicals couple to endgroup radicals on polymeric lignin. In effect, lignin polymers "grow" from phenolic endgroups. The growing ends may "fuse", and divide into two growing ends.

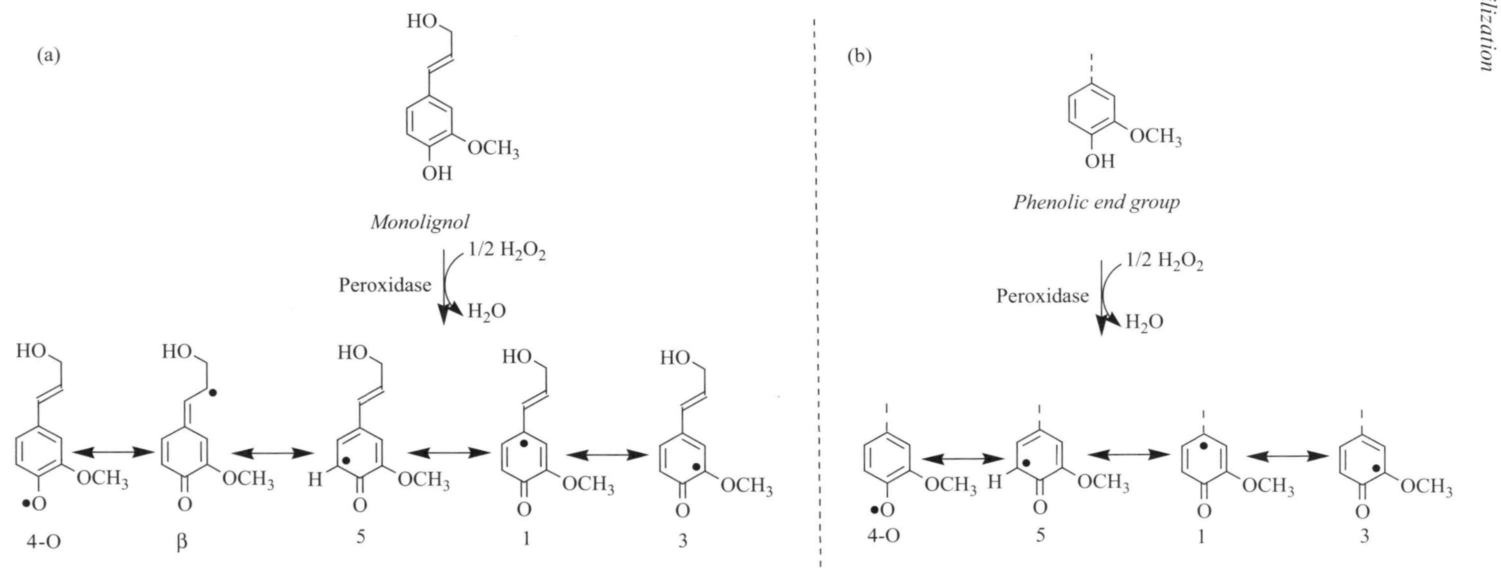

Resonance forms ordered after abundance. The position of the unpaired electron is marked.

Figure 9.6 Oxidation of monolignols (a) and phenolic endgroups (b) during lignin biopolymerizations. As shown, the type of radicals are similar, except that the endgroups normally cannot have unpaired electrons on the β-position.

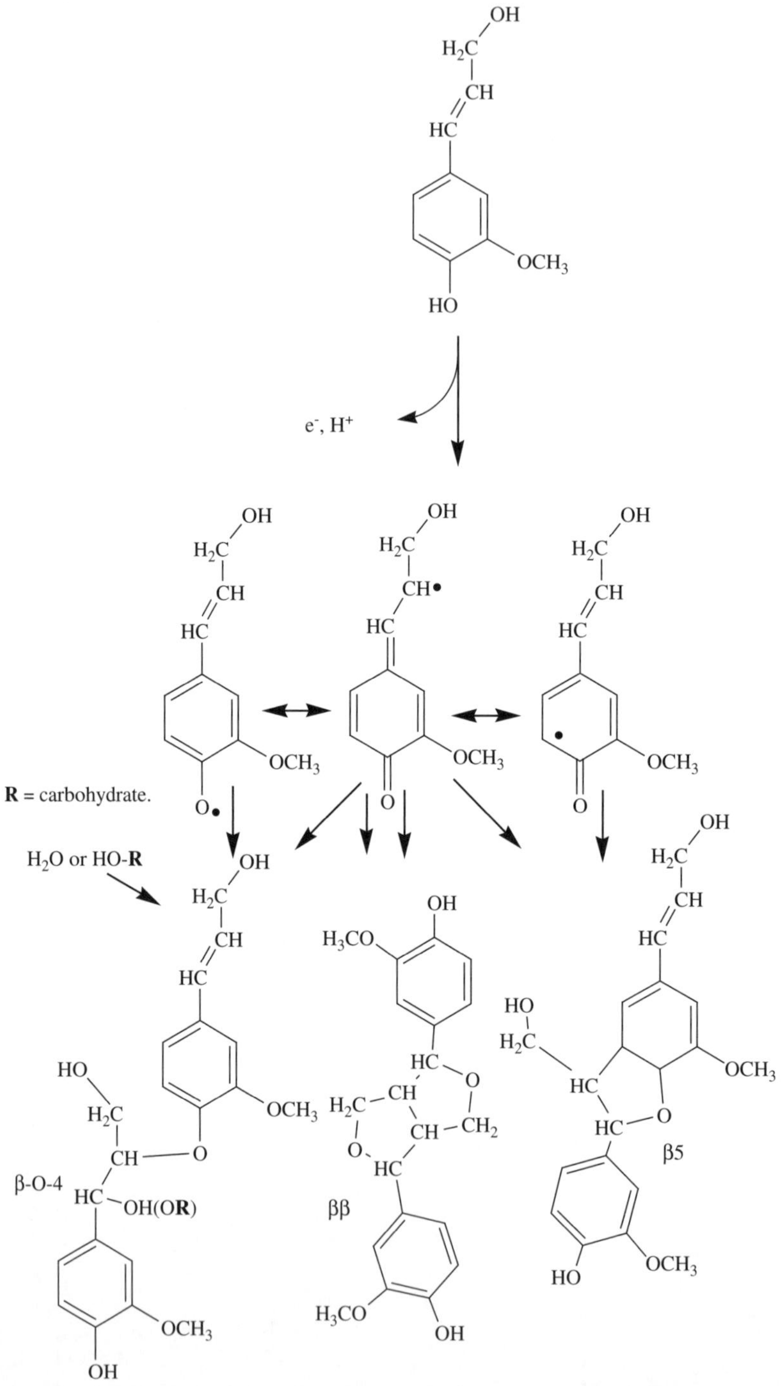

Figure 9.7 Example of pathways leading to creation of intermonolignol bonds.

Analysis of binding patterns in lignin suggests that the second process dominates – at least in wood.

The oxidation of monolignols and endgroups to radicals is often assumed to be carried out by peroxidases, and several candidates have been described,[35] while the copper-containing oxidase laccase has also been suggested.[40] However, the steric factor, as well as structural factors, may make it difficult for enzymes to introduce the radicals into lignin – especially in the late phase of lignification, when the structure of the cell wall is so compact that large molecules such as enzymes have difficulty diffusing and coming into contact with the phenolic endgroups. It has therefore been suggested that the enzymes receive help from a low molecular weight *redox shuttle* that is oxidized by the enzymes and in turn oxidizes phenolic endgroups to radicals, with the redox shuttle returning to its relaxed form. Mn(II)/Mn(III), superoxide and monolignol radicals have been suggested to play this role.[41–43] Other low molecular weight compounds, such as radical scavengers and reducing agents, may also play a role in lignin biopolymerization.[44,45]

Monolignols are synthesized from the amino acid phenyl alanine in the general propanoid pathway, which has been subjected to intensive studies. It has been shown that it is possible to change the proportions and even the structure of the monolignols used in lignification. This is due to the *metabolic plasticity of lignin*, *i.e.* the system for polymerizing monomers of lignin is subject to much less biological control than for other biopolymers, such as cellulose and DNA. Consequently, slightly modified monolignols can be included in the polymer, and a functional lignin still be produced. From this, it shall not be understood that lignin can be made of any phenolic compound, or that the properties of lignin are unchanged by the inclusion of modified monolignols. On the contrary, modified monolignols can give a lignin that reacts differently in, for instance, chemical pulping. There are currently large projects underway aimed at making such transgenic plants, which will be industrially very important for cultivation in plantations. This research has been thoroughly reviewed elsewhere.[46,47]

9.3 Reactions of Lignin During Technical Processes

During different types of technical processes involving wood, lignin can be obtained as a byproduct. Generally, the lignin undergoes rather intensive chemical modification during these treatments. In chemical pulping the goal of the process is to remove the lignin, and this is done by: (i) *depolymerizing the lignin*; (ii) *breaking bonds between lignin and polysaccharides*; and (iii) *making the lignin more soluble*. The latter can be done both by modifying the lignin, or using a good solvent for lignin.[19,48]

9.3.1 Reactions During Alkaline Pulping

Chemical pulping in alkaline solution is the dominant form of chemical pulping today. The method is based on the fact that lignin is degraded under alkaline conditions, and that strong alkali solutions are good solvents for fragmented lignin. The pulp produced is relatively strong and the method is, with modern techniques, easy to implement in very large units. The lignin produced as a byproduct of the process has so far found few high-value applications and is today mainly burned in the recovery boiler as part of the chemical regeneration of the pulping chemicals, and for generating heat.

Alkaline pulping can consist of simply heating of biomass in strong alkali (*soda pulping*) – today this is mainly used for nonwood plants – but in most cases additives are used that promote lignin degradation; these can be sulfide ions (*kraft pulping*), anthraquinone (*anthraquinone pulping*), or polysulfide (*polysulfide pulping*). Polysulfide pulping is done in the presence of sulfide ions and can be seen as a version of kraft pulping. Anthraquinone can also be added to the kraft cook to improve selectivity.[19,49,50]

The most important chemical reactions deal with the β–O–4 bonds. At high temperature and high alkali concentrations nonphenolic β–O–4 bonds can be broken, as described in Figure 9.8. This reaction is important in all alkaline pulping techniques. Note that the reaction normally creates a phenolic end-group. Phenolic β–O–4 bonds can be attacked with the different additives in pulping liquids, sulfide, anthraquinone and polysulfide, as described in Figures 9.9 and 9.10. These reactions are considerably faster than the reaction of nonphenolic lignin in Figure 9.8. The degradation of lignin can thus be described as the nonphenolic degradation (Figure 9.8) that cuts the lignin creating starting points for the fast reactions (Figure 9.9) that "peel" off monolignol units from the phenolic endgroup, until a condensed bond is reached. These types of reactions proceed until the lignin fragments are small enough to be soluble in the strong alkali solution.[19] However, covalent bonds between lignin and polysaccharides (LCC) are an important obstacle for the delignification process; it has been shown that almost all lignin molecules remaining attached on the pulp fiber are covalently attached to polysaccharides.[33] Some types of LCC can be broken as described in Figure 9.9, but covalent bonds between lignin and polysaccharides survive the kraft cook.[31,33]

Unfortunately, the specificity of the alkaline chemical pulping process is not 100%; unfavorable side reactions may occur. Figure 9.10 shows the most

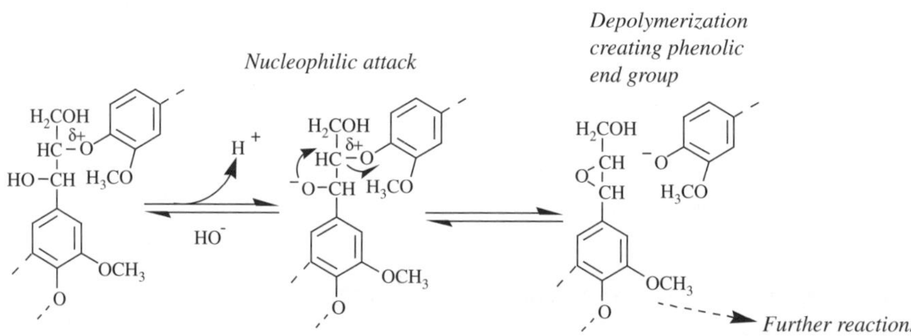

Figure 9.8 Alkali breakage of nonphenolic β–O–4 bond.

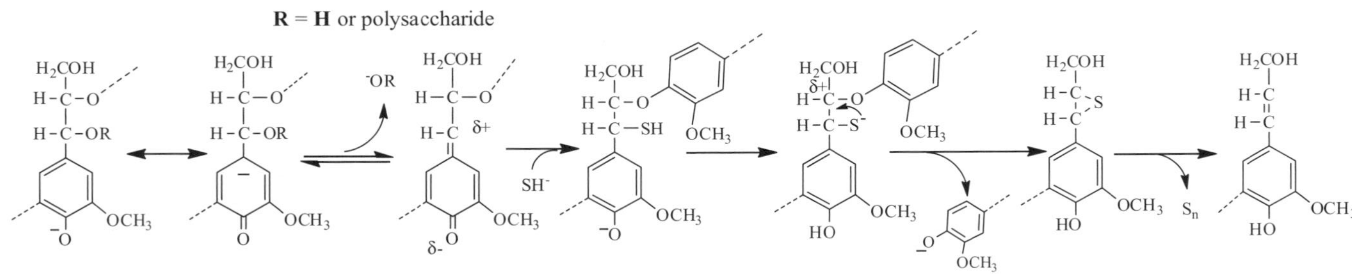

Deprotonated phenolic endgroups have a resonance form, where the negative charged is delocalized on the 1-position.

The hydroxyl (or LCC) of α-position can leave forming a quinomethide intermediate. This has a δ+ on α-position, that can undergo nucleophilic attack by sulphide ions.

The sulphide couples to the α-position, gets deprotonated and performs a nucleophilic attack at the β-position. This breaks the ether to the next monolignol residue and "peels off"one monolignol residue creating a new phenolic endgroup. The created thioepoxygroup eliminates elementary sulphur creating a double bond.

Figure 9.9 Sulfide-dependent breakage of β–O–4 bond.

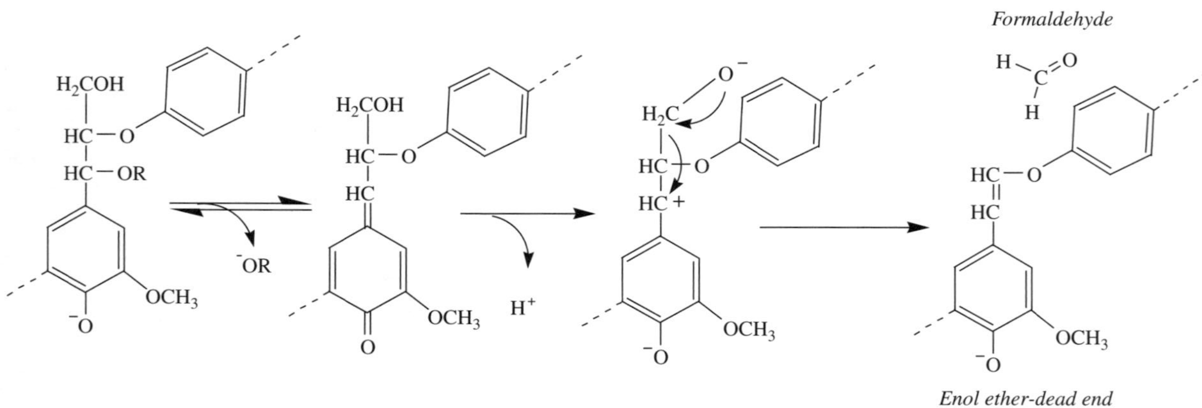

Figure 9.10 Formation of enol ether from β–O–4 bonds under alkaline conditions. This structure is a "dead end" in alkaline pulping delignification.

important, in which a phenolic endgroup is converted to an enol ether structure that is stable under alkaline conditions. In this way, a "dead end" is created that stops further degradation on this part of the lignin polymer. As the alkaline pulping proceeds, such stable structures and condensed bonds will be accumulated in the remaining lignin in the fiber. These structures, together with LCC (see above), may be part of the explanation why delignification proceeds more slowly in the latter stages of chemical pulping.[19,51] Practically, it is therefore not possible to remove all of the lignin from a fiber in kraft pulping or other alkali pulping methods. The degree of delignification varies between different pulp qualities. For instance, strong brown papers used for sack paper and the outer layer in corrugated board contain approximately 4–8% lignin (the wood used for these products contains around 27% lignin on wood), whereas pulp that is to be used for fine paper may contain around 2% lignin on wood. These pulp qualities are bleached for further lignin removal.

Both the lignin remaining in the fiber, and especially the lignin released in the process liquid, undergo various types of additional reaction that are not fully understood, although they may include radical coupling reactions, which modify the structure and create strongly colored structures. This makes it necessary to bleach kraft lignin fibers, if they are to be applied in bright products. Furthermore, the lignin released in the process liquid will become heterogenic and heavily chemically modified. Therefore, it can be separated into several fractions with different properties.[52]

Both polysulfide and antraquinone can take part in delignification reactions in similar ways to sulfide ions, but are regarded as being more efficient than sulfide ions. Polysulfide and anthraquinone can also oxidize endgroups on polysaccharides, which gives less carbohydrate degradation during alkaline pulping (Figure 9.11).[49,50,53]

9.3.2 Reactions During Neutral and Acidic Sulfite Pulping

Sulfite pulping was formerly the most important method for chemical pulping, but has now been replaced by mainly kraft pulping for most applications. Dissolving pulps, *i.e.* pulps used for the manufacture of cellulose derivatives, regenerated cellulose, *etc.*, are, however, still mainly made according to sulfite processes. Another advantage is that the lignin produced as byproduct is more valuable than the kraft lignin. The method is based on heating wood in a sulfite solution at low or neutral pH. Different counterions can be used to the sulfite ions, *e.g.*, Ca^{2+}, Na^{+}, Mg^{2+}, NH_4^{+}.

Sulfite pulping is carried out at lower pH than the methods discussed above, and consequently the solubility of lignin is much lower. However, the lignin is chemically modified *via* the introduction of anionic groups *via* sulfonation, as shown in Figure 9.12, making the lignin soluble at a wide pH range. In addition to this solubilizing reaction, lignin can be depolymerized by hydrolysis of β–O–4 bonds (as shown in Figure 9.13) in acidic sulfite pulping. During neutral

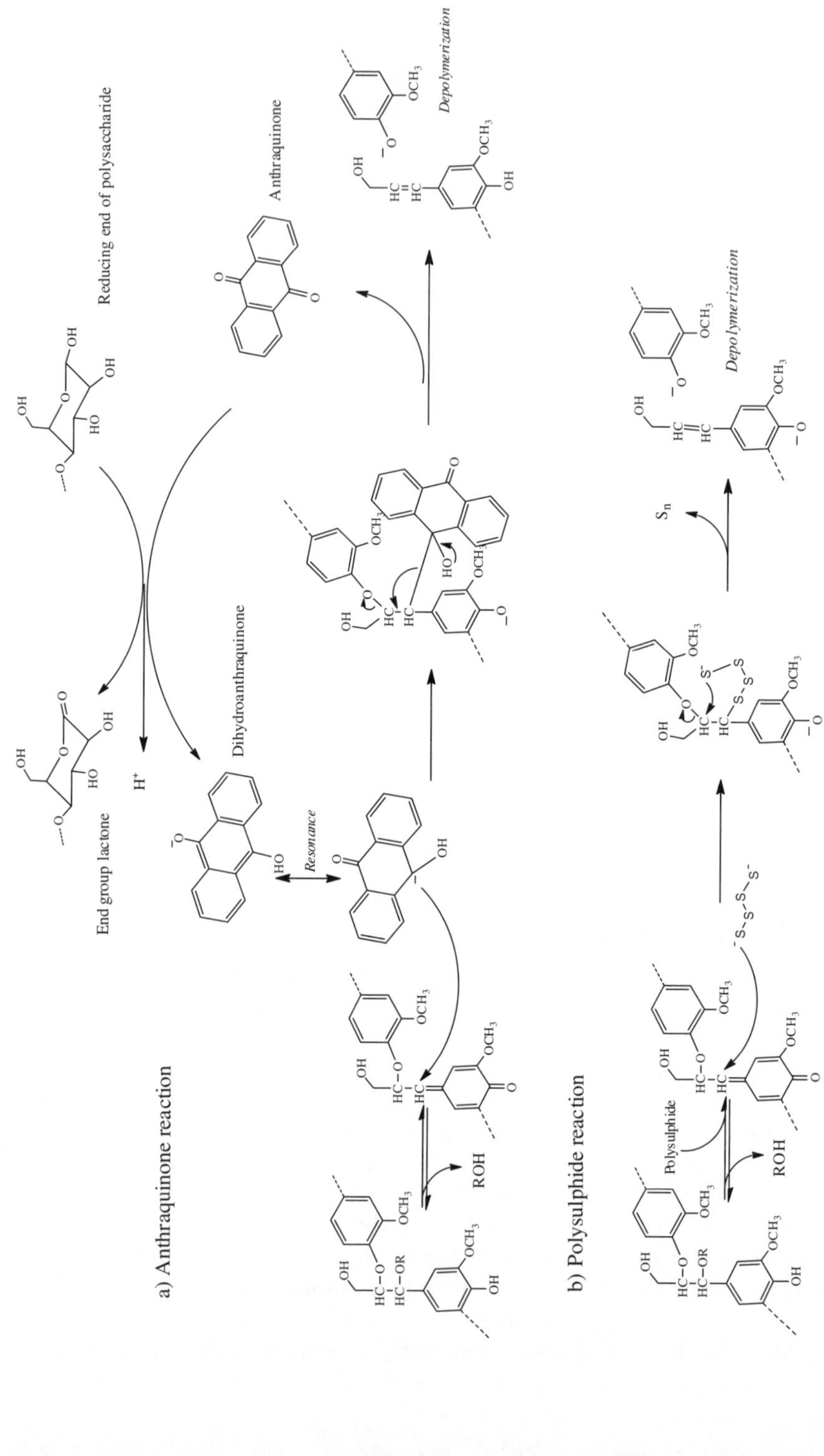

Figure 9.11 Pulping reactions with (a) anthraquinone and (b) polysulfide.

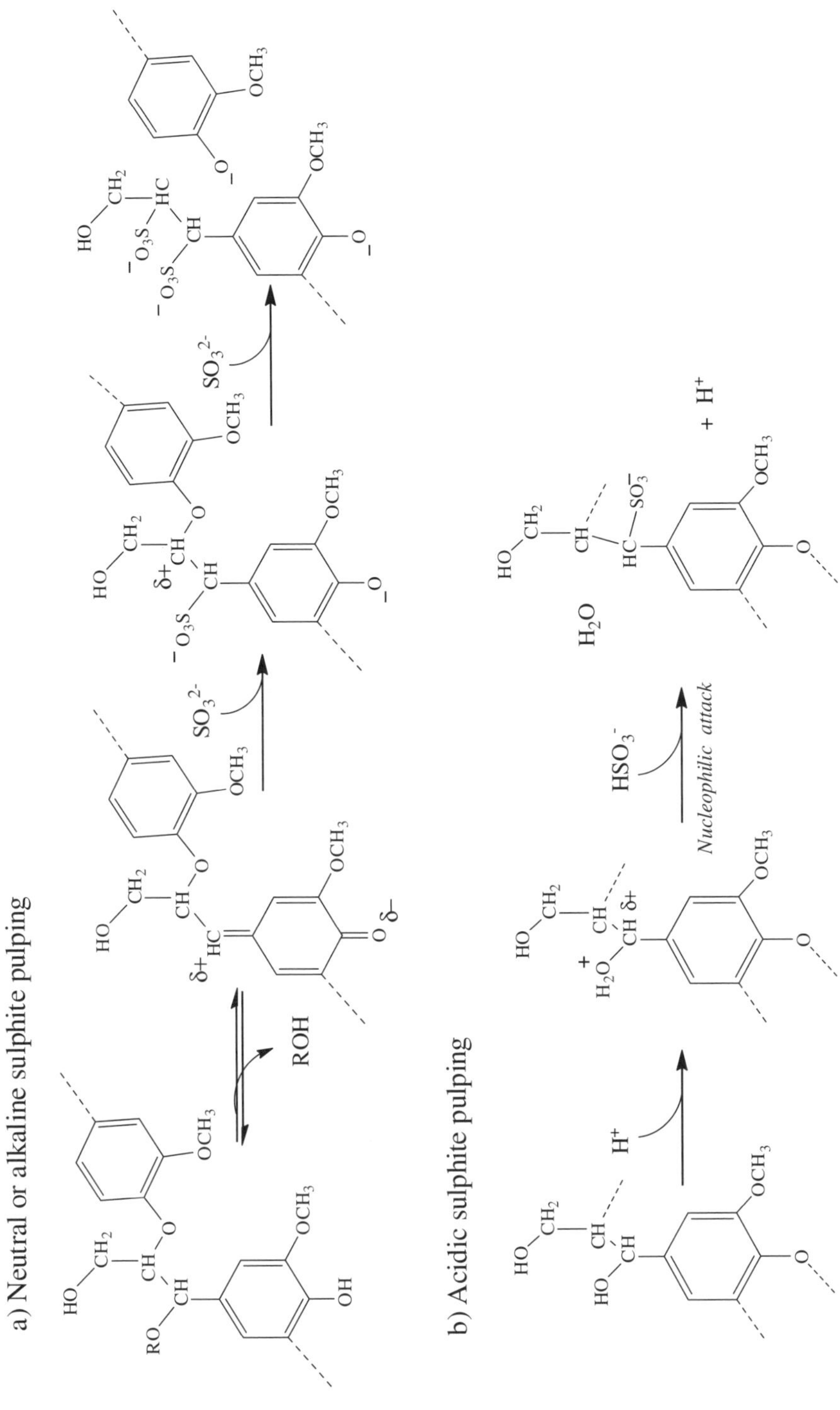

Figure 9.12 Sulfonation of lignin in (a) alkaline and neutral sulfite pulping, and (b) acidic sulfite pulping.

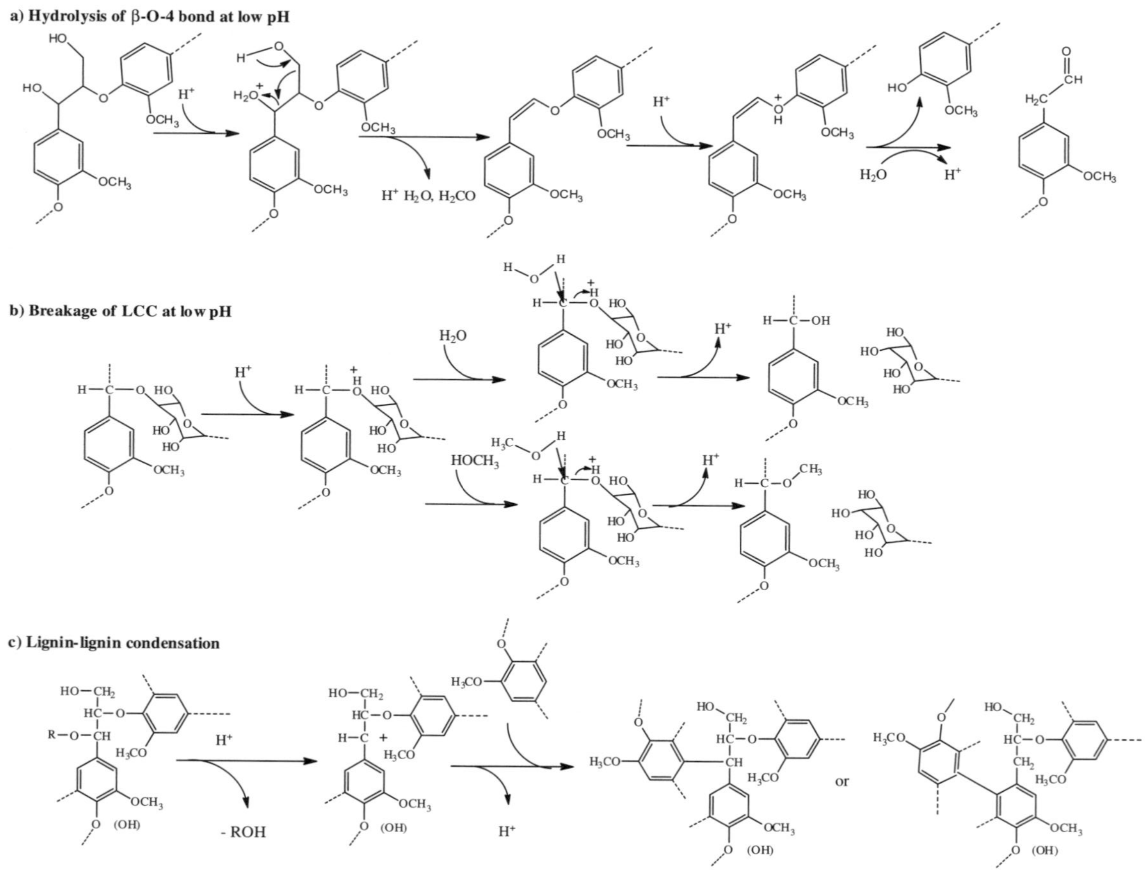

Figure 9.13 Important lignin reactions under acidic conditions, such as acidic sulfite pulping, steam explosion and acid hydrolysis. (a) Hydrolysis of β–O–4 bonds. (b) Hydrolysis/methanolysis of LCC bond. (c) Condensation reaction leading to formation of extra C–C bonds between phenylpropenyl (C9) units.

sulfite pulping, a peeling chemistry similar to the one in alkaline pulping occurs (Figure 9.12). LCC can be broken as shown in Figure 9.13. However, as in alkaline pulping, the breakage of LCC is far from complete in the pulping, and also here remaining lignin in the fiber is covalently attached to polysaccharides, although the binding pattern appears to be different. Technical data for sulfite pulping is shown in Table 9.2.

Also, in sulfite pulping various types of condensation reactions occur that lead to the formation of brown color in the lignin, both in solution and in the pulp fiber, although not to the same degree as in kraft pulping.[54,55]

9.3.3 Reactions During Organosolv Pulping

Organosolv pulping is carried out in an organic solvent, normally alcohols or carboxylic acids (or a mixture of organic solvent and water), which is a better solvent for lignin than water-based systems. In practice, the pulping can be described as an extraction, but in most cases it is combined with alkaline or acidic pulping. There are also multistage concepts. The chemical reactions in organosolv pulping are therefore principally identical with the ones described above, although bleaching chemicals have been tested in organosolv concepts. The main drawback with organosolv concepts has been that that the recovery of the cooking liquids has been relatively costly. Table 9.3 provides a summary of some concepts for organosolv pulping.[56–60]

An interesting property with organosolv pulping is that alcohols can replace water in hydrolysis reactions, giving methylated (if methanol is used) products. Consequently, such a lignin may be more hydrophobic than technical lignins obtained by kraft pulping.

9.3.4 Reactions During Bleaching of Pulp

Chemical fibers are, in general, colored brown after pulping and contain some remaining lignin, as described above. This lignin needs to be removed if the fibers are to be used in applications where a bright fiber is needed. Furthermore, many applications need pulps with low biological activity, high purity and uniformity (low sheaves content). The process for obtaining these goals is called *bleaching*, which removes the resistant remaining lignin by other methods than in pulping. The most important methods are based on oxidations that destroy the "sticky" aromatic groups in lignin and convert them to hydrophilic structures such as carboxylic acid, although there are also reactions that depolymerize the lignin, or convert the colored structures to bright structures (lignin-preserving bleaching). The chemistry of bleaching is thus very complex, and a complete description will not be given here. In Table 9.4 the most important bleaching methods are shown. In Figure 9.14 some important reactions in the three most important bleaching methods, oxygen delignification, alkaline peroxide bleaching and chlorine dioxide bleaching, are

Table 9.2 Technical data for chemical pulping.

Process	*pH*	*Active reagent*	*Counterion*	*Cooking temperature °C*	*Pulp yield %*	*Applications*
Alkaline sulfite	9–13.5	SO_3^{2-}, OH^-	Na^+	160–180	45–60	Similar to kraft pulp
Neutral sulfite (NSSC)[(2)]	5–7	HSO_3^-, SO_3^{2-}	Na^+, NH_4^+	160–180	75–90	Corrugated medium, semi-chemical pulp
Bisulfite (Magnefite)	3–5	HSO_3^-, H^+	Mg^{2+}, Na^+	150–170	50–65	Printing paper, tissue
Acid (bi)sulfite	1–2	SO_2*H_2O, H^+, HSO_3^-	Ca^{2+}, Mg^{2+}, Na^+	125–143	40–50	Dissolving pulp, tissue, printing paper, greaseproof paper and other special paper

Table 9.3 Examples of concepts in organosolv pulping.

Name	*Solvent used*	*pH and temp*	*Comment*
ALCELL	Ethanol/water (50/50)	Weakly acidic. The high temperature (200 °C) leads to liberation of acids like acetic acid from the wood, which work as catalysts.	Used for hardwood. Properties of pulp are similar to kraft pulping.
Organocell	Methanol	Alkaline 165–170 °C	Anthraquinone is added in catalytic amounts.
ASAM	Methanol	Alkaline	Anthraquinone and sodium sulfite are added to the solution.
MILOX	Formic acid	Low pH, 90–100 °C	Hydrogen peroxide is used that forms performic acid.
Acetosolv	Acetic acid	Low pH. Mineral acids are added as catalyst. 150–200 °C	Hydrogen peroxide is used that forms peracetic acid.
Acetocell	Acetic acid	Low pH	Autocatalyzed system.
Formacell	Acetic acid and formic acid (10–15%)	Low pH, 150–190 °C	Can be used as dissolving pulp.

Table 9.4 Common bleaching methods for chemical pulp.

Bleaching method	*Effects on lignin*	*Comment*
Oxygen delignification (O)	Ring opening of phenolic structures forming muconic acid. Hydroxyl radicals can also break β–O–4 bonds.	A cheap, environmentally friendly and robust method for removing lignin. Less good in brightness improvement. Often used as first bleaching stage. May have selectivity problems.
Chlorine dioxide bleaching (D)	Ring opening of phenolic structures. Chlorination of aromatic structure (including nonphenolic structures) and double bonds.	Very powerful and selective bleaching method both for lignin removal and brightness improvement. Chlorinates.
Hydrogen peroxide bleaching (P)	Degradation of quinones and other colored structures. Oxidize phenones.	Good especially for brightness improvement.
Alkaline extraction (E)	Extract fragmented lignins. Hydrolyze muconic acid esters, and decloronates.	Mainly used for extracting lignin after acidic bleaching stages.
Ozone bleaching (Z)	Oxidize phenolic and nonphenolic aromatics to muconic acids.	Very powerful bleaching method that can also attack nonphenolic lignin.

shown. The solubilized products are oxidized and therefore of lower energy content than from pulping techniques. The molecular weights of the released lignin fragments are also relatively low, and the aromatic structure of the lignin is partly lost (Figure 9.14).[61–65]

9.3.5 Reactions During Acid Hydrolysis and Steam Explosion

Steam explosion and acid hydrolysis are processes that "open" the structure so that material can be extracted. In both cases acidic hydrolysis reactions dominate, similar to the one described in Figure 9.13. Even if no acid is added, the pH tends to be low due to the formation of organic acids, such as acetic acid, from the hemicelluloses during the process. It is known that the acidic conditions applied during acid hydrolysis or encountered during steam explosion of biomass lead to formation of a carbenium ion by proton-induced elimination of water (ether, ester) from the benzylic position. In the presence of other electron-rich carbon atoms such as the C-2/C-6 present in aromatic rings, condensation reactions occur leading to formation of extra C–C bonds between lignin structures (Figure 9.13).[66,67]

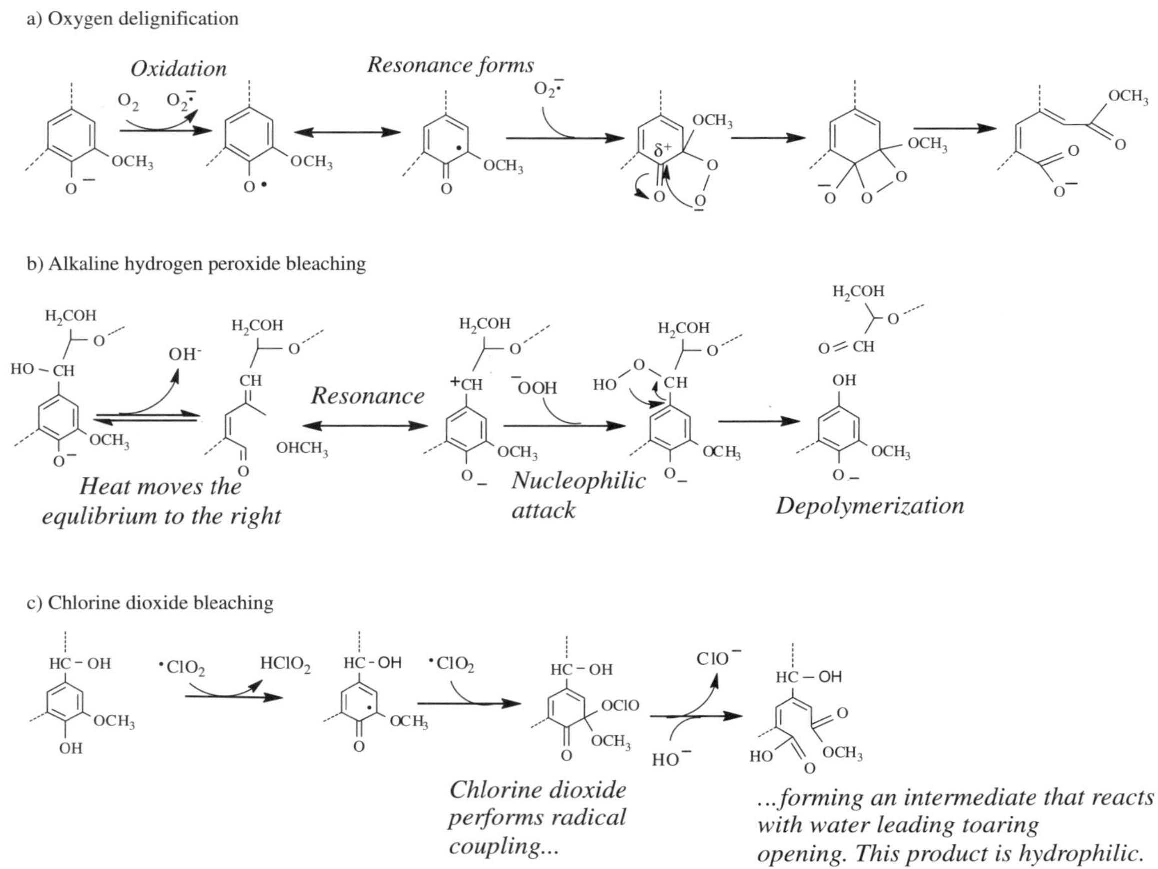

Figure 9.14 Examples of important reactions during different bleaching methods. (a) During oxygen delignification, phenolic endgroups are oxidized to phenolic radicals forming superoxide anion. These species react with each other forming muconic acid. (b) During alkaline peroxide bleaching deprotonated peroxide reacts with the α-carbon of phenolic endgroups leading to depolymerization. (c) During chloride dioxide bleaching muconic acid is created by a series of oxidations. Note that the chemical reactions in bleaching are very complex, and that many other reactions, including depolymerization of the lignin, may occur.

9.4 Preparation of Technical Lignins

Technical lignins are lignins obtained as a byproduct in chemical pulping, steam explosion, acid hydrolysis and bleaching. Currently, it is mainly the lignin obtained from chemical pulping that is of interest, and especially the water-soluble lignosulfonates. Lignin from bleaching processes is obtained in smaller amounts, and is also less interesting from an energy perspective, since it is more oxidized and thereby less energy rich. Such lignin is therefore normally discarded, or burned in a recovery boiler.

The black liquor obtained after alkaline pulping is a very complex mixture of organic components originating not only from lignin, but also from carbohydrates and extractives, as well as inorganic material. There are two main reasons for preparing lignin from pulping liquors; one is that the produced lignin has value as an energy source, as well as for other applications (see below). However, the dominant reason, at least in a short-term perspective, is that removal of organic material from the black liquor makes chemical recovery in the recovery boiler more efficient, this being in many cases a bottleneck with respect to increasing the production of chemical mills.[68]

9.4.1 Precipitation of Lignin

Lignin from alkali pulping methods such as the kraft process is soluble at high pH, but a majority of the lignin precipitates if the pH is lowered. This can be done with addition of sulfuric acid or by flushing the liquid with CO_2, so that carbonic acid is created, which lowers the pH. An advantage of the latter method is that the remaining liquids after precipitation can be "closed", *i.e.* used in a chemical recovery system in the kraft mill that regenerates the pulping liquids without disturbing the balance of sulfur and sodium content. The technique of flushing CO_2 in black liquor for precipitation of lignin has long been used for the removal of lignin by sedimentation in small-scale pulping in, for instance, China. The precipitate is then often burned and part of the smoke gases is used for the lignin precipitation. The sediment is, however, impure and wet, and for more advanced applications washing on filters is necessary. One major problem has been that the washing of technical lignins on filters is difficult to perform. The reason for this is that the ion strength decreases during the filtration and the lignin partly redissolves. This gives rise to gels, which clog the filter and filter cake, making large-scale filtration practically impossible. A recently developed and patented method, *lignoboost,* has presented a solution to this problem; the initial filter cake is resuspended in acidic water and filtered again. The pH and ion strength are now low enough for the production of a stable precipitate, which can be washed. Figure 9.15 presents the method. The produced lignin can be pelleted and used as fuel (it has a heating value of 26 GJ/tonne), and possibly also as raw material for chemicals production. The content of ash and sulfur is relatively low in lignoboost lignin.[69]

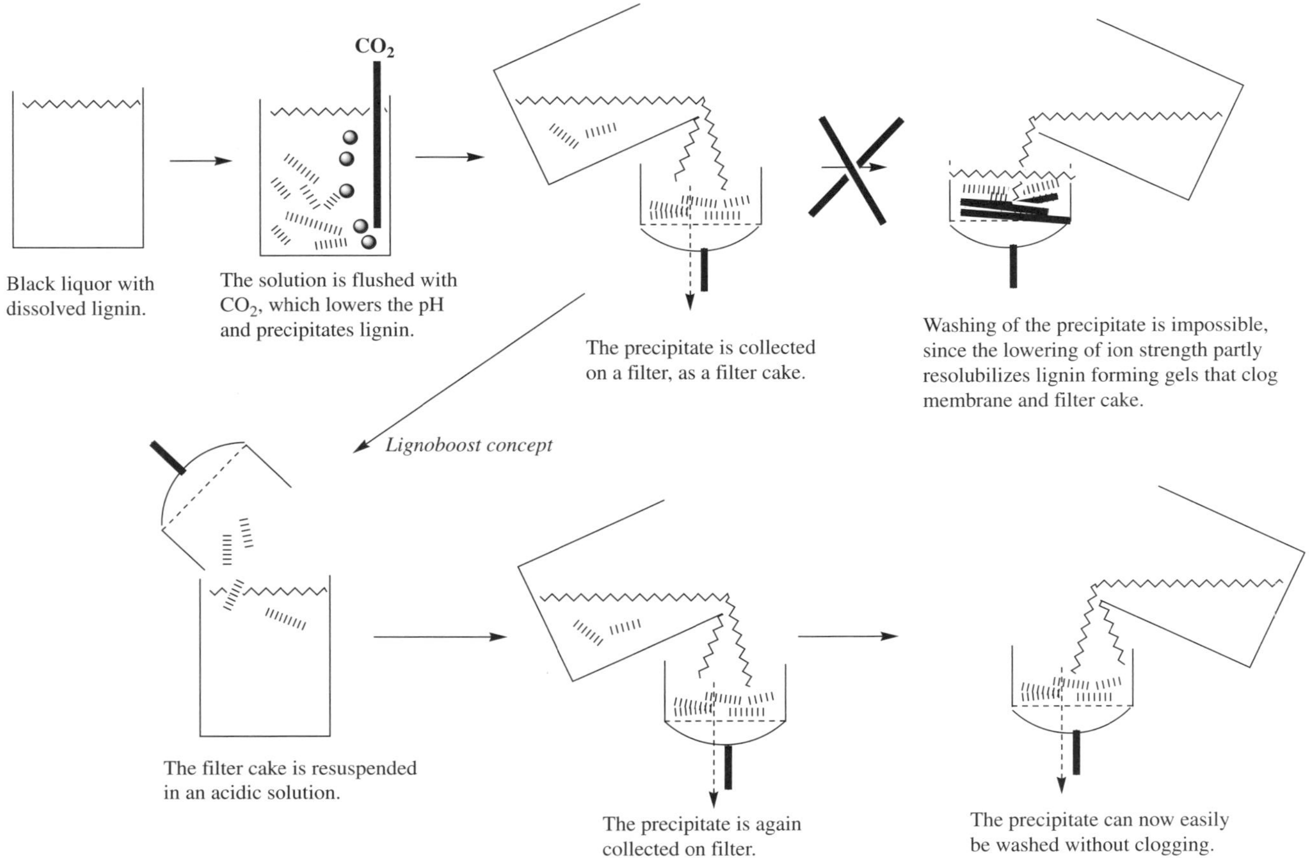

Figure 9.15 Schematic presentation of the Lignoboost process for production of solid lignin fuel. The black liquor is first concentrated by evaporation, and the strong black liquor is flushed with carbon dioxide for precipitation of the lignin, which is collected by filtration. The liquid goes to the recovery boiler, and the filter cake is resuspended in acidified water, re-filtered and washed.

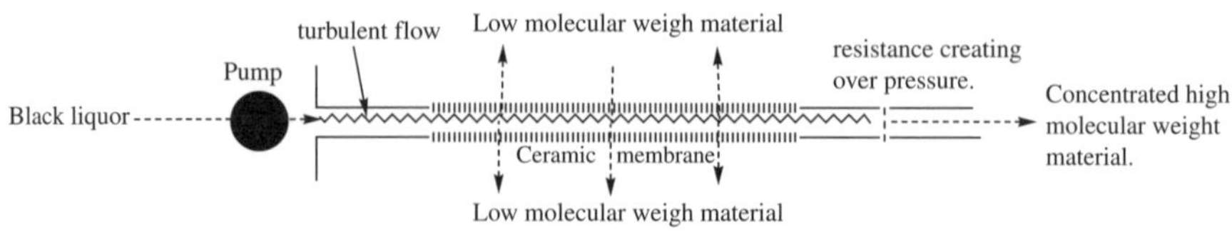

Figure 9.16 Filtration of black liquor on ceramic membranes.

9.4.2 Preparation with Ultrafiltration

An alternative to precipitation is to obtain lignin-enriched liquid fractions by means of ultrafiltration, which separates higher molecular weight material from the low molecular weight salts in "online" systems. By using a series of membranes with different cutoffs, it is possible to obtain material with relatively homogeneous composition. Ceramic filters that are stable at high temperature and a broad pH range can be used, and the flow is turbulent in the filtering direction, which minimize clogging tendencies (see Figure 9.16).[68,70,71]

9.4.3 Preparation of Lignosulfonates

Lignosulfonate is prepared from the pulping liquor of sulfite pulping by different methods, such as concentration by ion-exchange chromatography or ultrafiltration followed by spray drying.

9.5 Use of Technical Lignins in the Energy Sector

Lignin can be used in the energy sector either for simple burning as a solid fuel for generation of heat or electricity, or for conversion to liquid fuel for utilization in internal combustion engines. Lignin is rather well suited for the first application due to its relatively high energy content (discussed above). The combustion of lignin in recovery boilers for producing heat for chemical pulping and for electricity production has, for a long time, constituted one of the main examples of biofuel utilization. However, the second application represents a technical challenge due to the complex structure of lignin.

9.5.1 Combustion of Technical Lignins

Lignin combustion refers to severe thermal decomposition at high temperature and complete oxidation in the presence of sufficient amounts of oxygen. Like the combustion of other organic materials, carbon dioxide and water are the endproducts from lignin combustion. The process is highly exothermic.

$$\text{Lignin} + O_2 \longrightarrow CO_2 + H_2O + \text{heat} \quad (9.1)$$

In the pulp and paper industry, combustion of lignin and other organic materials present in black liquor is currently practiced to generate process heat in connection with chemical recovery in kraft pulping. In ethanol plants, lignin waste obtained after conversion of the carbohydrates into ethanol by hydrolysis and fermentation is also subjected to combustion to generate process heat and thereby improve the process economics. Lignin can, of course, also be used for heating in other locations such as district heating installations.

Among the three major components in lignocellulosic materials, *i.e.* lignin, cellulose and hemicellulose, lignin gives the highest amount of specific heat during combustion, as discussed above. The gross heat of combustion determined for different virgin woods is reported to correlate linearly with their lignin content.[72] In the kraft pulping process, the energy generated at the recovery boiler from combusting black liquor is commonly beyond the demand of the process heat. This creates a bottleneck, therefore, for the expansion of pulp production. Since lignin is the major contributor to the heat generation from black liquor, partial removal of the lignin in the black liquor before combustion is an effective solution for this problem. The content of organic bound sulfur in technical lignin is an important factor, since material with high sulfur content may need more intense flue gas purification for removing sulfur dioxide from the gases.

9.5.2 Gasification of Lignin

Gasification of lignin occurs in the recovery boiler as an integral part of kraft pulping, but gasification of technical lignin can also be done in black-liquor gasification where the temperature and pressure are higher, and the gases produced are recovered. Chemically, strong (concentrated) black liquor (or lignin) is dried, and thereafter subjected to heat, resulting in pyrolysis of the lignin, *i.e.* a series of homolytic cleavage reactions caused by the high temperature that are still not totally understood. The final products of the pyrolysis are gases, including CO, H_2, CH_4, *etc.* (Figure 9.17). Some of these pyrolysis gases are burned by addition of dioxygen or air, generating heat for the endothermic pyrolysis reactions. The primary aim of lignin gasification is to obtain carbon monoxide and hydrogen (syngas). Compared to the original solid lignin, the syngas obtained is easier and more versatile to use. For example, the gas can be directly applied in gas engines and gas turbines. It can also be converted into liquid fuels, for example, by applying Fisher-Tropsch

Black liquor containing: Lignin, carbohydrates, organic sulfur, extractives, organic sodium salts etc. --- **Pyrolysis** (*High temp. no O_2*) ---> H_2, CO, H_2O, CH_4, H_2S, CO_2, CH_3SH, C_2H_6 (g)

Na_2CO_3, Na_2S, Na_2SO_4, K_2CO_3, K_2S, K_2SO_4, elementary C (s)

Figure 9.17 Pyrolysis of black liquor with examples of major products.

(FT) technology to produce green diesel.[73] It is also possible to make dimethylether from syngas, which can be used in diesel engines. These technologies are discussed in Chapters 5 and 6 in this volume.

There are many studies that correlate hydrogen production and lignin content in biomass during biomass gasification. Lignin is found to produce four times more hydrogen than cellulose and almost three times more than hemicellulose. The higher content of aromatic rings and methoxy groups in lignin may explain the higher production of hydrogen, in addition to a higher methane release during gasification.

9.5.3 Oxygen-Free Pyrolysis of Lignin

If lignin pyrolysis is performed in the absence of air, the molecular structure of lignin is broken down into smaller units, corresponding to gaseous or liquid products, by homolytic cleavage reactions. The liquid product obtained is known as *pyrolysis oil* or *bio-oil*. For fast pyrolysis of biomass, the bio-oils obtained are generally quite unstable because of viscosity changes, liability to oxidation and so on, which makes their use for chemicals and fuels problematic. Therefore, the bio-oils from biomass must usually be upgraded before they can be used as replacements for diesel and gasoline fuels. Among others, hydrodeoxygenation using sulfided NiMo and CoMo catalysts and zeolite upgrading are two successful means for upgrading.[73]

Lignin can be converted directly into a transportation fuel by hydrodeoxygenation using the sulfided NiMo and CoMo catalysts or the zeolite upgrading methods mentioned above for bio-oil upgrading. The major products from the hydrodeoxygenation of lignin include phenols, cyclohexane, benzene, naphthalene and phenanthrene, typically with a liquid yield of around 60% of the original lignin. Zeolite upgrading of lignin results in liquid components that include toluene, benzene and xylene.[73]

9.5.4 General Considerations on Converting Technical Lignin into Transportation Fuels

Generally speaking, there is one common problem encountered in all types of thermal decomposition of lignin, namely the lack of product selectivity. In other words, a large variation of products is commonly obtained. To produce specially targeted products by promoting product selectivity and maximizing product yield, the reaction conditions and chemical reagents applied, including catalysts (if any), should be carefully designed and controlled based on current knowledge about the reaction mechanisms.

The van Krevelen diagram is a plot constructed using the molar ratio of hydrogen-to-carbon (H/C ratio) as the ordinate and the molar oxygen-to-carbon ratio (O/C ratio) as the abscissa, which is used extensively in the geochemistry literature to study the evolution of coals or oil samples. In this diagram, hydrocarbons that possess only hydrogen and carbon in an H/C ratio

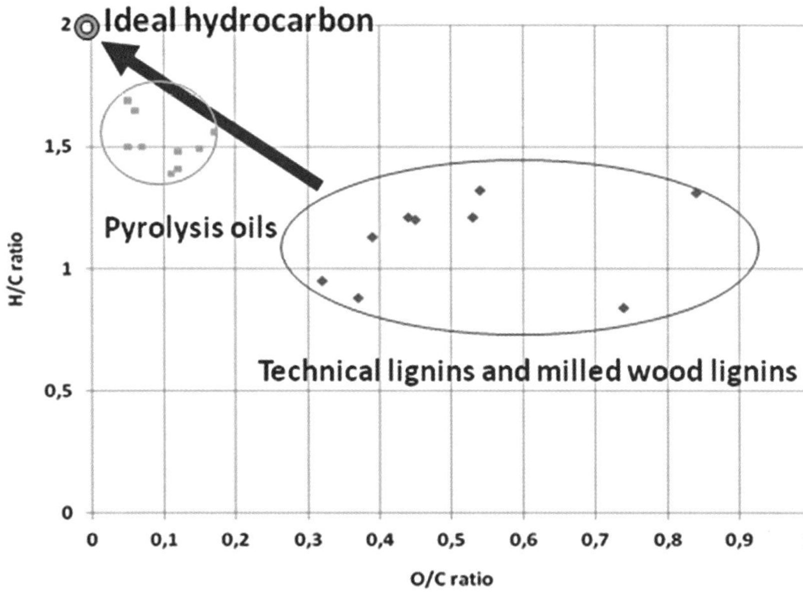

Figure 9.18 Van Krevelen diagram showing the H/C *vs.* O/C ratios of technical lignins and of pyrolysis oils obtained from technical lignins in the presence of formic acid (data taken from Tables 9.1 and 9.2). The arrow indicates the ideal pathway for thermal treatment of technical lignins leading to bioliquids with H/C and O/C ratios close to ideal hydrocarbon applicable as transportation fuels.

of around 2, are located at the top left corner, as shown in Figure 9.18. These hydrocarbons could be set as the targeted structures for lignin thermal conversion to obtain transportation or motor fuels ("ideal hydrocarbon"). Like milled wood lignins (representatives of native lignins), technical lignins contain substantial amounts of oxygen, the O/C ratio ranging between 0.3 and 0.85. Due to the high oxygen content and the high content of aromatic rings, the H/C ratio in technical lignins is around 0.8 to 1.3 (Table 9.5). In order to convert technical lignins into the "ideal hydrocarbon", the thermal treatment should be conducted in such a way that the structural conversions will follow the direction illustrated by the arrow in Figure 9.18. This means that oxygen must be to a large degree eliminated and at the same time, hydrogen needs to be introduced.

It is known that there is a general tendency during the heating of organic compounds for the O/C ratio to be lowered due to the elimination of stable oxygen containing molecules like CO_2 and water. Therefore, the introduction of hydrogen is desirable for the "ideal" conversion. This means that thermal treatments must be conducted under reducing conditions or in the presence of hydrogen donors.[74]

As discussed above, lignin is a large three-dimensional molecule composed of randomly branched phenylpropenyl (monolignol, C9) units with the β–O–4

Table 9.5 O/C and H/C molar ratios of technical and milled wood lignins (data adapted from reference 78 if not specified).

Lignin type	*Wood species*	*O/C ratio*	*H/C ratio*
Kraft lignin[a]	Birch	0.33	1.14
Sulfite lignin (lignosulfonate)	Spruce	0.84	1.31
Alkali lignin	Spruce	0.74	0.84
Organosolv lignin	Spruce	0.32	0.95
Acid-hydrolysis lignin	Spruce (strong acid)	0.37	0.88
	Spruce (weak acid)	0.54	1.32
Steam-explosion lignin	Birch	0.53	1.21
	Spruce	0.39	1.13
Milled-wood lignin	Aspen	0.45	1.20
	Spruce	0.44	1.21

[a]Data calculated from ref. 89.

bond as the most abundant linkage between the C9 units. During the conversion of lignin into transportation fuels, depolymerization of the lignin structure is also desired. Chemically, cleavage of the β–O–4′ linkage is one of the main reactions leading to depolymerization, either directly[75] or after a dehydration between Cα and Cβ.[76] In technical lignins, modified structures are also present (see above), which result from the reactions occurring during the processing. For example, condensed structures between the C9 units might be present after *e.g.*, steam explosion with extra bonding between lignin C9 units (Figure 9.13). Therefore, a technical process suitable for converting all types of technical lignins into transportation fuels should be able to cleave both the β–O–4′ linkage in native lignins and the extra C–C bonds in the modified lignins.

9.5.5 Pyrolysis of Technical Lignins in the Presence of Formic Acid

Recently, a pyrolysis process applying a strongly hydrogen-donating reaction medium consisting of formic acid and alcohol has been developed. This process, termed the "lignin-to-liquid" process, can utilize all types of technical lignin as the feed. After $\geq$ 8–10 h at 380 °C, technical lignins are converted into biofuels with O/C ratios of 0.05–0.17 and H/C ratios of 1.35–1.69 (Table 9.6), which are close to the ratios of the targeted hydrocarbons and are, in fact, in the same range as heavy oils (Figure 9.18). These biofuels are therefore suitable as blending components to be combined with conventional fossil fuels for motor-fuel applications.

At 380 °C, both lignin depolymerization and hydrodeoxygenation take place in one single step. The depolymerization of lignin takes place *via* cleavage of either C–O (*e.g.*, in the β–O–4′ linkage) or C–C bonds (*e.g.*, in condensed structures), forming C- or O-centered radicals. Under heating, different reactions of formic acid decomposition with and without water present generate

Table 9.6 O/C and H/C molar ratios of pyrolysis oils formed upon the pyrolysis of technical lignins in the presence of formic acid (data adapted from ref. 78).

Lignin type	*Wood species*	*O/C ratio*	*H/C ratio*
Steam explosion	spruce	0.12	1.48
	spruce	0.07	1.5
	spruce	0.11	1.39
	spruce	0.05	1.69
	spruce	0.05	1.5
Steam explosion	birch	0.06	1.65
	birch	0.17	1.56
Acid hydrolysis	spruce	0.15	1.49
Lignosulfonate	spruce	0.12	1.41

CO, CO_2 and H_2O.[77] At the high temperature applied, CO_2 and H_2 are considered as the main products. The reactions are probably slow and there is a continuous formation of very reactive hydrogen atoms. These hydrogen atoms can capture and cap the reactive C- or O-centered radicals once they are formed and prevent their recombination. The hydrogen also forms water with the oxygen released from the lignin.

Analysis of the biofuels obtained has shown that a complete degradation of the lignin takes place irrespective of its origin or if the structure contains condensed bonding or not. The resulting bio-oil has a low molecular-mass distribution with a preponderance of aliphatic hydrocarbon structures. A substantial number of phenolic compounds are also present and some of these also contain carboxyl groups. The reactions that take place are depicted in Figure 9.19. It should be noted here that the lignin-to-liquid process can completely demethoxylate the lignin aromatic rings as shown by the model study reported in ref. 78 (Figure 9.20). As a result, the phenol structures present in the biofuel are chemically more active than the original phenolic lignin structures with more free *ortho* positions, although methyl groups may be located in this position; they may be sufficiently reactive to be usable for a variety of applications such as the manufacture of phenol-biofuel-formaldehyde resins. Compared to the other processes reported in the literature for lignin thermal conversion, such a single-step process can be considered as a combination of lignin degradation (*e.g.*, fast pyrolysis) and bio-oil upgrading.

9.6 Other Technical Uses of Lignin

Lignin is used in, or has been tried for, many other types of technical applications. To date, lignosulfonates have been the most commercially successful of these.

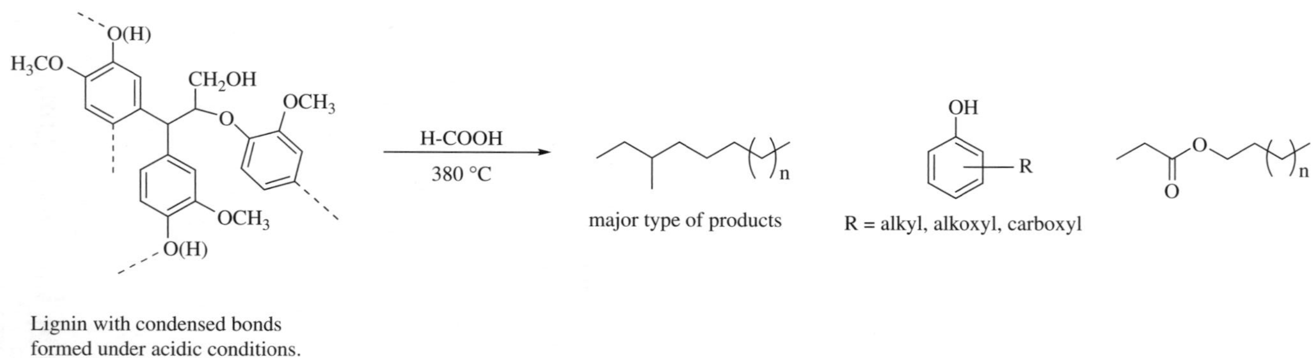

Figure 9.19 Schematic picture of the products formed upon the pyrolysis of lignin in the presence of formic acid, where both β–O–4′ linkage and C–C condensed bond are cleaved (adapted from ref. 90).

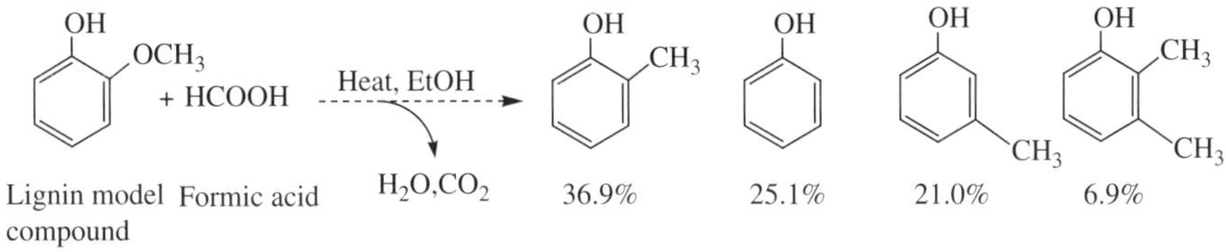

Figure 9.20 Completely demethoxylated products formed upon the pyrolysis of lignin model (guaiacol) in the presence of formic acid (adapted from ref. 78).

9.6.1 Lignosulfonates as Dispersing Agent

Lignosulfonates differ from other technical lignins by their good solubility properties and the relatively high degree of polymerization. The lignosulfate molecule includes both charged groups (sulfonic acid), and the relatively hydrophobic and "sticky" aromatic group. This makes it ideal as a dispersing agent, since the polymer can "carry" different types of particles efficiently. The main applications are in oil drilling, as an additive in concrete, a binder for dust and an additive for asphalt. By volume, the use of lignosulfonate in concrete is the largest. Addition of lignosulfonates to a concrete mix reduces the amount of water and thus lowers the viscosity, resulting in a substantial increase in concrete strength.[79]

9.6.2 Manufacture of Vanillin

Vanillin is the main chemical giving vanilla its pleasant smell and taste. Natural vanilla is relatively expensive and therefore synthetic vanillin is also produced. Its main use is as a spice in mostly sweet food (ice cream, *etc.*), but is also used in perfumes, for hiding the bad taste of medicines, and for the synthesis of other chemicals. Most synthetic vanillin is made from petroleum, but a minor fraction is obtained from lignosulfonate as the raw material. This product is often regarded as having higher quality than other synthesized vanillin, probably due to the fact that it contains acetovanillone as a contaminant.

The manufacture is done by an alkaline oxidation of lignosulfonate at high temperature (Figure 9.21).[80]

9.6.3 Soil-Improving Agent

Technical lignins have been tried as soil-improvement agents, where they help to bind metal salts and other minerals needed for plants. This application is not surprising given the role of microbially processed lignin, humus, in soil. Lignin is sometimes used together with other components. Apart from acting as a binder for minerals, lignin may affect the mechanical properties of soil by working as a dispersing agent for clay particles.[81]

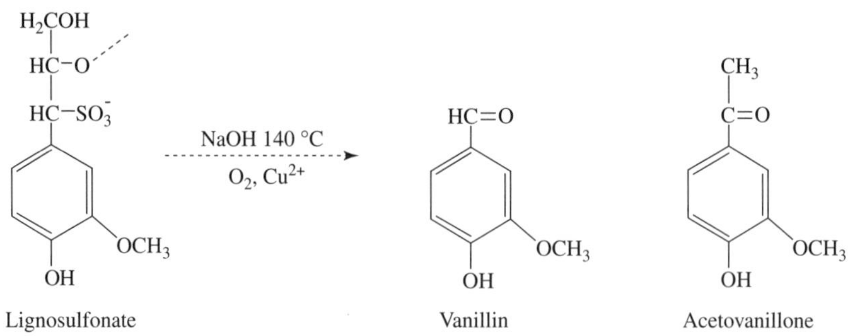

Figure 9.21 Principle for the manufacture of vanillin from lignosulfonate.

9.6.4 Manufacture of Carbon Fiber

It has been demonstrated that it is possible to make carbon fibers from lignin. In essence, the manufacture of carbon fiber involves subjecting a prefiber made of carbon-containing material to high temperature (up to 2000 °C!) under oxygen-free conditions. The noncarbon elements are thereby converted to gases, whereas the carbon is converted to graphite that form fibers with high strength and stiffness.[82]

Lignin has the advantage that as a raw material it is cheap, and that the carbon content is high for a biopolymer. A disadvantage is that it is difficult to make a prefiber of high quality from lignin, and the produced fibers are not of the same quality as carbon fibers made from the synthetic polymer polyacrylonitrile.

9.6.5 Lignin as Binder in Composite Material and Filler in Polymers

As discussed above, lignin has similarities with man-made phenolic plastics, and has a biological role to "glue" other wood polymers together. Therefore, it is not surprising that lignin has been tried as a filler in certain types of synthetic polymers – the most successful have been phenyl formaldehyde resins, where up to 50% of the resin can be replaced by kraft lignin. Particularly interesting is the use of technical lignins as a gluing substance in wood-based materials such as strand boards. Normally, these are manufactured by pressing wood pieces of different shapes together at high temperature. The concept is attractive, since lignin is a low-value renewable material, and wood materials made this way may have fewer problems with formaldehyde emissions than the presently employed urea-formaldehyde binder.[83–85]

9.6.6 Lignin for Dust Control

Lignosulfonates are also used for *dust control of gravel roads*, where they are applied as a water solution. Here too, it is the ability of lignins to bind to particles that is the important factor. The fact that it is a natural biopolymer is, of course, important for this application.

References

1. C. K. Boyce, G. D. Cody, M. L. Fogel, R. M. Hazen, C. M. O. D. Alexander and A. H. Knoll, *Int. J. Plant Sci.*, 2003, **16**, 691.
2. M. Baucher, B. Monties, M. Van Montagu and W. Boerjan, *Crit. Rev. Plant Sci.*, 1998, **17**(2), 125.
3. H. H. Johnston, *Plant and Soil*, 1964, **21**(2), 191.
4. J. M. Schopf, *J. Sediment. Petrol.*, 1952, **22**(2), 611.
5. J. Ralph, J. Peng, F. Lu and R. F. Hatfield, *J. Agric. Food Chem.*, 1999, **47**(8), 2991.
6. M. Westoby and B. Rice, *Evolution*, 1981, **35**(4), 713.
7. P. Kendrik and P. R. Crane, *Nature*, 1997, **389**, 33.
8. J. S. Sperry, U. G. Hacke and J. Pitterman, *Am. J. Botany*, 2006, **93**, 1490.
9. P. H. Raven, R. F. Evert and S. E. Eichhorn, *Biology of Plants*, 6th edn., W.H. Freeman and Company Worth Publishers, New York, NY, USA, ISBN 1-57259-041-6 (1999).
10. N. Wikström, V. Savolainen and M. W. Chase, *Proc. R. Soc. Lond. B*, 2001, **268**, 2211.
11. R. Hatfield and W. Vermerris, *Plant Physiol.*, 2001, **126**, 1351.
12. W. Boerjan, J. Ralph and M. Baucher, *Annu. Rev. Plant Biol.*, 2003, **54**, 519.
13. K. Freudenberg, *Angew. Chem.*, 1956, **68**, 508.
14. R. A. Dixon, F. Chen, D. Guo and K. Parvathi, *Phytochemistry*, 2001, **57**(7), 1069.
15. Z. Jin, Y. Matsumoto, T. Tange, T. Akiyama, M. Higuchi, T. Ishii and K. Iiyama, *J. Wood Sci.*, 2005, **51**, 424.
16. G. H. N. Towers and R. D. Gibbs, *Nature*, 1953, **172**, 25.
17. K. V. Sarkanen and H. L. Hergert, "Classification and distribution", pp 63–65 in K. V. Sarkanen and C. H. Ludwig (ed.) *Lignins: Occurence, formation, structure and reactions*. Wiley-Interscience, New York, NY, USA (1971).
18. E. Sjöström, *Wood Chemistry. Fundamentals and Applications, (2nd edn.)*, Academic Press Inc., San Diego, USA, (1993).
19. J. Gierer, *Wood Sci. Technol.*, 1980, **14**, 241.
20. L. Hildén, G. Henriksson, J. Li, P. Ljungquist and G. Johansson, *FEBS Lett.*, 2000, **477**, 79.
21. C. K. Mathews and van K. E. Holde, *Biochemistry*, The Bejamin/Cummings Publishing Company Inc., Rewood City, Carlifornia, USA, ISBN 0-8053-5015-2 (1990).
22. J. Wadenbäck, D. Clapham, G. Gleerstedt and S. von Arnold, *Holzforchung*, 2004, **58**(2), 107.
23. N. Terashima and K. Fukushima, *Wood Sci. Technol.*, 1988, **22**(3), 259.

24. U. Westermark, *Wood Sci. Technol.*, 1985, **19**(3), 223.
25. C. Grunwald, K. Ruel, Y. S. Kim and U. Schmitt, *Plant Biology*, 2002, **4**(1), 13.
26. G. Henriksson, L. Zhang, T. Nilsson, A.B. Ohlsson and T. Berglund, *Inhomogenity in lignin structure between different cell wall layers in conifers and hardwood*, Fifth Plant Biomechanics Conference Stockholm, Sweden, Vol **I**, pp 145–150 (2006).
27. K. E. Wolter, J. M. Harkin and T. K. Kirk, *Physiologia Plantarum*, 2006, **31**(2), 140.
28. N. Takahashi and T. Koshijima, *Wood Sci. Technol.*, 1988, **22**(2), 177.
29. T. Iversen and S. Wännström, *Holzforchung*, 1986, **40**(1), 19.
30. Z. F. Jin, K. S. Katsumata, T. B. T. Lam and K. Iiyama, *Biopolymers*, 2006, **83**(2), 103.
31. M. Lawoko, G. Henriksson and G. Gellerstedt, *Holzforchung*, 2003, **57**, 69.
32. M. Lawoko, G. Henriksson and G. Gellerstedt, *Holzforchung*, 2006, **60**, 156.
33. M. Lawoko, R. Berggren, F. Berthold, G. Henriksson and G. Gellerstedt, *Holzforchung*, 2004, **58**(6), 603.
34. J. Ralph, K. Lundquist, G. Brunow, F. Lu, H. Kim, P. F. Schatz, J. Marita, R. D. Hatfield, S. Ralph, A. Holst, J. Christensen and W. Boerjan, *Phytochem. Rev.*, 2004, **3**, 29.
35. K. Marjamaa, E. M. Kukkola and K. V. Fagerstedt, *J. Exp. Bot.*, 2009, **60**(2), 367.
36. K. Freudenberg and H. Schluter, *Chem. Ber.*, 1955, **88**, 617.
37. V. L. Chiang, *Envir. Chem. Lett.*, 2006, **4**(3), 143.
38. E. Ammalahti and G. Brunow, *Holzforchung*, 2000, **54**(6), 605.
39. M. Toikka and G. Brunow, *J. Chem. Soc. Perk. Trans.*, 1999, **13**, 1877.
40. P. R. LaFayette, K.-E. L. Eriksson and J. F. D. Dean, *Plant Mol. Biol.*, 1999, **40**(1), 23.
41. U. Westermark, *Wood Sci. Technol.*, 1982, **16**, 71.
42. H. Önnerud, L. M. Zhang and G. Henriksson, *Plant Cell*, 2002, **14**(8), 1953.
43. R. Hatfield, J. Ralph, G. Gellerstedt and J. H. Grabber, *Planta*, 2008, **228**(6), 919.
44. A. Holmgren, G. Henriksson and L. Zhang, *Biomacromolecules*, 2008, **9**(12), 3378.
45. A. Holmgren, G. Brunow, G. Henriksson, L. Zhang and J. Ralph, *Org. Biomol. Chem.*, 2006, **4**, 3456.
46. C. Chen, M. Baucher, J. H. Christensen and W. Boerjan, *Euphytica*, 2001, **118**, 185.
47. G. Pilate, E. Guiney, K. Holt, M. Petit-Conil, C. Lapierre, J.-C. Leplé, B. Pollet, I. Mila, E. A. Webster, H. G. Marstorp, D. W. Hopkin, L. Jouanin, W. Boerjan, W. Schuch, D. Cornu and C. Halpin, *Nature Biotechnol.*, 2002, **20**, 6107.
48. S. Aziz and K. Sakanen, *TAPPI J.*, 1989, **72**(3), 169.
49. L. Löwendahl and O. Samuelsson, *TAPPI J.*, 1978, **61**(2), 19.
50. O. Pekkala, *Paperi Ja Puu*, 1986, **68**, 385.

51. J. Gierer, S. Ljunggren, P. Ljungquist and I. Noren, *Svensk Papperstidning1980,* **83***(3), 75.*
52. G. Gellerstedt, A. Majtnerova and L. Zhang, *Comp. Rend. Biolog.*, 2004, **327**(9–10), 817.
53. F. Berthold, E. L. Lindfors and G. Gellerstedt, *Holzforchung*, 1989, **52**(5), 481.
54. G. Gellerstedt and J. Gierer, *Acta Chem. Scand.*, 1968, **22**, 2510.
55. G. Gellerstedt and J. Gierer, *Acta Chem. Scand. Ser. B Org. Chem. Biochem.*, 1975, **29**(5), 561.
56. M. C. Schroeter, *TAPPI J.*, 1991, **74**(10), 197.
57. T. J. McDonough, *TAPPI J.*, 1993, **76**(8), 186.
58. H. Sixta, H. Harms, S. Dapia, J. C. Parajo, J. Puls, B. Saake, H. P. Fink and T. Roeder, *Cellulose*, 2004, **11**(1), 73.
59. J. Sundquist, *Paperi ja Puu*, 1996, **78**(3), 92.
60. E. Muurinen, *Organosolv pulping. A review and distillation study related to peroxyacid pulping*. ISBN 951-42-5660-3. ISBN 951-42-5661-1 (2000).
61. J. Gierer, *Holzforchung*, 1990, **44**, 387.
62. J. Gierer, *Holzforchung*, 1990, **44**, 395.
63. S. Ljunggren, E. B. Gidnert and J. Kolar, *TAPPI J.*, 1996, **79**(12), 152.
64. M. Ek, J. Gierer and K. Jansbo, *Holzforchung*, 1989, **43**(6), 391.
65. C. Brage, T. Eriksson and J. Gierer, *Holzforchung*, 1991, **45**(2), 147.
66. J. Li, G. Henriksson and G. Gellerstedt, *Biores. Technol.*, 2007, **98**(16), 3061.
67. T. Dizhbite, G. Zakis, A. Kizima, E. Lazareva, G. Rossinskaya, V. Jurkjane, G. Telysheva and U. Viesturs, *Biores. Technol.*, 1999, **67**(3), 221.
68. A. Keyoumu, R. Sjödahl, G. Henriksson, M. Ek, G. Gellerstedt and M. E. Lindström, *Ind. Crop. Prod.*, 2004, **20**(2), 143.
69. http://www.afconsult.com/upload/CAMPAIGNS/cpb/2006/Tomani.pdf
70. A. S. Jonsson and O. Wallberg, *Delaination*, 2009, **237**, 254.
71. A. S. Jonsson, A. K. Nordin and O. Wallberg, *Chem. Eng. Res. Des.*, 2008, **86**(11A), 1271.
72. R. H. White, *Wood Fib. Sci.*, 1987, **19**(4), 446.
73. G. W. Huber, S. Iborra and A. Corma, *Chem. Rev.*, 2006, **106**(9), 4044.
74. T. Barth and M. Kleinert, *Chem. Eng. Technol.*, 2008, **31**(5), 773.
75. http://www.brdisolutions.com/pdfs/bcota/abstracts/17/z195.pdf
76. B. H. Hwang and J. R. Obst, *Proceedings of the International Conference on Forest Products (IAWPS2003)*, 2003, 1165.
77. J. Yu and P. Savage, *Ind. Eng. Chem. Res.*, 1998, **37**, 2.
78. M. Kleinert and T. Barth, *Energy & Fuels*, 2008, **22**, 1371.
79. L. H. Grierson, J. C. Knight and R. Maharaj, *Cem. Conc. Res.*, 2005, **35**(4), 631.
80. V. E. Tarabanko, N. A. Fomova, B. N. Kuznetsov, N. M. Ivanchenko and A. V. Kudryashev, *React. Kinet. Catal. Lett.*, 1995, **55**, 161.
81. G. Shulga, T. Betkers, J. Brovkina, B. Neiberte, A. Verovkins, O. Belous, D. Ambrazaitene and A. Zukauskaite, *Mol. Cryst. Liq. Cryst.*, 2008, **486**, 1333.

82. C. D. Warren, F. L. Paulauskas, F. S. Baker, C. C. Eberle and A. Naskar, *Sample J.*, 2009, **45**(2), 24.
83. M. Westin, R. Simonson and B. Östman, *Holz als Roh- und Werkstoff*, 2001, **58**, 393.
84. B. Danielson and R. Simonson, *J. Adhes. Sci. Technol.*, 1998, **12**(9), 923.
85. W. G. Glasser, *For. Prod. J.*, 1981, **31**(3), 24.
86. L. Zhang and G. Gellerstedt, *Magn. Reson. Chem.*, 2007, **45**(1), 37.
87. E. A. Capanema, M. Y. Balakshin and J. F. Kadla, *J. Agricult. Food Chem.*, 2004, **52**, 1850.
88. E. A. Capanema, M. Y. Balakshin and J. F. Kadla, *J. Agricult. Food Chem.*, 2005, **53**, 9639.
89. G. Gellerstedt, K. Gustafsson and R. A. Northey, *Nordic Pulp Pap. Res. J.*, 1988, **3**, 87.
90. G. Gellerstedt, J. Li, I. Eide, M. Kleinert and T. Barth, *Energy & Fuels*, 2008, **22**, 4240.

CHAPTER 10

Catalysts in Biomass Pyrolysis

ANGELOS A. LAPPAS, ELENI F. ILIOPOULOU AND K. KALOGIANNIS

Chemical Process Engineering Research Institute (CPERI), Center for Research & Technology Hellas (CERTH), 570 01, Thessaloniki, Greece

10.1 Introduction to Biomass Catalytic Pyrolysis

Biomass is today the main renewable-energy source available, producing *via* various conversion processes (biological and thermochemical) liquid, solid and gaseous fuels. Existing technologies for the conversion of biomass to 1st-generation biofuels use only a small part of the biomass (that sometimes competes with food) and give questionable CO_2 benefit. Lignocellulocic biomass is used for 2nd-generation biofuels, requiring, however, higher investment costs and producing very expensive fuels.

Today, biomass thermal pyrolysis has received special attention[1–3] and is considered an emerging technology, converting biomass directly into solid, liquid and gaseous products. It can be used either as an independent process for the production of fuels and valuable chemical products or as an initial step to gasification or combustion.[4] The biomass pyrolysis process is simple and inexpensive and of particular importance for agrarian countries with available biomass. Biomass pyrolysis is defined as the thermal degradation/depolymerization of lignocellulose in the absence of oxygen, with short residence times and at atmospheric pressure, which results in the production of charcoal (solid), bio-oil (liquid), and gaseous fuel products.[5–8] The liquid or oil fraction is commonly called pyrolytic oil or bio-oil and requires upgrading before utilization as a fuel or source of chemical feedstock. Produced char can be used as a

RSC Energy and Environment Series No. 1
Thermochemical Conversion of Biomass to Liquid Fuels and Chemicals
Edited by Mark Crocker

Published by the Royal Society of Chemistry, www.rsc.org

fuel or to make activated carbon. Moreover, the pyrolysis gaseous products are of high calorific value and may be used to supply the fuel for the pyrolytic process.[4,9–12] For many biomass systems, the products of pyrolysis can be controlled by regulating the pyrolysis conditions. Parameters that affect the decomposition of biomass are temperature, heating rate, gas flowrate, pressure, sample size, and inorganic salts. Depending on these operating conditions, the pyrolysis can be subdivided mainly into slow pyrolysis or fast pyrolysis.[13–15] Of these, the most important is the fast pyrolysis that maximizes bio-oil production. A detailed overview of the conventional thermal biomass pyrolysis process is given in Chapter 7 of this book.

Bio-oil is a complex mixture of water (ca. 20–25 wt%) and several hundred mainly oxygenated, polar organic compounds (ca. 75–80 wt%), that exhibit a wide range of chemical functionalities: ketones, aldehydes, sugars, carboxylic acids and aromatics.[16–20] The aromatics are mainly composed of phenols and guaiacols, which can represent one-fourth of the liquid fraction derived from lignocellulosic biomass.[21,22] In most cases, bio-oil consists of two phases, an aqueous phase containing oxygenated organic compounds of lower molecular weight and a non-aqueous phase, which contains heavier organic compounds (mainly aromatics). The components of bio-oil are different size molecules, derived from depolymerization and fragmentation reactions of the three key biomass building blocks: cellulose, hemicellulose, and lignin. The aqueous phase has high oxygen content, originating from the cellulose and hemicellulose fractions of biomass. The water-insoluble phase derives from the lignin fraction of biomass, and has lower oxygen content and high viscosity.[23,24] A detailed overview of the properties and quality characteristics of the bio-oil produced from the thermal biomass pyrolysis process is also given in Chapter 7 of this book. It should be emphasized that although bio-oil is a very attractive fuel when compared to the initial biomass, it has some disadvantageous characteristics (corrosiveness, instability under storage and heating conditions, immiscibility with petroleum fuels, high acidity, high viscosity, and low calorific value) that serve as primary obstacles for the further application and commercial scale-up of bio-oil utilization. Therefore, improvement of the bio-oil quality is a prerequisite before its upgrading and final utilization.[16,25,26]

One of the most promising methods to improve the quality of bio-oil by minimizing its undesirable properties (high viscosity, corrosivity, instability, *etc.*) or to increase the production of high added-value chemicals is the application of heterogeneous catalysis in biomass pyrolysis. This can be done using the catalyst in the pyrolysis reactor (*in situ* catalytic upgrading) or downstream using a catalyst bed to upgrade the pyrolysis vapors. In the *in situ* catalytic pyrolysis, the catalyst usually also plays the role of the heat carrier. Heterogeneous catalysis has been used for many years in conventional oil refineries to convert heavy hydrocarbons into suitable, lighter fuels and/or chemicals and this concept is currently transferred to the biomass catalytic pyrolysis process. An ultimate target of the research in catalytic biomass pyrolysis is the production of improved bio-oil for use within modern refinery infrastructure (as a biocrude). Moreover, there is a research trend to render bio-oils as feedstocks or co-feedstocks in petroleum streams, co-processing them with hydrocarbon fractions, such as vacuum gas oils, in FCC and/or HT processes of a conventional refinery.[27] Ideally, novel catalysts should produce high quality and yield of bio-oil with low amounts of oxygen and water, should affect bio-oil composition, minimizing undesirable oxygenated compounds (alcohols, ketones, acids, and carbonyls), and should exhibit both resistance to deactivation and hydrothermal stability. Research on the development and optimization of new catalysts includes controlled formation of appropriate catalyst particles and tailoring the porosity, acidity, basicity, and metal–support interactions in the candidate catalytic materials.[27–29]

10.2 Overview of the Biomass Catalytic Pyrolysis Process

Since biomass pyrolysis reactions are endothermic, a heat carrier should be available to provide heat to the reacting biomass particles. Different types of heating have been proposed, but the most effective one is direct heat transfer using a solid inert material (usually silica sand). The type of heat carrier dictates the type of the reactor and in general the process scheme of biomass pyrolysis. An overview of these process schemes is given in Chapter 7. The most efficient process technology for biomass pyrolysis is based on fluid-bed reactors. Biomass particles are introduced into a bed of hot sand, fluidized by a gas, which is usually a recirculated product gas. High heat-transfer rates from the fluidized sand result in rapid heating of the biomass particles. This process has been proven as very efficient for maximizing bio-oil yield and for this reason it has been scaled up by companies such as Dynamotive. This technology, although suitable for conventional biomass pyrolysis, is very difficult to apply to biomass catalytic pyrolysis. The reason is that catalyst deactivation usually takes place during catalytic pyrolysis that makes the catalyst inactive within a very short time. Thus, even by replacing the inert heat carrier by a catalytic material, the life of this material will be very limited as very rapidly the catalyst will be deactivated and will behave as an inert material. This technology can be used for downstream upgrading of bio-oil vapors by putting a fixed or fluid bed of a catalyst at the reactor exit. However, again catalyst deactivation in this second bed can be a serious issue except if the catalyst is continuously replaced. This downstream catalytic pyrolysis has been applied by various researchers in the open literature as will be described in a subsequent section.

The most appropriate process for biomass catalytic pyrolysis should be based on the circulating fluid-bed (CFB) technology that includes a regeneration step for the continuous catalyst regeneration. In this way, the catalyst is always mixed with the biomass particles in an active state. The technology is similar to the fluid catalytic cracking (FCC) process that is the biggest conversion unit in a refinery for the conversion of heavy hydrocarbons to lighter fuels. Commercial CFB technology for biomass catalytic pyrolysis is not available today and thus in the following paragraphs we will try to describe the process based on an existing pilot plant operated in the Chemical Process Engineering Research Institute (CPERI). This pilot-plant unit is described in Figure 10.1.[30] The entire process scheme is divided into five sections: biomass feed, catalyst regeneration, reactor/mixing zone, product stripper – solid/vapor separation and liquid product recovery. The biomass is loaded at ambient temperature into a feed hopper (D-61) that can be subsequently pressurized. Then, the biomass is fed from the feed hopper to the hot reaction section using a screw feeder (SF-61). The solid heat carrier (sand or catalyst) is loaded into heat carrier vessel or regenerator (D-101). The regenerator is a fluid-bed reactor aiming to burn off the coke that is deposited on the catalyst and provide the required heat for the cracking of biomass. The sand or catalyst in this vessel is fluidized with air. A sample flow is continuously taken from the flue gas stream and passed through a CO_2 analyzer to measure in real time the coke deposited on the heat carrier. Solids from the heat carrier vessel pass through a standpipe and slide valve (SV-101), which controls the flow of the hot catalyst towards the reactor.

The reactor usually consists of a mixing zone vessel (D-201) and the riser. The mixing zone should be designed to promote intimate contact of hot catalyst, hot fluidizing gas, and biomass in order to have very high heating rates of the biomass particles. Once mixed, the combination is transported out of the mixing zone into the riser (D-202), where the pyrolysis reactions continue to take place. The mixture of biomass pyrolysis vapors and solid heat carrier then enters the cyclonic head of the stripper tangentially. The function of the stripper is twofold. First, it allows the separation of oil vapor from biomass and catalyst solids *via* its cyclonic head. Secondly, steam or nitrogen is fed at the bottom of the stripper to remove trapped vapors within the solids in the stripper standpipe. The residence time of the solids in the stripper standpipe can be adjusted and controlled *via* a differential pressure transducer and slide valve (SV-301). Solids pass through the stripper standpipe to an injector (D-304) that helps recirculate the spent heat carrier back to the regenerator *via* the liftline (D-305). Downstream of the product stripper, the hot separation of entrained solids from the product vapor is accomplished by an external cyclone and filter housing (F-301). The final section of the unit is the product-recovery system, which encompasses a number of vessels and heat exchangers along with a stabilizer that recover the produced bio-oil. Finally, the process gases are passed through special devices for the online measurement of the flow of the cracked gases.

The factors of the above technology leading to an optimum bio-oil yield include (besides the type of catalyst) the design of the mixing zone to assure the very rapid mixing of biomass particles with the hot sand particles that come from the regenerator, the residence time of the vapor in the riser that should be low (<2 s) to avoid overcracking of the vapors, the fast separation of the solid and the vapors in the cyclone type stripper and the fast quench of the produced vapors in the heat exchangers that also favors a high liquid yield. Ensyn and CPERI have developed technology for CFB pyrolysis.[30–32] The rotating-cone technology developed by BTG also works in the circulating mode and thus, can also be used for catalytic pyrolysis.[2,33]

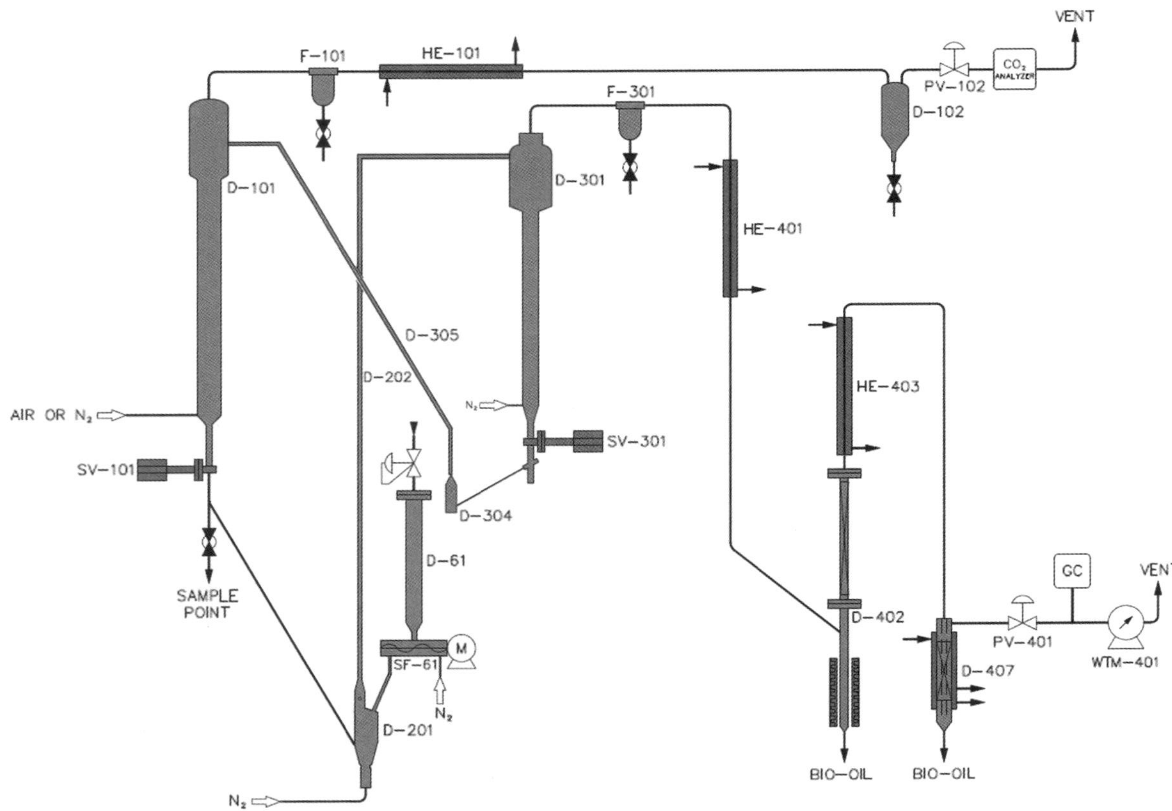

Figure 10.1 Schematic flow diagram of the CPERI biomass pyrolysis pilot-plant unit depicted in the recirculation of the heat-carrier configuration. The following sections of the BIOCAT unit are presented: heat carrier vessel or regenerator (D-101), biomass feed hopper (D-61), mixing zone (D-201), reactor/riser (D-202), stripper (D-301), lift line injector (D-304), lift line (D-305), bio-oil recovery vessels (D-402, D-407), heat exchangers (HE-101, HE-401, HE-403).

10.3 Catalyst Effects on Bio-Oil Yield and Quality

10.3.1 Microporous Catalysts

It is clear that since pyrolysis requires the breaking of carbon–carbon bonds, it can be catalyzed by solid acids, *e.g.*, zeolites, which are used in oil refineries for the same purpose. In this respect, several groups have studied the catalytic pyrolysis of biomass over acidic zeolite catalysts, such as zeolites Y and ZSM-5.[34–39] These zeolites are both very acidic, the latter being also very shape selective due to its two-dimensional channel-like pore system with a pore diameter of ~0.55 nm.

Studying the pyrolysis of wood-based biomass in the presence of H-ZSM-5 suggested that all of the deoxygenation, decarboxylation, and decarbonylation reactions of the bio-oil components, as well as cracking, alkylation, isomerization, cyclization, oligomerization, and aromatization reactions, are catalyzed by acidic sites of the zeolite by a carbenium-ion mechanism. However, tar and coke were also formed as undesirable byproducts.[40] In general, zeolite ZSM-5 has been the main zeolite investigated as a catalyst for biomass pyrolysis and has been found to dramatically change the composition of bio-oils by both reducing the amounts of oxygenated compounds in bio-oil *via* deoxygenation reactions and simultaneously increasing the aromatic species, producing a lighter fraction (gasoline-type fuel), while also decreasing the bio-oil molecular weight. Williams *et al.* used a combined fluidized-bed pyrolysis unit in which a fixed catalyst bed was packed on the freeboard of the fluid bed.[12,35,38,40–43] The pyrolysis oil was highly oxygenated and consisted of a single liquid phase with a significant aqueous content. After catalysis, gas and aqueous phases increased at the expense of the oil yield, while separate oil and aqueous phases were present. The presence of the zeolite catalyst reduced oxygen content in the bio-oil from 33 to 13%. Oxygen removal was found to take place as H_2O at lower temperatures and as CO and CO_2 at higher temperatures.[12,41] The latter case is preferable so that the hydrogen can participate in hydrocarbon forming reactions. When investigating Na-ZSM-5-, H-ZSM-5- and Y-type zeolites and activated alumina, the ZSM-5 catalysts gave the highest yields of hydrocarbon products.[38] The influence of ZSM-5 catalyst deactivation due to coke formation on the upgraded pyrolysis product was also examined.[44] Coke formation is reported to be extremely high during the initial catalytic stages, resulting in catalyst deactivation and thus reduced activity and selectivity and also caused reactor plugging problems. All these results confirm that the circulating fluid-bed technology is the best for catalytic biomass pyrolysis.

Atutxa *et al.*[37] studied the catalytic pyrolysis of pine biomass over H-ZSM-5 zeolite in a spouted-bed reactor. They observed that by increasing the amounts of acid sites in the reactor, the gas yield increases and the oil yield decreases, keeping though the char yield fairly constant but lower than that formed in the absence of any catalyst. The use of the catalyst drastically changed bio-oil composition, as the heavy organic fraction is transformed to an aqueous fraction *via* cracking, whereas the aqueous liquid fraction undergoes dehydration, decarboxylation and decarbonylation reactions. The application of HZSM-5 was successful in the deoxygenation of the bio-oil.[37] Fast pyrolysis of vegetable biomass in a fluidized-bed reactor was also successfully performed using Ni-H-ZSM-5 as catalyst.[45]

In a recent study[34] catalytic pyrolysis of pine biomass was successfully carried out in a fluidized-bed reactor using H-beta zeolites. The presence of H-beta zeolite catalysts decreased the yield of the organic oil (from ~27% to 12–17% depending on the silica to alumina ratio of the H-beta zeolite structure) and increased the percentage of char (from ~15% to >20%) as a result of catalyst coking. More water was formed (water yield increased from ~5% to ~13% in most cases) during the catalytic pyrolysis of biomass than in the non-catalytic case, the increase of the water content being probably a result of the formation of polyaromatic hydrocarbons. The gas yield was fairly constant in both the catalytic and non-catalytic experiments. H-beta zeolite catalyst used in the pyrolysis experiment was regenerated by calcination in air. The surface area and the acidity were regained after such treatment. Aldehyde formation was influenced by the different acidities of the zeolites, while formation of PAHs was significant only in the presence of the acidic zeolites. It was generally observed that higher acidities produced less organic oil and more water and PAHs.[34] The same group reports in a following study the influence of several zeolite structures, beta, Y, ZSM-5 and mordenite, on the yields and the chemical composition of the bio-oil obtained from the catalytic pyrolysis of biomass. H-ZSM-5 zeolite gave the highest liquid yield, while non-catalytic pyrolysis produced roughly the same percentage of oil but with lower water content. The formation of acids and alcohols over ZSM-5 was lower, while the formation of ketones was higher than over the other zeolites tested. It was possible to successfully regenerate the spent zeolites without changing their zeolitic structure.[46]

An increased formation of coke is reported when using Y-zeolites, attributed to the larger pores of these catalysts allowing larger coke precursors to enter the catalyst pore structure, thus initiating and contributing to coke formation. For all these catalysts the aqueous product is the main result of oxygen removal from biomass pyrolysis vapors. In addition, bio-oil upgrading resulted in an increase in aromatics and PAHs *via* a suggested dual mechanistic route, either through aromatization reactions of low molecular weight hydrocarbons or direct deoxygenation reactions.[35]

A very systematic and comprehensive study regarding the effect of zeolitic catalysts in biomass pyrolysis was performed in CPERI. From this work, we present in Figure 10.2 the raw liquid yield (wt% on biomass) as a function of the total surface area (TSA) of the catalytic materials. The experimental point at zero surface area represents tests with silica sand and it essentially comprises the conventional or thermal biomass pyrolysis case. All the other points represent catalytic biomass pyrolysis runs with the different zeolitic materials. Figure 10.2 shows that the highest raw total liquid yield (around 74 wt%) was produced from the silica sand. The use of a zeolitic catalytic material does not favor the production of the liquid (bio-oil). The more active the catalyst (higher

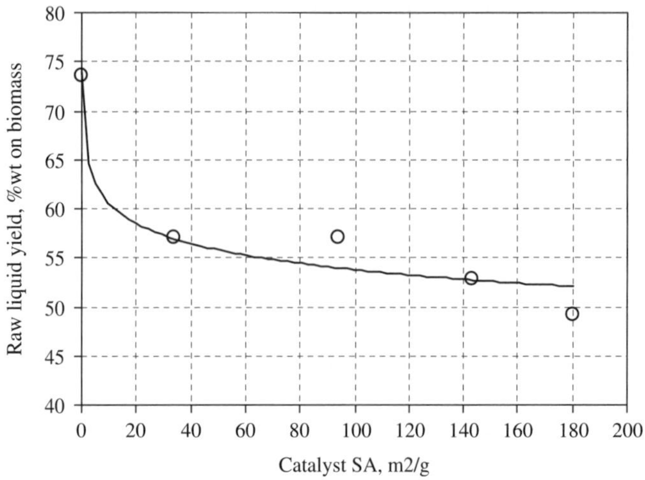

Figure 10.2 Effect of catalyst surface area on liquid yield.

surface area), the less liquid is produced. Nevertheless, the catalytic effect on the bio-oil yield is not linear. Even with the less-active catalyst (with TSA around 30 m^2/g) there is significant drop of 17 wt% in the liquid yield (from 74 wt% to 57 wt%). Catalysts with higher activity (higher TSA) decrease the bio-oil yield but to a lesser extent as compared to the low activity catalyst. Even with the more active catalyst (TSA = 180 m^2/g) the liquid yield only dropped from 57 wt% to 50 wt%. These experimental results show that the effect of catalyst surface area (or catalyst acidity) on the bio-oil yield seems to be logarithmic. This is a general effect, since it was found to be valid in all other conditions (temperature, solid/biomass ratio, type of biomass, *etc.*) tested in CPERI.

In Figure 10.3 the yield of the total gas is presented. The effect of the TSA on gas yield is also logarithmic and thus, by using a low-activity catalyst the gas yield increases considerably, reaching values of about 17.5 wt%. The use of a more active catalyst has an additional minor effect on gas yield by increasing it to only 18.5–19 wt%. In the pyrolysis gases the oxygen compounds (CO, CO_2) dominate, while the yields of light hydrocarbons are very low. With the silica sand the CO_2 was the component with the highest yield (6.9 wt% on biomass basis), while the CO was second with a yield of 3.8 wt%. By using a zeolitic catalyst, the distribution of the gases changes, favoring CO production rather than CO_2 production. It seems that the presence of a zeolitic catalyst favors secondary decarbonylation reactions rather than decarboxylation reactions, producing more CO and less condensable carbonylic compounds. Figure 10.4 depicts the remaining unconverted material (char) and the pyrolitic coke. The effect of the catalyst surface area on the char yield seems to be different as compared with that on liquid and gas yields. The increase of surface area causes a linear increase to the char yield. By increasing the surface area of the solid

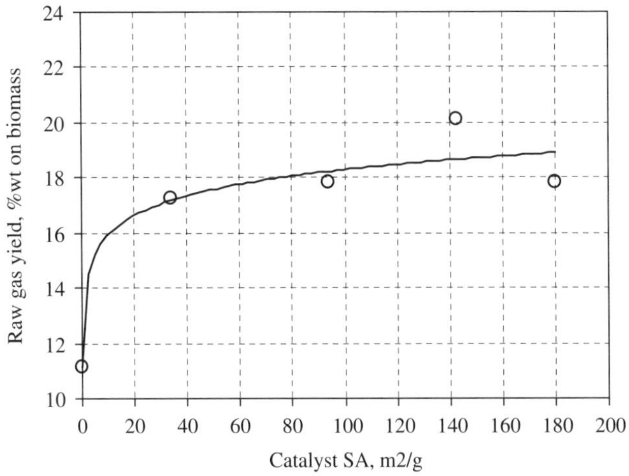

Figure 10.3 Effect of catalyst surface area on gas yield.

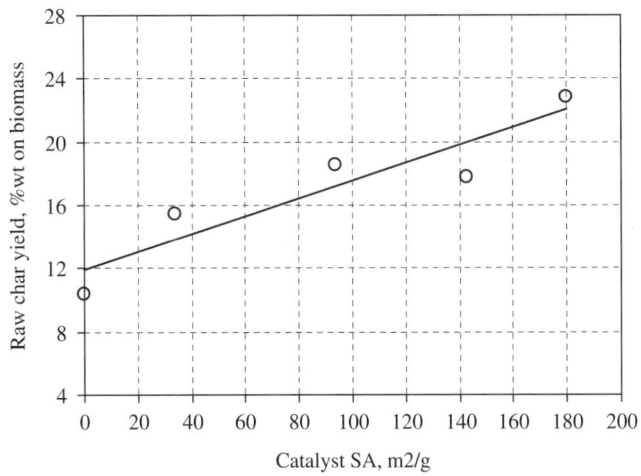

Figure 10.4 Effect of catalyst surface area on char yield.

material the char yield becomes higher and almost doubles (as compared with the silica sand) in the case of the highest activity catalyst. Figure 10.5 shows that the water yield is influenced by the catalyst activity in the same way as the liquid and gas yields. The water yield in the experiments with the inert material was around 17.5 wt%, while with the most active catalyst it can reach values up to 28 wt%. The pyrolytic water from the silica sand is a product of the primary dehydration reactions occurring during pyrolysis.[47] Even a low activity catalyst (30 m^2/g) can strongly promote secondary deoxygenation reactions, producing about 26 wt% water (the catalytically produced water is around 8.5 wt%).

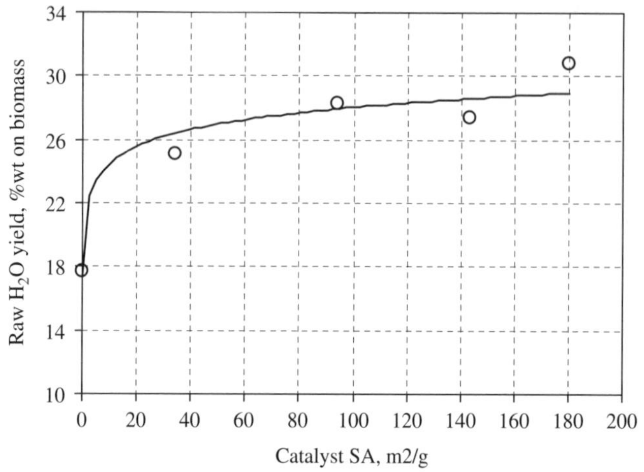

Figure 10.5 Effect of catalyst surface area on H_2O yield.

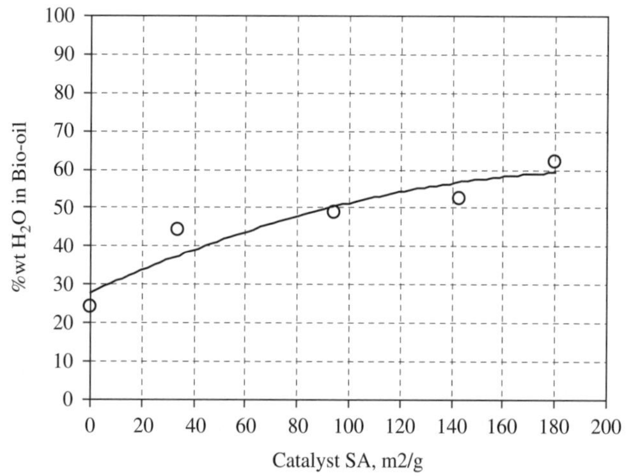

Figure 10.6 Effect of catalyst surface area on water in bio-oil.

From all previous CPERI results it can be concluded that the zeolite catalysts cause secondary reactions to the primary pyrolysis products, leading to lighter products and coke. These secondary reactions crack the bio-oil vapor compounds to lighter ones producing gases, coke and water.

Zeolitic catalysts play an important role not only in determining the various product yields but also in the quality of the produced bio-oil. The concentration of water in bio-oil is given in Figure 10.6 as a function of the catalyst total surface area. The water in the bio-oil produced from the inert material (silica sand) was around 24 wt%. This bio-oil consisted of a single phase. In the literature a wide range of water content in bio-oils has been reported (15–30 wt%) depending on the type of feedstock and process conditions.[18,47,48] Water comes from the original moisture in the feedstock and it is also a product of the primary dehydration reactions. Increasing the activity of the zeolitic catalyst significantly increases the concentration of water in the bio-oil, while for the most active catalyst (with the highest surface area) the water concentration was around 60 wt%. It must be pointed out that in most of the catalytic cases a two-phase bio-oil can be produced due to the high water content.

The presence of the catalyst causes a significant change to the bio-oil composition (as determined in CPERI by GC-MS techniques[37]). The bio-oil produced from the test with the silica sand contained mainly phenolic and carbonylic compounds (5.8 wt% and 4.0 wt%, respectively). Furans, acids and hydrocarbons were about 2.5 wt% each. Moreover, heavy oxygenates and unidentified compounds were present in significant amounts (14.7 and 9.7 wt%, respectively). When using a zeolite catalyst, the concentration of hydrocarbons increases with increasing catalyst surface area. When using the most active catalyst this concentration is almost double as compared to that from the non-active material. Carbonylic compounds were decreased. The production of light phenolic compounds was also influenced by the presence of the catalyst. The total concentration of the light phenols (anisole, phenol, m/p-anisol, o/p/m-cresol, guaiacol, xylenols, methyphenols) in the bio-oil produced from silica sand was only 10% of the total phenols. The majority of these phenolic compounds are heavy and are present as oligomers with high molecular weights.[48–50] In the catalytic pyrolysis experiments the fraction of light phenols in the total phenolic fraction reaches values of up to 30%. However, the most important effect of the catalyst is on the heavy compounds. The presence of the catalyst causes a drastic decrease in the production of these compounds that is depicted in Figure 10.7 as a function of catalyst surface area. In these series of experiments it should be clarified that as heavy fractions we consider organic compounds with molecular weight ranging from vanillin and higher, excluding PAHs.[28] It seems that by using a high-activity catalyst we can minimize these undesirable compounds. There is a clear trend which shows that in the presence of a catalyst we can produce a bio-oil with a micrometer carbon residue content (MCRT, that is an index of the heavy compounds in fuels) that is much better as compared with that from pyrolysis using an inert material. From the above results it can be concluded that the catalyst affects the primary heavy products, cracking them towards the production of coke and gases. In this way, although the yield of bio-oil is lower, the quality of the bio-oil is improved.

10.3.2 Mesoporous Acid Catalysts

As described in the previous section, the significant increase in the production of water and gases at the expense of the organics, the formation of undesired polycyclic aromatic hydrocarbons (PAHs) and the rapid catalyst deactivation

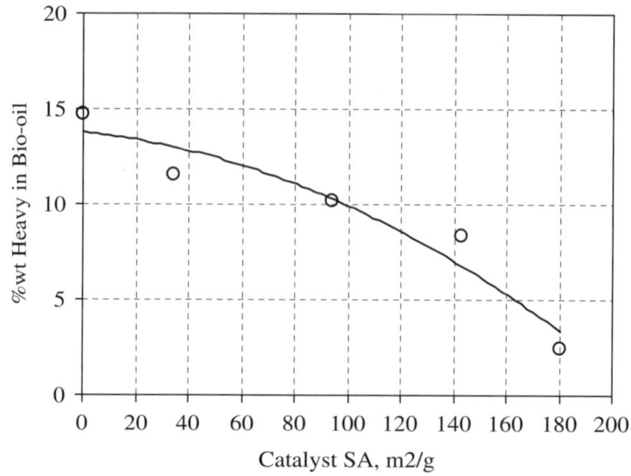

Figure 10.7 Effect of catalyst surface area on heavy compounds in bio-oil.

by coke deposition are serious drawbacks of the zeolitic materials. To overcome these problems, catalysts with a larger pore size than that of ZSM-5 have recently attracted much attention, as they are expected to allow larger pyrolysis products, particularly lignin-derived compounds, to enter, reformulate and exit the catalyst matrix with less chance of coke deposition and pore blocking.[51–56] The recent focus of catalyst development for the conversion of lignocellulosic biomass to valuable products is concentrated on mesoporous materials (MCM-41), one of the newest members of the mesoporous family of materials with a uniform pore-size distribution. The discovery of mesoporous molecular sieves of the M41S family in 1992, with the main representative being MCM-41, opened a new area in the application of these materials as supports for various metals and other catalytically active species.[56] These mesoporous molecular sieves exhibit extremely high surface areas ($\geq 1000\,m^2/g$) and large, but well-defined, pore sizes in the range of 20–300 Å. Showing a relatively narrow pore-size distribution, MCM-41 materials present a hexagonal arrangement of parallel mesopores and originally they were synthesized in the siliceous form. MCM-41, one of the latest members of the mesoporous family, possesses a hexagonal array of uniform mesopores whose dimensions can vary from 14 Å to greater that 100 Å in size.[56] Through proper control of preparation conditions, such as template, reaction temperature and time, and pH value, high quality MCM-41 materials with the following properties could be obtained: high surface area (more than $1000\,m^2/g$), large pore volume, uniform pore-size distribution and high thermal and chemical stability.[20] When they were first discovered, there were great expectations for applications in the petroleum industry, however, independent of the Al content in the framework, the MCM-41 material shows only weak acidity, which is comparable to amorphous aluminosilicates and therefore they are only promising for reactions that do not require strong acidity.[57] Very soon after their first synthesis, many laboratories incorporated various elements together with silicon into the framework of MCM-41 molecular sieves, for instance aluminum, vanadium, niobium, *etc.*, changing the physico-chemical properties of the material.[58] Mesoporous materials containing aluminium, such as Al-MCM-41, with improved acidic characteristics are considered to be of great importance, since the presence of aluminium in a large range of Si/Al ratios in the mesoporous material can generate the acidity and thereby the cracking characteristics of the catalysts.[59] With the incorporation of various metals (Ni, Al, Co, Mo, Fe, Cu) it was found that the catalytic properties of these materials could be improved.[60–62] When trivalent cations (like Al, B, Ga, Fe) substitute for Si in the walls of the mesoporous silica, the framework possesses negative charges that can be compensated by protons, and these solids can be used in acidic reactions. When other cations (Ti, V, Sn, Zr) are introduced, the electroneutrality is maintained and the corresponding mesoporous materials are used in specific reactions like oxidations.[63]

The one-dimensional mesopores of the mesoporous materials, in combination with their large surface area ($\geq 1000\,m^2/g$) and their mild acidity, seem highly advantageous for novel and probably useful catalytic materials for biomass pyrolysis. More significantly, their properties are expected to make them more accessible than microporous materials and for this reason more effective for the cracking of high molecular weight lignocellulosic molecules, thereby producing a liquid product with less heavy oxygenated compounds and possibly richer in high-value products. Factors such as the pore shape and size, the acidity and the redox properties of the transition metals can affect the catalytic pyrolysis product distribution, providing the desired environment for a controlled conversion of the high molecular weight lignocellulosic molecules. Besides testing the activity and selectivity of these materials in catalytic pyrolysis, research interest is also focused on the improvement of their hydrothermal stability by steam treatment, or by introduction of noble and/or transition metals to enhance the oxygen removal capacity and/or promote decarboxylation reactions.[28,29,52–55] Some results from the use of these mesoporous catalysts in biomass catalytic pyrolysis are given in the following paragraphs.

Samples of Al-MCM-41 materials with different Si/ Al ratios have been tested in a bench-scale pyrolysis unit, and compared with three metal-promoted mesoporous samples, Cu-Al-MCM-41, Fe-Al-MCM-41 and Zn-Al-MCM-41. The presence of the catalysts decreased the production of liquids in comparison to the non-catalytic runs, produced comparable or lower gas yield and increased the production of coke. Decreased organic fractions (7–17%) indicate an increased production of water due to the catalytic reactions in the presence of the mesoporous materials (Figure 10.8). All of the catalysts altered the quality of the pyrolysis products, decreasing almost all of the undesired oxygenates (Figure 10.9) and increasing the amount of phenolic compounds, important raw materials for the adhesives industry. A low Si/Al ratio of MCM-41 exhibited a positive effect on phenol yields and composition, while Fe- Al-MCM-41 and Cu-Al-MCM-41 are

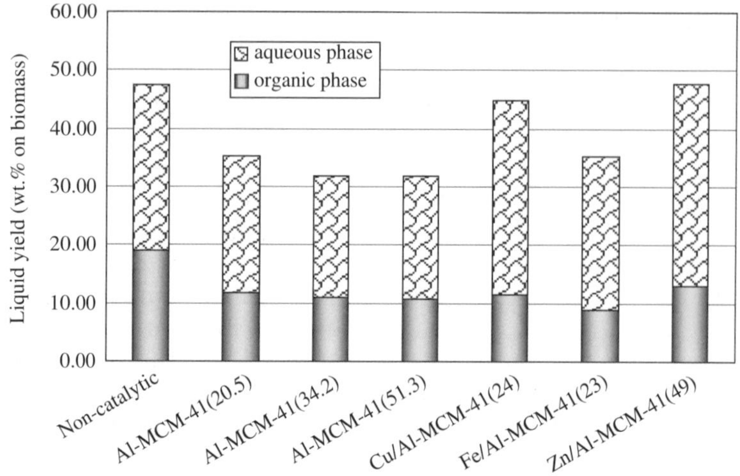

Figure 10.8 Effect of catalytic materials on the liquid yield produced during biomass catalytic pyrolysis (numbers in parenthesis refer to the Si/Al ratios of the Al-MCM-41 materials).

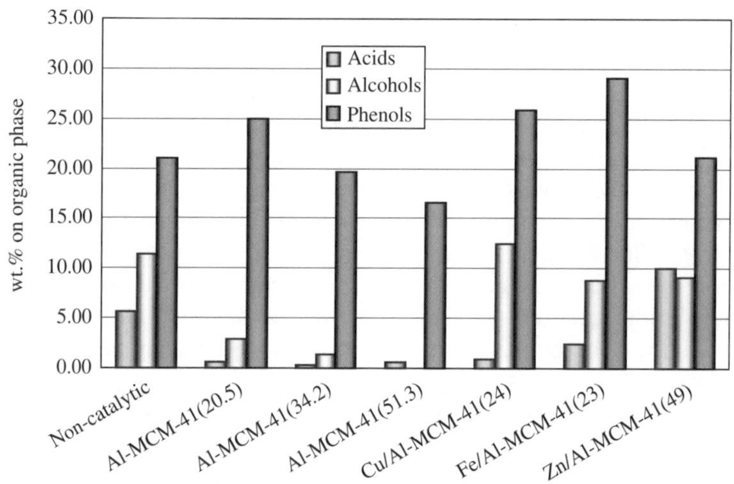

Figure 10.9 Acids, alcohols and phenols production from biomass catalytic pyrolysis (numbers in parenthesis refer to the Si/Al ratios of the Al-MCM-41 materials).

the best transition-metal-containing catalysts for the production of phenols (Figure 10.9). In addition, an increased concentration of hydrogen is reported in the gaseous pyrolysis products when using transition-metal-based catalysts, especially Cu/Al-MCM-41.[53]

However, as is well known in the literature,[64] Al-MCM-41 materials exhibit relatively low hydrothermal stability compared to zeolites, thus inhibiting their use in high-temperature processes in the presence of steam. Catalytic pyrolysis of biomass at 500–650 °C, which produces water vapor, and the intermediate regeneration of coked catalysts at 650–750 °C in a continuous-operation mode, can be considered as such processes. The hydrothermal stability and acidity of two mesoporous Al-MCM-41 (Si/Al ratios of 50 and 30) was thus examined in the *in situ* upgrading of lignocelluslosic biomass pyrolysis vapors, for comparison with a siliceous MCM-41 sample and with non-catalytic biomass pyrolysis.[29] The product yields and the quality of the produced bio-oil were significantly affected by the use of all of the MCM-41 catalytic materials, a behavior mainly attributed to the combination of the large surface area and the tubular mesopores (pore diameter ~2–3 nm) of the MCM-41 materials, which with their mild acidity leads to the desired environment for controlled conversion of the high molecular weight lignocellulosic molecules. Confirming results reported in the literature, the major improvement in bio-oil quality when using Al-MCM-41 catalytic materials was the increase of phenols concentration (useful chemicals) and the reduction of corrosive acids (undesirable in fuel bio-oils). Higher Si/Al ratios (*i.e.* lower Al content and lower number of acid sites) of the Al-MCM-41 samples enhanced the production of the organic phase of the bio-oil, while lower Si/Al ratios favored the conversion of the hydrocarbons of the organic phase into gases and coke. Moderate steaming of the Al-MCM-41 samples (at 550 and 750 °C, 20% steam partial pressure) decreased their surface area and number of acid sites by 40–60% depending on the Si/Al ratio of the samples and the steaming temperature. The main information provided by this study was that the relatively moderate steaming pretreatment of Al-MCM-41 samples resulted in active catalytic materials (no significant change in the activity exhibited before steaming), although they possessed relatively lower surface area and number of acid sites compared to the parent calcined samples. The product yields and composition of the bio-oil were mainly affected by the higher temperature steaming of the Al-MCM-41 samples. This effect is clearly evidenced in Figure 10.10, presenting the phenol concentration in the bio-oil produced using fresh or hydrothermally treated MCM-41 materials. As is apparent from Figure 10.10, moderate steaming has nearly no effect on the phenol production of the low Si/Al ratio Al-MCM-41 material. Fine tuning the acidity and porosity of the MCM-41-based materials seems to be a prerequisite for improving the bio-oil quality and the product selectivity.[29]

Two MSU-S mesoporous aluminosilicate materials that are synthesized from zeolite seeds have been also investigated as catalysts for the *in situ* upgrading of lignocellulosic biomass pyrolysis vapors, in comparison to a conventional Al-MCM-41 sample. The MSU-S samples exhibit high hydrothermal stability and increased acidity compared to Al-MCM-41. After steaming at 800 °C in a flow of N_2 containing 20% steam, the MSU-S materials retained significantly higher surface area and mesopore volume compared to conventional Al-MCM-41.[65,66] With regard to acidity, the MSU-S materials were more active in cumene

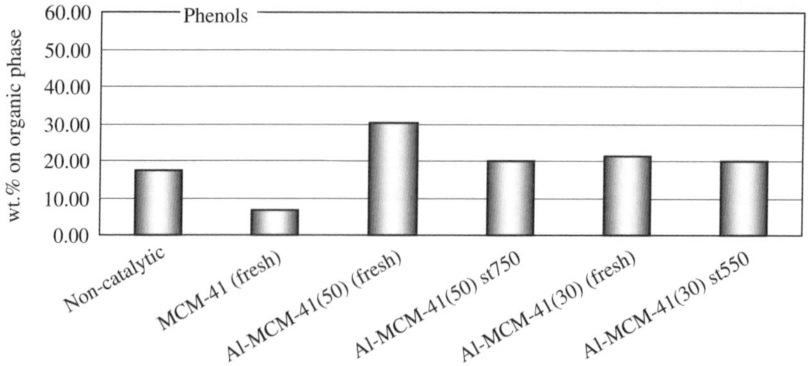

Figure 10.10 Biomass catalytic prolysis in the presence of (Al)MCM-41 materials: Effect of steaming on the selectivity of Al-MCM-41 materials to phenols (numbers in parenthesis refer to the Si/Al ratios of the Al-MCM-41 materials).

cracking compared to Al-MCM-41 with similar Al content, indicating the higher acid strength of the MSU-S materials. In addition to the above, the MSU-S samples exhibited higher gas-oil cracking activity after severe steaming compared to similarly steamed Al-MCM-41.[67] Based on the literature reported activity and hydrothermal stability of MSU-S materials, they can be considered as an effective choice for catalytic processes at high temperature in the presence of water (steam) or for high-temperature regeneration of coked catalysts, both cases being related to the *in situ* upgrading of biomass pyrolysis vapors.[29] Both of the two MSU-S catalysts tested led to a significantly lower quantity of organic phase in the obtained bio-oil, were quite selective towards polycyclic aromatic hydrocarbons (PAHs) and heavy fractions, and produced only small amounts of acids, alcohols, carbonyls, and phenols (Figure 10.11). Their stronger acid sites are claimed to be responsible for the enhanced yields of aromatics, PAHs, and coke, along with propene in the pyrolysis gases. Pyrolysis bio-oils with organic phases that are composed mainly of hydrocarbons, PAHs and heavy compounds, with a relatively low content of oxygenated compounds (phenols, acids, carbonyls, alcohols), such as those produced by the two MSU-S catalysts, can be of high value for refinery upgrading processes.

Future studies on the acidic properties of the mesoporous materials (type, strength, and number of acid sites) and on the pore architecture are required to gain information about the structure–property relationships of mesoporous materials as catalysts in the conversion of lignocellulosic biomass.[28] The performance of MCM-41 was also confirmed by a different research group investigating the catalytic upgrading of bio-oil using MCM-41 catalysts. Use of Al-MCM-41 catalysts produced a bio-oil of enhanced stability and with larger amounts of phenolic components.[68]

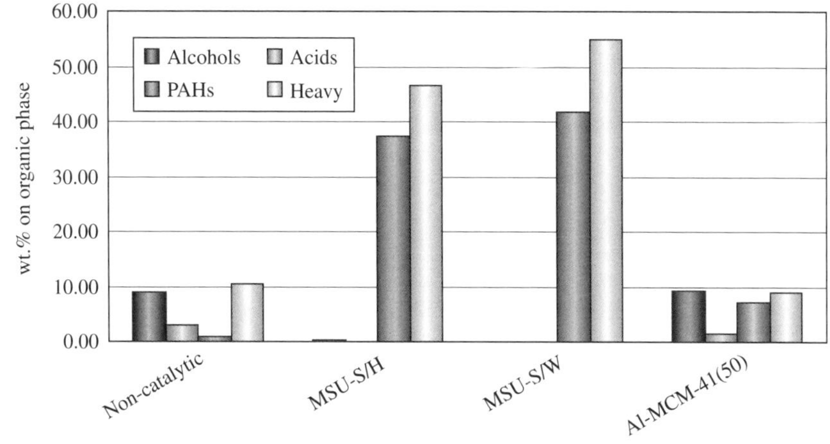

Figure 10.11 Effect of MSU materials on the composition of bio-oil produced from biomass feedstock (numbers in parenthesis refer to the Si/Al ratios of the Al-MCM-41 materials).

10.3.3 Basic Catalytic Materials

As previously described, solid acid catalysts are predominately tested in the effort to convert biomass to oils using pyrolysis. However, Nokkosmaki *et al.*[69] and Fabbri *et al.*[70] have also applied basic oxides in catalytic pyrolysis, ZnO and MgO, respectively. The first group pyrolyzed pine sawdust with three different zinc oxide (ZnO) catalysts at 600 °C, with 30 ms residence time, investigating their influence on the composition and stability of the bio-oil. They passed the pyrolysis vapors of pine biomass through a fixed bed of catalyst. They found that ZnO was a mild catalyst for producing bio-oils, showing only a small reduction in the liquid yield with only a 2% gas increase, while the stability of the catalytic oil was increased. The formation of anhydrosugars and formic acid was reduced when increasing the temperature of the catalyst. At 500 °C, the usual temperature of biomass pyrolysis, neither the former nor the latter products could be detected.[69] Catalytic pyrolysis of pinewood sawdust was also tested by microwave heating, applying inorganic additives with different basicity. All of the eight additives altered the yields and composition of the pyrolysis products. However, they caused a significant increase in the yield of the solid product and a decrease in the yield of gases. Yields of liquid products were not subject to dramatic change. All of the eight additives caused the gaseous products to be evolved earlier. The gases produced from pyrolysis consisted mainly of H_2, CH_4, CO and CO_2, while alkaline sodium compounds, NaOH, Na_2CO_3 and Na_2SiO_3, favored H_2 formation in the pyrolytic gases.[71] Recently, strongly basic zeolites have been synthesized from the reaction of zeolites with ammonia at elevated temperatures, providing unique activity and selectivity for base-catalyzed reactions.[72] Such basic zeolites, *e.g.* amine-substituted ZSM-5, are also suggested as a promising candidate catalyst for biomass conversion to hydrocarbons *via* catalytic pyrolysis.[73]

10.3.4 Catalysts for Biomass Conversion to Aromatics

The conversion of biomass compounds to aromatics by thermal decomposition in the presence of catalysts has been investigated using several zeolitic catalysts in both fluidized-bed and fixed-bed reactors, with characterization of the bio-oil effluent. When aromatics are produced (instead of oxygenated bio-oils) the wt% yield of the bio-oils decreases because oxygen is being removed as water, CO and CO_2. However, the produced aromatics have a higher heating value than the typical bio-oil because of their reduced oxygen content. An early study performed by researchers at Mobil[75] achieved a high degree of conversion (74%) of biomass feedstocks over ZSM-5 to form hydrocarbons (including aromatics) in hydrogen flow. They also showed that aqueous glucose feedstocks can be converted to aromatics. Dao and coworkers[76–80] carried out several studies on aqueous fructose and glucose feeds with ZSM-5 catalysts and metal-doped ZSM-5 catalysts (ZnZSM-5 and MnZSM-5), discovering increased yields of aromatics when using ZnZSM-5 and MnZSM-5 catalysts. Production of aromatics *via* biomass pyrolysis has been also reported. Olazar *et al.*,[74] pyrolyzing pine sawdust in a conical spouted-bed reactor, showed that a 30.8 wt% total organic yield (with a high percentage of aromatics: 12% carbon yield in the product oils) could be obtained with high catalyst (ZSM-5) to feed ratios. More recently, Bridgwater and coworkers[51] reported screening several microporous (HZSM-5) and mesoporous (Al-MCM-41, Al-MSU-F, alumina-stabilized ceria) catalysts for the fast pyrolysis of biomass to gasoline, producing in all cases aromatic hydrocarbons and reduced amounts of oxygenated lignin derivatives. Metal-exchanged zeolites used in the catalytic pyrolysis of biomass have been reported. Park *et al.*[81] used Ga-ZSM-5 in the catalytic pyrolysis of sawdust, producing a greater amount of aromatic hydrocarbons compared to H-ZSM-5 under the same conditions. Sulman and coworkers[82,83] studied catalytic pyrolysis in the presence of iron-impregnated zeolites (beta, mordenite, HY, ZSM-5). Catalyst modification with iron increased the hydrocarbon yields by 2- to 3-fold compared to catalysts without iron impregnation, under the same conditions. In a recent study, five catalysts were tested in the conversion of biomass compounds to aromatics by thermal decomposition, including ZSM-5, silicalite, beta, Y-zeolite and silica–alumina. ZSM-5 had the highest aromatic yields (30% carbon yield) and the least amount of coke.[82,83] In summary, further studies are required to develop or optimize a catalyst that maximizes the yield of gasoline-range hydrocarbons, focusing on the effect of acid strength and variation of the pore size (to facilitate transformation of a higher fraction of the larger biomass components, *e.g.*, lignin), the incorporation of metals in the zeolites, and finally the use of microporous basic catalysts.[73]

10.4 Recent Developments in Bio-Oil Upgrading

Recent developments in biomass catalytic pyrolysis concern biomass pre-treatment methods and bio-oil upgrading in downstream processes. Pretreated biomass can be used directly for *in situ* catalytic pyrolysis. In this area a patent was recently disclosed[84] for a hydrothermal treatment process for conversion of a carbon-based energy-carrier material (biomass particle). The process comprises a step for sensitizing or activating the carbon-based energy-carrier material to increase its susceptibility to hydrothermal conversion. As a result of the sensitization step, the hydrothermal conversion step itself may be carried out under relatively mild conditions (less than 300 °C).

Regarding downstream upgrading of the pyrolytic oils, the most important issue is deoxygenation. Currently, two technologies have been proposed for this process. The first one is a typical catalytic hydrotreatment with hydrogen under high pressure and/or in the presence of hydrogen donor solvents. This first route involves using conventional Co-Mo and Ni-Mo catalysts,[85–87] characterized by extensive catalyst deactivation due to coking. Hydrotreating is discussed in Chapter 11 of this book and so is not considered further here. The second technology utilizes cracking catalysts (zeolites, silica–alumina and molecular sieves) under atmospheric pressure without hydrogen.[88] Catalytic cracking exhibits several processing and economic advantages over hydrotreating: no H_2 is required, atmospheric processing reduces operating costs, and the temperatures employed are similar to those used in the production of bio-oil.[89]

10.4.1 Zeolite Upgrading – Catalytic Cracking

Catalytic upgrading of biomass pyrolytic oil is performed either in a single reactor (*in situ* upgrading of pyrolysis vapors during catalytic pyrolysis of biomass, described above) or in a separate process following bio-oil production. Conversion of bio-oils and model bio-oil compounds has been investigated by several groups using zeolites as catalysts.[90–97] The reactivity and reaction pathways for model bio-oil compounds using ZSM5 catalysts have been studied by Gayubo and coworkers. They suggested that alcohols convert into olefins at temperatures around 200 °C, then into higher olefins at 250 °C, and into paraffins and a small proportion of aromatics at 350 °C. Gayubo *et al.* have also recommended that the oil fractions that lead to thermal coking (such as aldehydes, oxyphenols, and furfurals) should be removed from the bio-oil prior to upgrading over zeolites.[95–97] Bakhshi and coworkers studied zeolite upgrading of wood-derived fast-pyrolysis bio-oils. In order to reduce coking during the upgrading process, Sharma and Bakhshi[98–100] studied the conversion of bio-oil in a dual-reactor system. The dual-reactor system was particularly effective in producing high amounts of organic liquid products. Thermal cracking of bio-oil in the first reactor, followed by catalytic upgrading (H-ZSM-5) in the second reactor, reduced catalyst coking in the second reactor, thus enhancing catalyst life.[94] Catalytic cracking using zeolites such as HZSM-5 is advantageous, avoiding pressure restrictions involved in hydrotreating (typically 10–20 MPa hydrogen) and product distributions restricted to the diesel fuel range.[101] Using shape-selective zeolitic catalysts gasoline-range hydrocarbons can be produced, though the overall yields tend to be lower compared to hydrotreating and they often suffer detrimentally from coking. In addition, hydrotreating tends to produce products usable only when blended with other fuels, whereas zeolite upgrading may produce premium chemicals.[102] In a recent study, sulfated zirconia was used as a potential low-temperature catalyst for upgrading biomass pyrolysis oil and was claimed to avoid problems due to polymerization of the biomass oils. Instead, sulfated zirconia affords both high product yields and lighter hydrocarbon distribution in the upgraded biomass oil.[102] A general concern involves char, coke and tar formation, representing a major problem during the upgrading of bio-oils; this should be minimized by selection of the appropriate catalyst.[85,90,103–104]

A series of catalysts including ZSM-5, silicalite, beta, Y-zeolite and silica-alumina were tested in the conversion of biomass compounds into aromatics. The first step in this process is the thermal decomposition of the biomass to smaller oxygenates that enter the catalyst pores, where they are converted to CO, CO_2, water, coke and volatile aromatics. The product selectivity depended on both the size of the catalyst pores and the nature of the active sites.[73]

10.4.2 Co-processing of Bio-Oil and Conventional Oil in Refinery Processes

Another way to transform bio-oils into good-quality fuels could be to co-feed the bio-oils with FCC feedstock.[105] Many studies have been performed in the area, mainly using model compounds. In a recent study, small amounts of phenol (0.01–4.0 wt.%) (biomass-derived oxygenate) were added to methylcyclohexane (representative of FCC feedstock) to simulate bio-oil co-feeding. Its influence on the activity and selectivity of an H-USY zeolite was investigated, as well as its contribution to zeolite deactivation.[106] A similar concept was studied by another group.[25] Catalytic cracking of glycerol and sorbitol, representative of biomass-derived oxygenates, was studied at 500–700 °C with six different catalysts (fresh fluid catalytic cracking (FCC) catalyst, equilibrium FCC catalyst, mesoporous Al_2O_3, USY zeolite, ZSM5-based FCC additive and an inert silicon carbide (SiC)). Glycerol was also converted when fed together with vacuum gas oil without significantly altering the product selectivity. This supports the co-feeding approach of biomass-derived products and petroleum-derived streams in an industrial FCC reactor. In addition, these experiments showed that mixtures of VGO with biomass-derived feedstocks can help to transfer hydrogen from the VGO to the biomass molecules, in agreement with the results discovered by Marinangeli *et al.*, who also presented VGO as a hydrogen donor.[107] Increased yields of ethylene and propylene when using mixtures of VGO and glycerol indicate a synergetic effect. One option for further improved olefins and aromatics yields is co-feeding glycerol and petroleum-derived feedstocks into an FCC reactor in the presence of a ZSM-5 additive producing more olefins and less coke.[89] A complex reaction pathway is involved in the catalytic cracking of biomass: dehydration reactions, hydrogen-producing reactions, and hydrogen-consuming reactions. A balance between these three pathways is needed to produce the desired olefins from biomass-derived molecules.[25]

Integration of bio-oil upgrading in the petrochemical refinery presents the advantage that refineries are already built and this provides a low capital cost route to make biofuels. Several oil companies are evaluating this possibility, considering at least three options for using petroleum refineries to convert biomass into fuels and chemicals: (1) utilization of biomass-derived syngas or hydrogen, (2) fluid catalytic cracking (FCC), and (3) hydrotreating–hydrocracking.[108]

10.5 Conclusions

Biomass catalytic pyrolysis is an emerging technology that results in the effective utilization of biomass in the production of clean and environmentally benign bio-oil for combustion purposes and for use as a feed in existing oil refineries (biocrude). In order to make biomass an economically viable alternative for the production of transportation fuels and chemicals, a low cost, large-scale operation such as biomass liquefaction *via* pyrolysis will be preferable to the gasification/gas-to-liquids (GTL) route. Gasification of biomass followed by Fischer–Tropsch synthesis is very capital intensive and the economics will require large-scale production ($\gg$1000 MW). A catalytic flash pyrolysis system is much less costly, and much smaller units can be used. In this way the pyrolysis oil can be produced in the region where the biomass is generated. This will also stimulate the local use of biomass for green fuels and chemicals. The process has been proven at the pilot scale, while currently demonstration scales are being investigated. A considerable amount of research is in progress, aimed at the development and evaluation of new catalysts for catalytic biomass pyrolysis, producing a bio-oil with less oxygen compared to conventional bio-oil. Research thus far has shown that the presence of catalysts in biomass pyrolysis can indeed significantly affect the bio-oil yield and quality. Zeolite catalysts seem to be very active for this process since they crack the bio-oil, giving lower yield but with a better quality. Similar results have been obtained for mesoporous materials although they also seem to produce some useful chemicals. Research on mesoporous materials should be continued at the pilot scale, since up to now they have mostly been tested at the bench scale. Basic catalytic materials also appear to be interesting. In general, research to this point has shown that there are catalysts capable of producing a bio-oil with very good quality, however, the bio-oil yields from these catalysts are usually very poor. We need to continue research on biomass catalytic pyrolysis in order to find optimum catalysts that do not sacrifice the bio-oil yield for its quality. However, even a small improvement in the quality of the bio-oil would be beneficial for any downstream bio-oil upgrading process.

References

1. S. Yaman, *Energy Convers. Manage.*, 2004, **45**, 651.
2. A. V. Bridgwater and G. V. C. Peacocke, *Renew. Sust. Energy Rev.*, 2000, **4**(1), 1.
3. A. Demirbas and M. Balat, *Energy Convers. Manage.*, 2006, **47**, 2371.

4. D. Mohan, C. U. Pittman and P. H. Steele, *Energy Fuel*, 2006, **20**, 848.
5. O. Onay and O. M. Kockar, *Biomass Bioenergy*, 2004, **26**(3), 289.
6. R. C. Brown, *Biorenewable Resources Engineering New Products From Agriculture,* Iowa State Press, Iowa, USA, 2003.
7. A. Demirbas, *Energy Sources*, 2009, **31**, 546.
8. R. K. Sharma, J. B. Wooten, V. L. Baliga, X. Lin, W. G. Chan and M. R. Hajaligol, *Fuel*, 2004, **83**, 1469.
9. M. N. Islam and M. R. A. Beg, *Bioresour. Technol.*, 2004, **92**, 181.
10. M. N. Islam, R. Zailani and F. N. Ani, *Renew. Energy*, 1999, **17**, 73.
11. A. Demirbas, *Energy Convers. Manage.*, 2001, **42**, 1357.
12. P. A. Horne and P. T. Williams, *Renew. Energy*, 1994, **5**, 810.
13. H. B. Goyal, D. Seal and R. C. Saxena, *Renew. Sustain. Energy Rev.*, 2008, **12**, 504.
14. B. V. Babu, *Biofuels, Bioprod. Biorefin.-Biofrp.*, 2008, **2**, 393.
15. A. Demirbas, *Fuel Process. Technol.*, 2007, **88**, 591.
16. G. W. Huber, S. Iborra and A. Corma, *Chem. Rev.*, 2006, **106**, 4044.
17. A. V. Bridgwater, *Chem. Eng. J.*, 2003, **91**(2–3), 87.
18. A. Oasmaa and S. Czernik, *Energy Fuels*, 1999, **13**, 914.
19. A. V. Bridgwater, *Catal. Today*, 1996, **29**, 285.
20. A. V. Bridgwater, *J. Anal. Appl. Pyrolysis*, 1999, **51**, 3.
21. R. Maggi and B. Delmon, *Biomass Bioenergy*, 1994, **7**, 245.
22. E. Furimsky, *Appl. Catal. A*, 2000, **199**, 147.
23. G.W. Huber, National Science Foundation-Division of Chemical, Bioengineering, Environmental, and Transport Systems, Washington, DC, 2008.
24. J. Piskorz, P. Majerski, D. Radlein, A. Vladars-Usas and D. S. Scott, *J. Anal. Appl. Pyrol.*, 2000, **56**, 145.
25. A. Corma, G. W. Huber, L. Sauvanaud and P. O'Connor, *J. Catal.*, 2007, **247**, 307.
26. A. V. Bridgwater and G. Grassi, *Biomass Pyrolysis Liquids Upgrading and Utilisation,* Elsevier Applied Science, London and New York, 1991.
27. M. Stocker, *Angew. Chem. Int. ed.*, 2008, **47**, 9200.
28. K. S. Triantafyllidis, E. F. Iliopoulou, E. V. Antonakou, A. A. Lappas, H. Wang and T. J. Pinnavaia, *Micropor. Mesopor. Mater.*, 2007, **99**, 132.
29. E. F. Iliopoulou, E. V. Antonakou, S. A. Karakoulia, A. A. Lappas, I. A. Vasalos and K. S. Triantafyllidis, *Chem. Eng. J.*, 2007, **134**, 51.
30. A. A. Lappas, V. S. Dimitropoulos, E. V. Antonakou, S. S. Voutetakis and I. A. Vasalos, *Ind. Eng. Chem. Res.*, 2008, **47**, 742.
31. Ensyn website 2009. http://www.ensyn.com/about.htm.
32. B. Freel and R Graham, Ensyn Technologies Inc., "Method and Apparatus for a Circulating Bed Transport Fast Pyrolysis Reactor System", US Patent 5,792,340, 1995.
33. BTG website 2009 Biomass Technology Group B.V.; Flash Pyrolysis. http://www.btgworld.com
34. A. Aho, N. Kumar, K. Eränen, T. Salmi, M. Hupa and D. Y. Murzin, *Process Safe. Environ. Protec.*, 2007, **85**, 473.

35. P. T. Williams and P. A. Horne, *J. Anal. Appl. Pyrolys.*, 1995, **31**, 39.
36. R. W. Thring, S. P. R. Katikaneni and N. N. Bakhshi, *Fuel Process. Technol.*, 2000, **62**, 17.
37. A. Atutxa, R. Aguado, A. G. Gayubo, M. Olazar and J. Bilbao, *Energy Fuel*, 2005, **19**, 765.
38. P. A. Horne and P. T. Williams, *Fuel*, 1996, **75**, 1043.
39. A. A. Lappas, M. C. Samolada, D. K. Iatridis, S. S. Voutetakis and I. A. Vasalos, *Fuel*, 2002, **81**, 2087.
40. S. Vitolo, B. Bresci, M. Seggiani and M. G. Gallo, *Fuel*, 2001, **80**, 17.
41. P. T. Williams and P. A. Horne, *Biomass Bioenergy*, 1994, **7**, 223.
42. P. A. Horne, N. Nugranad and P. T. Williams, *J. Anal. Appl. Pyrolysis*, 1995, **34**, 87.
43. P. T. Williams and P. A. Horne, *J. Anal. Appl. Pyrolysis*, 1995, **31**, 15.
44. P. A. Horne and P. T. Williams, *J. Analyt. Appl. Pyrolysis*, 1995, **34**, 65.
45. B. Valle, A. G. Gayubo, A. Atutxa, A. Alonso and J. Bilbao, *Int. J. Chem. React. Eng.*, 2007, **5**, A86.
46. A. Aho, N. Kumar, K. Eranen, T. Salmi, M. Hupa and D. Yu. Murzin, *Fuel*, 2008, **87**, 2493.
47. D. Chiaramonti, M. Bonini, E. Fratini, G. Tondi, K. Gartner, A. V. Bridgwater, H. P. Grimm, I. Soldaini, A. Webster and P. Baglioni, *Biomass Bioenergy*, 2003, **25**(1), 101.
48. S. Czernik and A. V. Bridgwater, *Energy and Fuels*, 2004, **18**(2), 590.
49. D. Meier and O. Faix, *Bioresour. Technol1999,* **68**, *71.*
50. K. Sipilä, K. Kuoppala, L. Fagernäs and A. Oasmaa, *Biomass Bioenergy*, 1998, **14**(2), 103.
51. A. Pattiya, J. O. Titiloye and A. V. Bridgwater, *J. Anal. Appl. Pyrolysis*, 2008, **81**, 72.
52. J. Adam, M. Blazsó, E. Mészáros, M. Stöcker, M. H. Nilsen, A. Bouzga, J. E. Hustad, M. Grønli and G. Øye, *Fuel*, 2005, **84**, 1494.
53. E. Antonakou, A. Lappas, M. H. Nilsen, A. Bouzga and M. Stöcker, *Fuel*, 2006, **85**, 2202.
54. J. Adam, E. Antonakou, A. Lappas, M. Stöcker, M. H. Nilsen, A. Bouzga, J. E. Hustad and G. Øye, *Micropor. Mesopor. Mater.*, 2006, **96**, 93.
55. M. H. Nilsen, E. Antonakou, A. Bouzga, A. Lappas, K. Mathisen and M. Stöcker, *Micropor. Mesopor. Mater.*, 2007, **105**, 189.
56. J. S. Beck, J. C. Vartuli, W. J. Roth, M. E. Leonowicz, C. T. Kresge, K. D. Schmitt, C. T. W. Chu, D. H. Olson, E. W. Sheppard, S. B. McCullen, J. B. Higgins and J. L. Schlenker, *J. Am. Chem. Soc.*, 1992, **114**, 10834.
57. U. Ciesla and F. Schüth, *Micropor. Mesopor. Mater.*, 1999, **27**, 131.
58. A. Lewandowska, S. Monteverdi, M. Bettahar and M. Ziolek, *J. Mol. Catal. A: Chem.*, 2002, **188**, 85.
59. F. A. Twaiq, R. A. Mohamed and S. Bhatia, *Micropor. Mesopor. Mater.*, 2003, **64**, 95.
60. A. Corma, *Chem. Rev.*, 1997, **97**, 2373.
61. M. L. Ocelli, S. Biz and A. Auroux, *Appl. Catal. A*, 1999, **183**, 231.

62. T. Takeguchi, J. B. Kim, M. Kang, T. Inui, W. T. Cheuh and G. L. Haller, *J. Catal.*, 1998, **175**, 1.
63. A. Tuel, *Micropor. Mesopor. Mat.*, 1999, **27**, 151.
64. A. Corma, M. S. Grande, V. Gonzalez-Alfaro and A. V. Orchilles, *J. Catal.*, 1996, **159**, 375.
65. Y. Liu, W. Zhang and T. J. Pinnavaia, *Angew. Chem., Int. ed.*, 2001, **40**, 1255.
66. Y. Liu and T. J. Pinnavaia, *J. Mater. Chem.*, 2004, **14**, 1099.
67. K. S. Triantafyllidis, A. A. Lappas, I. A. Vasalos, Y. Liu, H. Wang and T. J. Pinnavaia, *Catal. Today*, 2006, **112**, 33.
68. H. J. Park, J.-K. Jeon, K. Y. Jung, Y. S. Ko, J. M. Sohn and Y.-K. Park, *Korean Chem. Eng. Res.*, 2007, **45**, 340.
69. M. I. Nokkosmaki, E. T. Kuoppala, E. A. Leppamaki and A. O. I. Krause, *J. Anal. Appl. Pyrol.*, 2000, **55**, 119.
70. D. Fabbri, C. Torri and V. Baravelli, *J. Anal. Appl. Pyrol.*, 2007, **80**, 24.
71. M.-Q. Chen, J. Wang, M.-X. Zhang, M.-G. Chen, X.-F. Zhu, F.-F. Min and Z.-C. Tan, *J. Anal. Appl. Pyrol.*, 2008, **82**, 145.
72. S. Ernst, M. Hartmann, S. Sauerbeck and T. Bongers, *Appl. Catal. A*, 2000, **200**, 117.
73. T. R. Carlson, G. A. Tompsett, W. C. Conner and G. W. Huber, *Top. Catal.*, 2009, **52**, 241.
74. M. Olazar, R. Aguado, J. Bilbao and A. Barona, *AIChE J.*, 2000, **46**, 1025.
75. P. B. Weisz, W. O. Haag and P. G. Rodewald, *Science*, 1979, **206**, 57.
76. L.H. Dao Institut National de la Recherche Scientifique,"Converting biomass into hydrocarbons", Canadian Patent # 83-443162, 1986, p 10.
77. L. H. Dao, M. Haniff, A. Houle and D. Lamothe, *Preprints Am. Chem. Soc., Div. Fuel Chem.*, 1987, **32**, 308.
78. M. I. Hanniff and L. H. Dao, *Energy Biomass Wastes*, 1987, **10**, 831.
79. L. H. Dao, M. Haniff, A. Houle and D. Lamothe, *ACS Symp. Ser.*, 1988, **376**, 328.
80. M. I. Hanniff and L. H. Dao, *Appl. Catal. A*, 1988, **39**, 33.
81. H. J. Park, J.-I. Dong, J.-K. Jeon, K.-S. Yoo, J.-H. Yim, J. M. Sohn and Y.-K. Park, *J. Ind. Eng. Chem.*, 2007, **13**, 182.
82. V. V. Alferov, O. S. Misnikov, O. V. Kislitsa, E. M. Sulman, D. Y. Murzin and N. Kumar, *Kataliz v Promyshlennosti*, 2006, **6**, 42.
83. E. M. Sulman, V. V. Alferov, Y. Y. Kosivtsov, A. I. Sidorov, O. S. Misnikov, A. E. Afanasiev, N. Kumar, D. Kubicka, J. Agullo, T. Salmi and D. Y. Murzin, *Chem. Eng. J.*, 2007, **134**, 162.
84. D. Stamires, P. O'Connor, S. Daamen, BIOeCON International Holding N.V., "Process for the Conversion of Biomass to Liquid Fuels and Specialty Chemicals", European Patent 2021434, 2009.
85. D. C. Elliott, *Energy & Fuels*, 2007, **21**, 1792.
86. Y. H. E. Sheu, R. C. Anthony and E. J. Soltes, *Fuel. Proc. Technol.*, 1988, **19**, 31.

87. E. Churin, P. Grange and B. Delmon, in *Catalytic Upgrading of Pyrolysis Oils*, in *Biomass for Energy and Industry*, C. Crassi, C. Cosse, C. Santos, (eds.), Elsevier Appl. Sci., London, UK, 1990, p. 2616.
88. I. Demiral and S. Sensoz, *Bioresource Technol.*, 2008, **99**, 8002.
89. G. W. Huber and A. Corma, *Angew. Chem. Ind. ed.*, 2007, **46**, 7184.
90. J. D. Adjaye and N. N. Bakhshi, *Fuel Process. Technol.*, 1995, **45**, 185.
91. J. D. Adjaye and N. N. Bakhshi, *Fuel Process. Technol.*, 1995, **45**, 161.
92. J. D. Adjaye, S. P. R. Katikaneni and N. N. Bakhshi, *Fuel Process. Technol.*, 1996, **48**, 115.
93. R. K. Sharma and N. N. Bakhshi, *Energy Fuels*, 1993, **7**, 306.
94. S. T. Srinivas, A. K. Dalai and N. N. Bakhshi, *Can. J. Chem. Eng.*, 2000, **78**, 343.
95. A. G. Gayubo, A. T. Aguayo, A. Atutxa, R. Aguado and J. Bilbao, *Ind. Eng. Chem. Res.*, 2004, **43**, 2610.
96. A. G. Gayubo, A. T. Aguayo, A. Atutxa, R. Aguado, M. Olazar and J. Bilbao, *Ind. Eng. Chem. Res.*, 2004, **43**, 2619.
97. A. G. Gayubo, A. T. Aguayo, A. Atutxa, B. Valle and J. Bilbao, *J. Chem. Technol. Biotechnol.*, 2005, **80**, 1244.
98. R. K. Sharma and N. N. Bakhshi, *Bioresource Technol.*, 1993, **45**, 195.
99. R. K. Sharma and N. N. Bakhshi, *Biomass and BioEnergy*, 1993, **5**, 445.
100. R. K. Sharma and N. N. Bakhshi, *Fuel Proc. Technol.*, 1993, **35**, 201.
101. R. K. Sharma and N. N. Bakhshi, *Canadian J. Chem. Eng.*, 1991, **69**, 1071.
102. N. G. Wilson and P. T. Williams, *Int. J. Energy Res.*, 2003, **27**, 131.
103. P. Chantal, S. Kahaguine, J. L. Grandmaison and A. Mahay, *Appl. Catal.*, 1984, **10**, 317.
104. T. A. Milne, R. J. Evans and J. Filley, in *Research in Thermochemical Biomass Conversion,* A.V. Bridgwater, J.L. Kuester, (ed.), Elsevier Appl. Sci., London, 1988, p. 910.
105. M. C. Samolada, W. Balsauf and I. A. Vasalos, *Fuel*, 1998, **77**, 1667.
106. I. Grac, J.-D. Comparot, S. Laforge, P. Magnoux, J. M. Lopes, M. F. Ribeiro and F. Ramona Ribeiro, *Appl. Catal. A*, 2009, **353**, 123.
107. R. Marinangeli, T. Marker, J. Petri, T. Kalnes, M. McCall, D. Mackowiak, B. Jerosky, B. Reagan, L. Nemeth, M. Krawczyk, S. Czernik, D. Elliott, D. Shonnard, in "Opportunities for Biorenewables in Oil Refineries", Report No. DE-FG36- 05GO15085, UOP, 2006.
108. Y. C. Lin and G. W. Huber, *Energy Environ. Sci.*, 2009, **2**, 68.

CHAPTER 11

Hydrotreating for Bio-Oil Upgrading

MAIJA L. HONKELA, TUULA-RIITTA VILJAVA, ANDREA GUTIERREZ AND A. OUTI I. KRAUSE

Department of Biotechnology and Chemical Technology, Aalto University School of Science and Technology, P.O. Box 16100, FI-00076, Aalto, Finland

11.1 Introduction

Liquid bio-oils are produced from various biomass sources through processes such as pyrolysis and hydrothermal treatment. In this chapter we concentrate on wood-based pyrolysis oils. These oils are composed of several phases and contain water, oxygenated hydrocarbons, small carbonaceous particles and some minerals.[1] The organic phase, consisting of oxygenated hydrocarbons, is a mixture of aliphatic and aromatic acids, alcohols, esters, ethers, aldehydes, ketones and sugars.[1–3]

In contrast to fossil-based oils, bio-oils contain hardly any sulfur or nitrogen as shown in Table 11.1[4–6] in which the sulfur, nitrogen and oxygen contents of various synthetic and biomass-based liquids are presented. The table shows that while conventional oils contain less than 0.1 wt% oxygen, the oxygen content of bio-oils can approach 50 wt% depending on the biomass raw material, its pretreatment and the bio-oil production process.[4,7] The oxygen content of hydrothermal oil produced in the presence of water at supercritical conditions (553–653 K and 7–30 MPa)[6] is ower than that of pyrolysis oils. This is because during hydrothermal liquefaction the oxygen content is simultaneously

RSC Energy and Environment Series No. 1
Thermochemical Conversion of Biomass to Liquid Fuels and Chemicals
Edited by Mark Crocker

Published by the Royal Society of Chemistry, www.rsc.org

Table 11.1 Composition of different feeds.[4–6]

	Reference	*Sulfur content*	*Nitrogen content*	*Oxygen content*
Conventional crude	4	1.8	0.1	0.1
Coal-derived naphtha	4	0.1	0.5	4.7
Oil-shale crude	4	0.5	1.4	1.2
Tar-sand-based liquid	5	3.6	0.4	0.4
Pyrolysis oil	6	<0.1	<0.1	47
Pyrolysis oil	6	n.d.	n.d.	36
Hydrothermal oil	6	n.d.	n.d.	12

n.d. = not determined

decreased through the production of H_2O and CO_2. An example of the oxygen content of a hydrothermal oil is included in Table 11.1.

The high oxygen content of bio-oils causes several undesirable properties: high viscosity, nonvolatility, corrosiveness, immiscibility with fossil fuels, thermal instability and tendency to polymerize under exposure to air.[8–11] Thus, oxygen needs to be removed at least partly before bio-oil can be used as a fuel or processed further. The process of oxygen removal is also referred to as bio-oil upgrading.

For the purpose of removing oxygen, the hydrodeoxygenation (HDO) process has received the most attention in recent literature through its link with the refining of conventional oils. This process is similar to hydrodesulfurization (HDS) and hydrodenitrogenation (HDN). In the process, oxygen-containing molecules react to hydrocarbons in the presence of H_2 and a suitable catalyst and oxygen is removed mainly as water.

The characteristics of the HDO process and the results of HDO studies carried out with model components of wood-based pyrolysis oil will be presented in the following sections.

11.2 Early Hydrodeoxygenation Studies

HDO was expected to proceed simultaneously with HDS and HDN during catalytic hydrotreatment of mineral oils, but did not receive much attention because of the low oxygen content of the crudes and because oxygen was removed in a harmless form as water.[4,12] In the beginning of the 1980s research interest towards HDO increased because new, so-called synthetic liquids, *e.g.*, coal, oil shale and tar-sand-based liquids, contained more oxygen than traditional petroleum (Table 11.1).[5,12] In addition, the detrimental properties caused by the high oxygen content of biomass-based liquids provided an impetus to study HDO in more detail.

The early studies on HDO in our laboratory were concentrated on coal as the raw material.[13] Later on, even before the current emphasis on environmental protection, the focus was shifted to studying the HDO of the lignin fraction of wood[11,14,15] and recently to the HDO of vegetable oils as well.[16,17]

11.3 Hydrodeoxygenation

The HDO of wood-based pyrolysis oils is traditionally carried out at high pressures (7–20 MPa) in two steps: stabilization and hydrodeoxygenation.[4,18] The first step is performed at lower reaction temperature (below 573 K) with the aim of avoiding polymerization of oxygen-containing compounds and coking of the catalyst. During the first step only the most reactive oxygen-containing components are deoxygenated. In the second step oxygen is removed under higher reaction temperatures (623–673 K) from the highly resistant components. The first step, *i.e.* mild hydrotreating, consumes less hydrogen[18] and can be applied to produce only partially deoxygenated fuels suitable for the generation of electricity and heat or for co-feeding to oil refinery units.[18–20]

Both steps in the two-stage HDO process are carried out in the presence of a catalyst. Mainly sulfided CoMo and NiMo catalysts on γ-Al_2O_3 support have been used in the HDO of bio-oils, but alternative catalyst materials have also been developed and will be discussed in Sections 11.3.2 and 11.3.3. Many of the studies in the following sections have been carried out using model components to represent different types of functionality in the wood-based bio-oils, because of the complexity of real oils.

11.3.1 Sulfided CoMo and NiMo Catalysts

Sulfided CoMo and NiMo catalysts on γ-Al_2O_3 have been studied extensively in HDO due to their well-known and common industrial application in the HDS of fossil-based oils. At the laboratory scale these catalysts have been tested in their oxide form,[21,22] although it has been observed that only the sulfide form has any practical activity in oxygen-removal reactions. As an example of the use of sulfided CoMo and NiMo catalysts in HDO of real bio-oils, the composition of the bio-oil produced by the pyrolysis of olive oil wastes and that of the upgraded oil are shown in Table 11.2.[12] Both sulfided CoMo, as well as NiMo catalysts, were tested in the upgrading at 573–673 K and 12 MPa in a batch reactor. The table shows that with the NiMo catalyst a lower oxygen content was reached than with CoMo catalyst. On the other hand, less

Table 11.2 Composition of bio-oil produced by the pyrolysis of olive oil waste and the corresponding oil upgraded by HDO using either $CoMo/Al_2O_3$ or $NiMo/Al_2O_3$ as catalyst (573–673 K and 12 MPa in batch reactor). Reprinted with permission[12]. Copyright 2007 American Chemical Society.

	Bio-oil feed	$CoMo/Al_2O_3$	$NiMo/Al_2O_3$
Hydrogen, wt%	7.6	8.0	9.5
Oxygen, wt%	15.3	5.5	4.7
H/C atomic ratio	1.23	1.13	1.35
Oxygen removal,%		64	70

hydrogen was consumed and the saturation degree of the aromatic compounds in the oil was lower with CoMo.[23]

The sulfide structure of the catalyst is, however, not stable in the HDO of sulfur-free feeds. A decrease in the sulfur content of the catalyst[21,22] and formation of sulfur-containing HDO intermediates has been reported in model component studies.[14,15,24] In addition to the losses in the sulfur level, the sulfide structure of the catalyst may at least partially change into a less active form during HDO.[22,25] Therefore, sulfur is typically added to the system to maintain the activity of the catalyst.[4,26]

In oil refineries the CoMo and NiMo catalysts are typically activated by sulfidation pretreatment with carbon disulfide, mercaptans or dimethylsulfide before bringing the system to the reaction temperature.[27] In HDO model component studies higher temperatures, *e.g.*, 673 K,[28] and shorter presulfidation periods have often been used.

Preservation of the sulfidation degree of the catalysts by simultaneous HDO and HDS has been tested with model components,[29] however, the competition leads to a clear decrease in the HDO reaction rates. Cycling between HDO and HDS conditions has given more promising results.[21] Recently Kubička *et al.*[30] studied the effect of H_2S pulses on the activity of a NiMo catalyst during the HDO of vegetable oils, but did not obtain the same activity level for the catalyst as when feeding H_2S continuously. Apparently, the catalyst surface is permanently modified when H_2S is not present.[25,31]

Water is already present in the bio-oil feed and more is formed during the HDO. The water in the system, together with the oxygen-containing compounds, cause problems during the HDO. For example, they can affect the activity of the sulfided hydrotreatment catalysts by competitive adsorption on the active sites.[32] The adsorption tendency of water on the sulfide phase is, however, low and thus water causes only weak inhibition of the HDO of phenolic, ketonic and carboxylic groups.[32] Still, water can bind strongly on the support (γ-Al_2O_3) and interfere with the reactions that take place on the support. Recently, Popov *et al.*[33] observed that phenolic components adsorb on the support and hinder the accessibility of the sulfide-phase sites.

Water and oxygenates can also alter the surface structure and the chemical composition of γ-Al_2O_3-supported CoMo and NiMo catalysts. Water vapor has been reported to cause the recrystallization of the γ-Al_2O_3 support into hydrated boehmite, as well as the partial oxidation of the nickel sulfide phase into oxidized nickel species, resulting in decreased activity.[32]

11.3.2 Noble-Metal Catalysts

Addition of sulfur to the sulfur-free HDO feed to maintain the activity of the sulfide catalysts might not be industrially attractive because of contamination of the product by sulfur species. Thus new nonsulfided metallic catalysts have recently been developed. It is expected that they will work at lower temperatures and thus deactivation of the catalyst by coke formation can at least partially be avoided.[34] Furthermore, if the catalysts are produced on supports other than alumina, the problem of support instability (caused by water) can also be avoided.[34]

The products on noble-metal catalysts differ from those formed on sulfide catalysts. For example, activated carbon-supported Ru catalysts have been tested in the HDO of a mixture of four model compounds representing the most significant components of wood-based bio-oil.[34] Experiments performed in the temperature range 423–473 K at 13.6 MPa showed that different products are obtained from those with the sulfided catalysts. The major products with the Ru catalyst were saturated cyclic alcohols at lower temperatures and cyclic hydrocarbons at higher temperatures.

Gutierrez *et al.*[15] studied ZrO_2-supported mono- and bimetallic Rh, Pd and Pt catalysts (0.5 wt% total metal loading) in the HDO of guaiacol (2-methoxyphenol) and compared their performance to the performance of a sulfided alumina-supported CoMo catalyst. At 373 K the main reaction was hydrogenation over all of the catalysts, while at 573 K deoxygenation took place. In the experiments at 373 K carbon deposition was found to be significantly less with the noble-metal catalysts than with the sulfided catalyst. At 573 K with Rh, RhPd and sulfided CoMo, low O/C ratios similar to gasoline and diesel were obtained.

The noble-metal catalysts are, however, more sensitive towards poisoning than sulfide catalysts. Titania- and carbon-supported Ru catalysts were tested in the HDO of wood-based pyrolysis oil.[34] The composition of the white wood bio-oil and the hydrogenated oil are presented in Table 11.3. Experiments with the bio-oil showed that even very low levels of contaminants such as iron and sulfur deactivate the catalyst significantly.

11.3.3 Other Catalysts

The acidic γ-Al_2O_3 support is active in coke formation and polymerization and thus several neutral supports, such as carbon and silica, have been studied as supports for sulfided CoMo and NiMo catalysts.[35] Carbon was found to be a promising support when experiments were carried out with mixtures of model components;[35] in particular, the guaiacol methoxy group was selectively eliminated. In addition, modification of the support acidity by potassium

Table 11.3 Properties of white wood bio-oil and hydrogenated oil (454–511 K, 13.3–14.2 MPa, Ru catalyst). Reprinted with permission.[34]

Composition wt% (dry base)	*White wood bio-oil*	*Hydrogenated white wood bio-oil*
Carbon	48.3	65.6–74.0
Hydrogen	7.4	8.5–10.1
Oxygen	44.4	16.7–25.8
Hydrogen/carbon atomic ratio	1.82	1.49–1.68
Water content	30.0	9.7–15.7

addition to decrease coking was found promising. The impregnation order of the active metals has also been found to have some effect on the HDO activity of the activated-carbon-supported CoMo catalysts and selectivities were better when molybdenum was impregnated first.[36]

Catalysts containing a combination of Rh and Co, as well as various Ni and Ni-Cu catalysts, have been tested in the HDO of anisole and biodiesel (523-673 K, 0.5-2.0 MPa, fixed-bed flow reactor).[37] Catalyst supports varied from SiO_2, Al_2O_3, ZrO_2, CeO_2 to CeO_2-ZrO_2. Ni catalysts were found to be promising in the HDO. Ni-Cu catalysts were found to be even more active in the HDO of anisole than Ni as such. CeO_2 and ZrO_2 supports were found to be most suitable supports due to additional activation of the reacting components.

Centeno *et al.*[38] added noble metals (Pt, Pd, Rh and Ru) to Mo and CoMo catalysts supported on active carbon and tested them in the HDO of model components with various oxygen groups. Bimetallic noble-metal catalysts were observed to be very active in the hydrogenation of C=O bonds, but not in decarboxylation.

11.4 Studies with Wood-Based Pyrolysis Oil Model Components

Due to their high chemical stability, phenol and its derivatives are the most resistant components with respect to oxygen removal.[2,4] Guaiacol and substituted guaiacols, that are present in wood-based bio-oils in relatively large amounts (2–5 wt%),[39] are chemically very reactive on the other hand, and cause instability and polymerization of the oil.[2] Studying the HDO of both phenols and guaiacols can reveal some essential aspects of the HDO of bio-oils.

Table 11.4 shows the main components of wood-based bio-oil.[39] Phenol, anisole, guaiacol and substituted guaiacols are often used as model components to represent the lignin fraction of the wood-based bio-oils. Recently, the HDO

Table 11.4 Main components of pyrolysis liquids. Reprinted with permission.[39]

Component	*Mass %*
Water	20–30
Lignin fragments: Insoluble pyrolytic lignin	15–30
Aldehydes: Formaldehyde, acetaldehyde, hydroxyacetaldehyde, glyoxal, methylglyoxal	10–20
Carboxylic acids: Formic, acetic, propionic, butyric, pentanoic, hexanoic, glycolic, (hydroxy acetic)	10–15
Carbohydrates: Cellobiosan, α-D-levoglucosan, oligosaccharides, anhydroglucofuranose	5–10
Phenols: Phenol, cresols, guaiacols (GUA), syringols	2–5
Furfurals	1–4
Alcohols: Methanol, ethanol	2–5
Ketones: Acetol (1-hydroxy-2-propanone), cyclopentanone	1–5

of carbohydrates and acids has drawn increasing attention, because they are reactive and polymerize at HDO conditions to coke precursors.[40]

11.4.1 Phenol

Phenol is the simplest model component that can be used to study the reactivity of the bond between aromatic carbon and oxygen. This bond is clearly stronger than the aliphatic carbon–oxygen bond.[41,42] Phenol and its derivatives are typically formed as intermediates in the reactions of more complicated aromatic oxygen-containing molecules. Thus, it is important to know their reaction network.

The low reactivity of phenol in HDO can clearly be seen in studies in which reactions of phenol were compared to the reactions of aliphatic esters (methyl heptanoate and ethyl heptanoate).[14,24] The difference in the reactivities between phenol and the esters demonstrates the difference in oxygen removal from wood-based pyrolysis oils and from vegetable oils. The conversions as a function of time for the model components as such and as a mixture are presented in Figure 11.1.

In earlier studies concerning the HDO of phenol, oxygen removal was suggested to proceed *via* two parallel routes: the aromatic oxygen-free product, benzene, is formed in a hydrogenolysis reaction and the aliphatic products, cyclohexane and cyclohexene, in a combined hydrogenation-hydrogenolysis

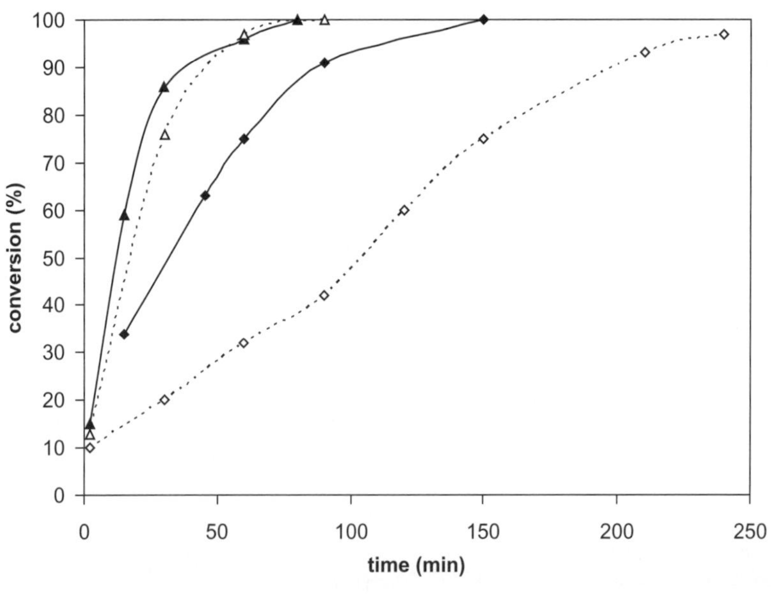

Figure 11.1 Conversion of pure phenol (◆), pure methyl heptanoate (▲), phenol in the binary mixture (◇) and methyl heptanoate in the binary mixture (Δ) in HDO experiments over sulfided NiMo at 523 K and 7.5 MPa.[24]

route.[11,43] Gevert *et al.*[44] proposed similar routes for the reactions of methyl-substituted phenols on a sulfided CoMo catalyst. They suggested that partially hydrogenated phenols, *i.e.* cyclohexanol and cyclohexenols, are formed as intermediates in the hydrogenation-hydrogenolysis path although these products were not detected due to their high reactivity under the HDO conditions. Correspondingly, Laurent and Delmon suggested that 4-methylcyclohexanol is an intermediate in the HDO of 4-methylphenol, although it was not detected in the liquid product.[45]

Senol *et al.*[14] studied the HDO of phenol on sulfided catalysts both in the gas phase in a plug-flow reactor and in the liquid phase in a batch system. The main oxygen-free products on sulfided NiMo were cyclohexane and cyclohexene, while on CoMo benzene was the main endproduct. Cyclohexanone and diphenyl ether were formed as oxygen-containing intermediates. Cyclohexanol was not detected in the products. Separate reaction tests with cyclohexanone and cyclohexanol confirmed that these intermediates are very reactive and that the only reaction route is that to alicyclics. In addition, even though the conversion profile for benzene was not typical for intermediates, it gave trace amounts of alicyclics in separate reaction tests. Later, in a study by Ryymin *et al.*,[24] the reaction scheme was revised on the basis of additional experiments carried out with phenol on sulfided NiMo in a batch system. With more detailed analysis, a conversion profile typical for intermediates was observed for benzene. A new hydrocarbon endproduct, cyclohexyl cyclohexane, was also detected in small amounts. The detailed overall reaction scheme is shown in Scheme 11.1.

As mentioned before in Section 11.3.1, a decrease in the sulfur content of presulfided catalysts has been reported in HDO studies with model components in the absence of sulfur additives.[21,22] As explained previously, the sulfur content of real bio-oils is negligible and therefore with sulfided catalysts a resulfiding agent is added to the feed to maintain the sulfide structure of the catalyst. In model component studies this component is often H_2S. However, the presence of a sulfur additive negatively affects the HDO reactions of phenol: on NiMo catalyst sulfur addition suppressed the pathways to cyclohexane and benzene similarly,[14] while over CoMo catalyst the pathway to benzene was suppressed more than the one to cyclohexane.[22] It was also observed that hydrogenation was suppressed over NiMo and practically unaffected over CoMo. The opposite effect of H_2S on the HDO of aliphatic esters has been reported:[46] addition of sulfur to the feed during HDO of methyl and ethyl heptanoate on sulfided NiMo in a flow reactor had a promoting effect on the deoxygenation reactions, indicating differences in the HDO reaction mechanisms between aromatic and aliphatic carbon–oxygen bonds.

Coordinatively unsaturated sites (CUS) associated with MoS_2 phase and sulfur-saturated sites are generally accepted to be the active sites for hydrotreating reactions.[4,47] Reductive reactions of phenol and its intermediates have been concluded to occur on CUS.[24,46] The inhibiting effect of H_2S is caused by the competitive adsorption of phenol and H_2S on these sites. Sulfur-saturated sites are needed for acid-catalyzed reactions.[24] Acid-catalyzed reactions play a

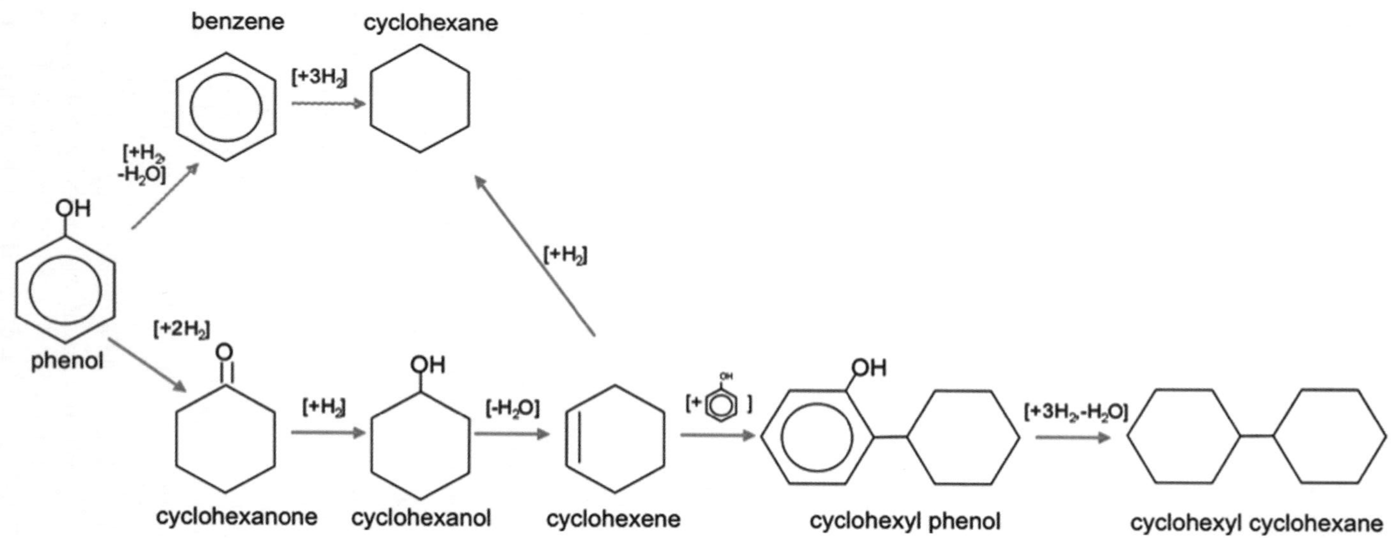

Scheme 11.1 HDO reaction pathways of phenol over sulfided NiMo catalyst (473–523 K, 7.5 MPa, batch reactor).[24]

major role in the reactions of aliphatic esters and contribute strongly to their high reactivity.[16,24]

Senol *et al.*[14] reported formation of trace amounts of a sulfur-containing compound, cyclohexanethiol, during HDO of phenol in the presence of H_2S. The study was carried out in a flow reactor on sulfided CoMo and NiMo catalysts. Cyclohexanethiol is highly reactive and reacts very quickly to cyclohexene and H_2S. It was concluded that the compound was formed in an acid-catalyzed nucleophilic substitution reaction in which the SH^- groups of the catalyst may act as nucleophiles. In the study cyclohexanethiol was detected only in the presence of H_2S.

11.4.2 Anisole

In anisole there are two kinds of oxygen–carbon bonds and the bond between the alicyclic carbon and oxygen is weaker than the bond between the aromatic carbon and oxygen.[41,42] There are more reaction products compared to phenol, which makes the reaction scheme more complicated. Viljava *et al.*[22] studied the reactions of anisole on a presulfided CoMo catalyst in a batch reactor and on both oxide (nonsulfided) and presulfided CoMo catalyst, as well as on the alumina support, in a flow reactor. The cleavage of the aliphatic carbon–oxygen bond giving phenol as the product proceeded effectively on the oxide catalyst and on the alumina support, while the sulfided catalyst was required for the oxygen-removal reactions. The direct splitting of the methoxy group of anisole was already ruled out in some earlier studies.[48–50] Several ring-methylated reaction products were formed, *e.g.*, *ortho*-cresol, *ortho*-methyl anisole and 2,6-xylenol. In addition to the main oxygen-free reaction products that were formed *via* phenol, toluene was formed from anisole. The simplified reaction network for anisole is shown in Scheme 11.2.

The HDO reactions of anisole were suppressed and the demethylation reaction to phenol enhanced in the presence of H_2S.[22] Added H_2S did not improve the stability of the sulfide catalyst. Though deactivation of the catalyst was concluded to be mainly due to coke formation at high temperatures, desulfurization of the sulfide structure could not be totally ruled out as the sulfur content of the catalyst decreased by about 16% during the HDO run. Desulfurization of the catalyst was not detected in HDS experiments.[21,22]

11.4.3 Guaiacols

Guaiacol has both a phenolic and a methoxy group. Bui *et al.*[51] studied the HDO reaction scheme with sulfided CoMo as catalyst and guaiacol as the model component. The methoxy group is removed through demethylation or demethoxylation. Formed phenol-like molecules react through direct hydrogenolysis of the C–O bond between the aromatic carbon and the

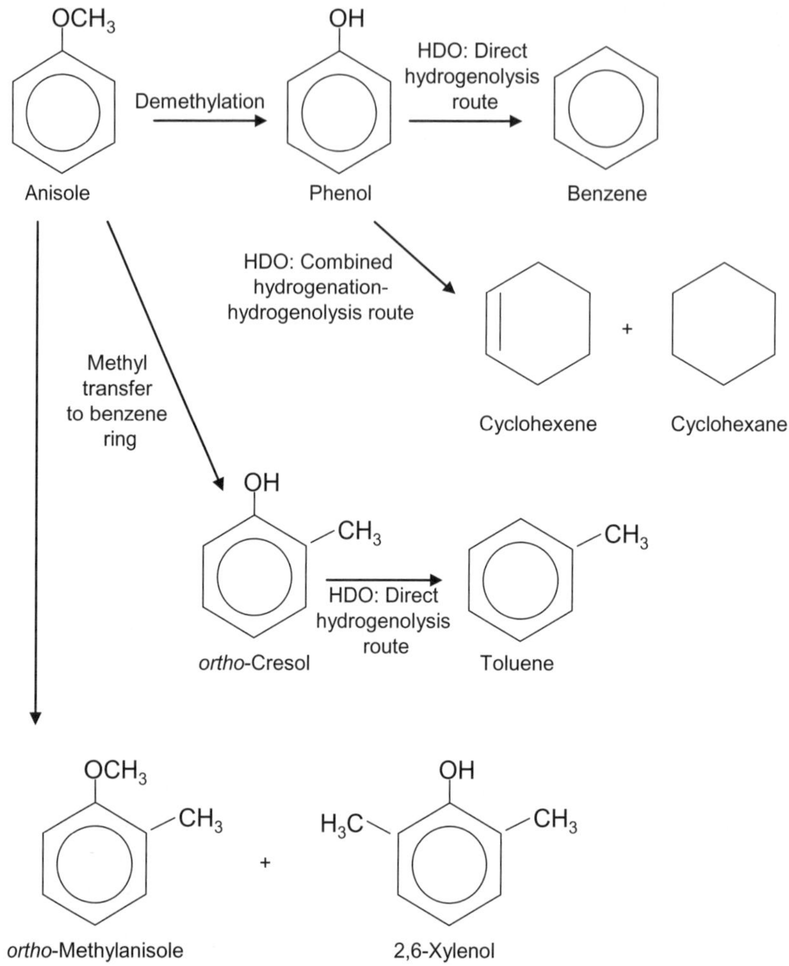

Scheme 11.2 HDO reaction pathways of anisole over sulfided CoMo catalyst (473–573 K, 1.5 MPa, flow reactor). Reprinted with permission from Elsevier.[22]

OH group or preliminary hydrogenation of the aromatic ring before elimination of the OH groups. Some heavier products resulting from condensation of two rings were also observed. A simplified reaction scheme without the heavier products and omitting some of the methylated compounds is presented in Scheme 11.3.[51]

Sulfur-containing intermediates have also been reported to form in the reactions of guaiacol on sulfide catalysts. In the HDO of guaiacol (573 K, 8 MPa, batch reactor) on sulfided CoMo, Gutierrez *et al.*[15] detected methanethiol and dimethyl sulfide in the gaseous product and cyclohexanethiol and methylthiocyclohexane in the liquid product.

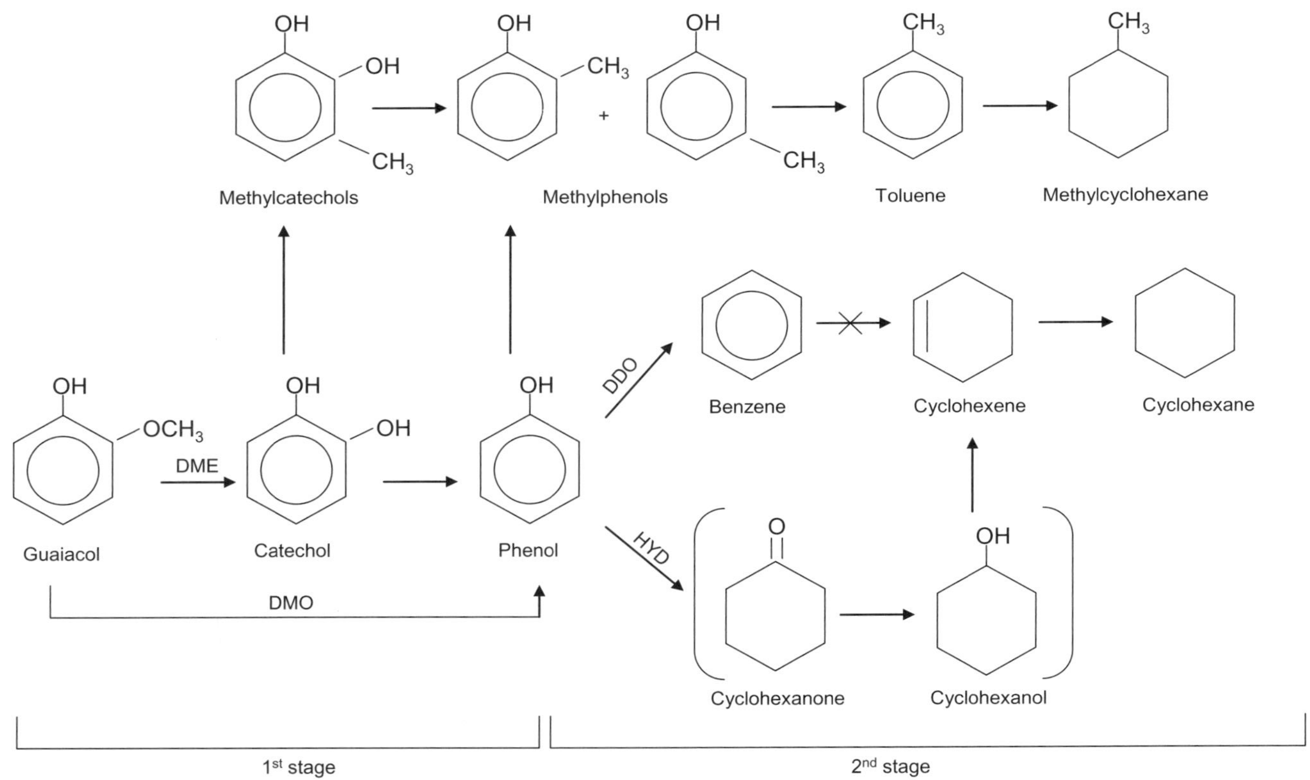

Scheme 11.3 HDO reaction pathways of guaiacol over sulfided CoMo catalyst at 573 K and 4 MPa. DME = demethylation, DMO = demethoxylation, DDO = direct hydrogenolysis and HYD = hydrogenation. Reprinted with permission from Elsevier.[51]

11.4.4 Ketones, Aldehydes, Acids and Carbohydrates

The instability of wood-based pyrolysis oils at high temperatures is attributed to the high concentration of unsaturated double bonds, such as the ones present in the alkenes, ketones and aldehydes in the oil. At high temperatures these components react through condensation reactions. Therefore, these functionalities need to be eliminated to avoid the formation of high molecular weight components. Thus, as explained in Section 11.3, the HDO is carried out in two steps. Oxygen removal from ketones and aldehydes, as well as reactions of unsaturated hydrocarbons (alkenes), aliphatic alcohols and esters, occur in the first step at low temperatures.[52] Then, higher temperatures are needed for the elimination of phenolic and furanic oxygen and the conversion of heavy molecules. The combination of the low- and high-temperature procedures enables complete oxygen removal and production of a high-grade hydrocarbon product.

Laurent and Delmon[26] observed that the ketonic group is easily deoxygenated on sulfided CoMo and NiMo at temperatures higher than 473 K, affording alkanes as the reaction products. This was thought to result partly from the ability of the γ-Al_2O_3 support to catalyze dehydration, resulting in fast conversion of the ketone group through alkenes to alkanes.

Aldehydes are also easily reduced to alcohols or even to the corresponding alkane in the presence of hydrogen. When the hydrogenation of linear aldehydes is performed with sulfided catalysts ($NiMo/Al_2O_3$) two types of reactions can be identified, namely, hydrogenation of the aldehyde to the corresponding alcohol and the formation of undesired heavy products. The alcohol formation is correlated to the concentration of CUS, while the formation of heavy products is mainly related to the surface concentration of OH^- and possibly SH^- groups.[53]

For ketones and aldehydes the transformation of the carbonyl group into methyl group proceeds according to three general mechanisms:[54]

(a) hydrogenation-dehydration: the carbonyl is transformed into an alcohol, followed by the dehydration of the alcohol to the corresponding alkene and the hydrogenation of the alkene to alkane (C–C=O → C–C–OH → C=C → C–C);
(b) hydrogenation-hydrogenolysis: first hydrogenation of the carbonyl to alcohol occurs, followed by hydrogenolysis of the C–O bond, producing the corresponding alkane (C–C=O → C–C–OH → C–C);
(c) direct hydrogenolysis of the C=O bond (C–C=O → C–C).

In addition to the main reaction paths presented above, cracking reactions take place under the experimental conditions. In the case of branched aldehydes cracking reactions are more important than in the case of linear aldehydes.[53]

Acetic acid can be used as a model compound for the acidic phase of wood-based bio-oils. Under HDO conditions the main hydrogenation product of acetic acid is ethanol and secondary products are condensation products such as ethyl acetate.[55] However, experiments performed with Ru/C demonstrated that this catalyst is not active in the hydrogenation of acetic acid to ethanol.[55] At temperatures where the conversion of the acid was significant, mainly gasification reactions were catalyzed. With Pd catalyst on the other hand, acetic acid can be hydrogenated to ethanol at 573 K.

Carbohydrates are known to follow several reaction pathways and especially undergo polymerization at high temperatures.[40] These reactions are affected by temperature and by the pH of the reaction media. With Ru/C catalyst at HDO conditions (523 K, 100 bar) mainly polyols and gas-phase components, such as methane, are produced.[40] Carbohydrates may also be selectively reacted to furfurals and further to C_7 to C_{15} alkanes under conditions similar to HDO on bifunctional catalysts (containing both acid and metal sites).[56]

11.5 Catalyst Comparison and Coking with Model Components and with Pyrolysis Oil

Gutierrez *et al.*[57] compared the results obtained with guaiacol as model component to those with wood-based pyrolysis oil. They studied mono- and bimetallic noble-metal catalysts and observed that with guaiacol (temperature 573 K and pressure 8 MPa), Rh and RhPd catalysts were the most active for oxygen-removal.[15] In the HDO of pyrolysis oil (temperature 623 K and pressure 20 MPa), however, no clear difference could be seen between the noble-metal catalysts as shown in Figure 11.2. In the figure the O/C and H/C molar ratios of the product mixture are presented. The H/C molar ratio close to 1 indicates that the product is highly aromatic. In this case study[57] it was concluded that the results with the model components are not sufficiently representative and hence real feed or different oil fractions also need to be tested.

Carbon deposition is one of the main challenges in the HDO of wood-based pyrolysis oils. Thus, it is of interest to identify the chemical species having the highest tendency to form carbon deposits, as well as the reaction pathways and the operation conditions leading to carbon formation. The temperature-programmed oxidation (TPO) technique is a useful method for the characterization of carbon deposits. In TPO, the carbon deposited on the catalyst is burnt with oxygen. Both oxygen content and heating rate can be varied.

The CO_2 thermograms for the carbon deposited on the Rh catalyst in guaiacol and pyrolysis oil experiments are shown in Figure 11.3.[57] The temperature ramp was 5 K/min from room temperature to 973 K. There was no correlation between the CO_2 maxima obtained in the burning of the carbon deposited in the HDO of guaiacol and the HDO of the pyrolysis oil, suggesting that the structure of the carbon is different. This could also indicate that in the case of the pyrolysis oil other chemical components are producing more carbon deposits during the HDO than guaiacols.

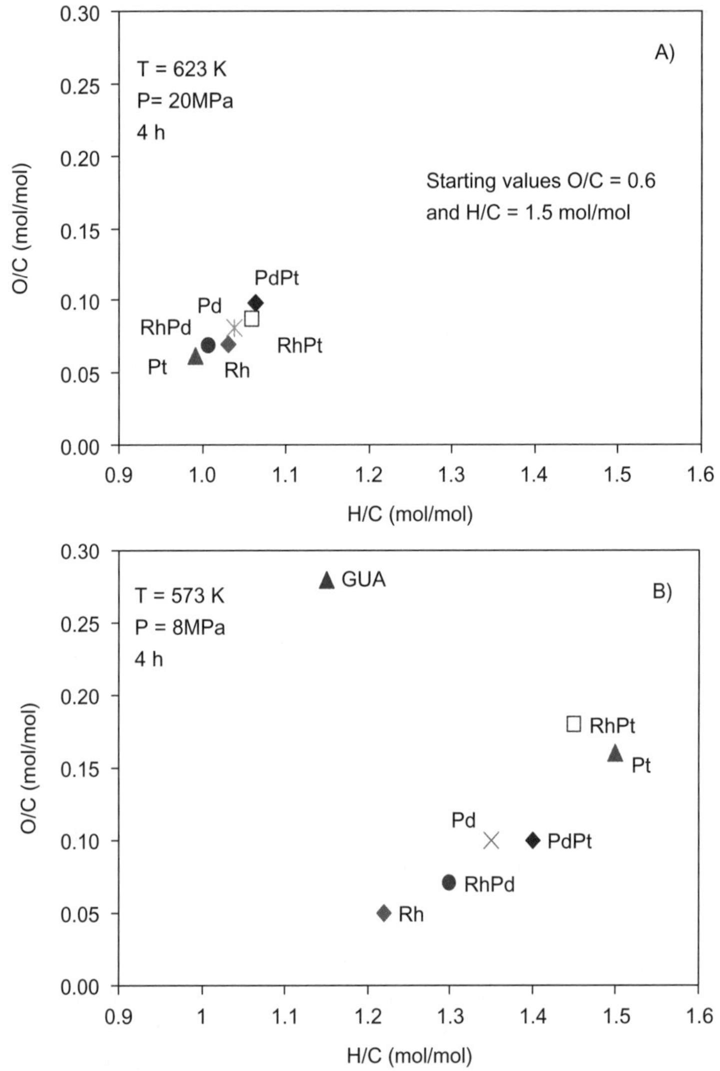

Figure 11.2 A) O/C and H/C molar ratios obtained in the HDO of pyrolysis oil (PO)[57] and B) in the HDO of guaiacol (GUA)[15] on ZrO_2-supported Rh, Pd, and Pt mono- and bimetallic catalysts (batch reactors).

11.6 Outlook

For the large-scale HDO processing of wood-based bio-oils, challenges still remain: how to produce the required hydrogen and where to get the energy to pressurize it. A recent study has shown that the use of a noncatalytic high-pressure thermal treatment (HPTT) process before HDO minimizes the hydrogen consumption.[58] Partial removal of oxygenates is already achieved in

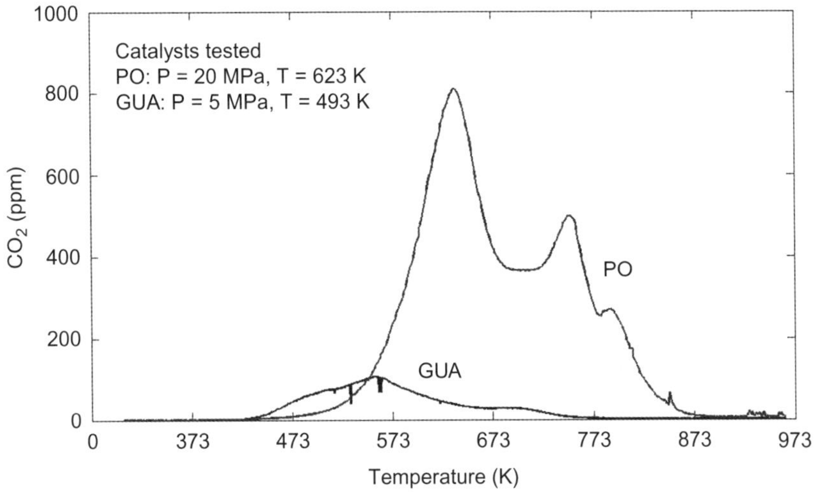

Figure 11.3 TPO for catalysts tested in the HDO of pyrolysis oil (PO) and guaiacol (GUA).[57]

this process and thus the properties of the oil are improved. In the future, it is also possible that hydrogen will be produced *in situ*.[59]

Additionally, the selection of the catalyst for these harsh conditions (high temperatures and pressures and acidic media) is challenging. Noble-metal catalysts have been shown to be promising in HDO[15] as replacements for sulfide catalysts, and some cheaper transition-metal alternatives have also shown their potential,[37] but need to be studied further.

The removal of oxygen is a crucial step in the processing of biomass into liquid fuels and HDO is one of the most promising options.

11.7 Summary

Given that HDO reaction networks have been found to be very complex, the use of model components has become a general approach for the study of HDO. The selection of the model component is important, because although HDO has been studied in connection with HDS and HDN, it has been observed that oxygen is generally more difficult to remove than other heteroatoms. Recent studies have, however, shown that careful study of the real reaction mixture needs to be carried out in addition to model component studies. In addition, comparing HDO results from different studies is challenging, because the experimental equipment as well as the reaction conditions have significant effects on the results.

HDO is often carried out with sulfided CoMo and NiMo catalysts on alumina support at high hydrogen pressures (7–20 MPa) and temperatures (573–673 K). Deactivation of these catalysts is mainly caused by carbon deposition and the loss of sulfur. New catalytic materials, such as noble and transition metal catalysts, have been identified as promising alternatives to the sulfided catalysts.

References

1. S.R.A. Kersten, W.P.M. van Swaaij, L. Lefferts and K. Seshan, in *Catalysis for Renewables – From Feedstock to Energy Production*, G. Centi and R.A. van Santen, (ed.), Wiley-VCH, Weinheim, 2007, p. 119.
2. R. Maggi and B. Delmon, *Stud. Surf. Sci. Catal.*, 1997, **106**, 99.
3. Z. Su-Ping, Y. Yong-Jie, R. Zhengwei and L. Tingchen, *Energy Sources*, 2003, **25**, 57.
4. E. Furimsky, *Appl. Catal. A*, 2000, **199**, 147.
5. E. Furimsky, *Catal. Rev.-Sci. Eng.*, 1983, **25**, 421.
6. A.A. Peterson, F. Vogel, R.P. Lachance, M. Fröling, M.J. Antal Jr and J.W. Tester, *Energy Environ. Sci.*, 2008, **1**, 32.
7. D. Mohan, C.U. Pittman Jr and P.H. Steele, *Energy & Fuels*, 2006, **20**, 848.
8. R.K. Sharma and N.N. Bakhshi, *Can. J. Chem. Eng.*, 1991, **69**, 1071.
9. R. Maggi and B. Delmon, *Biomass Bioenergy*, 1994, **7**(1–6), 245.
10. A.V. Bridgwater and M.-L. Cottam, *Energy & Fuels*, 1992, **6**, 113.
11. T.-R., Viljava, Doctoral Thesis, Helsinki University of Technology, Espoo Finland, 2001.
12. D.C. Elliott, *Energy & Fuels*, 2007, **21**, 1792.
13. A. Vuori, Doctoral Thesis, Helsinki University of Technology, Espoo, Finland, 1986.
14. O.İ. Şenol, E.-M. Ryymin, T.-R. Viljava and A.O.I. Krause, *J. Mol. Catal. A: Chem.*, 2007, **277**, 107.
15. A. Gutierrez, R.K. Kaila, M.L. Honkela, R. Slioor and A.O.I. Krause, *Catal. Today*, 2009, **147**, 239.
16. O.İ. Şenol, Doctoral Thesis, Helsinki University of Technology, Espoo Finland, 2007.
17. E.-M. Ryymin, M.L. Honkela, T.-R. Viljava and A.O.I. Krause, *Appl. Catal. A*, 2009, **358**, 42.
18. A.V. Bridgwater, *Appl. Catal. A*, 1994, **116**, 5.
19. D.C. Elliott and G.G. Neuenschwander, in *Developments in Thermochemical Biomass Conversion*, A.V. Bridgwater and D.G.B. Boocock, (ed.), Blackie Academic & Professional, London, 1997, p. 611.
20. R.E. Maggi and D.C. Elliott, in *Developments in Thermochemical Biomass Conversion*, A.V. Bridgwater and D.G.B. Boocock, (ed.), Blackie Academic & Professional, London, 1997, p. 575.
21. T.-R. Viljava, S. Komulainen, T. Selvam and A.O.I. Krause, *Stud. Surf. Sci. Catal.*, 1999, **127**, 142.
22. T.-R. Viljava, S. Komulainen and A.O.I. Krause, *Catal. Today*, 2000, **60**, 83.
23. D.C. Elliott and E.G. Baker, *Biotechnology and Bioengineering Symp.*, 1984, **14**, 159.

24. E.-M. Ryymin, M.L. Honkela, T.-R. Viljava and A.O.I. Krause, manuscript in preparation.
25. E. Furimsky and F.E. Massoth, *Catal. Today*, 1999, **52**, 381.
26. E. Laurent and B. Delmon, *Appl. Catal. A*, 1994, **109**, 97.
27. J.H. Gary and G.E. Handwerk, *Petroleum Refining*, 3rd edn., Marcel Dekker, New York, 1994.
28. R.P. Silvy, P. Grange, F. Delannay and B. Delmon, *Appl. Catal.*, 1989, **46**, 113.
29. T.-R. Viljava and A.O.I. Krause, *Stud. Surf. Sci. Catal.*, 1997, **106**, 343.
30. D. Kubička, J. Horáček and P. Šimáček, Proceedings of the 9th EuropaCat conference, Salamanca, Spain, 2009, P3–45.
31. M.W. Vogelzang, C.-L. Li, G.C.A. Schuit, B.C. Gates and L. Petrakis, *J. Catal.*, 1983, **84**, 170.
32. E. Laurent and B. Delmon, *J. Catal.*, 1994, **146**, 281.
33. A. Popov, E. Kondratieva, J.M. Goupil, J. El Fallah, L. Mariey, A. Travert, F. Maugé, J.P. Gilson, X. Portier, Y. Romero, F. Richard and S. Brunet, Proceedings of the 9th EuropaCat conference, Salamanca, Spain, 2009, P6–11.
34. D.C. Elliott, G.G. Neuenschwander, T.R. Hart, J. Hu, A.E. Solana and C. Cao, in *Science in Thermal and Chemical Biomass Conversion*, A.V. Bridgwater and D.G.B. Boocock, (ed.), **vol. 2**, CPL Press, Newbury, 2006, p. 1536.
35. A. Centeno, E. Laurent and B. Delmon, *J. Catal.*, 1995, **154**, 288.
36. M. Ferrari, B. Delmon and P. Grange, *Carbon*, 2002, **40**, 497.
37. V.A. Yakovlev, S.A. Khromova, O.V. Sherstyuk, V.O. Dundich, D.Yu. Ermakov, V.M. Novopashina, M.Yu. Lebedev, O. Bulavchenko and V.N. Parmon, *Catal. Today*, 2009, **144**, 362.
38. A. Centeno, R. Maggi and B. Delmon, *Stud. Surf. Sci. Catal.*, 1999, **127**, 77.
39. A.V. Bridgwater, S. Czernik and J. Piskorz, in *Fast Pyrolysis of Biomass: A Handbook Volume 2*, A.V. Bridgewater, (ed.), CPL Press, Newbury, 2002, p. 1.
40. J. Wildschut, J. Arentz, C.B. Rasrendra, R.H. Venderbosch and H.J. Heeres, *Environ. Prog. Sustainable Energy*, 2009, **28**, 450.
41. R.T. Sanderson, *Polar Covalence,* Academic Press, New York, 1983.
42. S.W. Benson, *Thermochemical Kinetics,* John Wiley & Sons, New York, 1976.
43. O. Weisser and S. Landa, *Sulfide Catalysts, Their Properties and Applications,* Pergamon Press, Oxford, 1973.
44. B.S. Gevert, J.-E. Otterstedt and F.E. Massoth, *Appl. Catal.*, 1987, **31**, 119.
45. E. Laurent and B. Delmon, *Ind. Eng. Chem. Res.*, 1993, **32**, 2516.
46. O.İ. Şenol, T.-R. Viljava and A.O.I. Krause, *Appl. Catal. A*, 2007, **326**, 236.
47. H. Topsøe, B.S. Clausen and F.E. Massoth, *Hydrotreating Catalysis,* Springer-Verlag, Berlin, 1996.
48. M.K. Huuska, *Polyhedron*, 1986, **5**, 233.
49. M. Huuska and J. Rintala, *J. Catal.*, 1985, **94**, 230.

50. S.J. Hurff and M.T. Klein, *Ind. Eng. Chem. Fundam.*, 1983, **22**, 426.
51. V.N. Bui, G. Toussaint, D. Laurenti, C. Mirodatos and C. Geantet, *Catal. Today*, 2009, **143**, 172.
52. P. Grange, E. Laurent, R. Maggi, A. Centeno and B. Delmon, *Catal. Today*, 1996, **29**, 297.
53. X. Wang, R.Y. Saleh and U.S. Ozkan, *J. Catal.*, 2005, **231**, 20.
54. D. Procházková, P. Zámostný, M. Bejblová, L. Červený and J. Čejka, *Appl. Catal. A*, 2007, **332**, 56.
55. D.C. Elliott and T.R. Hart, *Energy & Fuels*, 2009, **23**, 631.
56. G.W. Huber, J.N. Chheda, C.J. Barrett and J.A. Dumesic, *Science*, 2005, **308**, 1446.
57. A. Gutierrez, A. Ardiyanti, M. Honkela, E. Heeres and O. Krause, oral presentation, Abstracts of papers of 21st North American Catalysis Society Meeting, San Francisco, 2009.
58. F. de Miguel Mercader, J. A. Hogendoorn, R.H. Venderbosch, S.R.A. Kersten and M.J. Groeneveld, submitted for publication.
59. C.A. Fisk, T. Morgan, Y. Ji, M. Crocker, C. Crofcheck and S.A. Lewis, *Appl. Catal. A.*, 2009, **358**, 150.

CHAPTER 12

Ionic Liquids for the Utilization of Lignocellulosics

SAMUEL A. MORTON III[a] AND LAUREL A. MORTON[b]

[a] Center for Applied Energy Research, University of Kentucky, Lexington, Kentucky 40511, USA; [b] Department of Chemistry, Eastern Kentucky University, Richmond, Kentucky 40475, USA

12.1 Introduction to Lignocellulosics

Given the ever-increasing world-wide demand for energy, coupled with the instability of current petroleum resources, the development of sustainable, biologically derived energy and commodity chemical feedstock sources is critical. A great deal of interest and effort is currently being focused on the development of technologies to convert lignocellulosic material ("woody" biomass) into renewable replacements and augments for current petroleum-derived platform chemicals and fuels.[1–8] To this end, both the United States and the European Union have adopted policy directives and research initiatives directed at replacing a portion of currently petroleum-derived products with biomass-derived alternatives. The United States Department of Energy has established goals to derive 25% of industrial organic chemicals and 30% of transportation fuels from renewable biomass sources by the year 2025.[9,10] Additionally, a number of states have adopted "Renewable Portfolio Standards" that set statewide goals aimed at the use of renewable/alternative energy sources that include the use of biomass.[11] The European Union has, through the Biofuels Research Advisory Council, set a goal of replacing 25% of transportation fuels from biomass sources by 2030.[12] These are long-range

RSC Energy and Environment Series No. 1
Thermochemical Conversion of Biomass to Liquid Fuels and Chemicals
Edited by Mark Crocker

Published by the Royal Society of Chemistry, www.rsc.org

targets and as such are prone to acceleration, deceleration, and/or replacement; however, it is evident that some appreciable portion of energy and chemical feedstocks will be produced from biomass in the future. Therefore, energy-efficient and environmentally friendly methods of processing of lignocellulosic materials yielding valuable bioproducts (biopolymers, biofuels, commodity chemicals, *etc.*) are needed. These new methods for the full utilization of lignocellulosic biomass must address the challenge of "biomass recalcitrance" – the natural resistance plant matter exhibits to biodegradation.[8] The recalcitrance of biomass is generally attributed to cellular construction and the arrangement of substructures, the heterogeneity of lignin, the insolubility of the cellular biopolymers, and the presence or formation of inhibiting compounds during degradation.[8]

At its most basic, lignocellulosic biomass is composed of three primary components: lignin, cellulose, and hemicelluloses (also known as polyose).[13,14] Lignin comprises roughly 20–35% of the dry mass of woody plants, and is a randomly structured amorphous polyphenolmeric material derived from three basic monomers, coniferyl alcohol, sinapyl alcohol, p-coumaryl alcohol. Lignin serves a variety of purposes: (1) controlling the cellular transport of water, nutrients, and other metabolites, (2) binding cells in a resin-like material that imparts resistance to physical damage (compression, impact, and bending), and (3) confounding biological degradation through the heterogeneous makeup of the lignin polymer. Cellulose comprises roughly 40–50% of the dry mass of woody plants, and is the most abundant of the cellular sugars present in biomass as well as the best utilized. Cellulose is, in general, assumed to be composed of a polymeric chain of glucose molecules. Cellulose imparts a great deal of resilience to plant matter as it is mostly insoluble in the natural environment and its fibers are remarkably strong. When released from the lignin matrix, cellulose can be readily degraded by a range of enzymes either by cleavage at the terminal polymer ends or at locations nearer the center of the molecule. Hemicelluloses are the least utilized of the primary plant cellular components and comprise approximately 20–30% of the dry mass of woody plants. Hemicellulose is a rather deceptive descriptor for what are essentially the noncellulose plant polysaccharides composed of a variety of C5 and C6 sugar monomers (xylose, mannose, galactose, rhamnose, and arabinose). The hemicelluloses primarily provide structural benefits to the lignocellulose polymer matrix. A representative set of chemical structures for lignin, cellulose, and hemicellulose can be seen in Figure 12.1.

While not all plant-derived material contains the same ratio of lignin to cellulose to hemicelluloses, there are trends that exist between general plant groupings (softwoods, hardwoods, grasses, *etc.*). Additionally, the three primary fractions are sometimes accompanied by various quantities of starch, ash, pectins, lipids, and other trace species-specific compounds. The relative quantities and compositions for a range of common biomasses are given in Table 12.1.[13,15]

A significant hurdle to the ready adoption of lignocellulosic-based fuels and chemicals is that the economically viable approaches are only able to utilize a portion of the total available plant matter. Most lignocellulosic processing

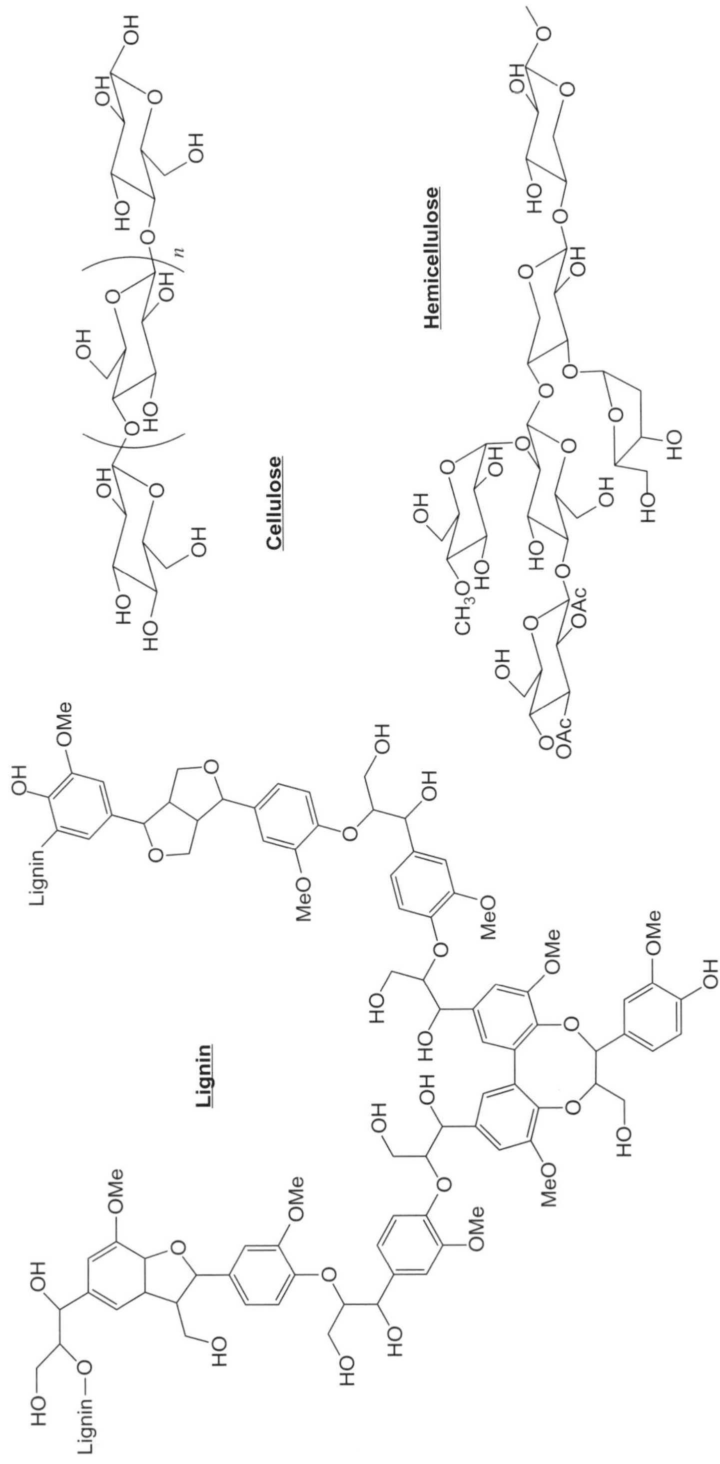

Figure 12.1 Lignocellulosic biomass constituent compound structures.

Table 12.1 Observed distribution of common constituent fractions in a range of common lignocellulosic biomasses.

	Biomass	*Cellulose*	*Lignin*	*Hemicellulose*	*Pectin, Starch, Ash, Etc.*
Hardwoods	Red Maple (*Acer rubrum*)	45	24	29	2
	Paper Birch (*Betula papyrifera*)	42	19	38	1
	American Beech (*Fagus grandfolia*)	45	22	29	4
	Quaking Aspen (*Populus tremuloides*)	48	21	27	4
	American Elm (*Ulmus americana*)	51	24	23	2
Softwoods	Balsam Fir (*Abies balsamea*)	42	29	27	2
	White Spruce (*Picea glauca*)	41	27	31	1
	Eastern White Pine (*Pinus strobus*)	41	29	27	3
	Eastern Hemlock (*Tsuga canadensis*)	41	33	23	3
	Northern White Cedar (*Thuja occidentalis*)	41	31	26	2
Agricultural	Corn Stover	36	17	23	16
	Sugarcane	22	11	15	52
	Sweet Sourghum	35	17	17	28
	Switchgrass	43	9	33	14

systems seek only to convert the cellulose fraction to glucose followed by fermentation to an alcohol product. This leaves the lignin and hemicellulose fractions to be discarded or merely burned as a heat-generation fuel source.[13] An economic analysis performed by Zhang, summarized in Figure 12.2, demonstrates the significant improvements in the economic potential of lignocellulosic biomass that would result from improvements in the recovery of these underutilized fractions.[16]

Current methods for separating the primary biomass fractions (*i.e.* delignification) include oxidation, hydrogenolysis, mild hydrolysis, and acidolysis, and involve economically and energetically unsustainable processes utilizing high thermal conditions, aggressive chemical routes, or costly enzymes that generate performance, economic, and environmental challenges. The most prevalent strategy for lignocellulosic processing is the kraft pulping process. This process is directed at the removal of lignin for the purpose of leaving a sufficiently robust but flexible cellulose dominant fiber for the production of various fiber products, papers being the primary example. The chemistry of the pulping

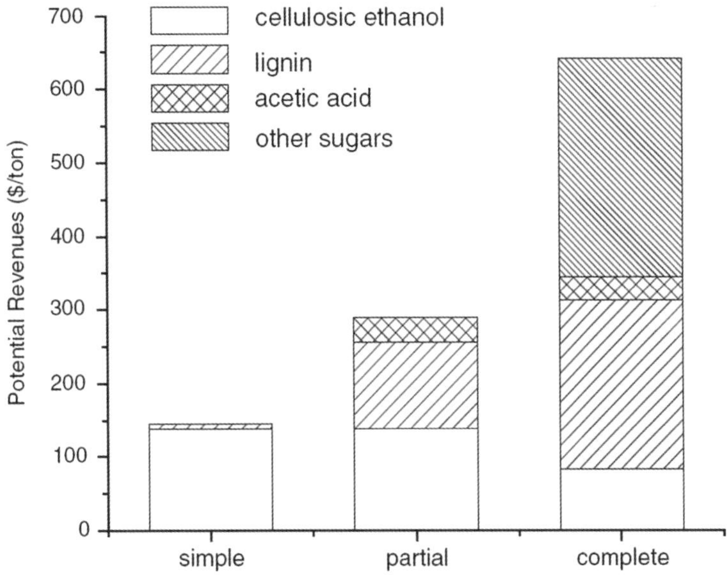

Figure 12.2 Economic assessment for various extents of biomass utilization in a lignocellulosic biorefinery. [With kind permission from Springer Science + Business Media.][16]

process requires the use of an aqueous solution of sodium hydroxide and sodium sulfide under elevated pressures and temperatures. During this extended "cook" the lignocellulosic material is depolymerized into smaller soluble polymer fragments from which roughly 90–95% of the lignin is removed. The residual removable lignin is finally extracted during the pulp bleaching process. Consequences of the need to remove such a significant amount of lignin are that the residual fibers are weakened significantly and there is potential for environmental damage from the release of the chemicals used and the byproducts produced.[13,14,17]

This thermochemical route to delignification is effective at removing the lignin; however, it is not effective at delivering the lignin in a form that can be utilized as a broad chemical feedstock. Another method for chemical deconstruction of lignocellulosic material involves the development of selective oxidation or depolymerization catalysts.[18–22] While these novel catalysts have been proposed and demonstrated, no large industrial-scale processes have been developed utilizing the technology, in part because as of yet no real market exists to use the specific feedstock chemicals that could be produced from lignocellulosic material. The remaining mechanism for lignocellulosic degradation is that of biological processes present in nature.[23] Although lignocellulosic materials can be transformed *via* enzymatic degradation, the enzymes required have limited functionality under the conditions presently common in industrial delignification to effectively dissolve lignin and cellulose.[13] A potential alternative to costly and temperamental enzymes is the use of biomimetic complexes. Fortunately, as delignification has been a source of long-standing research interest, the biological pathways utilized by a range of organisms, the chief of which are the fungi, have been well studied. From the studies of active sites in lignin-specific enzymes, biomimetic catalytic compounds, such as metalloporphyrins and metallophthalocyanines, have been developed.[13,24–26] Despite the level of interest and effort expended, there remain challenges to the implementation of alternatives to kraft pulping, primarily revolving around issues with process development cost and complexity related to the dual problems of solubility and reactivity of lignocellulosic biomass. These linked problems are traditionally resolved independent of one another, by the selection of an effective dissolution strategy followed by the separation of the various biomass fractions.

Recently, a novel class of compounds, ionic liquids, has been presented as a possible solution to the challenge of biomass recalcitrance.[27–32] These compounds have demonstrated the ability to dissolve either the entirety of the lignocellulosic biomass or the individual fractions selectively. While the potential for ionic liquids to serve as solvents for the chemical and biochemical processing of biomass is high, the body of scientific knowledge is still in the early stages of exploration and many challenges remain to be resolved. The purpose of this chapter is to provide a "state-of-the-art" review of the scientific literature and to identify trends that are developing and issues that remain.

12.2 Ionic Liquids

12.2.1 Classification and Nomenclature

Ionic liquids are essentially nonvolatile liquid/organic salts.[33] Ionic liquids can be more broadly defined to include compounds traditionally identified as molten salts, such as cryolite, which require elevated temperatures to exist in a liquid state. To distinguish these high-temperature molten salts from room-temperature ionic liquids (RTIL), a definition revolving around melting point and liquid phase range is used in the literature. By this definition, a room-temperature ionic liquid exhibits a melting point temperature typically less than 100 °C as well as having a liquid range that can span from below 0 °C to upwards of 400 °C.[33–41] While this definition is sufficient for our current level of development in the field of ionic liquids, it could omit ionic liquid-like compounds yet to be identified that melt at temperatures above the arbitrarily selected 100 °C, and so the final definition of an ionic liquid remains to be settled.

Ionic liquids are novel solvent-like compounds that, based on their physical and chemical properties as well as their selectable chemical composition, have demonstrated a significant potential for use in a broad range of applications. There appear to be a multitude (on the order of 10^{18})[38] of potential ionic liquids that can be derived from various combinations of currently available anions and cations. Most of the published work to date has employed a limited

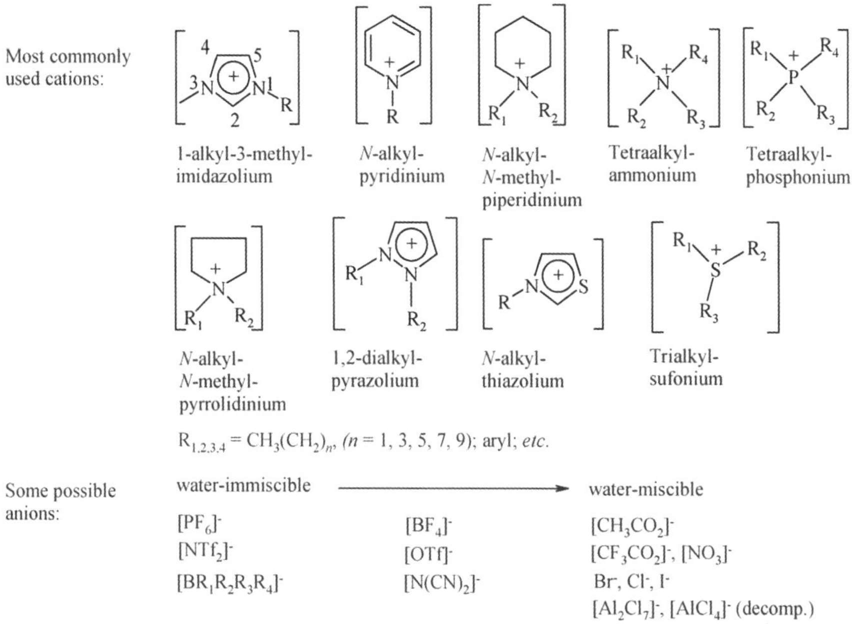

Figure 12.3 Typical ionic-liquid compositional cations and anions [Reproduced by permission of The Royal Society of Chemistry.][33]

number of ionic liquids. These compounds are typically composed of a few select cations and anions as can be seen in Figure 12.3.

While there are increasingly large numbers of unique ionic liquids described in the literature, the available research has dealt with only a small number of ionic liquid "families", with the bulk of the studies investigating compounds classified as alkyl-imidazolium ionic liquids. This is likely due to the simple fact that until recently there was limited accessibility to quality ionic liquids. In 2000, a research group interested in ionic liquids would likely have synthesized the required compounds in their own laboratory, whereas now there are approximately 300 commercially available compounds in various purities and quantities from both large-scale and small-scale chemical vendors.[33]

Ionic liquids have a long yet obscure history; a truncated timeline is shown in Figure 12.4. The first literature-verified reference to what we now call ionic liquids was published by Paul Walden in 1914 in the *Bulletin de l'Academie Imperiale des Sciences de St.-Petersbourg* regarding the synthesis and properties of ethylammonium nitrate $[EtNH_3][NO_3]$.[42] Following this seminal work, ionic liquids disappeared from the scientific literature until a few years after the end of the Second World War with the work of Hurley and Wier related to the use of a simple pyridinium ionic liquid for the electrodeposition of aluminum.[43] As with the work of Walden, ionic liquids again fell into obscurity until the research work of Robert Osteryoung and coworkers in 1975.[44] It is at this point that many consider the field of ionic liquid research to have truly begun.

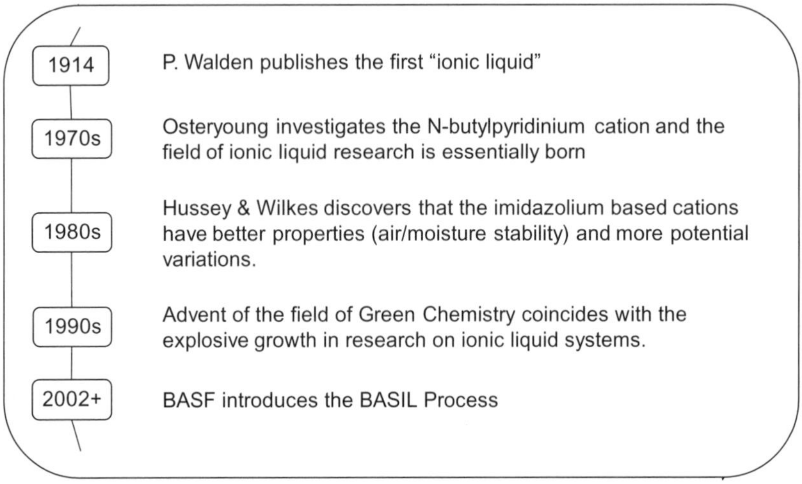

Figure 12.4 Time-line of the major milestones in the history of ionic liquids.

Following this initial work, Osteryoung continued to explore the electrochemical properties of other ionic liquids, primarily of the chloroaluminate variety, for use in electrodeposition processes.[41] A critical shortcoming of the ionic liquids used at that time was poor air and moisture stability. This instability severely limited the potential applications of these compounds until the work of Wilkes and Hussey into the use of the imidazolium family of ionic liquids.[45] Wilkes *et al.* recognized that in addition to improved air/moisture stability, significantly more possible compounds could be produced based on the imidazolium-type ionic liquids than the previously favored chloroaluminate compounds, due to the ability to control both the functional composition of the cation and the anion. Following the discovery of the imidazolium family of compounds, ionic liquids were included in the burgeoning field of green chemistry in large part due to their negligible vapor pressure (early work described them as nonvolatile; however, recent work published by Earle *et al.* has demonstrated the reasonable distillation of some ionic liquids).[46] The first industrial process specifically utilizing an ionic liquid, the biphasic acid scavenging utilizing ionic liquids (BASIL™) Process, was developed by BASF in 2002; however, the process generated an ionic liquid as a product for recycle rather than using the ionic liquid as a solvent. Recently, compounds not initially identified as ionic liquids have been found in a number of chemical processes that predate the BASIL process; however, none of these processes are currently in use.

Despite the more than 30 years of consistently published scientific work, ionic liquids have only recently found significant and sustained interest in the scientific literature. Figure 12.5 gives a snapshot of this growing interest using a "Web of Science" citation search of articles containing keywords relevant to the subject of this article. This is not an exhaustive listing as both patents and

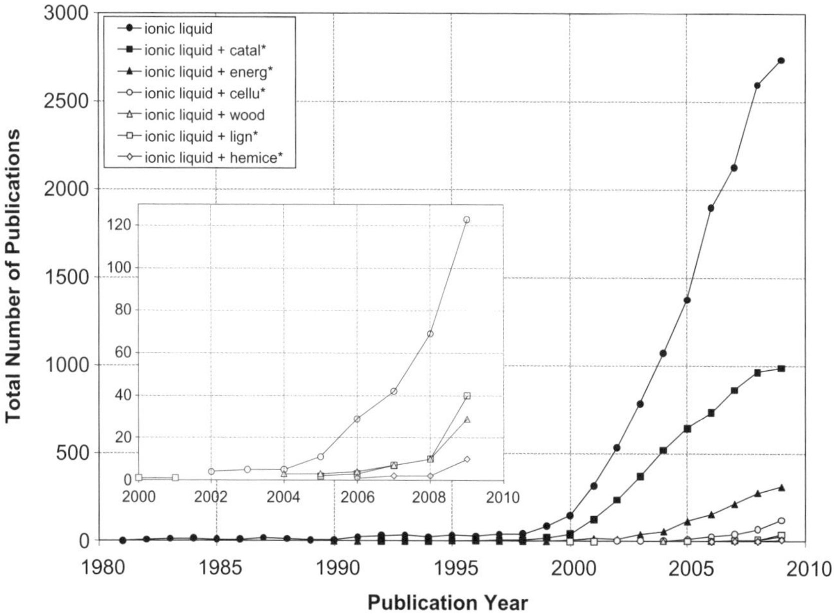

Figure 12.5 The growth in publications/citations relevant to the use of ionic liquids in lignocellulosic utilization.

new articles included in the citation system after the completion of this chapter are not included. As a comprehensive review of the long history of ionic liquids is not the purpose of this current work, we refer interested readers to the recent publication by Plechkova and Seddon for greater depth and detail.[33] In the following sections, we will provide a brief introduction to the general premise of ionic liquids before addressing the primary subject of this article, the intersection of ionic liquids and lignocellulosic materials.

12.2.2 Synthesis and Purification

There are a number of acceptable synthetic routes to produce ionic liquids.[47,48] These routes are dependent on the family of ionic liquid to be produced and should be selected based on the final purpose the ionic liquid will serve. A range of articles have been published outlining the synthesis of several of the primary ionic liquid families (chloroaluminate,[41] imidazolium,[49] phosphonium,[37,50–52] pyridinium,[53] pyrrolidinium,[54] *etc.*). As the number of various ionic liquid families is continuously expanding, it is not feasible to comprehensively outline all the synthetic pathways available. Therefore, we will employ the synthesis of a pair of imidazolium family ionic liquids, 1-butyl-3-methylimidazolium chloride ([$c_4$mim][Cl]) and 1-butyl-3-methylimidazolium hexafluorophosphate ([$c_4$mim][PF_6]), in our discussion of ionic liquid synthesis, the rationale for

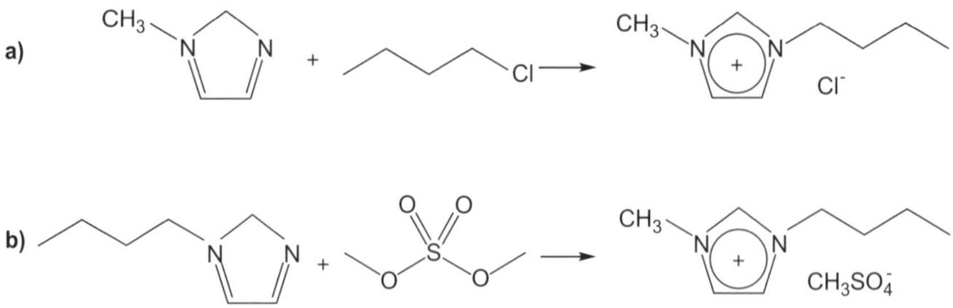

Figure 12.6 Synthetic pathways for the formation of the primary ionic liquids, [$c_4$mim][Cl] and [$c_4$mim][CH_3SO_4]

choosing one synthetic route over another, and issues related to the purity of the initial ionic liquid.

The synthesis of [$c_4$mim][Cl], which we will term a "primary" ionic liquid as it can later be used as a feedstock for subsequent or "secondary" ionic liquids, is traditionally based on the alkylation of 1-methylimidazole with 1-chlorobutane, in which one mole each of the reagents yields one mole of the target ionic liquid, as shown in Figure 12.6(a).[49] A drawback to the use of an alkyl halide is that the ionic liquid product will be contaminated with residual halide that can impact physical properties, catalytic performance, product purity, and long-term stability. While trace quantities of halides may not be of concern for some process situations, other processes may be significantly impacted.

An alternative synthetic route, shown in Figure 12.6(b), is the selection of an alternative imidazole and replacement of the alkyl halide with either an alkyl sulfate or an alkyl sulfonate.[55,56] In this example, the synthesis of 1-butyl-3-methylimidazolium methyl sulfate ([$c_4$mim][CH_3SO_4]) proceeds *via* the reaction of 1-butylimidazole with dimethyl sulfate. There are significant advantages for the use of this alternative route, not the least of which are a comparatively increased rate of reaction and less-expensive reagents.[55] Thus, we have prepared two alternative primary ionic liquids that can serve either as our target compounds or as reagents for the synthesis of ([$c_4$mim][PF_6]). The preparation of this ionic liquid can also occur by a variety of mechanisms, two of these schemes being shown in Figure 12.7.

In the first route, the primary ionic liquid of choice is mixed with either a salt or acid with the desired anion. Figure 12.7(a) shows the metathesis of [$c_4$mim][Cl] with ammonium hexafluorophosphate using water as a solvent to yield the desired ionic liquid [$c_4$mim][PF_6]. Separation of the final product from the solvent is spontaneous, after which multiple rinses are performed with an appropriate solvent, such as diethyl ether, followed by drying under vacuum until the ionic liquid is sufficiently dried. A consequence of this process is that there exists a significant requirement for solvent consumption, thereby reducing the greenness of this reaction. An alternative to the above metathesis, column-based ion exchange, is shown in Fig 12.7(b).[57] In this process an ion-exchange resin, for example Amberlyst A-26(OH), is placed in a column and the primary

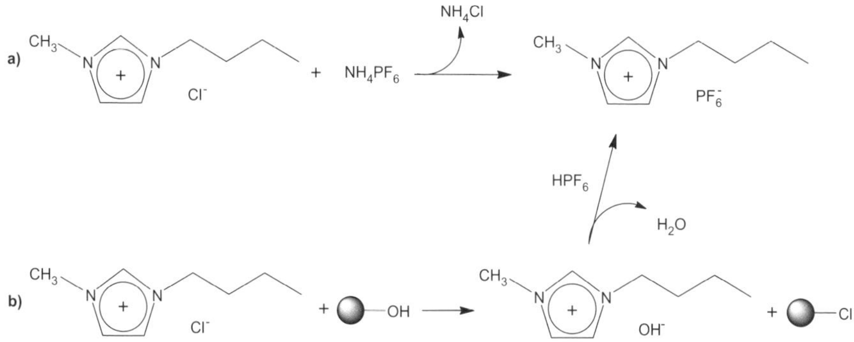

Figure 12.7 Synthetic pathways for the formation of the secondary ionic liquid, [c$_4$mim][PF$_6$].

ionic liquid, [c$_4$mim][Cl], is dissolved in a solvent (water will suffice) and repetitively passed through the ion exchange column. Assuming complete ion exchange between the resin and the reagent ionic liquid, the resulting solution contains the solvent (water) and an intermediate ionic liquid [c$_4$mim][OH]. An acid or salt containing the desired anion is stirred into this solution yielding the desired secondary ionic liquid, [c$_4$mim][PF$_6$] and water or a soluble salt. As before, this particular ionic liquid is insoluble in water as separation is spontaneous with the final ionic liquid only requiring drying under vacuum to remove residual water.

In addition to the traditional stirring at elevated temperature, there have been reports of other novel methods for the synthesis of these ionic liquids. These reports highlight either the effects of solvent on reaction rates,[58] the impacts of the physical and operation nature of the reactor on rapid production,[59] and the use of ultrasonic or microwave reactors.[60] One important feature of all these alternatives to the traditional use of solvents under reflux is the reduction or outright avoidance of thermally induced ionic-liquid discoloration.[61] Most ionic liquids should be nearly clear in nature; however, the general trend in the literature is a pale yellow to dark amber color. As with the halide impurities, if the ionic liquids are to be used in situations demanding high clarity,[62] the general advice is to dissolve the darkened ionic liquid and pass it through a packed column of silica gel or activated carbon. While the resulting ionic liquids are returned to a clear state, there is the potential for significant losses of product as well as indications that the sorbent material is retained by the ionic liquid.[63] Therefore, anyone considering an ionic liquid for a specific use must consider the impact of such impurities and balance these impacts with the economic impacts of attaining high purity from the outset. Regardless of the method for acquiring ionic liquids for research (be it in house synthesis or a commercial source), a recent publication by Smiglak *et al.* outline a useful framework for the selection, design, and use of ionic liquids.[64]

12.2.3 Potential Uses

The first widely studied ionic liquids were those containing chloroaluminate moieties developed for use as battery electrolytes.[41] Ionic liquids have been proposed as green extraction agents, thus reducing costs and environment impacts for many energy-intensive separation activities.[47,65–67] However, the primary focus for ionic liquid research has been to develop replacement solvents for use in catalytic reaction systems.[68–71] It is on this last application that the vast bulk of lignocellulosic/ionic liquid published research has focused, and we will return to it later in this chapter. There are a large number of high-quality review articles attending to the various uses of ionic liquids, both academic and industrial. One should note that most of the published ionic liquids were not designed for the processes for which they are proposed; in fact the vast bulk of ionic liquid research typically employs whatever ionic liquid is at hand without deliberate consideration for the appropriateness of the unique physical and chemical properties that ionic liquids might possess. To address this unfortunate fact, a category of ionic liquids, the task-specific ionic liquid (TSIL), has been developed.[72,73] TSILs are essentially ionic liquids that were prepared for a specific purpose with conscious thought to the impacts of their chemical and physical properties. The eventual adoption of ionic liquids to uses in the broader chemical industry is by the accounts of many inevitable; however, a number of challenges remain that must be overcome before this can occur.[33,74–76] Among these challenges, the requirements for recycle and reuse, as well as the development of a robust understanding of the environmental impacts from the inevitable release of ionic liquids into the environment are critical to the economic future of any ionic liquid-based process. While these two challenges are linked in many ways, we will address them individually in the following sections for the sake of simplicity.

12.2.4 Reuse, Recycle, and Stability

While ionic liquids do have exciting possibilities as alternative green solvents, the current cost of production for these compounds is such that their use is not feasible for a once-through-type process. In the recent ionic liquid review article by Plechkova and Seddon, a discussion of ionic liquid pricing is presented in which an ionic liquid, choline chloride, was determined to cost only five times the purchase price of a traditional industrial solvent, dimethyl sulfoxide (DMSO).[33] We can repeat this analysis for the primary ionic liquid used in our discussion of ionic-liquid synthesis, [$c_4$mim][Cl], for which ACROS lists 500 g at \$637.50 (US) or \$223/mole. The two reagent chemicals, 1-methylimidazole and 1-chlorobutane cost approximately \$15.2 and \$10.7 per mole, respectively. Thus, we can see that the ionic liquid is nearly an order of magnitude more expensive than its starting materials; this most likely reflects the scale of production more than the cost of said production. DMSO is comparable in cost to 1-chlorobutane, therefore on a mass basis the ionic liquid is nearly 25–30 times as expensive as DMSO. Even assuming the purchase price is reduced at large scales by a factor of approximately five, which has been demonstrated for an existing ionic liquid produced on the multiton scale,[48] the ionic liquid solvent remains significantly more expensive than traditional molecular solvents. However, this becomes less of a factor if the ionic liquid can be recycled multiple times; Plechkova and Seddon state that 10–20 recycles could bring ionic liquids into cost parity with conventional solvents.[33] It is therefore critical to develop recycling strategies in tandem with the development of potential ionic-liquid applications.

There are a number of possible recycle and purification strategies that could be employed in a large-scale process using ionic liquids, such as traditional solvent extraction, membrane separation, and supercritical extraction.[48,75] Regardless of the method employed, the long-term stability of the ionic liquid should be monitored closely. It is well known that some ionic liquids can be subject to decomposition, involving either hydrolysis[77] or thermolysis,[78,79] that could negatively impact the economic and environmental performance of these solvents. It is now becoming customary in the ionic liquid relevant literature to report some effort at solvent recycle. The reliability of this data for process condition relevant consideration is not always clear, as often the ionic liquids are extensively and, in an industrial context, unrealistically purified between cycles. However, there remains a great deal of work to be performed before the issue of ionic-liquid recycle can be resolved.

12.2.5 Toxicity and Biodegradation

The subject of ionic liquid toxicity is one of significant importance in our modern era of industrial chemical safety and environmental regulation. It is also an unfortunate fact of life that there will be chemical releases, be they accidental, incidental, and intentional. Therefore, we must assume that steps are required to understand the consequences of these inevitable releases should ionic liquids be incorporated into the chemical enterprise. Unfortunately, while the potential uses for ionic liquids are continually evolving, a similarly robust investigation of the environmental consequences of these compounds is not occurring.[74] While many classify ionic liquids as green solvents, based almost exclusively on their negligible vapor pressure at standard conditions, there is a growing body of evidence calling this into some degree of question. A number of recent reports have been published focused on the impacts of ionic liquids on freshwater organisms,[80–83] their migration through the soil,[84] biological uptake into micro-organisms,[85] mutagenicity,[86] and general toxicological evaluation.[87] It should be evident from the previously mentioned studies that toxicology is a very broad subject area and despite the importance of such studies on the future of ionic-liquid utilization, there appear to only be a few scattered research programs with a specific focus on this topic. According to one of these groups, the development of sustainable chemicals (and ionic liquids in particular) requires that we must strive to understand the true environmental impacts of chemical release to the environment.[88] They propose five essential ecotoxicological indicators: release, range, bioaccumulation, biological activity, and uncertainty, that must be determined for any ionic liquid selected for use in a large-scale process. At present such a limited set of usable data exists that the level of uncertainty dictates the assumption that ionic liquids are not environmentally benign and should not be casually treated as such. It is encouraging then that interest is increasingly related to the impacts of ionic liquids on wastewater-treatment system organisms[89,90] (the first line of effluent related environmental protection) and the potential for the design of biodegradable ionic liquids.[50,83,90–94] It is important to note that in many instances ionic liquids are being proposed as process components in the biological conversion of lignocellulosic material. Thus, it is critical that increased emphasis should be focused on studying the impact of these compounds on biofuel and biochemical relevant organisms *via* short-term and long-term studies.

It was our intention that this basic introduction to ionic liquids would provide the reader, regardless of their level of expertise in this subject, with a common foundation of knowledge for the subsequent discussion of the potential application of ionic liquids in the processing of lignocellulosic material. As shown in Figure 12.4, we are in the early days of the use of ionic liquids in the processing of such materials (although a patent issued in 1934 proposed the use of an ionic liquid for cellulose dissolution)[95] and it is our intention that this review of the existing literature will function as a catalyst for increasing interest in the research and use of ionic liquids.

12.3 Ionic Liquids and Lignocellulosics

As stated earlier, ionic liquids are of specific interest in the production of biofuels and biochemicals from lignocellulosic materials because of the potential to overcome, or at least reduce, the issue of recalcitrance by resolving the dual problems of solubility and reactivity. Therefore, in the following sections we will separately address these two problems for the whole processing of lignocellulosic material, as well as each of the three primary components, cellulose, hemicellulose, and lignin.

12.3.1 Unrefined Lignocellulosics

The currently dominant prototypical lignocellulosic-based biorefinery, for which a simple flowsheet is shown in Figure 12.8, is conceptually defined as having three primary stages: (1) pretreatment to fractionate the biomass, (2) enzymatic or catalytic processing of the various fractions, and (3) production of endproducts from the refined and separated biomass components. As indicated earlier there are significant economic advantages to utilizing a larger portion of the original lignocellulosic biomass for purposes other than the generation of onsite energy. There are a great many approaches to achieving this improved economic outlook, of which processing with ionic liquids is a recent and rapidly expanding possibility.

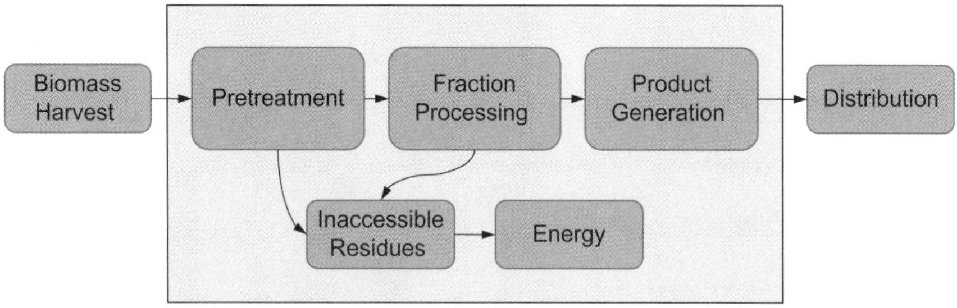

Figure 12.8 Conceptual flow diagram for a lignocellulosic biomass refinery.

Often, when the topic of lignocellulosic biomass arises, the general assumption is that the discussion is focused on "woody" biomass from trees or residual material from the corn-based ethanol industry. This obscures the fact that lignocellulosic biomass can encompass a great range of biomass, from tropical grasses to temperate trees. To avoid this trap we will focus our conversation more broadly to that of unrefined lignocellulosic biomass, wherein the biomass is directly interacting with the ionic liquids without prior processing, other than that required to introduce said biomass to the biorefinery process.

Though the modern intersection of lignocellulosic biomass and ionic liquids in the scientific literature occurred only recently with the seminal work of Swatloski *et al.* concering the dissolution of cellulose in the ionic liquid [$c_4$mim][Cl] (published in 2002)[29] the range of interest and ongoing scientific efforts are phenomenal. Despite the exponential growth in publications, seen in the inset of Figure 12.5, the research applications of ionic liquids with respect to lignocellulosics remain few. The current applications can generally be divided into five categories: (1) dissolution in ionic liquids as pretreatment, (2) dissolution combined with reaction in ionic liquids, (3) ionic liquids as analytical tools, (4) the use of ionic liquids as a product-relevant additive, and (5) the use of ionic liquids to recover waste volatile organics to reduce environmental impacts.[32,96]

Of these five applications, the dissolution of unrefined lignocellulosic biomass in ionic liquids has received the greatest amount of interest in the available scientific literature. The general goal of these works has been to verify the dissolution of whole biomass as a means to delignify and fractionate the components. There are a number of significant advantages to this: (1) the cellulose, once released from the lignin matrix, is much more accessible to enzymatic and catalytic conversion, (2) the hemicelluloses become accessible for utilization at significantly reduced cost and with less degradation, and (3) the lignin is depolymerized but remains in a minimally degraded form from which a wide array of chemical feedstock structures can in principle be extracted. While there have been a number of efforts to expand on the fact that certain ionic liquids dissolve cellulose, the most authoritative works focused on unrefined lignocellulosics are those of Fort *et al.*,[27] Kilpelainenn *et al.*,[28] and

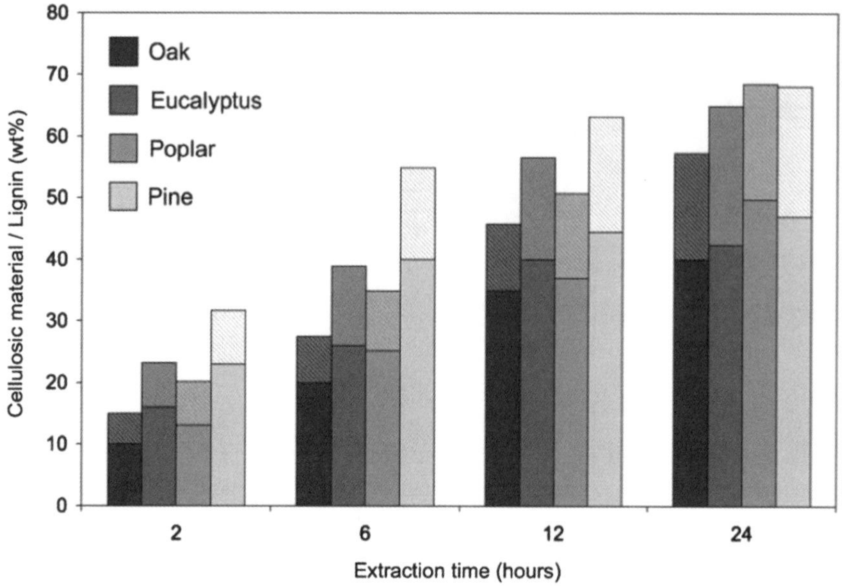

Figure 12.9 Extraction of cellulosic material and lignin from a range of woody biomasses over time. [Reproduced by permission of The Royal Society of Chemistry.][27]

Sun *et al.*[97] These studies focused on developing a better understanding of the specific process parameters that impact solubility. The work of Fort and coworkers investigated the solubility of a range of woods in a single ionic liquid, [$c_4$mim][Cl].[27] A summary of the timeframe and extent of dissolution is presented in Figure 12.9.[27] From this figure it is apparent that, regardless of wood type, highly effective dissolution was obtained after very long contacting at elevated temperatures. Additionally, it was determined that the reconstituted material contained sufficient quantities of hemicellulose and/or lignin to inhibit the formation of cellulose hydrogels. They describe this work as preliminary and outline a range of items that remain to be resolved, including the effects of time and temperature on the recovered cellulose, the implications of toxicity relative to the ionic liquid, and the need to mitigate problems with viscosity. Kilpelainenn and coworkers investigated the potential for a range of imidazolium-based ionic liquids (primarily with a chloride anion) for the dissolution of a selection of softwood biomass.[28] They determined that [amim][Cl] was at least as effective as [$c_4$mim][Cl] at solubilizing the biomass. Unlike the previous work of Fort *et al.* in which wood chips were used as the biomass source, Kilpelaninenn *et al.* investigated a range of preprocessed wood samples and noted improvements in dissolution time. It was noted that hazy solutions were obtained and by inference the presence of residual lignin was assumed. This is likely due to the ability of the ionic liquid to effectively disrupt the hydrogen-bonding characteristic of cellulose but not to fully solvate the highly aromatic

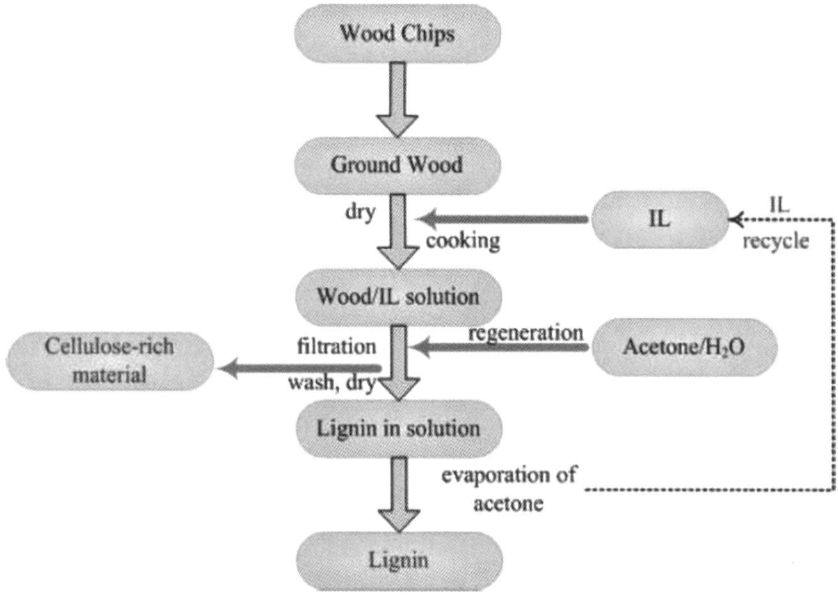

Figure 12.10 Conceptual flow diagram for the dissolution, fractionalization, and recovery of lignocellulosic materials in ionic liquids. [Reproduced by permission of The Royal Society of Chemistry.][97]

lignin. A major revelation from their work was the determination that particle size and the water content of the biomass were significant parameters impacting the dissolution of wood.[28] Sun *et al.* extended the work of Fort *et al.* to address some of the earlier unresolved issues as well as investigating a new ionic liquid, [$c_2$mim][OAc].[97] This work is effectively the most recent and comprehensive evaluation of the parameters affecting the solubility of wood in an ionic liquid. They reported that [$c_2$mim][OAc] was a better solvent for wood compared to [$c_4$mim][Cl], that hardwoods dissolve more rapidly and to a greater extent than softwoods, and that dissolution time could be positively impacted by appropriate milling and/or the use of microwave or ultrasonic pretreatment. Another significant outcome of this work was the development of a process-flow diagram, shown in Figure 12.10, representing the use of ionic liquids in dissolving and fractionating woody biomass. This was significant in that it proposed a reasonably inexpensive method for extracting the lignin from the ionic liquid, thereby allowing for the recycle of the ionic liquid solvent, a feature absent from the previously mentioned works.

Two other studies merit notation before we proceed. The first, that of Xie and Shi, focused on the liquefaction of wood in a novel imidazolium ionic liquid and was the first literature report of the dissolution of unrefined woody biomass.[31] This study demonstrated that ionic liquids were effective candidates for the whole liquefaction of lignocellulosic biomass, as well as effective catalytic liquefaction using an ionic-liquid/aluminum chloride system. The second recently published study is that of Miyafuji *et al.*, which focused on providing greater detail with respect to the process of dissolution.[98] They demonstrated the order and relative rate of the dissolution and conversion of the various lignocellulosic components. The order of liquefaction was determined to be the polysaccharides, celluloses and hemicelluloses, followed by the remaining lignin. Additionally, for all but the highest temperatures tested, the lignin was essentially unaffected until late in the liquefaction process. They were able to determine that glucose is liquefied first, followed by the polysaccharides present in hemicellulose.

Beyond these studies, other groups have specifically evaluated the use of ionic liquids for the selective extraction of either lignin[99] or cellulose[100] from lignocellulosic biomass. Tan and coworkers determined that an imidazolium alkylbenzenesulfonate ionic liquid, [$c_2$mim][ABS], was an effective candidate for recovering lignin from dissolved lignocellulosic biomass.[99] They used higher temperatures than previous studies and consequently were able to dramatically reduce the required processing time. In addition to evaluating the thermochemical properties of the recovered lignin, Tan *et al.* demonstrated the ability to recover and recycle the ionic liquid. It should be noted that although they utilized an industrial-grade anion source, the recovery process for the recycled ionic liquid required significant time, solvent, and energy and bears further attention. Lee *et al.* provided contemporaneous evidence to Sun *et al.* that [$c_2$mim][OAc] was superior to the prevailing ionic-liquid candidates relative to biomass dissolution and cellulose recovery.[100] Additionally, they also demonstrated a decrease in the efficacy of their ionic liquid during recycle, attributing these results to retention and accumulation of lignin.

It should be noted here that the vast bulk of ionic liquid/lignocellulosic work has focused on an extremely limited portfolio of ionic liquids, mainly of the imidazolium family. The sheer number of potential ionic liquids that could still be evaluated is daunting to consider; however, a recent high-throughput screening technique has been developed by Zavrel *et al.*[101] While this is an encouraging development, the method requires a specific instrument that may limit its acceptance and adoption. This study, which tested a robust set of candidate ionic liquids (mostly of the imidazolium family group), provided further evidence that both [$c_2$mim][OAc] and [amim][Cl] were ideal candidates for further development.

A last addition to the subject of ionic-liquid-induced lignocellulosic dissolution is contained in a study on the preparation of lignocellulosic aerogels. As for most previous studies, the work of Aaltonen and Juhiainen utilized only one ionic liquid, [$c_4$mim][Cl], and tested a range of lignocellulosic characteristic components and one softwood, spruce.[102] While the aerogels were successfully formed, they proved to be structurally susceptible to mechanical degradation, and, when lignin was present, brown in color; however, the premise of developing aerogel materials from dissolved lignocellulosic biomass was validated.

The second application of ionic liquids for the utilization of lignocellulosic biomass is the dissolution and *in situ* reaction of the biomass without extraction from the ionic liquid. The focus of this application is twofold: (1) dissolution of the lignocellulosic biomass and (2) the conversion of the liberated polysaccharides into desirable feedstock chemicals. Nearly all the work falling into this category is focused on the *in situ* production of 5-hydroxymethylfurfural (HMF) and/or furfural. HMF has been identified as one of the top candidate platform chemicals derivable from biomass that can be converted into a range of useful industrially relevant chemicals and the desirable fuel alternative 2-dimethylfuran (DMF).[103,104] A simplified reaction scheme can be seen in Figure 12.11.

A primary limitation inherent to the vast bulk of the work on deriving HMF from biological sources is the reliance on edible-food-grade compounds, such as glucose and fructose.[105] There have been several recent publications that seek to remedy this "food *versus* fuel" conundrum through the use of ionic liquids to overcome the recalcitrant nature of lignocellulosic biomass. The first article to direct its attention to this problem was that of Li *et al.*[106] They dissolved a variety of biomass samples in a range of ionic liquids, again only of the imidazolium family, and introduced several known industrial acids to determine the impacts on the hydrolysis of the biomass. They did not seek to process the biomass through to the generation of HMF or furfural in any significant amounts, instead focusing on the evolution of the polysaccharides from cellulose and hemicellulose by means of the analysis of total reducing sugars (TRS). The results reported were very encouraging with the best-performing system, [$c_4$mim][Cl] + hydrochloric acid (HCl), exhibiting a TRS evolution comparable or better than that reported for traditional enzymatic or acid hydrolysis processes. An additional revelation was that the use of sulfuric acid was detrimentally impacted by the presence of the lignin fraction, which required the use of higher sulfuric-acid loadings to achieve the expected performance. A subsequent study by Vanoye *et al.* focused on developing a kinetic understanding of the hydrolysis of cellulose and the subsequent decomposition of glucose in [$c_2$mim][Cl],[107] the extent of which is a concern for the overall efficiency of acid hydrolysis. While the bulk of this study focused on cellobiose as a proxy for cellulose, it concluded with a study of miscanthus grass hydrolysis. The results were surprising, as the optimal conditions for the production of glucose determined using the cellobiose model were not the same as for the untreated biomass. They proposed that this was due to the inefficient dissolution of the unrefined biomass, a fact alluded to by other researchers,[28] and the consequential heterogeneous conditions of the ensuing reactions. While the aforementioned studies have focused on the preliminary steps to the production of HMF, only a few published studies have reported work on the complete (or nearly complete) reaction pathway outlined in Figure 12.12.[105,108,109] The work of Seivers *et al.* focused on the conversion of a well characterized loblolly pine biomass by acid hydrolysis with trifluoroacetic acid (TFA) in the proven cellulose solvent [$c_4$mim][Cl].[108] They verified that the reaction would proceed through to the near-complete conversion of the carbohydrate components into water-soluble compounds, and that the performance of the ionic liquid system was superior in many ways to the more traditional aqueous-based systems. The inhibition of the conversion by lignin

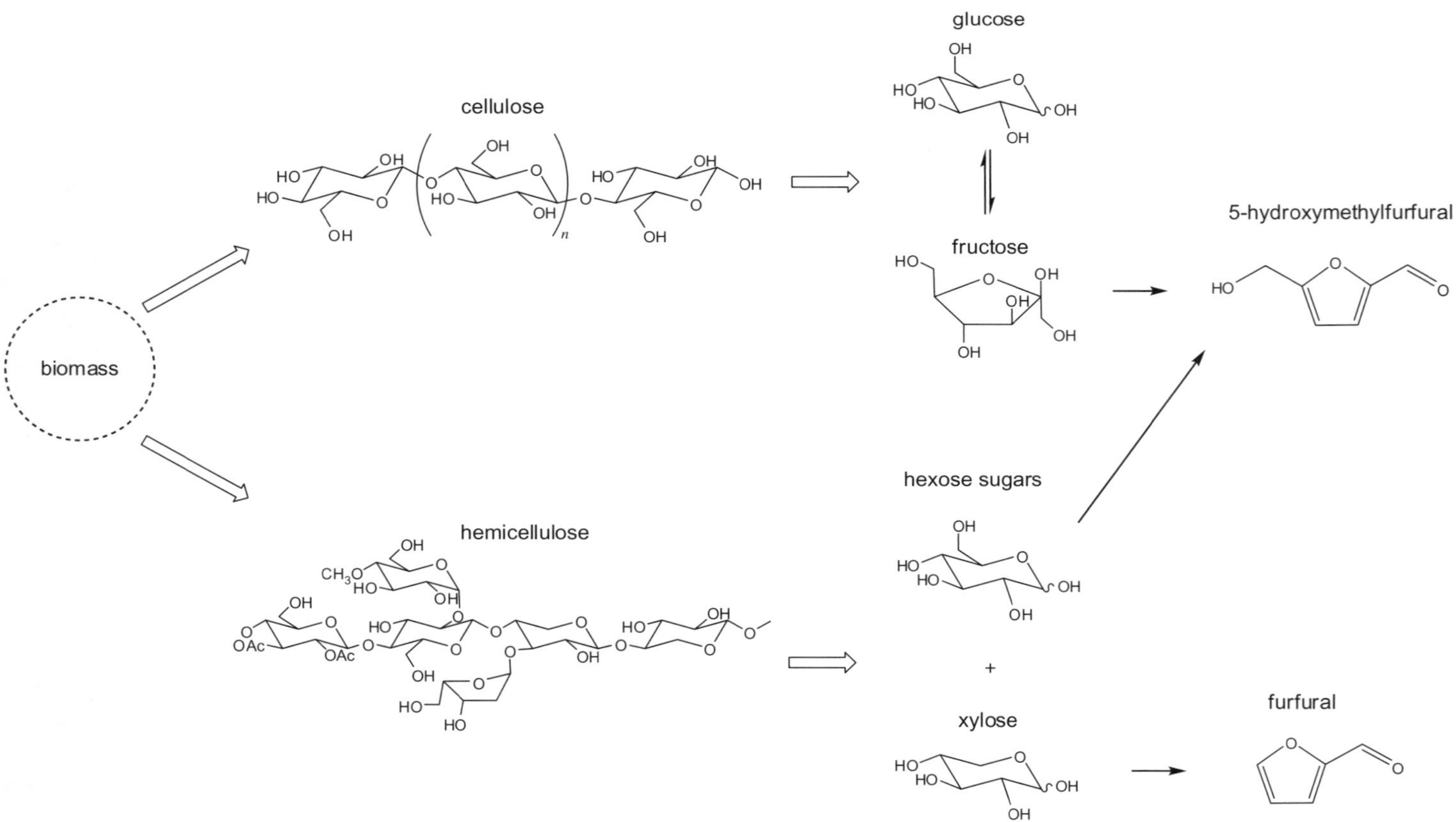

Figure 12.11 Production of 5-hydroxymethylfurfural and furfural from cellulose and hemicellulosic sugars.

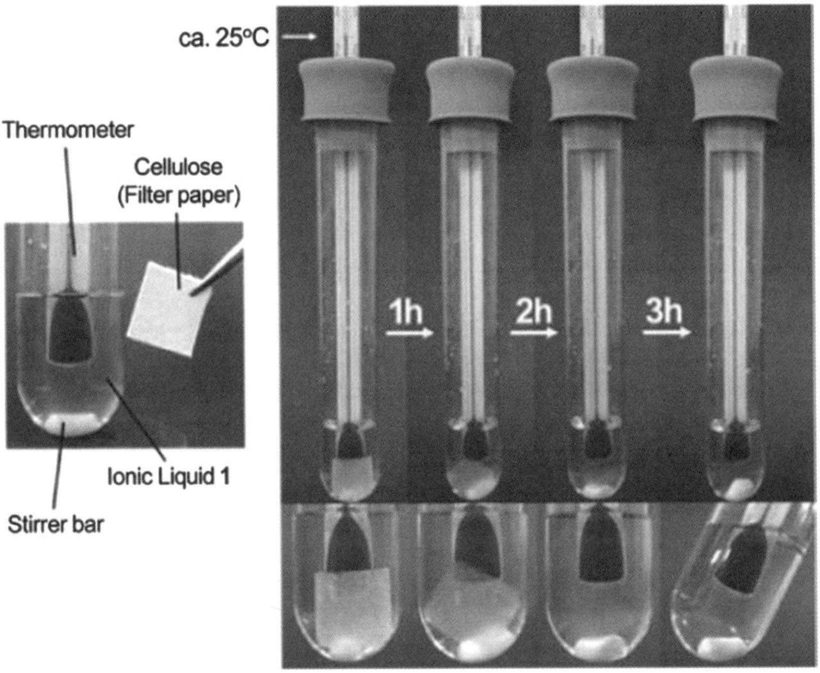

Figure 12.12 Dissolution of cellulose in an ionic liquid at 25 °C. [Reproduced by permission of The Royal Society of Chemistry.][126]

was not observed as would be expected with TFA as the acid catalyst. An interesting observation was made that while the hexose sugars reacted as expected to form HMF, the pentose sugars liberated from hemicellulose did not yield the expected quantities of furfural. This was attributed to the possibility of significantly different relative reactivities and they suggest that future work should focus on this, as well as improving the selectivities for various products. The remaining two studies of merit are those of Binder and Raines[105] and Zhang and Zhou.[109] We grouped these two reports mostly because they employed a chromium chloride ($CrCl_2$) acid catalyst that brings with it a whole host of environmental and toxicological issues. Binder and Raines did not focus their work on the use of ionic liquids as the primary solvent, but instead employed the ionic liquid, [$c_2$mim][Cl], as a reaction additive. It should be noted that in some systems the mass fraction of ionic liquid *versus* all other components was significant, with peak performance realized at ionic liquid weight fractions of 60%, calling into question the description of the ionic liquid as an additive. The results were impressive, adding credibility to the use of ionic liquids in low-temperature acid hydrolysis of biomass. The authors acknowledge the issues related to the use of chromium-based catalysts and have begun work directed at reducing (or eliminating) their requirement. Overall, this study provides the best justification for the accelerated investigation of ionic liquids for the conversion of lignocellulosic biomass. Zhang and Zhou also employed a chromic salt, $CrCl_3$, as the catalyst; however, they used the ionic liquid [$c_4$mim][Cl] primarily as a solvent. The innovative aspect of this work is that the reactions were performed in a microwave-reaction environment and the biomass required no pretreatment. They reported dramatic improvements in the time required for maximal conversion, down from several hours in a 200 °C oil bath to less than 5 min in the microwave reactor. Similar results have been reported previously for the conversion of cellulose and glucose.[110] While the bulk of the work concerning the direct dissolution/reaction of lignocellulose has focused on the HMF/furfural conversion pathway, a pair of reports by Xie *et al.* has demonstrated the acetylation, benzoylation and carbanilation of Norway spruce and southern pine sawdust at moderate temperatures for potential use in wood-based thermoplastics.[30,111] The reaction products have thermal, chemical and mechanical properties that are very distinct from those of the original starting materials. Additionally, it was clear from their results that hydroxyl groups from cellulose, hemicelluloses, and lignin were functionalized; however, the precise distribution and extent in a given fraction was not able to be determined. With the exception of this last study, there is little indication of the products resulting from the nonsugar components, or of the likely sugar degradation products resulting from undesired reactions. A justification for this omission could be that the product solutions are too heterogeneous for standard analytical methods. This could be remedied, or at a minimum alleviated, by the third application of ionic liquids with lignocellulosics.

As indicated in many of the previous works, it is apparent that both lignocellulosic materials and their degradation/reaction products do not lend themselves to easy analytical evaluation. The very recalcitrant nature of the material in question results in the significant destruction of the finer polymeric structure during fractionation, a typical requirement for the solvents employed in standard analytical methods (GC, HPLC, NMR, *etc.*). The potential for ionic liquids to dissolve lignocellulosic material for processing also holds for their use as analytical solvents. In fact a number of groups have used traditional NMR[27,28,97] or FTIR[112] analysis to observe the changes in lignocellulosic structure during dissolution. Another recent article by Jiang *et al.* reports a significant improvement in the development and utilization of predeuterated pyridinium ionic liquids for NMR analysis of the dissolution of poplar and switchgrass biomass. These results are in line with nonlignocellulosic-centered work utilizing ionic liquids as either enabling or enhancing solvents for complex analytical measurements.[113]

The fourth application of ionic liquids in the utilization of lignocellulosic biomass concerns the incorporation or infusion of ionic liquids into post-processing consumer products. To date, ionic liquids have been proposed as wood preservatives, antielectrostatic agents, and fire retardants. These applications, *in toto*, rely on the hydrophobicity of certain ionic liquids, thus preventing the leaching of the ionic liquid from the wood product; however, long term studies have yet to be performed in adverse conditions to determine if these ionic liquids will truly remain in the wood product. For wood preservation where biological degradation is the target, the toxicity of the ionic liquid under consideration toward the target degradation organism must be determined.[114–117] For antielectrostatic applications the wood is either soaked or brushed with the ionic liquid, resulting in an effective treatment method.[118] Applications related to fire retardation rely on the known nonflammability of ionic liquids; however, the article that presents the bulk of the recent information on this subject is not currently accessible and there is only the statement in the brief review by Han *et al.* as evidence.[96]

The last application of ionic liquids for the utilization of whole lignocellulosic biomass is less related to the biomass itself than to the current pulping industry. It is a known and accepted fact of modern chemical processing that certain modes of processing release low concentration of volatile organic compounds (VOCs) and other hazardous air pollutants (HAPs). As air-quality standards grow increasingly more restrictive regarding the uncontrolled emission of VOCs and HAPs, effective yet economic capture strategies are needed to meet (and hopefully exceed) these requirements. It is well known that ionic liquids can act as extractants for the removal of volatilized compounds from gas streams, thus it is reasonable to assume that ionic liquids would have potential in controlling the emission of VOCs and HAPS from the wood-processing industry. A series of articles by Wang *et al.* and Milota *et al.* detail the development of an ionic-liquid-based gas scrubber for specific removal of volatilized methanol and formaldehyde.[119–121] They demonstrated that the ionic liquid could be recycled and reused effectively to capture and concentrate the targeted compounds either for future combustion or production as a saleable byproduct. While this application is not directly tied to the production of fuels and chemicals from lignocellulosics, it could have a positive impact on the overall economic and environmental performance of future biorefineries.

12.3.2 Cellulose

Cellulose is by far the most investigated and utilized fraction of the lignocellulosic energy portfolio. This is due to the fact that cellulose is essentially a linear polymer of glucose molecules that when released from cellulose become available for biological conversion to various fungible biofuels. While this can sound simple in concept, cellulose is difficult to depolymerize due in large part to its general insolubility in natural solvents (such as water) or those common to industrial processes (ethanol, methanol, acetone). The solvent systems that are currently used on the industrial scale have a range of environmental and process-related issues, and, while effective, are generally not predicted to be competitive for use in future lignocellulosic biorefineries.

Cellulose was the first of the lignocellulosic fractions to be the focus of ionic-liquid research studies. The modern era of these studies can be traced to the seminal work of Swatloski *et al.* in 2002 and their initial foray into the dissolution of cellulose with a range of ionic liquids and processing conditions.[29] They reported the ability of [$c_4$mim][Cl] to dissolve up to 10 wt% cellulose after heating at 100 °C. Since this initial study, there has been an increasing interest in the dissolution, pretreatment, enzymatic processing, chemical conversion, and recovery of cellulose and cellulose-derived products using ionic liquids, which is evident in the inset of Figure 12.5.

A number of review articles have recently been published outlining the growth and current state of cellulose/ionic-liquid research.[122–125] In light of the existence of these reviews, we have elected to focus our chapter more toward the noncellulose research pertaining to ionic liquids. Before we move on to the hemicellulose and lignin fractions, a few articles merit discussion. These articles can be fairly consistently divided into sets focusing on: (1) the dissolution of cellulose in ionic liquids, (2) the chemical conversion of cellulose in ionic liquids, and (3) the biochemical reaction of the reconstituted cellulose/glucose molecules.

Addressing the issue of cellulose solubility in ionic liquids has been a central focus to most of the work present in the literature. Recently, Fukaya *et al.* published a study of the dissolution of cellulose in a set of novel polar ionic liquids.[126] This work presents two interesting findings. The first is that ionic liquids containing phosphorus oxide anions, of the general type [$c_2$mim][(MeO)(R)PO_2], were observed to have low viscosity, good thermal stabilities, and the ability to dissolve up to 10 wt% cellulose at temperatures lower than 70 °C. Figure 12.12 provides one of the best visual demonstrations of cellulose solubility in an ionic liquid in the open scientific literature. The work of Mazza *et al.* is important in that it stretches beyond the common imidazolium chloride ionic liquids typical of most ionic liquid/cellulose work.[127] In their study they sought to incorporate the effects of water on the dissolution of cellulose in the target ionic liquids. Not surprisingly, [$c_4$mim][Cl] was one of the only two ionic liquids able to truly solubilize the wet cellulose. The second ionic liquid, 1,3-dimethylimidazolium dimethylphosphate, while able to dissolve the cellulose, was not nearly as effective and suffered from significant viscosity issues. The work of Vitz *et al.* focused on studying dissolution of cellulose in a wide variety of imidazolium family ionic liquids.[128] They report the identification of a new cellulose dissolving ionic liquid, [$c_2$mim][Et_2PO_4], with little degradation of the polymer structure.

It is well understood that the ability of an ionic liquid to dissolve cellulose opens up the potential for the further chemical processing of the material. As indicated previously, the direct conversion of the glucose released from the cellulose to HMF or similar furan products is a potential high value target. Yu *et al.* present a one-step ionic-liquid enabled, $CrCl_2/CuCl_2$ catalyzed, chemical conversion of cellulose to HMF.[129] They report success and high levels of conversion; however, they leave the elucidation of the catalytic mechanism to a future study. On a different thread, the work of Rinaldi *et al.* demonstrated the catalytic depolymerization of cellulose in ionic liquids.[130] This is notable in that if the depolymerization reactions can be better controlled, improvements to the subsequent acid hydrolysis can be realized. Alternatively, if oligomers rather than sugars can be produced, new routes to novel chemical feedstocks could be developed.

If the ionic liquid is used to dissolve and recrystallize the cellulose as a pretreatment step leading to enzymatic conversion of the modified cellulose, significant improvements in reaction rates and product yields might be realized. Kamiya *et al.* demonstrated the interesting phenomenon that a 1:4 ionic liquid:water ratio had significant positive effects on the enzymatic saccharification of cellulose to glucose and cellobiose.[131] The significance of this result is the potential it presents for the elimination of the reconstitution step in the traditional enzymatic process.

12.3.3 Hemicellulose

As mentioned previously, hemicellulose is the "catch-all" grouping for the noncellulosic polysaccharides and is known to account for 20–30% of the lignocellulosic biomass of woody plants. The noncellulosic polysaccharides common to hemicellulose, shown in Figure 12.13, can be grouped into essentially three main categories: (1) pentoses or 5-carbon sugars (xylose, arabinose), (2) hexoses or 6-carbon sugars (mannose, glucose, galactose), and (3) sugar acids.[21,132] These sugars are incorporated into polymeric structures, xylans for

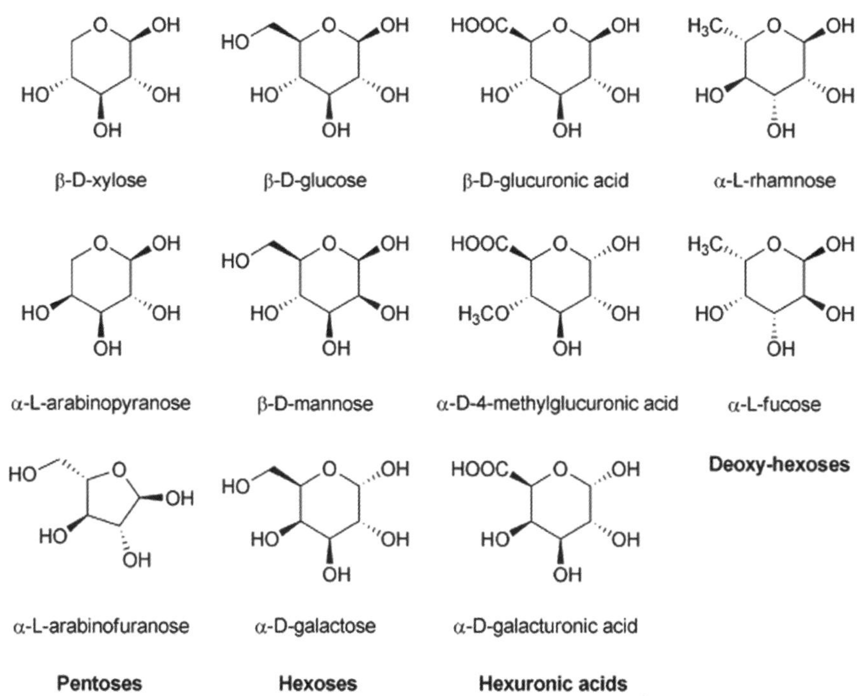

Figure 12.13 Constituent sugars present in hemicellulosic material. [Reproduced by permission of The Royal Society of Chemistry.][21]

the pentose sugars and glucomannans for the hexose sugars. The relative abundances of the hemicellulosic polymeric structures vary widely across the spectrum of available lignocellulosic biomass, with hardwoods predominantly containing xylans and softwoods primarily glucomannans.[132] A number of factors are at play in the general lack of effective use of the hemicellulosic sugars, namely, the issue of biomass recalcitrance, the elevated reactivity and degradation during hydrolysis, and the general inability of most pentose-fermenting organisms to tolerate economically viable levels of ethanol. Thus, an opportunity for ionic liquids is readily apparent in the greater utilization of the hemicellulosic fraction of lignocellulosic biomass.

Unlike the large body of work available on the ionic-liquid-enhanced processing of cellulose, hemicellulose has seen relatively little interest in the broader literature. This interest takes two distinct forms: (1) dissolution of the hemicellulose fraction and (2) the reactive processing of the solubilized hemicellulose by traditional chemical methods. It has been noted that the selective extraction of the hemicellulose sugars before either the enzymatic or catalytic conversion of the cellulose fraction would have significant practical and economic benefits.[133]

There are a few general references to the retention of the hemicellulosic fraction during ionic-liquid dissolution of unrefined biomass, but they are mostly made in passing and are not the primary focus of the research. One of the articles concerning the dissolution of lignocellulosic biomass does discuss the dissolution of the hemicellulose as well as the relative order of dissolution of nonglucose sugars to glucose.[98] Unfortunately this article is of a general nature and does not provide any significant insight into the fate of the individual hemicellulosic sugars themselves. Another study, by Fukaya *et al.*, explored the solubility of several polysaccharides in a range of imidazolium formate ionic liquids.[134] This work demonstrated that the formate ionic liquids were superior to one of the favored cellulose dissolving ionic liquids, [amim][Cl], and also discussed the effects of temperature. Fukaya *et al.* were able to achieve a xylan solubility of nearly 20 wt% at a temperature of roughly 90 °C, which was comparable to the observed solubility of cellulose. Unfortunately, it is not clear that the ionic liquids tested would exhibit selectivity for cellulose over xylan as mixed-sugar systems were not evaluated. A third group, Wu *et al.*, focused their attention on the ternary equilibrium in an ionic liquid, [$c_4$mim][BF_4], water, sugar (xylose, glucose, fructose, or sucrose) system.[135] The result was the determination that the order of solubility was sucrose$<$glucose$<$xylose$<$fructose. Additionally, they proposed and tested an ionic-liquid-recycling scheme utilizing ethyl acetate; however, the reported recovery percentages for the ionic liquid were dismal.

Much as with the dissolution of hemicellulose, there are few direct and specific references to targeted reactions in ionic liquid solvents. These studies fall into two categories, production of furans and the reactive production of functionalized hemicellulosic sugars. Sievers *et al.* focused on the acid-catalyzed conversion of sugars to HMF and furfural, as well as the acid-catalyzed hydrolysis of pine wood.[108,136] Both studies utilized the ionic liquid [$c_4$mim][Cl], but employed different acids, sulfuric acid and trifluoroacetic acid, respectively.

The work of Lima *et al.* also focused on the acid-catalyzed generation of furan compounds. This study was primarily different in that it analyzed two ionic liquids, [$c_4$mim][Cl] and [$c_2$mim][HSO_4], where the second functioned as both solvent and acid catalyst.[137] All three studies demonstrated that the conversions of hexoses and pentoses were possible; however, significant hurdles remain relative to the selectivity of the reaction system and the performance and recovery of the ionic liquids. The final study, by Ren *et al.*, was focused on the acetylation of wheat-straw hemicellulose in the ionic liquid, [$c_4$mim][Cl], with iodine acting as the catalyst.[138] As with many of the previous studies on lignocellulosic acetylation the results were satisfactory with high degrees of hydroxyl group replacement with reasonable reaction times and temperatures.

12.3.4 Lignin

Lignin is an amorphous polyphenolmeric resin-like material and is the primary agent responsible for the recalcitrance of woody biomass. Lignin's amorphous nature makes it difficult to characterize in simple terms or to develop a precise understanding of lignin specific chemical reactions. Traditionally, this has been alleviated by the use of model compounds that represent the predominant bonds and structures present in the lignin matrix. These linkages can be seen in Figure 12.14, while the relative general distribution of each in hardwoods and softwoods is presented in Table 12.2.[139]

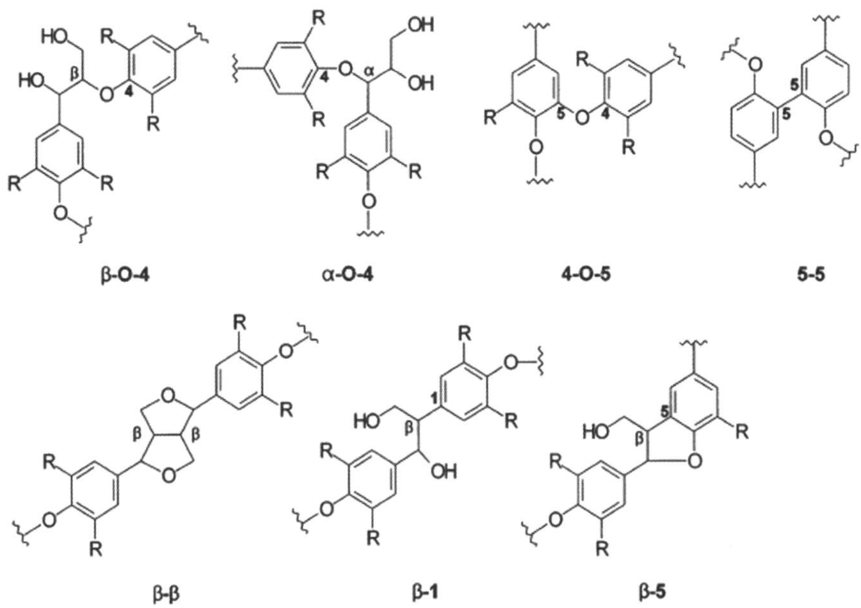

Figure 12.14 Structural linkages present in lignin. [Reprinted with permission from Elsevier.][139]

Table 12.2 Distribution (in %) of structural linkages present in lignin. Reprinted with permission from Elsevier.[139]

Type of Linkage		*Softwood Lignin*	*Hardwood Lignin*
Arylglycerol-β-aryl ether	β-O-4	46	60
Phenylcoumaran	β-5	11	6
Noncyclic benzyl aryl ether	α-O-4	7	7
Biphenyl	5-5	10	5
Diaryl ether	4-O-5	4	7
1,2-Diarylpropane	β-1	7	7
Resinol	β-β	2	3
Others		13	5

A database of identified lignin model compounds has been developed and maintained by the United States Department of Agriculture.[140] This database can be used to identify specific model compounds for use in reactive studies focused on the cleavage of the lignin matrix or the development of catalytic or enzymatic approaches to lignin modification.

In much the same manner as hemicellulose, the interaction of ionic liquids and the lignin fraction of lignocellulosic biomass has seen a very limited exposure in the scientific literature when compared to cellulose.

Unlike the simpler sugar components of the cellulose and hemicellulose fraction, the complex chemical structures that comprise lignin do not lend themselves to easily accessible biofuels or feedstock chemicals. This is in part due to the unique and primarily random generation of the lignin matrix, the variation of lignin structure with plant type, and the difficulty in extracting the matrix while retaining component structural integrity. A recent report by the Office of the Biomass Program of the United States Department of Energy provides a highly detailed review of the challenges and potential for deriving platform chemicals from lignin.[141] In the context of the immediate generation of biorefineries, lignin has little purpose other than to provide a fuel source for the generation of refinery heat. This is a shame, as lignin is essentially the sole renewable source of aromatic compounds that, if they could be selectively removed from the lignin matrix, would provide significant value to future generation biorefineries.[16]

The bulk of ionic-liquid/lignin-specific publications have only been published in the last few years, the earliest in 2007. These reports cover much the same areas as previously indicated for lignocellulosics, (1) dissolution in ionic liquids, (2) reactions in ionic liquids, and (3) analysis of lignin using ionic liquids. Most of the studies that discuss the dissolution of lignin in ionic liquids do so for the general purpose of delignifying the biomass, focusing on the cellulose for further conversion and doing little with the lignin residue. The work of Miyafuji *et al.*,[98] Sievers *et al.*,[108] and Sun *et al.*[97] are in this vein and have been discussed earlier. While the work of Lee *et al.* also deals with the displacement of lignin from cellulose-containing solutions, there are two major conclusions that merit discussion.[100] The first conclusion of note is that the ionic liquid [$c_2$mim][OAc] was extremely effective at extracting both cellulose and lignin from maple wood flour and was easily recovered after precipitation of the dissolved lignin using water. The second conclusion was that the degree of cellulose hydrolysis did not require complete extraction of the lignin, in fact more than 90% cellulose hydrolysis was achieved with just a little more than 40% of the lignin extracted. Unfortunately, no indication that the lignin was improved or available for further chemical treatment was given. The remaining lignin-centric studies focused on the dissolution and retention of the lignin fraction specifically. The work of Tan *et al.* has been discussed previously; however, it is important to note that they were able to extract lignin from lignocellulosic biomass and their ionic liquid, [$c_2$mim][ABS], was recyclable, although their process for doing so did not appear energetically efficient.[99] The last study was in fact the first published analysis of the use of ionic liquids specifically for the purpose of dissolving lignin. Pu *et al.* demonstrated, for the set of ionic liquids they tested, that the methyl sulfate anion containing ionic liquids were the best solvents for lignin and that increased chain length did not have a dramatically negative effect.[142] This study did not provide much additional insight other than an interesting approach to the assignment of chemical shifts in ^{13}C NMR measurements.

The articles focusing on reactions of lignin in ionic liquids can be further subdivided into (1) work employing lignin model compounds that represent the linkages present and (2) the nonspecific functionalization of the lignin complex without extraction. As with the studies of cellulose conversion, lignin reactions can be subdivided into chemical and biochemical. The first work to specifically evaluate the conversion of a lignin model compound in an ionic liquid was performed by Kumar *et al.*[26] Their work focused on the oxidation of veratryl alcohol in the ionic liquid, [$c_4$mim][PF_6], using both a metalloporphyrin catalyst and the horseradish peroxidase enzyme. They observed that the highest yields occurred in the ionic liquid. Additionally, the ionic liquid was recycled at least five times, although, there was a slight decrease in the overall yield during these recycles. There is no significant discussion of the reason for this decrease, but the authors do indicate that in the typical aqueous system, the enzyme is destroyed after the second cycle, whereas it survived through all five recycles in the ionic liquid. Another study focused on lignin model compounds was that of Binder *et al.* in which eugenol, 4-ethyguaiacol, and 2-phenylethyl phenyl ester were converted using a large range of acid catalysts in a wide array of ionic liquids.[143] While the conversions of eugenol proceeded as expected, 4-ethylguaiacol and to some degree 2-phenylethyl phenyl ester resisted conversion. Following the determination of an optimal set of conditions from the eugenol experiments, organosolv lignin was tested in the ionic liquid [$c_2$mim][OTf]. While it was apparent that the organosolv lignin was soluble in the ionic liquid, none of the monomeric products were observed, leading to the conclusion that either the alkyl linkages resisted conversion or the lignin resisted depolymerization by a self-condensation reaction. Kubo *et al.* directed their attention to the dehydration of a β–O–4 model compound, guaiacylglycerol-*β*-guaiacyl (GG), in the three ionic liquids predominantly studied for lignocellulosic dissolution.[144] The most interesting result was that a traditionally unstable enol-ether intermediate was formed as the primary stable product in all the ionic liquids studied. The primary result of this work was the suggestion that ionic liquids might be suitable solvents for the production of certain lignin model compounds. Finally the work of *Xie et al.*, discussed earlier, reported the acylation and carbanilation of lignin; however, the complex nature of the lignocellulosic material studied did not allow for specific analysis of the type and degree of lignin functionalization.[30]

12.4 Conclusions (Issues that Remain to be Resolved)

Ionic liquids show great potential for the improved processing of lignocellulosic biomass into fuels and/or platform chemical feedstocks. While these compounds have a lengthy history as proposed alternatives to various traditional chemical solvents, little of this promise has been realized beyond the laboratory. Before this transition can occur there are a number of issues that remain to be resolved:

Ionic Liquids

- There are a multitude of synthetic pathways that can be used for the preparation of laboratory-scale quantities of a given ionic liquid. Many of these methods are not efficient with respect to the energetic and material costs of production and consequently are anything but green. It is therefore imperative that new industrial-scale ionic-liquid production capabilities be developed that can produce lower-cost ionic liquids with minimal reagent compound contamination and reduced environmental impacts.
- It is nearly guaranteed that chemical process leaks will occur over the operational lifespan of a given process. If ionic liquids are to be adopted into the future lignocellulosic biomass-based industries, the evaluation of toxicity should be made a priority together with the potential for biodegradation as well as remediation strategies should a process leak result in environmental contamination.
- Many of the physical and chemical properties of ionic liquids that are attractive for their process related uses are significant challenges when recycle, purification, and reuse are considered. Most of the recycle and recovery strategies employed in the literature involve significant quantities of very volatile solvents as well as long timeframes in a vacuum environment. Consequently, lower impact methods for the effective reuse of ionic liquids must be developed before adoption into the larger chemical enterprise.
- Ionic liquids are often described as being thermally stable solvents. While this is essentially true, there are indications that long-term exposure to temperatures well below the known thermal decomposition temperature can, nevertheless, still result in degradation of the ionic

liquid and the generation of undesirable byproducts. Additionally, the impacts of realistic processing equipment on the long-term performance and stability of ionic liquids have not been studied effectively to date and merit significant future evaluation.

Lignocellulosics

- While a number of research groups have discussed the processing of unrefined biomass, there remain many questions related to the recovery of the biomass components, the analysis and description of the resulting ionic liquid/lignocellulosic systems, and the characterization of products resulting from direct or indirect chemical conversion of the biomass.
- Many of the discussions concerning the solubility of lignocellulosic biomass in ionic liquids return to a common portfolio of ionic liquids, mainly [$c_4$mim][Cl] and [$c_2$mim][OAc]. The sheer number of possible ionic liquids implies that other possible and potentially superior biomass-processing solvents must exist, and efforts should be made to expand future research away from the comfortable favorites of the past.
- The body of cellulose-related research is significantly larger than that of hemicellulose and lignin. As these two fractions typically comprise more than 50% of lignocellulosic biomass, there needs to be a greater share of the attention directed toward developing task-specific ionic-liquid systems. Such attention could focus on selective extractions of subunits of these fractions or the selective conversion of the pentose sugars without the requirement of prior separation.
- Regardless of the short-term success of the lignocellulosic chemical platform, the need to develop technologies and markets to consume these platform chemicals must be developed in tandem with the improvements in the efficiency and performance of next-generation biorefineries.

As stated earlier, these are the early days of ionic liquid research into the utilization of lignocellulosic biomass and significant developments are probable. It is likely that ionic liquids will have a place in the modern chemical industry, although the details and sector of first adoption remain to be identified. However, their considerable potential for the improved utilization of lignocellulosics could be a herald of things to come.

References

1. L. Ponti and A. P. Gutierrez, *Bull. Sci. Technol. Soc.*, 2009, **29**, 493.
2. G. Resch, A. Held, T. Faber, C. Panzer, F. Toro and R. Haas, *Energy Policy*, 2008, **36**, 4048.
3. A. J. Ragauskas, C. K. Williams, B. H. Davison, G. Britovsek, J. Cairney, C. A. Eckert, W. J. Frederick, J. P. Hallett, D. J. Leak, C. L. Liotta, J. R. Mielenz, R. Murphy, R. Templer and T. Tschaplinski, *Science*, 2006, **311**, 484.

4. L. Petrus and M. A. Noordermeer, *Green Chem.*, 2006, **8**, 861.
5. S. Pacala and R. Socolow, *Science*, 2004, **305**, 968.
6. M. I. Hoffert, K. Caldeira, G. Benford, D. R. Criswell, C. Green, H. Herzog, A. K. Jain, H. S. Kheshgi, K. S. Lackner, J. S. Lewis, H. D. Lightfoot, W. Manheimer, J. C. Mankins, M. E. Mauel, L. J. Perkins, M. E. Schlesinger, T. Volk and T. M. L. Wigley, *Science*, 2002, **298**, 981.
7. G. Berndes, C. Azar, T. Kaberger and D. Abrahamson, *Biomass Bioenergy*, 2001, **20**, 371.
8. M. E. Himmel, S. Y. Ding, D. K. Johnson, W. S. Adney, M. R. Nimlos, J. W. Brady and T. D. Foust, *Science*, 2007, **315**, 804.
9. Office of the Biomass Program-US Department of Energy, Biomass Multi-Year Program Plan, July 2009.
10. Energy Information Administration (EIA), Annual Energy Outlook 2009 with Projections to 2030, DOE/EIA-0383(2009), 2009.
11. E. Doris, J. McLaren, V. Healey and S. Hockett, State of the States 2009: Renewable Energy Development and the Role of Policy, NREL/TP-6A2-46667, 2009.
12. Biofuels Research Advisory Council, Biofuels in the European Union-A vision for 2030 and beyond, EUR 22066, **2006**.
13. D. N. S. Hon and N. Shiraishi, *Wood and cellulosic chemistry,* Marcel Dekker, New York, 2001.
14. R. E. Kirk and D. F. Othmer, *Kirk-Othmer Encyclopedia of Chemical Technology*, John Wiley & Sons: Wiley Interscience (Online Service), 2001.
15. G. W. Huber, S. Iborra and A. Corma, *Chem. Rev.*, 2006, **106**, 4044.
16. Y. H. P. Zhang, *J. Ind. Microbiol. Biotechnol.*, 2008, **35**, 367.
17. F. S. Chakar and A. J. Ragauskas, *Ind. Crop and Prod.*, 2004, **20**, 131.
18. D. S. Argyropoulos, *Oxidative delignification chemistry: fundamentals and catalysis,* American Chemical Society, Washington, DC, 2001.
19. A. R. Gaspar, J. A. F. Gamelas, D. V. Evtuguin and C. P. Neto, *Green Chem.*, 2007, **9**, 717.
20. B. N. Kuznetsov, S. A. Kuznetsova, V. G. Danilov, I. A. Kozlov, V. E. Taraban'ko, N. M. Ivanchenko and N. B. Alexandrova, *Catal. Today*, 2002, **75**, 211.
21. R. Rinaldi and F. Schuth, *Energy Environ. Sci.*, 2009, **2**, 610.
22. K. Ruuttunen and T. Vuorinen, *Ind. Eng. Chem. Res.*, 2005, **44**, 4284.
23. A. T. Martinez, F. J. Ruiz-Duenas, M. J. Martinez, J. C. del Rio and A. Gutierrez, *Curr. Opin. Biotechnol.*, 2009, **20**, 348.
24. C. Crestini, A. Pastorini and P. Tagliatesta, *J. Mol. Catal. A: Chem.*, 2004, **208**, 195.
25. C. Crestini, R. Saladino, P. Tagliatesta and T. Boschi, *Bioorganic Med. Chem.*, 1999, **7**, 1897.
26. A. Kumar, N. Jain and S. M. S. Chauhan, *Synlett*, 2007, 411.
27. D. A. Fort, R. C. Remsing, R. P. Swatloski, P. Moyna, G. Moyna and R. D. Rogers, *Green Chem.*, 2007, **9**, 63.
28. I. Kilpelainen, H. Xie, A. King, M. Granstrom, S. Heikkinen and D. S. Argyropoulos, *J. Agri. Food Chem.*, 2007, **55**, 9142.

29. R. P. Swatloski, S. K. Spear, J. D. Holbrey and R. D. Rogers, *J. Am. Chem. Soc.*, 2002, **124**, 4974.
30. H. Xie, A. King, I. Kilpelainen, M. Granstrom and D. S. Argyropoulos, *Biomacromolecules*, 2007, **8**, 3740.
31. H. L. Xie and T. J. Shi, *Holzforschung*, 2006, **60**, 509.
32. S. D. Zhu, *J. Chem. Technol. Biotechnol.*, 2008, **83**, 777.
33. N. V. Plechkova and K. R. Seddon, *Chem. Soc. Rev.*, 2008, **37**, 123.
34. S. G. Cull, J. D. Holbrey, V. Vargas-Mora, K. R. Seddon and G. J. Lye, *Biotechnol. Bioeng.*, 2000, **69**, 227.
35. J. H. Davis and P. A. Fox, *Chem. Commun.*, 2003, 1209.
36. S. A. Forsyth, J. M. Pringle and D. R. MacFarlane, *Aust. J. Chem.*, 2004, **57**, 113.
37. K. J. Fraser and D. R. MacFarlane, *Aust. J. Chem.*, 2009, **62**, 309.
38. J. D. Holbrey and K. R. Seddon, *Clean Prod. Process.*, 1999, **1**, 223.
39. K. N. Marsh, J. A. Boxall and R. Lichtenthaler, *Fluid Phase Equil.*, 2004, **219**, 93.
40. K. N. Marsh, A. Deev, A. C. T. Wu, E. Tran and A. Klamt, *Korean J. Chem. Eng.*, 2002, **19**, 357.
41. R. A. Osteryoung, *J. Electrochem. Soc.*, 1987, **134**, C506.
42. P. Walden, *Bull. de l'Academie Imperiale des Sciences de St.-Petersbourg*, 1914, **8**, 405.
43. F. H. Hurley and T. P. Wier, *J. Electrochem. Soc.*, 1951, **98**, 207.
44. H. L. Chum, V. R. Koch, L. L. Miller and R. A. Osteryoung, *J. Am. Chem. Soc.*, 1975, **97**, 3264.
45. J. S. Wilkes, J. A. Levisky, R. A. Wilson and C. L. Hussey, *Inorg. Chem.*, 1982, **21**, 1263.
46. M. J. Earle, J. Esperanca, M. A. Gilea, J. N. C. Lopes, L. P. N. Rebelo, J. W. Magee, K. R. Seddon and J. A. Widegren, *Nature*, 2006, **439**, 831.
47. R. D. Rogers and K. R. Seddon, *Ionic liquids as green solvents: progress and prospects,* American Chemical Society, Washington, DC, 2003.
48. P. Wasserscheid and T. Welton, *Ionic liquids in synthesis,* Wiley-VCH, Weinheim, 2003.
49. P. Bonhote, A. P. Dias, N. Papageorgiou, K. Kalyanasundaram and M. Gratzel, *Inorg. Chem.*, 1996, **35**, 1168.
50. F. Atefi, M. T. Garcia, R. D. Singer and P. J. Scammells, *Green Chem.*, 2009, **11**, 1595.
51. C. J. Bradaric, A. Downard, C. Kennedy, A. J. Robertson and Y. H. Zhou, *Green Chem.*, 2003, **5**, 143.
52. R. E. Del Sesto, C. Corley, A. Robertson and J. S. Wilkes, *J. Organomet. Chem.*, 2005, **690**, 2536.
53. S. H. Zhao, E. L. Zhao, P. Shen, M. G. Zhao and J. Y. Sun, *Ultrason. Sonochem.*, 2008, **15**, 955.
54. M. Anouti, M. Caillon-Caravanier, Y. Dridi, H. Galiano and D. Lemordant, *J. Phys. Chem. B*, 2008, **112**, 13335.
55. J. D. Holbrey, W. M. Reichert, R. P. Swatloski, G. A. Broker, W. R. Pitner, K. R. Seddon and R. D. Rogers, *Green Chem.*, 2002, **4**, 407.

56. C. C. Cassol, G. Ebeling, B. Ferrera and J. Dupont, *Adv. Synth. Catal.*, 2006, **348**, 243.
57. I. Dinares, C. G. de Miguel, A. Ibanez, N. Mesquida and E. Alcalde, *Green Chem.*, 2009, **11**, 1507.
58. J. C. Schleicher and A. M. Scurto, *Green Chem.*, 2009, **11**, 694.
59. M. A. Gonzalez and J. T. Ciszewski, *Org. Process Res. Dev.*, 2009, **13**, 64.
60. G. Cravotto, L. Boffa, J. M. L'Eveque, J. Estager, M. Draye and W. Bonrath, *Aust. J. Chem.*, 2007, **60**, 946.
61. J. Karkkainen, J. Asikkala, R. S. Laitinen and M. K. Lajunen, *Zeit. Natur. B*, 2004, **59**, 763.
62. M. J. Earle, C. M. Gordon, N. V. Plechkova, K. R. Seddon and T. Welton, *Anal. Chem.*, 2007, **79**, 758.
63. B. R. Clare, P. M. Bayley, A. S. Best, M. Forsyth and D. R. MacFarlane, *Chem. Commun.*, 2008, 2689.
64. M. Smiglak, A. Metlen and R. D. Rogers, *Acc. Chem. Res.*, 2007, **40**, 1182.
65. H. Zhao, S. Q. Xia and P. S. Ma, *J. Chem. Technol. Biotechnol.*, 2005, **80**, 1089.
66. A. Berthod, M. Ruiz-Angel and S. Carda-Broch, *J. Chromatogr. A*, 2008, **1184**, 6.
67. X. Han and D. W. Armstrong, *Acc. Chem. Res.*, 2007, **40**, 1079.
68. N. Jain, A. Kumar, S. Chauhan and S. M. S. Chauhan, *Tetrahedron*, 2005, **61**, 1015.
69. T. Welton, *Coord. Chem. Rev.*, 2004, **248**, 2459.
70. J. S. Wilkes, *J. Mol. Catal. A: Chem.*, 2004, **214**, 11.
71. Z. Yang and W. B. Pan, *Enzyme Microb. Tech.*, 2005, **37**, 19.
72. J. H. Davis, *Chem. Lett.*, 2004, **33**, 1072.
73. S. G. Lee, *Chem. Commun.*, 2006, 1049.
74. P. J. Scammells, J. L. Scott and R. D. Singer, *Aust. J. Chem.*, 2005, **58**, 155.
75. M. Wagner, *Chim. Oggi-Chem. Today*, 2005, **23**, 32.
76. M. Wagner, *Chim. Oggi-Chem. Today*, 2007, **25**, 14.
77. R. P. Swatloski, J. D. Holbrey and R. D. Rogers, *Green Chem.*, 2003, **5**, 361.
78. R. E. Del Sesto, T. M. McCleskey, C. Macomber, K. C. Ott, A. T. Koppisch, G. A. Baker and A. K. Burrell, *Thermochim. Acta*, 2009, **491**, 118.
79. M. Kosmulski, J. Gustafsson and J. B. Rosenholm, *Thermochim. Acta*, 2004, **412**, 47.
80. C. W. Cho, Y. C. Jeon, T. P. T. Pham, K. Vijayaraghavan and Y. S. Yun, *Ecotox. Environ. Safe.*, 2008, **71**, 166.
81. K. J. Kulacki and G. A. Lamberti, *Green Chem.*, 2008, **10**, 104.
82. C. Pretti, C. Chiappe, I. Baldetti, S. Brunini, G. Monni and L. Intorre, *Ecotoxicology Environ. Safety*, 2009, **72**, 1170.
83. A. S. Wells and V. T. Coombe, *Organic Process Res. Dev.*, 2006, **10**, 794.

84. W. Mrozik, C. Jungnickel, T. Ciborowski, W. R. Pitner, J. Kumirska, Z. Kaczynski and P. Stepnowski, *J. Soils Sediments*, 2009, **9**, 237.
85. R. J. Cornmell, C. L. Winder, G. J. T. Tiddy, R. Goodacre and G. Stephens, *Green Chem.*, 2008, **10**, 836.
86. K. M. Docherty, S. Z. Hebbeler and C. F. Kulpa, *Green Chem.*, 2006, **8**, 560.
87. M. Rebros, H. Q. N. Gunaratne, J. Ferguson, K. R. Seddon and G. Stephens, *Green Chem.*, 2009, **11**, 402.
88. J. Ranke, S. Stolte, R. Stormann, J. Arning and B. Jastorff, *Chem. Rev.*, 2007, **107**, 2183.
89. M. A. Azimova, S. A. Morton and P. D. Frymier, *J. Environ. Eng.-ASCE*, 2009, **135**, 1388.
90. K. M. Docherty, J. K. Dixon and C. F. Kulpa, *Biodegradation*, 2007, **18**, 481.
91. A. Esquivel-Viveros, F. Ponce-Vargas, P. Esponda-Aguilar, L. A. Prado-Barragan, M. Gutierrez-Rojas, G. J. Lye and S. Huerta-Ochoa, *Rev. Mex. Ing. Quim.*, 2009, **8**, 163.
92. N. Gathergood, P. J. Scammells and M. T. Garcia, *Green Chem.*, 2006, **8**, 156.
93. J. R. Harjani, R. D. Singer, M. T. Garciac and P. J. Scammells, *Green Chem.*, 2009, **11**, 83.
94. S. Morrissey, B. Pegot, D. Coleman, M. T. Garcia, D. Ferguson, B. Quilty and N. Gathergood, *Green Chem.*, 2009, **11**, 475.
95. *US Patent*, 1943176, 1934.
96. S. Q. Han, J. L. Li, S. D. Zhu, R. Chen, Y. X. Wu, X. Y. Zhang and Z. N. Yu, *Bioresources*, 2009, **4**, 825.
97. N. Sun, M. Rahman, Y. Qin, M. L. Maxim, H. Rodriguez and R. D. Rogers, *Green Chem.*, 2009, **11**, 646.
98. H. Miyafuji, K. Miyata, S. Saka, F. Ueda and M. Mori, *J. Wood Sci.*, 2009, **55**, 215.
99. S. S. Y. Tan, D. R. MacFarlane, J. Upfal, L. A. Edye, W. O. S. Doherty, A. F. Patti, J. M. Pringle and J. L. Scott, *Green Chem.*, 2009, **11**, 339.
100. S. H. Lee, T. V. Doherty, R. J. Linhardt and J. S. Dordick, *Biotechnol. Bioeng.*, 2009, **102**, 1368.
101. M. Zavrel, D. Bross, M. Funke, J. Buchs and A. C. Spiess, *Bioresource Technol.*, 2009, **100**, 2580.
102. O. Aaltonen and O. Jauhiainen, *Carbohyd. Polym.*, 2009, **75**, 125.
103. T. A. Werpy, J. E. Holladay and J. F. White, "Top Value Added Chemicals From Biomass: I. Results of Screening for Potential Candidates from Sugars and Synthesis Gas", PNNL-14808 Office of the Biomass Program, United States Department of Energy, 2004.
104. J. N. Chheda, G. W. Huber and J. A. Dumesic, *Angew. Chem. Int. ed.*, 2007, **46**, 7164.
105. J. B. Binder and R. T. Raines, *J. Am. Chem. Soc.*, 2009, **131**, 1979.
106. C. Z. Li, Q. Wang and Z. K. Zhao, *Green Chem.*, 2008, **10**, 177.

107. L. Vanoye, M. Fanselow, J. D. Holbrey, M. P. Atkins and K. R. Seddon, *Green Chem.*, 2009, **11**, 390.
108. C. Sievers, M. B. Valenzuela-Olarte, T. Marzialetti, D. Musin, P. K. Agrawal and C. W. Jones, *Ind. Eng. Chem. Res.*, 2009, **48**, 1277.
109. Z. H. Zhang and Z. B. K. Zhao, *Bioresource Technol.*, 2010, **101**, 1111.
110. C. Z. Li, Z. H. Zhang and Z. B. K. Zhao, *Tetrahedron Lett.*, 2009, **50**, 5403.
111. H. B. Xie, P. Jarvi, M. Karesoja, A. King, I. Kilpelainen and D. S. Argyropoulos, *J. Appl. Polymer Sci.*, 2009, **111**, 2468.
112. S. Singh, B. A. Simmons and K. P. Vogel, *Biotechnol. Bioeng.*, 2009, **104**, 68.
113. R. J. Soukup-Hein, M. M. Warnke and D. W. Armstrong, *Annu. Rev. Anal. Chem.*, 2009, **2**, 145.
114. J. Pernak, I. Goc and A. Fojutowski, *Holzforschung*, 2005, **59**, 473.
115. J. Pernak, M. Smiglak, S. T. Griffin, W. L. Hough, T. B. Wilson, A. Pernak, J. Zabielska-Matejuk, A. Fojutowski, K. Kita and R. D. Rogers, *Green Chem.*, 2006, **8**, 798.
116. J. Pernak, J. Zabielska-Matejuk, A. Kropacz and J. Foksowicz-Flaczyk, *Holzforschung*, 2004, **58**, 286.
117. M. Stasiewicz, A. Fojutowski, A. Kropacz and J. Pernak, *Holzforschung*, 2008, **62**, 309.
118. X. Li, Y. Geng, J. Simonsen and K. C. Li, *Holzforschung*, 2004, **58**, 280.
119. M. Milota, P. Mosher and K. Li, *For. Prod. J.*, 2008, **58**, 97.
120. M. Milota, P. Mosher and K. C. Li, *For. Prod. J.*, 2007, **57**, 73.
121. F. J. Wang, M. Milota, P. Mosher, K. C. Li and M. Yankus, *Wood Fiber Sci.*, 2007, **39**, 434.
122. Y. Cao, J. Wu, J. Zhang, H. Q. Li, Y. Zhang and J. S. He, *Chem. Eng. J.*, 2009, **147**, 13.
123. L. Feng and Z. I. Chen, *J. Mol. Liquids*, 2008, **142**, 1.
124. H. Ohno and Y. Fukaya, *Chem. Lett.*, 2009, **38**, 2.
125. S. D. Zhu, Y. X. Wu, Q. M. Chen, Z. N. Yu, C. W. Wang, S. W. Jin, Y. G. Ding and G. Wu, *Green Chem.*, 2006, **8**, 325.
126. Y. Fukaya, K. Hayashi, M. Wada and H. Ohno, *Green Chem.*, 2008, **10**, 44.
127. M. Mazza, D. A. Catana, C. Vaca-Garcia and C. Cecutti, *Cellulose*, 2009, **16**, 207.
128. J. Vitz, T. Erdmenger, C. Haensch and U. S. Schubert, *Green Chem.*, 2009, **11**, 417.
129. S. Yu, H. M. Brown, X. W. Huang, X. D. Zhou, J. E. Amonette and Z. C. Zhang, *Appl. Catal. A: Gen.*, 2009, **361**, 117.
130. R. Rinaldi, R. Palkovits and F. Schuth, *Angew. Chem. Int. ed.*, 2008, **47**, 8047.
131. N. Kamiya, Y. Matsushita, M. Hanaki, K. Nakashima, M. Narita, M. Goto and H. Takahashi, *Biotechnol Lett.*, 2008, **30**, 1037.
132. B. C. Saha, *J. Ind. Microbiol. Biotechnol.*, 2003, **30**, 279.

133. H. J. Huang, S. Ramaswamy, U. W. Tschirner and B. V. Ramarao, *Separ. Purif. Technol.*, 2008, **62**, 1.
134. Y. Fukaya, A. Sugimoto and H. Ohno, *Biomacromolecules*, 2006, **7**, 3295.
135. B. Wu, Y. M. Zhang and H. P. Wang, *J. Phys. Chem. B*, 2008, **112**, 6426.
136. C. Sievers, I. Musin, T. Marzialetti, M. B. V. Olarte, P. K. Agrawal and C. W. Jones, *Chemsuschem*, 2009, **2**, 665.
137. S. Lima, P. Neves, M. M. Antunes, M. Pillinger, N. Ignatyev and A. A. Valente, *Appl. Catal. A: Gen.*, 2009, **363**, 93.
138. J. L. Ren, R. C. Sun, C. F. Liu, Z. N. Cao and W. Luo, *Carbohydrate Polymers*, 2007, **70**, 406.
139. E. Dorrestijn, L. J. J. Laarhoven, I. Arends and P. Mulder, *J. Anal. Appl. Pyrol.*, 2000, **54**, 153.
140. S. A. Ralph, J. Ralph and L. L. Landucci, USDA Forest Products Laboratory, 2004, URL: http://ars.usda.gov/Services/docs.htm?docid=10491
141. J. E. Holladay, J. F. White, J. J. Bozell and D. Johnson, "Top Value-Added Chemicals from Biomass-Volume II-Results of Screening for Potential Candidates from Biorefinery Lignin", PNNL-16983, 2007.
142. Y. Q. Pu, N. Jiang and A. J. Ragauskas, *J. Wood Chem. Technol.*, 2007, **27**, 23.
143. J. B. Binder, M. J. Gray, J. F. White, Z. C. Zhang and J. E. Holladay, *Biomass Bioenergy*, 2009, **33**, 1122.
144. S. Kubo, K. Hashida, T. Yamada, S. Hishiyama, K. Magara, M. Kishino, H. Ohno and S. Hosoya, *J. Wood Chem. Technol.*, 2008, **28**, 84.

CHAPTER 13

Conversion of Cellulose to Sugars

SAMAR KUMAR GUHA, HIROKAZU KOBAYASHI AND ATSUSHI FUKUOKA

Catalysis Research Center, Hokkaido University, Kita 21, Nishi 10, Sapporo 001-0021, Kita-ku, Japan

13.1 Introduction

The amount of atmospheric CO_2 generated from the consumption of fossil fuels is increasing day by day. It is the main reason for the present global warming, as CO_2 is a greenhouse gas. Therefore, the production of energy, fuels and chemicals from renewable biomass is important to prevent global warming by decreasing atmospheric CO_2. Biomass is a sustainable resource that is constantly being formed by the interaction of air, water, soil and sunlight. It includes the following: waste materials (agricultural wastes, wood wastes, urban wastes), forest products (trees, logging residues, shrubs), food crops (starch crops such as rice, corn, wheat, potato, barley, sugar crops, vegetable oils, *etc.*), energy crops (grasses, woody crops, hydrocarbon plants), and aquatic biomass (algae, water wood, water hyacinth). In a biorefinery system, biomass is converted into necessary products (plastics, pharmaceuticals, adhesives, paints, dyes, pigments, ink, detergents, *etc.*), organic chemicals (solvents, phenolic compounds, furfural, fatty acids, acetic acid, carbon black, *etc.*), energy (electricity, heat) and fuels (ethanol, renewable diesel, methanol, hydrogen).

RSC Energy and Environment Series No. 1
Thermochemical Conversion of Biomass to Liquid Fuels and Chemicals
Edited by Mark Crocker

Published by the Royal Society of Chemistry, www.rsc.org

Carbohydrates such as starch, cellulose and hemicellulose are produced by plants during photosynthesis. Starch is the main constituent of rice, corn and potato, whereas cellulose (40–50%), hemicellulose (20–30%) and lignin (15–25%) are the constituents of wood. It is known that *ca.* 40% of the annual net yield of photosynthesis is cellulose. As a result, cellulose is the most abundant biological polymer on earth. It is currently regarded as a promising alternative for fossil fuels as it cannot be digested by human beings. Thus, unlike starch, its use will not exert a negative impact on food supplies.[1,2]

Both starch and cellulose are polymers of glucose with a molecular formula $(C_6H_{10}O_5)_n$ (Figure 13.1). Starch has a linkage of D-glucose with 1,4′-α- (amylose) (Figure 13.1(a)) and 1,6′-α-glycosidic bonds (amylopectin), while the chemical composition of cellulose is simple, consisting of D-glucose residues linked by 1,4′-β-glycosidic bonds to form linear polymeric chains (Figure 13.1(b)).

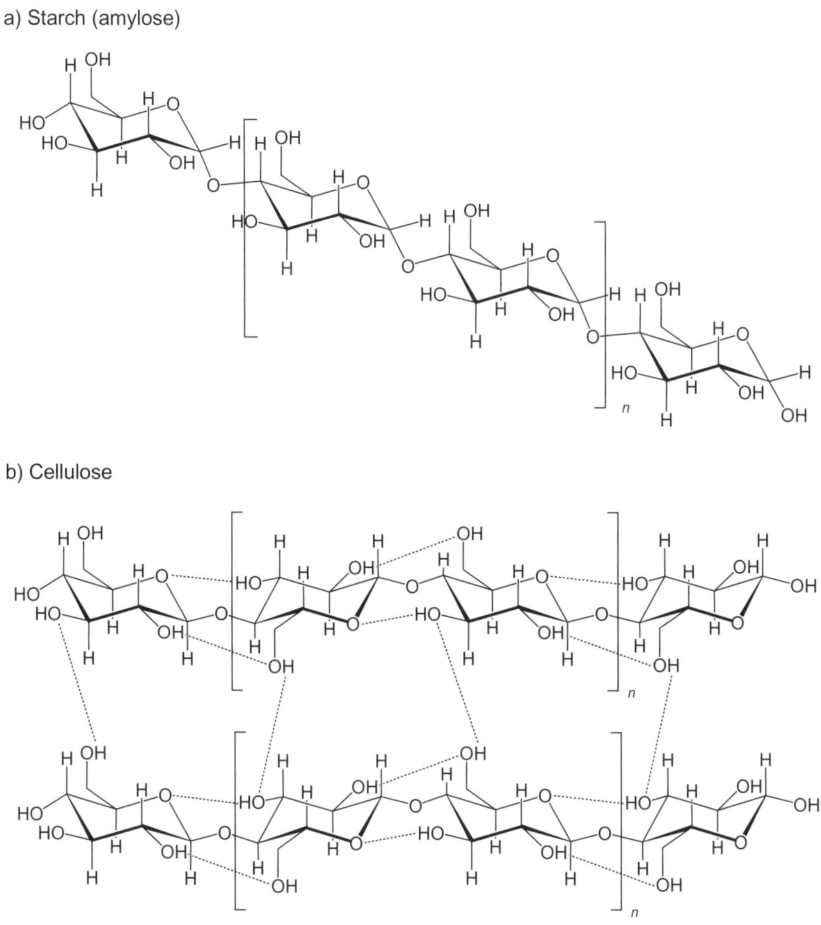

Figure 13.1 Structures of starch and cellulose.

In a chain, the number of glucose units may vary from 300 (rayon) – 15 000 (native cotton).[3] The difference in the bonding of glucose units between starch and cellulose gives rise to different properties. For example, starch is soluble in hot water and produces glucose as a hydrolysis product, but cellulose is insoluble in water or usual organic solvents. Cellulose is only soluble in concentrated aqueous zinc chloride, ammoniacal copper hydroxide[4] and ionic liquids.[5–7] Degradation of starch to chemicals can easily proceed over acids, enzymes and catalysts under relatively mild conditions, whereas it is hard to degrade cellulose to chemicals due to its robust crystal structure with extensive intramolecular and intermolecular hydrogen bonds.[4,8–10] As a result, the conversion of starch into chemicals is well studied based on the biorefinery concept.[10,11] However, it is believed that its main use as food should be secured preferentially. Being abundant and outside the human food chain makes cellulosic materials relatively inexpensive feedstocks for the production of fuels and chemicals.

In this chapter, we will briefly describe the conversion of cellulose to sugars. The options for the hydrolysis of cellulose, the use of catalysts for the conversion of cellulose to sugars and utilization of sugars will be described.

13.2 Structure and Properties of Cellulose

13.2.1 General Properties

Cellulosic materials are principally comprised of lignin, hemicellulose and cellulose, and are thus sometimes called lignocellulosic materials. The primary function of lignin is to provide structural support for the plant. Therefore, in general, the lignin contents in trees are higher than in grasses. Unfortunately, lignin, which does not contain any sugar, encloses the cellulose and hemicellulose molecules, making them difficult to reach.

Hemicelluloses are also complex polysaccharides and are comprised of various monosaccharides such as xylose, arabinose, mannose, glucose and glucuronic acid, which vary significantly among different woody and herbaceous biomass species. Many of the hemicelluloses have the generic formula $(C_5H_8O_4)_n$ and usually consist of 50 to 200 monomeric units and a few simple sugar residues.

In cellulose, the glucose molecules are linked in the form of D-anhydroglucopyranose units through 1,4′-β-D-glycosidic ether bridges (sharing one oxygen) to form a long chain (Figure 13.1(b)).[12] The repeating unit in the chain is anhydrocellobiose, and so the alternate glucose units in the chain are rotated by 180°. Cellulose has an average molecular weight ranging from 300 000 to 500 000.[1] The more generally known form of cellulose is called α-cellulose. It is soluble in 18% NaOH solution. α-Cellulose, which is dissolved in this solution, can be precipitated partially in neutral media, and this precipitate is termed β-cellulose. The remaining dissolved part is termed γ-cellulose.

Due to the intramolecular hydrogen bonds, the cellulose molecule has a rigid structure, whereas the intermolecular hydrogen bonds between adjacent cellulose molecules make it a highly water insoluble material. For the same reason, it acquires a flexible but stable configuration, which enables plants to build long stems with small diameters. Molecular simulations predict that there are eight hydrogen bonds per repeat unit in crystalline cellulose, while in the amorphous form this number is 5.3.[13] These results explain why the amorphous form of cellulose is more easily hydrolyzed than the crystalline form. The bond energy of a hydrogen bond between cellulose and water is nearly 25 kJ/mol.[14] Due to this high value, cellulose shows insoluble properties in water.

In nature, most cellulose occurs in the crystalline state and is defined as cellulose I. It has been reported that native cellulose from almost every source exists as a mixture of two crystalline forms, I_α and I_β, which can be found alongside each other and are part of cellulose type I.[15,16] The I_α/I_β ratio depends on the origin of the cellulose. The physical properties of cellulose I_α and I_β differ from each other due to differences in their crystal packing, molecular conformation and hydrogen bonding.[17] Cellulose I_β is more stable in chemicals than I_α because of the intramolecular and intermolecular hydrogen bonding. Sugiyama *et al.* reported that cellulose from some algae and bacteria is found to be rich in I_α, while cellulose from cotton, wood and tunicates is rich in I_β.[18] Cellulose I_α is metastable and can be converted totally to the stable I_β phase by annealing. The thermodynamically less-stable cellulose I can also be converted into other types of cellulose, namely II, III and IV, upon particular treatments, of which cellulose II is the most stable structure.[8,19,20] Cellulose I can be converted to cellulose II by treating with aq. NaOH solution or by dissolution of the cellulose and subsequent precipitation. Apart from the crystalline states, cellulose also exists in a noncrystalline state. Noncrystalline cellulose has been observed to be present along with cellulose I crystallites in cellulose microfibrils. A large number of crystalline forms of cellulose are obtained by physical and chemical treatments of cellulose postsynthesis. Crystalline cellulose can be converted to the more reactive amorphous cellulose by various physical and chemical methods. Recently, Deguchi *et al.* reported that crystalline cellulose undergoes transformation to the amorphous form in water at around 320 °C and under a pressure of 25 MPa,[21] in which the interaction between water and cellulose plays an essential role.

13.2.2 Crystallinity Index

The crystallinity index (CrI) of cellulose is defined as the percentage of the crystalline part present in the total cellulose. It is an important factor with respect to the reactivity of this substrate. X-ray diffraction (XRD) and NMR spectroscopy are generally used to determine the CrI of cellulose. The most commonly used method to measure the crystallinity of dried cellulose samples is based on XRD,[8] as it enables structural changes to be monitored during different treatments. The crystallinity index of cellulose I is calculated using the following formula:

$$\mathrm{CrI} = 1 - h_{\mathrm{am}}/h_{\mathrm{cr}} = 1 - h_{\mathrm{am}}/(h_{\mathrm{tot}} - h_{\mathrm{am}}) \qquad (13.1)$$

where, h_{am} and h_{cr} represent the corresponding height of amorphous cellulose and crystalline cellulose, respectively, in the (002) reflection at $2\theta = 22.5°$, and $h_{tot} = h_{am} + h_{cr}$.[9] Furthermore, the crystallinity index can also be calculated from ^{13}C NMR spectroscopy using the following equation:

$$\mathrm{CrI} = A_{86-92\,\mathrm{ppm}} / \left(A_{79-86\,\mathrm{ppm}} + A_{86-92\,\mathrm{ppm}}\right) \times 100\% \tag{13.2}$$

In this case, the crystallinity index is determined from the peak areas of the C4 signals. The peak at 86–92 ppm is assigned to crystalline cellulose, and the peak at 79–86 ppm is assigned to amorphous cellulose.[22,23]

13.2.3 Degree of Polymerization

In general, the degree of polymerization (DP) is defined as the number of repeating monomer units in an average polymer chain. In carbohydrate chemistry, the degree of polymerization (DP) is defined as the average number of glucosyl residues per cellulose molecule.[24] It is a measure of molecular weight (MW) and can be defined in terms of number average (DP_N), weight average (DP_W) or viscosity (DP_V).[8,9] It is possible to measure DP by the following methods: (i) by dissolution of cellulose in an aqueous solution of cupric ammonium hydroxide,[25] (ii) by the formation of cellulose derivatives with organic solvents,[26] (iii) by dissolving it in ionic liquids such as *N,N*-dimethylacetamide (DMA)/LiCl.[27]

13.3 Options for Cellulose Saccharification

13.3.1 Cellulose Conversion

In 1819, Braconnot discovered that upon boiling a dilute solution of canvas in vitriol oil (otherwise known as sulfuric acid), a fermentable sugar was formed that was identical with the sugar from grapes (glucose) or starch. This process of breaking cellulose into glucose and oligosaccharide is called "cellulose saccharification."[28]

In order to efficiently convert cellulose, it is necessary to separate cellulose from other lignocellulosic components such as hemicellulose (xylans, mannans, galactans) and lignin (phenolic polymer) as much as possible. Several physical and/or chemical methods can be used for this purpose. By applying these processes, the surface area of the cellulose crystallite is increased by size reduction and swelling.[29] These various methods include the following: physical pretreatment (milling and grinding, steaming, extrusion and expansion, high-energy radiation, pyrolysis, *etc.*), chemical methods (alkali, acid, gas treatment, oxidizing agents, swelling agents, solvent extraction, *etc.*), biological methods (lignin-consuming micro-organisms, cellulose-attacking micro-organisms, lignin and/or cellulose-attacking insects, *etc.*). Some enzymatic methods are also known for this purpose.

13.3.2 Methods for the Saccharification of Cellulose

Various methods are known for the conversion of cellulose and hemicellulose into chemicals. Saccharification of cellulose involves two reactions: hydrolysis to reducing sugar and consecutive decomposition of the sugar formed. The hydrolysis reaction for cellulose conversion into sugar polymers is shown as

$$(C_6H_{10}O_5)_n + nH_2O \rightarrow nC_6H_{12}O_6 \quad (13.3)$$

Cellulose hydrolysis to soluble sugars can be catalyzed either by acids or enzymes. Other approaches include thermochemical transformation and recently developed hydrothermal methods in sub- and supercritical water. A short discussion about these methods is given below.

13.3.2.1 Acidic Hydrolysis

There are two basic types of acid hydrolysis processes that are based on the use of either dilute or concentrated acid. The most widely used acids for this purpose are HCl and H_2SO_4.[14,30–32] Dilute acid processes are conducted at high temperatures of 120 to 200 °C and pressures of 0.1 to 0.5 MPa. The major advantage of dilute acid processes is the short reaction time, which is in the range of seconds or minutes for continuous processes.[33,34] The disadvantage is their low sugar yield. Most dilute acid processes are limited to a sugar-recovery efficiency of around 60%. On the other hand, relatively mild temperatures, moderate pressures and longer reaction times are necessary in concentrated acid processes. The advantage of this process is the high efficiency of sugar recovery, which can reach over 90% for both xylose and glucose sugars.

The decomposition of the formed sugar and the hydrolysis of the cellulose are both first-order reactions and possess rates of similar magnitude. Constants are determined to show the dependence of reaction rate on acid concentration and temperature. On the basis of kinetic studies, Saeman *et al.* observed that the efficiency of cellulose conversion to reducing sugars increases with increasing temperature and increasing acid concentration.[30] On the other hand, the precaution reported by Bouchard *et al.* should also be kept in mind, namely that the structure of cellulose can be modified and frequently becomes more difficult to degrade at 200 °C and above in the presence of acids.[35]

13.3.2.2 Enzymatic Hydrolysis

Another basic method for cellulose saccharification is enzymatic hydrolysis. Enzymes are naturally occurring plant proteins that cause certain chemical reactions to occur. Cellulases of fungal or bacterial origin are in fact a mixture of enzymes that act in concert and synergistically,[36] and are very specific in their action. However, hydrolysis of cellulose to glucose in aqueous system catalyzed by the cellulase enzyme suffers from slow reaction rate because cellulose is a water-insoluble crystalline biopolymer, whereas the cellulase is a water-soluble compound. This makes it difficult for the enzymes to come into close contact with the substrate due to the low surface area available.[37,38] Moreover, Gregg *et al.* reported that enzymatic hydrolysis of cellulosic substrates depends on a number of factors, which include the crystallinity of the cellulose, the degree of polymerization, the lignin content and the surface area accessible to the cellulases.[39] For effective enzymatic hydrolysis, a pretreatment process is thus necessary to break the crystalline structure of the lignocellulose and to remove the lignin for exposure of the cellulose and hemicellulose molecules. Recently, Kumar *et al.* have reported a list of pretreatment methods and the yields of the resulting sugars obtained from the enzymatic hydrolysis of various feedstocks.[40] In addition to the above methods, microbial pretreatment for enzymatic hydrolysis has also been reported.[41]

The disadvantages of the enzymatic approach are low activity, the high cost of the enzyme, the need to use buffers to keep the pH in a range in which the enzymes work and the difficulties in the separation of the product from the enzyme as both are soluble in water. However, in the hydrolysis of a pretreated cellulosic substrate the use of a less-expensive enzyme has several advantages, including high efficiency, low byproduct formation, low construction cost for the set up and low process energy requirements.[33]

13.3.2.3 Thermochemical Transformation

Another route for the saccharification of cellulose is thermochemical treatment, in which the thermochemical transformation of cellulose occurs at high temperature to give various chemicals. The process can be divided into two categories: gasification and pyrolysis (Figure 13.2). In gasification, cellulose is decomposed at high temperature ($>700\,^{\circ}C$) in the absence/presence of oxygen but presence of steam to give syngas (mainly $CO + H_2$).[42,43] On the other hand, in the pyrolysis process the decomposition of cellulose is performed at relatively moderate temperature ($\leq 700\,^{\circ}C$) in the absence of oxygen to yield volatile

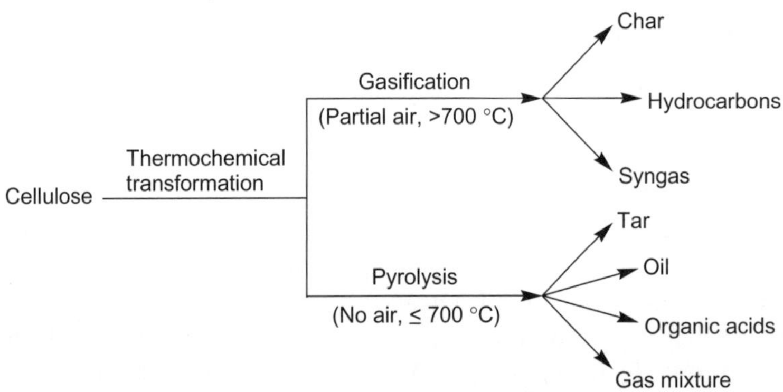

Figure 13.2 Model figure for the thermochemical transformation of cellulose.

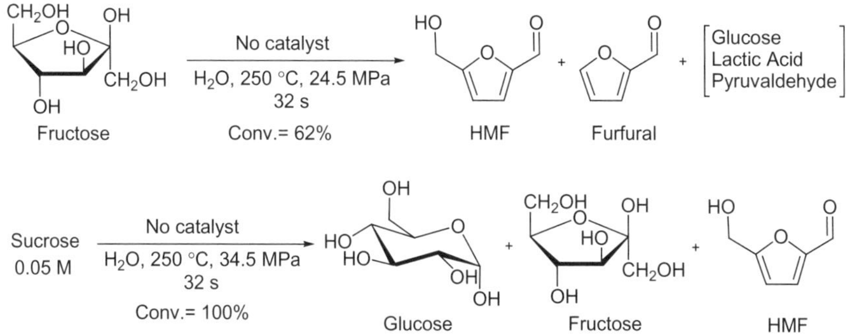

Figure 13.3 Hydrolysis of fructose and sucrose in H_2O at 250 °C without any catalysts.

gases, organic acids, oils, tar and char.[42,44] The advantage of these processes is that they can use a wide variety of feedstocks, especially for the gasification route. The major disadvantage is the lack of product selectivity.

13.3.2.4 Sub- and Supercritical Water

In 1990, Antal *et al.* reported the formation of 5-hydroxymethylfurfural (HMF) from the noncatalyzed conversion of D-fructose and sucrose in water at 250 °C (Figure 13.3).[45] Subsequently, Adschiri *et al.* investigated the conversion of cellulose in sub- and supercritical water at a pressure of 25 MPa over a temperature range of 200–400 °C.[46] They found that cellulose was rapidly hydrolyzed by *in situ* generated acid to form water-soluble species and that the yield of glucose increased with increasing temperature. It was also explained later why the formation of hydrolysis products in supercritical water was higher than in subcritical water.[47–49] The disadvantage of this method is the low selectivity to any one product. It also requires heavy investment as the reactor material is prone to corrode due to the superacidity of supercritical water.

13.4 Catalytic Conversion of Cellulose

13.4.1 Solid Acid-Catalyzed Reaction

Taking into account the drawbacks of the above methods, there is a need for a new process for the conversion of cellulose into monomers using chemical catalysts. Our group reported that sulfonated mesoporous silicas can be efficiently used as solid acid catalysts in the hydrolysis of soluble sugars.[50] The sulfonated mesoporous silicas were synthesized by anchoring propylsulfonyl groups onto mesoporous silicas. The sulfonated mesoporous silicas gave higher glucose yields than Amberlyst-15 and Nafion SAC-13. Both are ion-exchange resins with SO_3H groups. This higher catalytic activity may be due to the high surface area and efficient use of acid sites in the mesoporous silicas.[51] Abbadi *et al.* reported

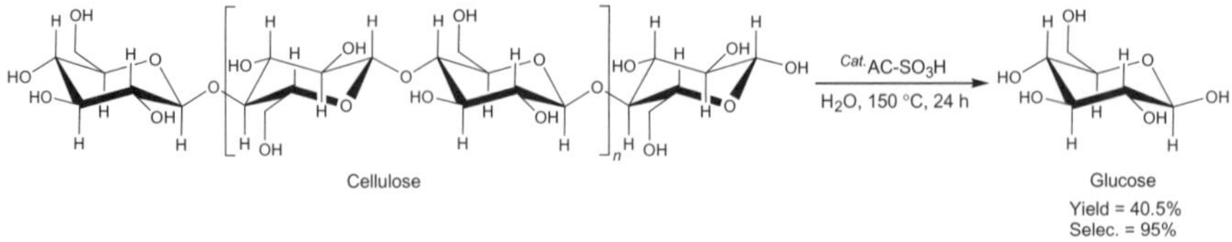

Figure 13.4 Solid acid-catalyzed conversion of cellulose into glucose.

the hydrolysis of starch into glucose in good selectivity over solid acids such as ion-exchange resins as catalyst.[52] Glucose was obtained in 35% yield by using ion-exchange resin as catalyst at 130 °C and 1 atm of pressure for 24 h.

Amberlyst resins are stable enough for use in ionic liquids, in which they are powerful catalysts for the selective depolymerization of cellulose.[53] Microcrystalline cellulose, which cannot be hydrolyzed by conventional dilute-acid processes, is easily hydrolyzed using Amberlyst 15DRY in 1-butyl-3-methylimidazolium chloride ([BMIM]Cl), although the molecular catalyst *p*-toluenesulfonic acid was found to have better catalytic activity than Amberlyst 15DRY. However, the work-up for the sugar-extraction process may become difficult because of the high solubility of sugars in ionic liquids.[54]

Several new solid acid catalysts have been prepared that are active for cellulose hydrolysis. Hara *et al.* reported that the amorphous carbon-bearing SO_3H group showed higher catalytic activity than other solid acids in the conversion of cellulose into 1,4-β-glucan and glucose.[55] In this case, the carbon materials were prepared from microcrystalline cellulose by pyrolysis, followed by sulfonation with fuming sulfuric acid. The higher activity of the carbon catalyst is due to the adsorbing ability of 1,4-β-glucan which is not adsorbed on other solid acids and the tolerance of SO_3H groups to hydration in the carbon material. Onda *et al.* reported the solid acid-catalyzed hydrolysis of cellulose into glucose in H_2O (Figure 13.4).[56] Among a series of tested catalysts, the sulfonated activated carbon (AC–SO_3H) prepared from commercially available activated carbon showed significantly higher catalytic activity and glucose was obtained in 41% yield with 95% selectivity. The high activity is due to the strong acidity of SO_3H groups and the highly hydrophobic surface.

13.4.2 Supported Metal-Catalyzed Reaction

Various supported-metal catalysts have been used for the conversion of starch to polyhydric alcohols by hydrogenation at high temperature and pressure.[57] By applying hydrogenation conditions, our group reported the first catalytic degradation of cellulose into sorbitol with high selectivity (Figure 13.5).[58–60] Before this report, the known route for sorbitol synthesis was based on the hydrolysis of cellulose by mineral acids to form glucose, followed by hydrogenation. To achieve a sufficient concentration of *in situ* generated acid sites, hydrogen gas was pressurized in an autoclave with a supported-metal catalyst, an approach that until then had not been applied for the conversion of

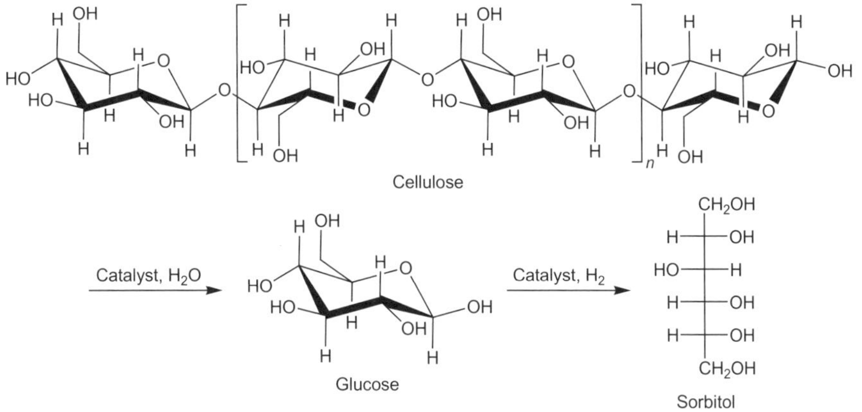

Figure 13.5 Catalytic conversion of cellulose to sorbitol.

cellulose. Sugar alcohols with 88% selectivity were obtained by using Pt/Al_2O_3 as a catalyst, and the high selectivity is probably due to the high thermal stability of sorbitol over glucose at high temperature. The reaction is believed to proceed in 2 steps: hydrolysis of cellulose to glucose by the acid sites produced *in situ* and reduction of the C=O group in glucose on the metal with H_2 to form sorbitol. Subsequently, Liu *et al.* studied this process using a Ru/C catalyst.[61] In their work, the reaction proceeded through two steps: hydrolysis of cellulose to glucose by H^+ ions that are reversibly formed *in situ* from water at elevated temperatures (above 200 °C) and the subsequent hydrogenation reaction catalyzed by Ru/C. The sugar alcohols were obtained in relatively high yield, which was attributed to the higher reaction temperatures and the well-known high efficiency of Ru/C in hydrogenation reactions.[62,63] At 240 °C, the sugar alcohols (hexitols) were obtained with a maximum selectivity of 61%. The conversion of the model substrate cellobiose into sugar alcohols was also studied, and 100% conversion was found to be possible at pH 2.[64]

Zhang *et al.* showed that a tungsten carbide catalyst gave a high yield of ethylene glycol (EG) relative to other polyols from the degradation of cellulose.[65] The advantages of this procedure are the low price of tungsten carbide and the high yield of EG. They also showed that the reaction can be promoted with a small amount of nickel to give EG in 61% yield. Ni promotes the formation of tungsten carbide phase, as well as effecting a remarkable change in the particle-size distribution.

13.4.3 Other Catalytic Conversions of Cellulose

Recently, Zhang *et al.* has described a single-step conversion of cellulose to HMF catalyzed by a variety of metal chlorides in an ionic liquid, [EMIM]Cl (Figure 13.6).[66] The catalytic activity of the individual metal chlorides was found to be poor, whereas a combination of metal chlorides ($CuCl_2$ and $CrCl_2$)

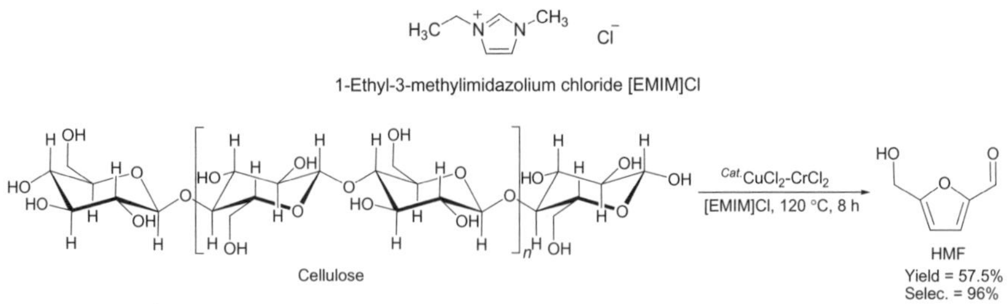

Figure 13.6 Structure of [EMIM]Cl and hydrolysis of cellulose to HMF in [EMIM]Cl.

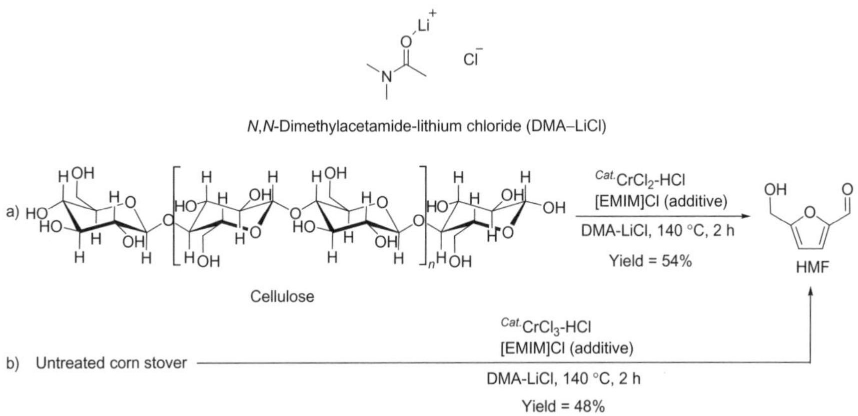

Figure 13.7 Structure of DMA-LiCl and hydrolysis of cellulose and corn stover into HMF in DMA-LiCl.

showed good catalytic activity in [EMIM]Cl. At 120 °C for 8 h in the presence of $CuCl_2/CrCl_2$, HMF was obtained in 58% yield with 96% purity. On the other hand, glucose was obtained as a major product when the temperature or the reaction time was reduced. In an another report, Raines *et al.* showed a single-step conversion of cellulose to HMF by using $CrCl_2$ or $CrCl_3$ as a catalyst in *N,N*-dimethylacetamide (DMA)-LiCl solvent system.[67] In some cases, the ionic liquid [EMIM]Cl was used as an additive to improve the product yield. HMF was obtained from cellulose hydrolysis in 54% yield within 2 h at 140 °C (Figure 13.7(a)). Untreated lignocellulosic biomasses such as corn stover or pine sawdust can also be hydrolyzed to produce HMF in up to 48% yield under similar conditions (Figure 13.7(b)).

13.5 Overview of Routes for Utilization of Sugars

As described above, various sugars and sugar alcohols are obtained from cellulose degradation. These sugars and sugar alcohols can be converted to other value-added chemicals and fuels.[68,69] Some of the utilization routes are described below.

13.5.1 Utilization of Glucose

Glucose has been the focus of much attention due to its wide range of uses, from the food industry to the manufacture chemicals. Routes exist for the conversion of glucose to value-added chemicals such as HMF, sorbitol, gluconic acid, glucaric acid, *etc.* (Figure 13.8). HMF is a potential precursor to fuels and plastics. Other value-added chemicals, 2,5-furandicarboxylic acid and levulinic acid, are obtained from HMF by oxidation and dehydration, respectively.[69] Dumesic *et al.* introduced the idea of using HMF as a key intermediate to produce liquid alkanes from glucose (Figure 13.9).[70] It can be

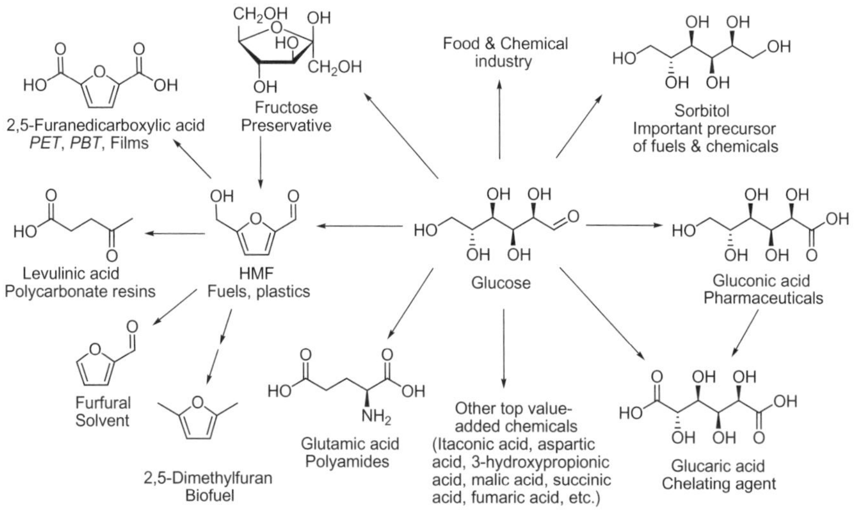

Figure 13.8 Use of glucose as a feedstock for the production of value-added chemicals.

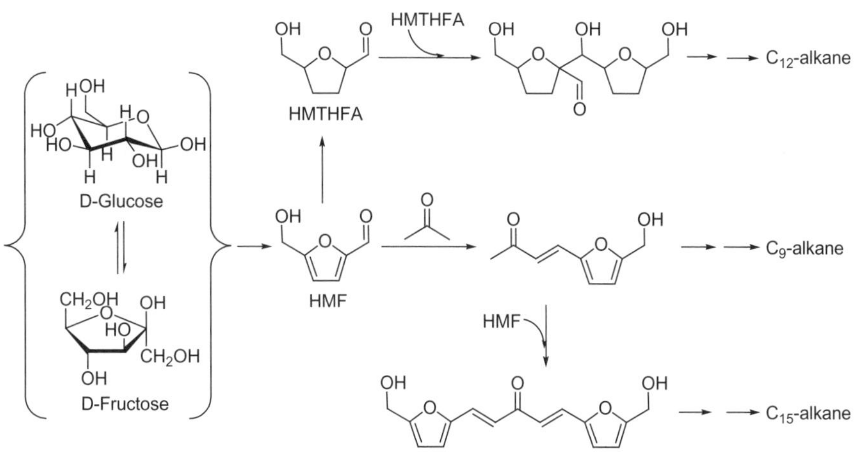

Figure 13.9 Formation of liquid alkanes from glucose (or fructose) *via* HMF.

synthesized from glucose by dehydrocyclization reactions under acidic conditions, and various solid acids and zeolites can be used for this purpose. Water is widely used as a solvent for this reaction due to the insolubility of HMF in water, which makes the separation easier.[71–73] Amberlyst-15, Nafion and Nb_2O_5 have also been used as solid catalysts for the conversion of glucose to HMF.[74,75] Zhang *et al.* reported the $CrCl_2$ catalyzed conversion of glucose to HMF in [EMIM]Cl solvent with a yield of about 70%.[76] They propose that $CrCl_2$ reacts with [EMIM]Cl to give the $[CrCl_3]^-$ anion, which plays a role in proton transfer. Another mild procedure is the $CrBr_3$-catalyzed conversion of glucose in which the yield of HMF is increased to 80%.[67] In this case, LiBr was used as an additive together with the catalyst in DMA at 100 °C.

As described above, supported-metal-catalyzed hydrogenation of the –CHO group of glucose produces sorbitol. The uses of sorbitol as a precursor to fuels and chemicals are described in Section 13.5.3. In contrast, oxidation of glucose with heterogeneous catalysts produces gluconic acid,[77–79] which is widely used in the pharmaceutical and food industries. Rossi *et al.* used gold nanoparticles (Au NPs) on activated carbon (AC) as the catalyst and demonstrated its high catalytic activity and selectivity in the oxidation of glucose to gluconic acid (Figure 13.10, route a),[77] whereas Baatz *et al.* used Au supported on Al_2O_3 for this purpose (Figure 13.10, route b).[78] In the latter case, addition of base was found to be essential to push the reaction towards product formation. Haruta *et al.* used relatively small Au NPs on a variety of support materials.[79] They found that the reaction between a volatile organogold complex, $[Me_2Au(acac)]$ (acac = acetylacetonate), and Al_2O_3 or TiO_2 formed Au clusters that showed excellent catalytic activity in the oxidation of glucose. There are also reports of the aerobic oxidation of glucose to gluconic acid catalyzed by transition-metal catalysts such as Pt/C,[80] Pd/C,[81] Pd–Bi/C,[81,82] *etc.*

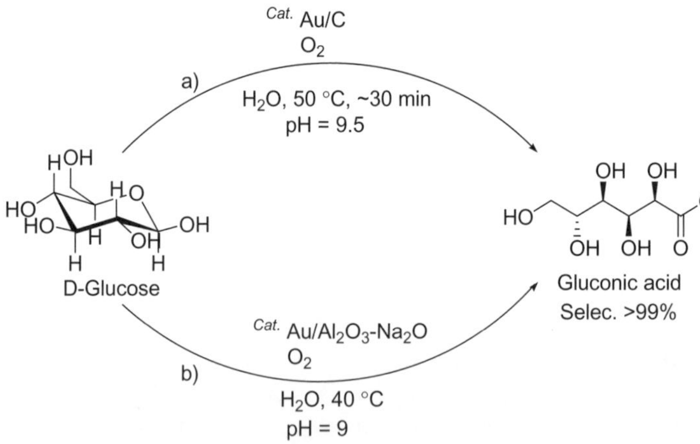

Figure 13.10 Oxidation of glucose to gluconic acid catalyzed by Au/C or Au/Al_2O_3-Na_2O.

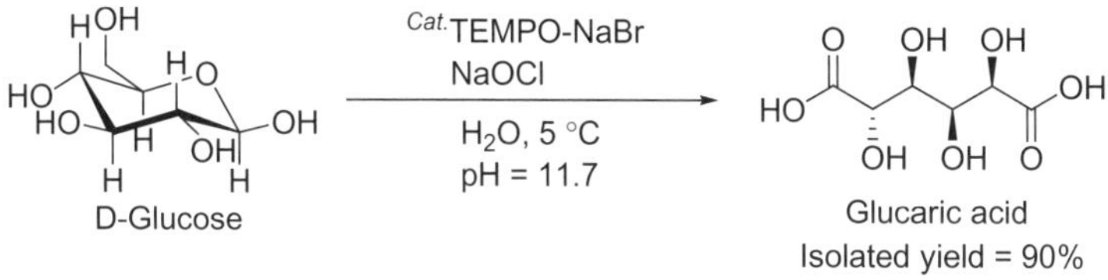

Figure 13.11 Oxidation of glucose to glucaric acid using TEMPO-NaBr-NaOCl.

Glucaric acid is another value-added chemical and is widely used as a chelating agent, in ionic/nonionic surfactants, as a carbohydrate polymer building block, *etc*. It can be prepared from glucose by oxidation. Besson *et al.* reported the formation of glucaric acid from the oxidation of glucose using Pt/C catalysts.[83] Catalyst promotion with bismuth or gold was examined and the yield of glucaric acid was around 40%. Other groups reported that glucaric acid was obtained from glucose oxidation using sodium nitrite,[84] TEMPO-mediated oxidation,[85,86] *etc*., and glucaric acid was also isolated from the oxidation of glucose in a TEMPO-NaBr-NaOCl system (Figure 13.11).[86]

Glucose is also an important precursor for the formation of other value-added chemicals, such as, 1,4-diacids (succinic, fumaric and malic), 3-hydroxypropionic acid, aspartic acid, glutamic acid and itaconic acid.[69]

13.5.2 Utilization of Fructose

Fructose is a simple sugar, found naturally in honey and most fruits. It is combined with glucose to form sucrose (table sugar). Fructose is an isomer of glucose and is 73% sweeter than sucrose at room temperature. It is used in the pharmaceutical and food industries, especially in foods for diabetics. It is absorbed in the body more slowly than glucose and is converted to glycogen in the liver. The uses of fructose are shown in Figure 13.12. Fructose undergoes dehydrocyclization reactions to yield HMF in the presence of acid catalysts.[87–89] Higher rates and selectivities for the dehydration of fructose to HMF are obtained than for glucose. Purification of the formed HMF becomes difficult in acid-catalyzed reactions as the acid catalysts cause many side reactions. Recently, procedures have been developed to prepare HMF from fructose with good yield and selectivity in ionic liquids (Figure 13.13).[67,76] Fructose is hydrogenated by $RuHCl(PPh_3)$ to mannitol and sorbitol.[90] Glycerol can be prepared from fructose *via* hydrogenation catalyzed by $H_2Ru(PPh_3)_4$.[91]

13.5.3 Utilization of Sorbitol

Sorbitol, also known as glucitol, is a sugar alcohol that the body metabolizes slowly. Industrially, it is obtained by reduction of glucose over Raney Ni catalyst, the aldehyde group being converted to an alcohol group.[68,92] Due to the fact that it contains two-thirds of the calories of sugar, it is used as a substitute

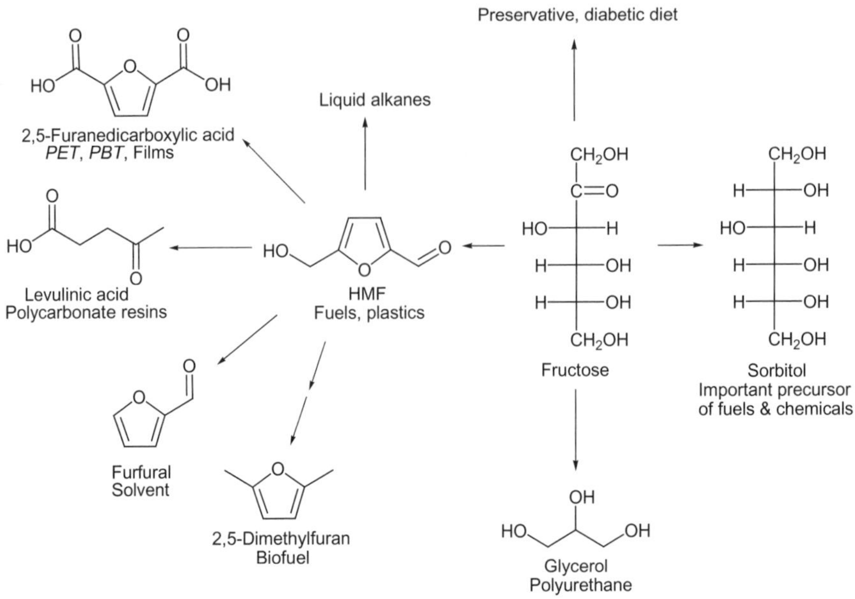

Figure 13.12 Uses of fructose.

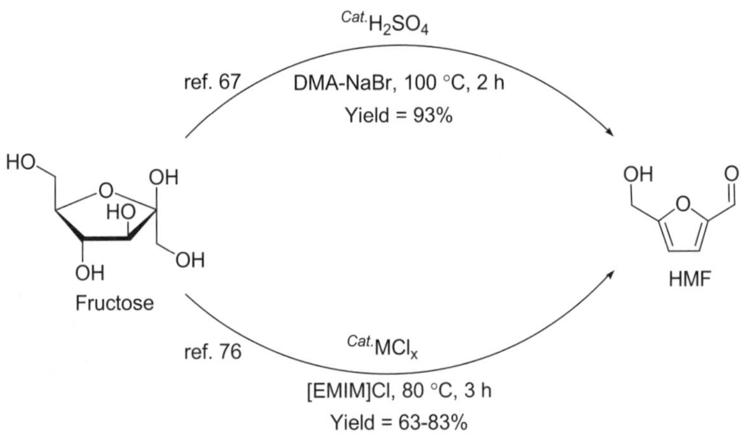

Figure 13.13 Preparation of HMF from fructose.

for sucrose in various foods, especially in diet foods and pharmaceutical tablets. It does not promote tooth decay. As shown in Figure 13.14, it is potentially an important raw material for the synthesis of value-added chemicals such as isosorbide, 1,4-sorbitan, glycols, glycerol, lactic acid, L-sorbose, *etc.* Depending on the reaction conditions, it is a good source for the production of H_2 gas and CH_4,[93,94] as well as higher alkanes.[95]

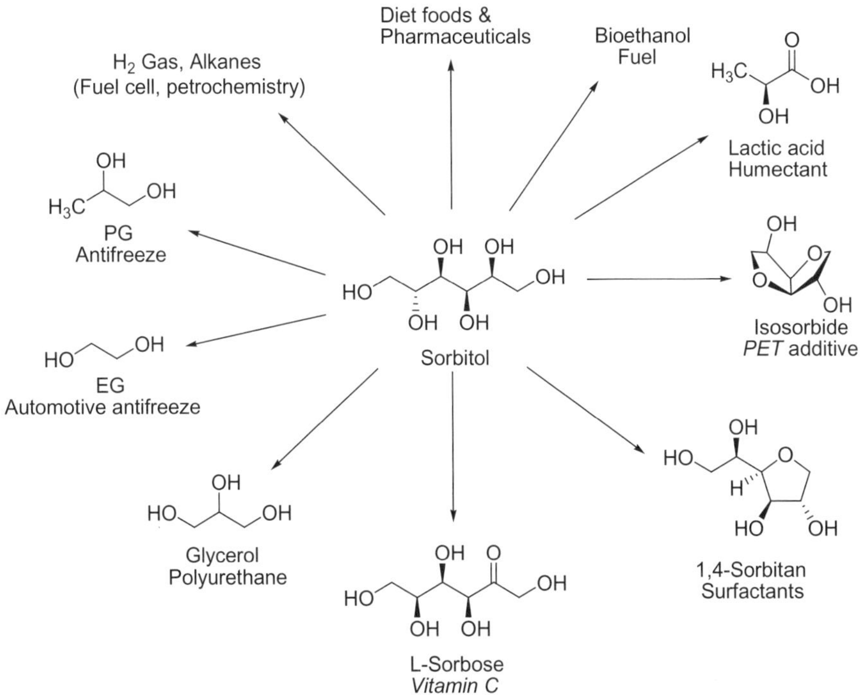

Figure 13.14 Use of sorbitol as a feedstock for the production of chemicals.

Barbier *et al.* investigated the Ru-catalyzed hydrogenolysis of sorbitol and found the formation of ethylene glycol, propylene glycol and glycerol.[96] In a basic medium, the formation of the glycols was independent of temperature, whereas in a neutral medium an increase in reaction temperature led to an increase in the selectivity to glycerol. Glycerol is used for oral care products, in pharmaceuticals, for foods and beverages and for the production of polyurethane. By hydrogenolysis, glycerol is converted into lower polyols and alcohols. Propylene glycol is used as an antifreeze, as a humectant reagent, in the preparation of unsaturated polyester resins, and in liquid detergents, paints, coatings, *etc*. Ethylene glycol is widely used as an automotive antifreeze and as a precursor to polyesters. C_2 and C_3 diols were also found to be formed in a CuO-ZnO catalyzed hydrogenolysis reaction of sorbitol at a temperature of 240 °C.[97] Upon decreasing the reaction temperature, the amount of C_2 and C_3 diols was decreased and the selectivity to deoxyhexitol was increased. The best yield of deoxyhexitols (63%) was obtained from an aqueous solution of sorbitol catalyzed by CuO-ZnO under H_2 pressure at 180 °C. Hydrogen and light alkanes (mainly methane) can be produced by aqueous-phase reforming of glycerol.[93,94]

Isosorbide (1,4:3,6-dianhydro-D-sorbitol) is used as an industrial solvent, plasticizer, and as an additive for increasing the glass transition temperature of PET-like polymers such as polyethylene isosorbide terephthalates (to enable the storage of hot drinks in PET bottles). It is prepared from sorbitol by dehydration reactions. Gadelle *et al.* reported that treatment of sorbitol with anhydrous hydrogen fluoride, catalyzed by formic or acetic acid, produces isosorbide in 57% isolated yield.[98] Formic acid was found to show better performance than acetic acid for this reaction. They also showed that the 3,6-anhydride (3,6-sorbitan) is exclusively formed in the initial stages of the reaction and then converted to isosorbide with increasing reaction time. Barbier *et al.* prepared isosorbide with 69% selectivity from sorbitol using a bimetallic Cu-Ru or Cu-Pt catalyst.[99] It is known that Raney Cu does not catalyze hydrogenolysis of C–C or C–O bonds. In order to increase the catalytic activity, Ru or Pt was deposited on Raney Cu by oxidoreduction. As a result, C–C and C–O bond cleavage was observed on Cu through the nucleophilic action of water or adsorbed hydroxyls. Cu-Ru showed better catalytic activity than Cu-Pt, and the yield of isosorbide was increased when the temperature and the hydrogen pressure was increased to 260 °C and 7 MPa. The reaction rate obtained with the Cu-Ru bimetallic catalyst was also significantly higher than for a dilute acid catalyst ($pH = 3$).

Sorbitan can be synthesized from sorbitol. It is used in the production of surfactants such as polysorbates, and is an intermediate in the synthesis of isosorbide from sorbitol. Another important derivative of sorbitol is L-Sorbose, which is a precursor to Vitamin C (ascorbic acid).[100] Sorbitol can also be fermented to ethanol *via* dehydrogenation using an enzyme.[101]

13.6 Conclusion

Lignocellulosic biomass, the most abundant renewable resource on earth, is a sustainable source of carbon for the production of fuels and chemicals. Various methods are well known for the hydrolysis of cellulose. In almost all cases, a pretreatment is necessary to obtain a good conversion and yield of the resulting products. At present, acid- or enzyme-catalyzed hydrolysis of cellulose is planned for use in practical applications, although these approaches are not without limitations: multiple steps, low activity, formation of neutralization wastes, corrosion hazard, and so on. In addition to traditional cellulose saccharification processes, the recently developed catalytic conversion of cellulose has been described in this chapter. Environmentally friendly solid acid catalysts and supported-metal catalysts have drawn particular attention from researchers in the field. Due to the low cost and easier separation of the products after the reaction, the area of catalytic conversion of biomass is growing rapidly. Here we have summarized recent work concerning the conversion of biomass to chemicals by heterogeneous catalysts. The importance of the sugar and sugar alcohols obtained from the conversion of cellulose is also illustrated. Heterogeneous catalysts can also be used in the conversion of sugar compounds to fuels and other chemicals. Glucose, fructose and sorbitol are converted to a number of value-added chemicals. Based on the foregoing, we believe that heterogeneous catalysis represents a new methodology for the conversion of cellulose and related compounds in biorefineries.

References

1. D. L. Klass, in *Biomass for Renewable Energy, Fuels, and Chemicals,* Academic Press, San Diego, 1998.
2. G. W. Huber, S. Iborra and A. Corma, *Chem. Rev.*, 2006, **106**, 4044.
3. D. Fengel, in *Polysaccharides,* W. Burchard, (ed.), Springer, Berlin, 1985, pp.15.
4. L. T. Fan, M. M. Gharpuray and Y. H. Lee, *Cellulose Hydrolysis,* Springer, Berlin, 1987.
5. R. P. Swatloski, S. K. Spear, J. D. Holbrey and R. D. Rogers, *J. Am. Chem. Soc.*, 2002, **124**, 4974.
6. H. Zhang, J. Wu, J. Zhang and J. He, *Macromolecules*, 2005, **38**, 8272.
7. S. Zhu, Y. Wu, Q. Chen, Z. Yu, C. Wang, S. Jin, Y. Ding and G. Wu, *Green Chem.*, 2006, **8**, 325.
8. H. A. Krässig, *Cellulose: Structure, Accessibility and Reactivity,* Gordon and Breach, Yverdon, 1993.
9. Y. P. Zhang and L. R. Lynd, *Biotechnol. Bioeng.*, 2004, **88**, 797.
10. H. Danner and R. Braun, *Chem. Soc. Rev.*, 1999, **28**, 395.
11. http://www1.eere.energy.gov/biomass/integrated_biorefineries.html.
12. C. E. Wyman, S. R. Decker, M. E. Himmel, J. W. Brady, C. E. Skopec and L. Viikari, in *Polysaccharides*, S. Dumitriu, (ed.), Marcel Dekker, New York, 2005.
13. K. Mazeau and L. Heux, *J. Phys. Chem., B*, 2003, **107**, 2394.
14. O. Bobleter, *Progr. Polym. Sci.*, 1994, **19**, 797.
15. R. H. Atalla and D. L. VanderHart, *Science*, 1984, **223**, 283.
16. M. Jarvis, *Nature*, 2003, **426**, 611.
17. Y. Nishiyama, J. Sugiyama, H. Chanzy and P. Langan, *J. Am. Chem. Soc.*, 2003, **125**, 14300.
18. J. Sugiyama, J. Persson and H. Chanzy, *Macromolecules*, 1991, **24**, 2461.
19. G. L. Klein and W. R. Snodgrass, in *Encyclopedia of Food Science, Food Technology and Nutrition*, R. Macrae, R. K. Robinson and M. J. Saddler, (ed.), Academic Press, London, 1993, pp. 758.
20. A. C. O'Sullivan, *Cellulose*, 1997, **4**, 173.
21. S. Deguchi, K. Tsujii and K. Horikoshi, *Chem. Commun.*, 2006, 3293.
22. H. Zhao, J. H. Kwak, Y. Wang, J. A. Franz, J. M. White and J. E. Holladay, *Energy Fuels*, 2006, **20**, 807.
23. H. Kono, S. Yunoki, T. Shikano, M. Fujiwara, T. Erata and M. Takai, *J. Am. Chem. Soc.*, 2002, **124**, 7506.
24. K. K-. Leyer, E. Agosin, A. H. Conner and T. K. Kirk, *Appl. Environ. Microbiol.*, 1992, **58**, 1266.
25. D. Klemm, B. Philipp, T. Heinze, U. Heinze and W. Wagenknecht, in *Comprehensive Cellulose Chemistry. I. Fundamentals and Analytical Methods,* Wiley-VCH, Weinheim, 1998.
26. T. K. Ng and J. G. Zeikus, *Anal. Biochem.*, 1980, **103**, 42.
27. A. M. Striegel, *Carbohyd. Polym.*, 1997, **34**, 267.

28. H. Braconnot, *Annales de chimie et de physique (Ann. Chim. Phis/A. ch.)*, 1819, **12**, 172.
29. L. T. Fan, in *Adv. Biochem. Eng.*, A. Fiechter, (ed.), Springer, Berlin-Heidelberg-New York, 1982, **vol. 23**.
30. J. F. Saeman, *Ind. Eng. Chem.*, 1945, **37**, 43.
31. W. S. Mork and M. J. Antal Jr, *Ind. Eng. Chem. Res.*, 1992, **31**, 94.
32. R. W. Torget, J. S. Kim and Y. Y. Lee, *Ind. Eng. Chem. Res.*, 2000, **39**, 2817.
33. P. C. Badger, in *Trends in New Crops and New Uses*, J. Janick and A. Whipkey, (ed.), ASHS Press, Alexandria, VA, 2002.
34. K. H. Kim, M. P. Tucker and Q. A. Nguyen, *Biotechnol. Prog.*, 2002, **18**, 489.
35. J. Bouchard, N. Abatzoglou, E. Chornet and R. P. Overend, *Wood. Sci. Technol.*, 1989, **23**, 461.
36. N. S. Mosier, P. Hall, C. M. Ladisch and M. R. Ladisch, *Adv. Biochem. Eng. Biotechnol.*, 1999, **65**, 23.
37. A. P. Dadi, S. Varanasi and C. A. Schall, *Biotechnol. Bioeng.*, 2006, **95**, 904.
38. D. N. Thompson, H. C. Chen and H. C. Grethlein, *Bioresour. Technol.*, 1992, **39**, 155.
39. D. J. Gregg and J. N. Saddler, *Biotechnol. Bioeng.*, 1996, **51**, 375.
40. S. Kumar, S. P. Singh, I. M. Mishra and D. K. Adhikari, *Chem. Eng. Technol.*, 2009, **32**, 517.
41. J. Shi, R. R. S-Shivappa, M. Chinn and N. Howell, *Biomass and Bioenergy*, 2009, **33**, 88.
42. B. Kamm, P. R. Gruber and M. Kamm, in *Biorefineries-Industrial Processes and Products,* Wiley-VCH, Weinheim, 2006, **vol. 1**.
43. M. Asadullah, K. Fujimoto and K. Tomishige, *Ind. Eng. Chem. Res.*, 2001, **40**, 5894.
44. L. Garcia, M. L. Salvador, J. Arauzo and R. Bilbao, *Ind. Eng. Chem. Res.*, 1998, **37**, 3812.
45. M. J. Antal Jr and W. S. L. Mok, *Carbohydr. Res.*, 1990, **199**, 91.
46. T. Adschiri, S. Hirose, R. Malaluan and K. Arai, *J. Chem. Eng. Jpn.*, 1993, **26**, 676.
47. M. Sasaki, B. Kabyemela, R. Malaluan, S. Hirose, N. Takeda, T. Adschiri and K. Arai, *J. Supercrit. Fluid.*, 1998, **13**, 261.
48. M. Sasaki, Z. Fang, Y. Fukushima, T. Adschiri and K. Arai, *Ind. Eng. Chem. Res.*, 2000, **39**, 2883.
49. T. Sakaki, M. Shibata, T. Sumi and S. Yasuda, *Ind. Eng. Chem. Res.*, 2002, **41**, 661.
50. P. L. Dhepe, M. Ohashi, S. Inagaki, M. Ichikawa and A. Fukuoka, *Catal. Lett.*, 2005, **102**, 163.
51. A. Fukuoka and P. L. Dhepe, *Stud. Surf. Sci. Catal.*, 2006, **38**, 197.
52. A. Abbadi, K. F. Gotlieb and H. van Bekkum, *Starch*, 1998, **50**, 23.
53. R. Rinaldi, R. Palkovits and F. Schüth, *Angew. Chem. Int. ed.*, 2008, **47**, 8047.

54. Q. Liu, M. H. A. Janssen, F. Rantwijk and R. A. Sheldon, *Green Chem.*, 2005, **7**, 39.
55. S. Suganuma, K. Nakajima, M. Kitano, D. Yamaguchi, H. Kato, S. Hayashi and M. Hara, *J. Am. Chem. Soc.*, 2008, **130**, 12787.
56. A. Onda, T. Ochi and K. Yanagisawa, *Green Chem.*, 2008, **10**, 1033.
57. P. Jacobs and H. Hinnekens, *EP Pat.*, 0329923, 1989.
58. A. Fukuoka and P. L. Dhepe, *Angew. Chem. Int. ed.*, 2006, **45**, 5161.
59. H. Hattori and T. Shishido, *Catal. Surv. Jpn.*, 1997, **1**, 205.
60. T. M. Yurieva, L. M. Plyasova, O. V. Makarova and T. A. Krieger, *J. Mol. Catal. A: Chem.*, 1996, **113**, 455.
61. C. Luo, S. Wang and C. H. Liu, *Angew. Chem. Int. ed.*, 2007, **46**, 7636.
62. B. Kusserow, S. Schimpf and P. Claus, *Adv. Synth. Catal.*, 2003, **345**, 289.
63. B. W. Hoffer, E. Crezee, P. R. M. Mooijman, A. D. Langeveld, F. Kapteijn and J. A. Moulijn, *Catal. Today*, 2003, **79**, 35.
64. N. Yan, C. Zhao, C. Luo, P. J. Dyson, H. Liu and Y. Kou, *J. Am. Chem. Soc.*, 2006, **128**, 8714.
65. N. Ji, T. Zhang, M. Zheng, A. Wang, H. Wang, X. Wang and J. G. Chen, *Angew. Chem. Int. ed.*, 2008, **47**, 8510.
66. Y. Su, H. M. Brown, X. Huang, X.-d. Zhou, J. E. Amonette and Z. C. Zhang, *Appl. Catal., A*, 2009, **361**, 117.
67. J. B. Binder and R. T. Raines, *J. Am. Chem. Soc.*, 2009, **131**, 1979.
68. T. Werpy and G. Petersen, in "Top value-added chemicals from biomass, 2004, vol 1: results of screening for potential candidates from sugars and synthesis gas", US Department of Energy, Energy Efficiency and Renewable Energy, Battelle, www.nrel.gov/docs/fy04osti/35523.pdf.
69. J. N. Chheda, G. W. Huber and J. A. Dumesic, *Angew. Chem. Int. ed.*, 2007, **46**, 7164.
70. G. W. Huber, J. N. Chheda, C. J. Barrett and J. A. Dumesic, *Science*, 2005, **308**, 1446.
71. K. Lourvanij and G. L. Rorrer, *Ind. Eng. Chem. Res.*, 1993, **32**, 11.
72. K. Lourvanij and G. L. Rorrer, *J. Chem. Technol. Biotechnol.*, 1997, **69**, 35.
73. C. Moreau, R. Durand, F. Alies, M. Corillon, T. Frutz and M. Teholeyre, *Ind. Crops Prod.*, 2000, **11**, 237.
74. D. Mercadier, L. Rigal, A. Gaset and J. P. Gorrichon, *J. Chem. Technol. Biotechnol.*, 1981, **31**, 497.
75. P. Carniti, A. Gervasini, S. Biella and A. Auroux, *Catal. Today*, 2006, **118**, 373.
76. H. Zhao, J. E. Holladay, H. Brown and Z. C. Zhang, *Science*, 2007, **316**, 1597.
77. S. Biella, L. Prati and M. Rossi, *J. Catal.*, 2002, **206**, 242.
78. C. Baatz and U. Prüβe, *Catal. Today*, 2007, **122**, 325.
79. T. Ishida, N. Kinoshita, H. Okatsu, T. Akita, T. Takei and M. Haruta, *Angew. Chem. Int. ed.*, 2008, **47**, 9265.
80. A. Abbadi and H. van Bekkum, *J. Mol. Catal. A: Chem.*, 1995, **97**, 111.
81. M. Besson and P. Gallezot, *Catal. Today*, 2000, **57**, 127.

82. M. Besson, F. Lahmer, P. Gallezot, P. Fuertes and G. Flèche, *J. Catal.*, 1995, **152**, 116.
83. M. Besson, G. Fleche, P. Fuertes, P. Gallezot and F. Lahmer, *Recl. Trav. Chim. Pays-Bas*, 1996, **115**, 217.
84. I. A. Grigor'eva, S. S. Chernaya and S. R. Trusov, *Russ. J. Appl. Chem.*, 2001, **74**, 2021.
85. M. Ibert, F. Marsais, N. Merbouh and C. Brückner, *Carbohydr. Res.*, 2002, **337**, 1059.
86. J.-F. Thaburet, N. Merbouh, M. Ibert, F. Marsais and G. Queguiner, *Carbohydr. Res.*, 2001, **330**, 21.
87. F. S. Asghari and H. Yoshida, *Ind. Eng. Chem. Res.*, 2006, **45**, 2163.
88. B. F. M. Kuster, *Starch Starke*, 1990, **42**, 314.
89. Y. Róman-Leshkov, J. N. Chheda and J. A. Dumesic, *Science*, 2006, **312**, 1933.
90. W. M. Kruse and L. W. Wright, *Carbohydr. Res.*, 1978, **64**, 293.
91. M. A. Andrews and S. A. Klaeren, *J. Am. Chem. Soc.*, 1989, **111**, 4131.
92. P. Gallezot, P. J. Cerino, B. Blanc, G. Flèche and P. Fuertes, *J. Catal.*, 1994, **146**, 93.
93. G. W. Huber, J. W. Shabaker and J. A. Dumesic, *Science*, 2003, **300**, 2075.
94. R. D. Cortright, R. R. Davda and J. A. Dumesic, *Nature*, 2002, **418**, 964.
95. G. W. Huber, R. D. Cortright and J. A. Dumesic, *Angew. Chem. Int. ed.*, 2004, **43**, 1549.
96. D. K. Sohounloue, C. Montassier and J. Barbier, *React. Kinet. Catal. Lett.*, 1983, **22**, 391.
97. B. Blanc, A. Bourrel, P. Gallezot, T. Haas and P. Taylor, *Green Chem.*, 2000, **2**, 89.
98. J. Defaye, A. Gadelle and C. Pedersen, *Carbohydr. Res.*, 1990, **205**, 191.
99. C. Montassier, J. C. Ménézo, J. Moukolo, J. Naja, L. C. Hoang, J. Barbier and J. P. Boitiaux, *J. Mol. Catal.*, 1991, **70**, 65.
100. P. D. Wulf, W. Soetaert and E. J. Vandamme, *Biotechnol. Bioeng.*, 2000, **69**, 339.
101. A. T. Brown and C. E. Patterson, *Archs. Oral. Biol.*, 1973, **18**, 127.

CHAPTER 14

Conversion of Carbohydrates to Liquid Fuels

GEOFFREY AKIEN, LONG QI AND ISTVÁN T. HORVÁTH

Department of Biology and Chemistry, City University of Hong Kong, Kowloon, Hong Kong SAR, China

14.1 Introduction

The development of the liquid fuel-based transportation system was perhaps one of the most important and enabling technologies of the industrial revolution. Just imagine the fuel delivery/distribution and the solid/liquid waste collection/removal problems of a major city of several millions, mainly using horses for transportation.[1] Luckily, the fossil-fuel-based transportation fuels such as gasoline, diesel, and kerosene are liquids, which can be easily stored and distributed, and the final products of their combustion, carbon dioxide and water, are readily dispersed in the environment. Similarly, the production of electricity from fossil fuels is also convenient, because it can respond readily to rapidly changing energy demands. In general, fossil fuels have been used either directly, or through the generation of electricity, to improve the quality of life by controlling the temperature of living, travelling, and working places, securing efficient refrigeration to maintain healthy foods all the way from the source to consumption, providing electricity to equipment for work and entertainment, and finally, allowing us to travel anywhere in the air, on water and on land. One of the most challenging impacts of the rapidly growing global population is the depletion of the fossil resources[2] that provide 90% of all of our energy needs.[3] Therefore, the development of alternative energy sources is

RSC Energy and Environment Series No. 1
Thermochemical Conversion of Biomass to Liquid Fuels and Chemicals
Edited by Mark Crocker

Published by the Royal Society of Chemistry, www.rsc.org

crucial to the sustainability of mankind.[4] While it is difficult to predict the exact date of the depletion of fossil fuels, the transition to renewable resources should be accelerated because of the frequently and unexpectedly changing political/economical environments, resulting in limited access to and rising costs of fossil fuels. Since the application of sustainable technologies for energy, including the efficient conversion of solar energy to electricity or hydrogen, might take longer than is expected,[5] the development of alternative liquid fuels has been considered a key research area in the last two decades.[6] In addition, shifting the production of carbon-based chemicals from fossil fuels to renewable raw materials is also a key issue of sustainability.

14.2 Carbohydrate-Containing Biomass

Photosynthesis utilizes the energy of the sun to convert carbon dioxide and water into 170 billion tons of biomass each year, of which 75% are carbohydrates with an empirical formula of $C_6H_{12}O_6$:[7]

$$6CO_2 + 6H_2O \xrightarrow{h\nu} C_6H_{12}O_6 + 6O_2 \tag{14.1}$$

The simplest carbohydrates are the naturally occurring 5- and 6-carbon monosaccharides, the pentoses and hexoses, respectively, which can be linked together by glycosidic bonds in various combinations to form di-, oligo-, and polysaccharides.[8] Sucrose, ordinary table sugar, is the derivative of glucose and fructose (Figure 14.1) and is the largest volume single chemical in the world – 161 million tons in 2008/2009.

More than 100 countries produce sucrose, 80% of which is made from sugarcane grown in the tropical and subtropical areas of the world. Generally, the costs of sugar production from sugarcane are lower than from sugar beets. Starch consists of 300–600 glucose units linked to form linear amylose (20–25%) (Figure 14.2) and branched amylopectin (75–80%) (Figure 14.3). Starch is the most important component of rice, wheat, corn, potatoes, and cassava.

The largest proportion of the biomass is lignocellulose (typically containing 40-50 wt% cellulose, 25 wt% hemicellulose, and 25 wt% lignin), which provides

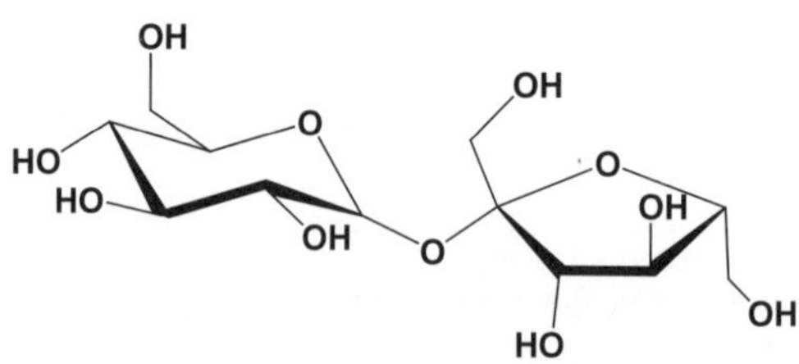

Figure 14.1 Sucrose.

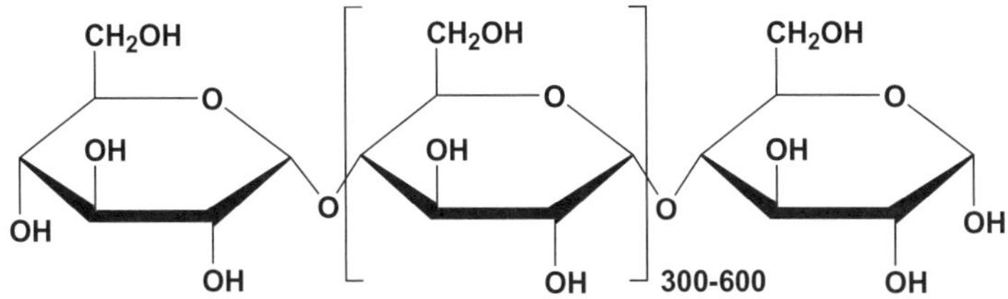

Figure 14.2 Amylose.

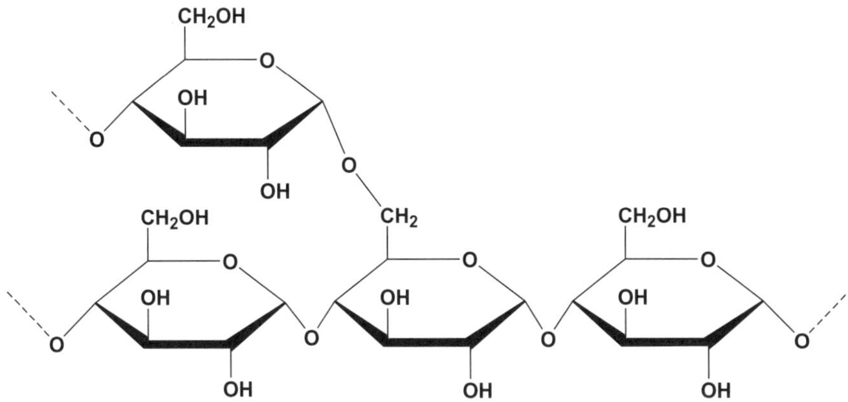

Figure 14.3 Typical branching in amylopectin.

strength to trees and plants. Cellulose is a mixture of linear polymers of 7000–15 000 glucose units (Figure 14.4), depending on the source.

While the linear chains are stabilized by intramolecular hydrogen bonds, adjacent parallel chains are bonded to each other by intermolecular hydrogen bonds, resulting in a crystalline and strong material. Most mammals cannot digest cellulose as a food, and this part of the vegetable intake is eliminated largely unchanged. However, cows and termites have intestinal micro-organisms that can convert cellulose to monosaccharides by the use of β-glycosidase enzymes. Cellulose is generally accompanied by hemicellulose, which is a lower molecular weight amorphous polymer containing mostly pentoses (500–3000 units) and is characterized by a branched structure. Consequently, hemicellulose is readily hydrolyzed by enzymes, acids or bases. Polysaccharides composed of mannose, galactose, xylose and arabinose are also known. Chitin,[9] a linear polymer of N-acetyl-D-glucose-2-amine monomer, is another polysaccharide that is readily available in large quantities (Figure 14.5), especially among coastal communities.

Finally, agricultural and food wastes could be another important feedstock, improving not only the sustainability of fuels, but also eliminating the negative environmental impacts of the used wastes.[10]

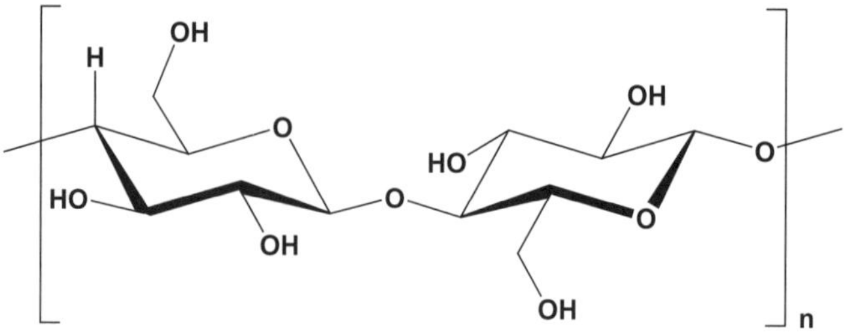

Figure 14.4 Cellulose.

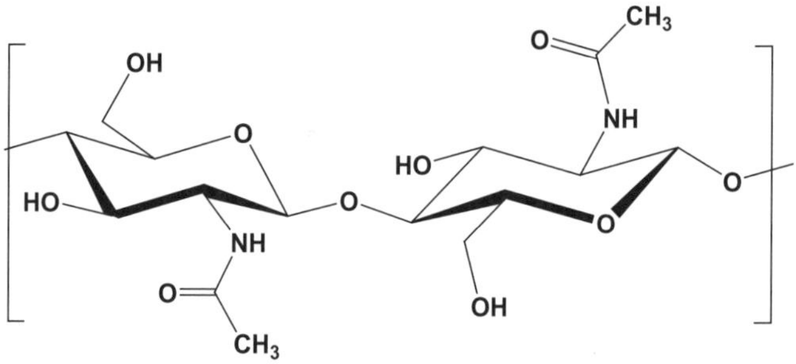

Figure 14.5 Chitin.

Carbohydrates are also present in algae as energy storage and structural materials[11] and the percentage of carbohydrates in dry algae matter can vary from 8 to 64% (Table 14.1).[12a] It should be noted that two species of algae in particular cultured in reduced-nutrient media had significantly greater dry weights and contained more carbohydrates and less proteins than those grown in the usual culture media.[12b] The obvious advantage of using algae as biomass for liquid fuels is the use of saltwater-based technologies, which will not compete for freshwater with human consumption and food production, provided the algae farms are ocean- or sea-based.

14.3 Potential Biomass-Based Liquid Fuels

The purpose of any liquid fuel is to provide energy for the user on demand. The sustainability of a liquid fuel requires that it can be produced from renewable resources by requiring less energy than it can provide. Thus, the energy-out – energy-in ratio must be higher, preferably much higher, than one. In addition, a sustainable liquid fuel should be continuously available, safe to use, store and transport, and have environmentally and health-friendly chemical and physical properties. As the primary users of liquid fuels have been the transportation industry, engine technologies have dominated the properties of the preferred

Table 14.1 General composition of different algae (% of dry matter).

Algae	*Protein*	*Carbohydrates*	*Lipids*
Anabaena cylindrica	43–56	25–30	4–7
Aphanizomenon flos-aquae	62	23	3
Chlamydomonas rheinhardii	48	17	21
Chlorella pyrenoidosa	57	26	2
Chlorella vulgaris	51–58	12–17	14–22
Dunaliella salina	57	32	6
Euglena gracilis	39–61	14–18	14–20
Porphyridium cruentum	28–39	40–57	9–14
Scenedesmus obliquus	50–56	10–17	12–14
Spirogyra sp.	6–20	33–64	11–21
Arthrospira maxima	60–71	13–16	6–7
Spirulina platensis	46–63	8–14	4–9
Synechococcus sp.	63	15	11

fuels, and only the environmental issues have forced the development of much more efficient engines and advanced fuels with designer properties. A variety of carbohydrates containing biomass-based organics including methanol,[13] ethanol,[14] butanol,[15] gamma-valerolactone (GVL),[16] and 2-methyl-tetrahydrofuran[17] have been proposed to either replace some of the conventional fuels such as gasoline, diesel, and kerosene, or to blend them in different proportions. While drop-in biofuels would be the best solution, the physical and chemical properties of these molecules are unique enough that their large-scale use might require significant changes in engine design.

14.4 Conversion of Carbohydrates to Liquid Fuels

14.4.1 Strategies

The strategies for sustainable conversion of the carbohydrates from biomass depend on, and are frequently limited by, the structures/properties of both the target fuels and the starting carbohydrates. Although the production and use of single chemical entities (*e.g.*, ethanol from glucose), instead of a mixture of compounds (*e.g.*, oxygenates from cellulose), could significantly simplify their worldwide monitoring and regulations, most biomass consists of a complex mixture of mono-, oligo- or polysaccharides incorporated or just included with other organic compounds. In the case of oligo- and polysaccharides, the depolymerization of the glycosidic bonds by catalytic hydrolysis using homogeneous or heterogeneous catalysts, including enzymes, seems to be the preferred route to produce monosaccharides for subsequent conversion; all in one pot or in separate steps.[18] The conversion of monosaccharides generally involves the stepwise removal of oxygen atoms by dehydration to generate double-bond-containing intermediates and products. For example, the dehydration of glucose or fructose in the presence of acids results in the formation of 5-hydroxymethyl-2-furfural (Scheme 14.1).

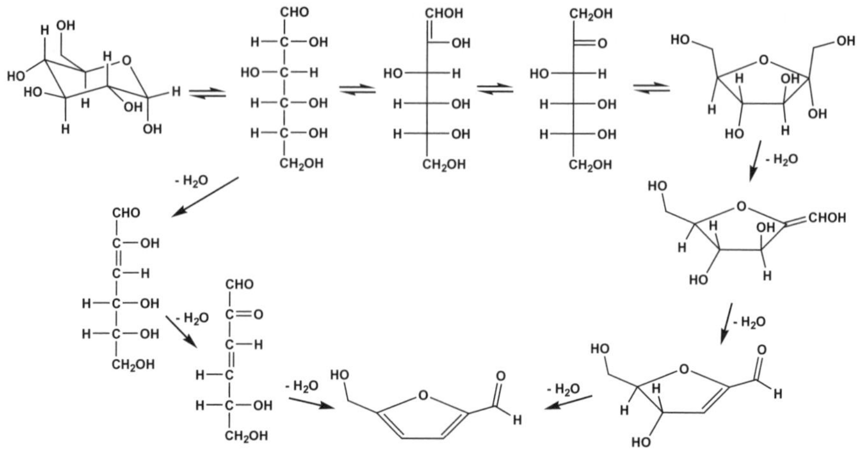

Scheme 14.1 Formation of 5-hydroxymethyl-2-furfural from glucose or fructose.

Since water is the product of the dehydration, hydration of the intermediates or the products is frequently an intrinsic and unavoidable step. While the intermediates and/or products with lower numbers of oxygen atoms could be dimerized, oligomerized, and even polymerized or reacted with other molecules (such as hydrogen, aldehydes, ketones, *etc.*) *in situ*, the stepwise removal of all of the oxygen atoms can also be carried out to produce hydrocarbons, though frequently leading to less carbon-containing chains than the starting carbohydrates. Another approach targets the production of a single intermediate, which could serve as a platform molecule to access different products including various liquid fuels.[19] Since drop-in technologies have been more attractive for the end-users of the fuels, the generation of "gasoline- or diesel-like" products has been preferred, especially because of the enormous number of operating engines in the world. Alternatively, the large-scale conversion of a biomass based product with "nongasoline- or nondiesel-like" combustion properties could be more feasible commercially, but replacement of the existing engine technologies with a new fuel-friendly system may take a very long time or may never happen at all.

14.4.2 Methanol

A methanol-based future was proposed by Olah and the concept of the methanol economy has been summarized.[20] While methanol has attractive properties for fuel applications[20,21] its high miscibility with water combined with its acute toxicity if ingested[22] should cast serious doubts on its global use. Since the technology to produce methanol from CO/H_2 has been established,[23] its availability from carbohydrates depends on the conversion of carbohydrates to CO/H_2 (see Chapter 4).

14.4.3 Ethanol

The use of ethanol as a liquid fuel preceded gasoline as some of the earliest combustion engines were run on ethanol.[14] Indeed, ethanol is a key liquid fuel produced from biomass, with the largest producers being the United States using corn, and Brazil using sugarcane, producing 34 and 22.5 billion liters per year, respectively.[24] European production is much less at 2.7 billion liters per year, largely from sugar beet, and is substantially more expensive, so will not be discussed further here. A comparison of the different crops used for sugar production is shown in Table 14.2.[25]

The ethanol in the United States is almost entirely produced from corn starch, where it comprises ~65 wt% of the grain on a dry basis.[18] The corn starch is isolated from the grain by two methods: wet milling and dry grinding.[30] In wet milling, the grains are soaked in dilute sulfuric acid for 1–2 days to separate the different components. The resulting slurry is then passed through a series of grinders, a screen, and centrifugal separators to obtain the starch in dilute solution form. In dry milling, the entire corn kernel is ground and water added to form the slurry. Once the starch is separated, glucoamylase and α-amylase enzymes are used to convert the starch to D-glucose, and ammonia is used to control the pH, as well as acting as a nutrient for the yeast used in later steps for fermentation.

$$\{C_6H_{12}O_6\}_n + nH_2O \rightarrow nC_6H_{12}O_6 + \text{heat} \quad (14.2)$$

The resulting sugar solution is processed at high temperatures to reduce the bacteria levels, then transferred and cooled in fermenters, where yeast (usually *Saccharomyces cerevisiae*, more commonly known as baker's yeast) is added to start the conversion. In contrast, the main soluble carbohydrate form in sugarcane is the disaccharide sucrose, the hydrolysis of which is promoted by the conditioning of the cane juice. The fermentation yeast is also capable of hydrolyzing sucrose to glucose and fructose.[31] Therefore, a discrete additional hydrolysis step is not necessary for sugarcane, and the cane juice extracted by crushing the milled stalks can be fermented directly.[32] The fermentation process takes approximately 2 days, with the broth being continually agitated and

Table 14.2 Comparison of bioethanol production processes.

	Sugarcane (in Brazil)	*Corn (in the US)*
Energy balance[a] [26]	8.1–10	0.77–1.4[b] [27]
Production cost (€/100 l)[28]	14.48	24.83
CO_2 reduction compared to gasoline[29]	84%	30%
Total production (billion l)	22.5	34
Area cultivated (million hectares)	3.4	8.13
Yield (l per hectare)	6471	4182

[a]Ratio of the energy output from ethanol to the ratio of fossil fuel inputs; higher than 1 is a saving.
[b]This value depends on the parameters of the assessments, though most recent studies tend to be more than 1.

$$C_6H_{12}O_6 + 2\,ADP + 2\,PO_4^{3-} + 2\,NAD^+ \longrightarrow 2\,CH_3COCH_2^- + 2\,ATP + 2\,NADH + 2\,H_2O + 2\,H^+$$

$$2\,CH_3COCH_2^- + 2\,H^+ \longrightarrow 2\,CH_3CHO + 2\,CO_2$$

$$2\,CH_3CHO + 2\,NADH \longrightarrow 2\,C_2H_5O^- + 2\,NAD^+$$

$$2\,C_2H_5O^- + 2\,H^+ \longrightarrow 2\,C_2H_5OH$$

Scheme 14.2 Fermentation of glucose to ethanol, *via* pyruvate and acetaldehyde intermediates.[33]

cooled to dissipate the heat generated.

$$nC_6H_{12}O_6 \rightarrow 2nC_2H_5OH + 2nCO_2 \tag{14.3}$$

Two-thirds of the carbon from the sugar is lost as carbon dioxide. It should be noted that one molecule of water is required for two molecules of ethanol in processing, not including the additional process water. Conveniently, *Saccharomyces cerevisiae* prefers fermentation to respiration (*i.e.* to CO_2 and H_2O), in both aerobic and anaerobic conditions, in contrast with many other microorganisms. A more detailed mechanism is described in Scheme 14.2.

The prospects of improving the 1st-generation sugarcane process are good, with an estimated doubling of production possible using genetic modification of the sugarcane, and a tenfold increase in production from increasing the area cultivated.[32]

A potentially more sustainable production method for bioethanol uses cellulose-containing biomass. In the corn-ethanol processes, only the grain, and indeed, only part of that, is used for bioethanol manufacture. In the sugarcane process, only the extracted cane juice is used for ethanol, with the bagasse residue usually burnt to power the sugar-refining plants. The second-generation cellulosic bioethanol processes can in principle use any cellulose-containing biomass, be it waste or otherwise. If the ethanol from lignocellulosic biomass replaces gasoline, then greenhouse-gas emissions can be cut by up to 85%.[34] The key difference between current bioethanol technologies and cellulosic ethanol is the need for the sugars to be released from the lignocellulose. At present, this is very expensive, amounting to as much as 40–45% of the projected total cost,[35] the pretreatment step alone accounting for 18–20% of the total cost. Reducing this expense will be possibly one of the most significant challenges in commercializing cellulosic ethanol technologies. Despite this, commercial quantities of cellulosic ethanol are due to be available in 2010.[36]

Biomass is mainly composed of lignocellulose, which is in turn composed of cellulose, hemicelluloses, and lignin. The composition of different potential feedstocks has been described.[31] As there is not enough waste to produce the required amounts of ethanol, other energy crops will need to be grown. The ideal energy crop is fast growing, requires minimal inputs (*e.g.* water and fertilizer) and can grow on poor-quality land. Crops that have been considered include switchgrass, poplar,[31] and *Miscanthus giganteus* (a large perennial grass).

The cellulose and hemicellulose can be broken down into their sugar constituents using cellulases. However, the rigid structure of cellulose prevents the cellulases accessing most of the cellulose, so it must be pretreated first. The first pretreatment is a physical one; the biomass must be ground or crushed into smaller particles to increase the surface area : volume ratio. The following step can take many forms: acid hydrolysis, steam explosion, ammonia fiber expansion, alkaline wet oxidation, ozonation; the particular type of pretreatment depends on the type of lignocelluloses, it not being uncommon for two or more pretreatments to be required. An effective pretreatment strategy should (a) break the lignocellulosic complex, (b) decrease the cellulose crystallinity, (c) preserve the hemicelluloses sugars, (d) limit the formation of degradation products that inhibit hydrolysis and fermentation, (e) minimize energy inputs and use of extraneous chemicals, (f) be simple to implement, (g) generate a high-value lignin coproduct, (h) minimize the production, if any, of toxic and hazardous wastes, and (i) generate the minimum amount of wastewater.[35] As an example, one of the more common chemical pretreatments is acid hydrolysis, as it can access 80–90% of the sugars available from the hemicelluloses. The acid acts to disrupt the hydrogen bonding within the lignocellulosic matrix, effectively decrystallizing it so it can be hydrolysed. Dilute acid hydrolysis (0.7–3.0% aqueous acid) requires high temperatures (200–240 °C), and produces inhibitors such as furfural, HMF, organic acids, and phenols. These inhibitors act against the yeasts used in the following fermentation step, and must be removed, *e.g.*, by using activated carbon. On the other hand, concentrated acid hydrolysis ($>70\%$) allows the use of lower temperatures, and does not produce inhibitors. The key issue with the concentrated acid process is that the consumption of acid is high, and recovery of the acid is difficult, requiring membrane or chromatographic techniques to remove it. The resulting liquor from the acid hydrolysis must also be neutralized, which is expensive and generates large amounts of waste.

Biological pretreatments using cellulases are at an early stage.[37] The cellulases are usually isolated from fungi like *Trichoderma reesei* and bacteria such as *Cellulomonas fimi*, and are actually a complex of three different classes of enzymes. This complexity, in combination with the heterogeneous nature of the process, *i.e.* soluble enzymes acting on insoluble substrates, makes cellulases challenging to understand, and therefore more difficult to rationally design.[38] One line of research involves the generation of artificial cellulase complexes by mixing the different components of cellulases from different organisms.

Fermentation is then carried out in another step using similar technology to first generation bioethanol. However, since the products from hemicellulose degradation are pentoses that *Saccharomyces cerevisiae* cannot natively ferment to ethanol,[39] the fermenting micro-organisms must be changed.[39]

It is possible to combine the cellulolysis and fermentation steps together in the simultaneous sacharification and fermentation (SSF) process.[31,40] The products from the hydrolysis step are consumed by the fermenting micro-organisms, so that problems with inhibition of the hydrolysis step are avoided. The capital cost can also be lower as one unit operation can be eliminated, compared to conducting the hydrolysis and fermentation in separate steps.

The ethanol produced from fermentation is in aqueous solution (for example, 39% ethanol in a cellulosic process[41]), so must be distilled first. Ethanol–water mixtures form a low-boiling azeotrope that limits the condensate purity to 95.6% ethanol. However, to use ethanol in a gasoline blend, the water content must then be reduced further. Older processes added benzene or cyclohexane to break the azeotrope, but many modern plants use molecular sieves to remove the water instead. Hot, pressurized, ethanol vapor is passed over a bed of molecular sieves. Two beds are run in parallel, with each bed being periodically regenerated by the application of vacuum. This method is much more energy efficient than the older azeotropic (or extractive) distillations.

Ethanol can be mixed in any proportion with gasoline, as long as the ethanol is anhydrous. The blends are named, *e.g.*, E10, where the fuel is a blend of 10% ethanol with 90% gasoline. Hydrous ethanol (usually containing 4.4% water) will phase separate with gasoline below 30% ethanol, causing problems in fuel lines, *etc.* The energy density of ethanol is 66% that of gasoline,[42] so a larger volume of fuel needs to be consumed for the same energy output. Ethanol is also more corrosive than gasoline, so cannot be distributed in existing pipelines. Most gasoline engines can tolerate E10 without modification, although performance can be affected due to the slightly different properties of the blend. Most ethanol–gasoline blends on the market are either low (E5–E15), or high (E85) ethanol. High-ethanol blends almost always require engine modification, but mean that the performance is reduced if using pure gasoline; this can be an issue where the availability of high-ethanol blends is intermittent.

Ethanol has found wide usage as an oxygenate in fuels, replacing the methyl *tert*-butyl ether (MTBE) being phased out due to groundwater-contamination issues. The oxygenates in transport fuels act to reduce the emissions of byproducts from partial combustion, *e.g.*, carbon monoxide, hydrocarbons, and particulates.[31] Since biomass does not possess the sulfur impurities present in crude oil, sulfur emissions are reduced, but oxygenated fuels tend to have increased nitrogen-oxide emissions.

14.4.4 Butanol

Butanol is an "advanced" biofuel. There are two major large-scale attempts at commercialization: DuPont and BP in the UK with British Sugar using the ABE process,[43] and Gevo using *iso*-butanol.[44] Both have pursued strategies involving the retrofitting of existing bioethanol facilities. Research on butanol from lignocellulosic feedstocks is also under way.[45]

Butanol is perhaps a more appropriate biofuel because it is more similar to gasoline than ethanol, with no modification of engines appearing to be required.[46,47] The energy density of butanol is only 10–20% less than gasoline, and it is also less corrosive than ethanol so can be transported using existing gasoline pipelines. An optimal blend of 30% butanol with 70% gasoline was found to reduce the unburnt hydrocarbons in the exhaust by 26%.[48] The vapor pressure of butanol is lower than ethanol, but because its latent heat of vaporization is lower too, cold starts are less of an issue than with ethanol. Existing bioethanol plants can easily be retrofitted to butanol because only the enzymes and distillation require minor changes. As butanol is not very hygroscopic, it can also be blended with gasoline at the source.[47] However, consumer acceptance of the smell of butanol may be a problem.

Typical enzymes used for fermentation include *Clostridium acetylbutylicum* and *Clostridium beijerinckii*, with both of them producing acetone and ethanol (among others) as coproducts in the ABE process.[49] Interestingly, compounds that inhibit the fermentation to ethanol, *e.g.*, furfural and HMF, actually stimulated production of butanol with *Clostridium beijerinckii*. Examples of genetically engineered *Escherichia coli* are known, but are in a much earlier stage of development.[50] It should be noted that the Clostridia micro-organisms are capable of metabolizing both hexoses and pentoses,[45] something that the most common ethanol fermentation enzymes cannot. A remarkable recent achievement has been the production of genetically engineered cyanobacteria capable of producing *iso*butyraldehyde and iso-butanol, directly from CO_2.[51]

The metabolic pathways of *Clostridium acetobutylicum* are described,[47] and the genome has been sequenced,[52] paving the way for genetic modification of the micro-organisms. The Clostridia metabolic pathways are similar to that of ethanol-producing micro-organisms, but there is an additional competing pathway at the pyruvate intermediate, going *via* butyraldehyde to produce the butanol.[47,52] Clostridia metabolism ceases in the presence of 20 g/L of solvents,[53] but research efforts are continuing to improve the tolerance.[54,55]

Like ethanol from fermentation, the butanol is in a dilute aqueous solution so must be separated. Current processes involve distillation, but the lower vapor pressure means that the distillation is more expensive, so other methods of separation are being researched, such as gas stripping, adsorption, liquid–liquid extraction, perstraction, pervaporation, and reverse osmosis.[53,45]

14.4.5 Methyl and Ethyl *Tertiary*-butyl ether

Methyl *tertiary*-butyl ether (MTBE) was originally used in gasoline as an octane booster and antiknock agent in 1979. Although it was later introduced to the gasoline market to improve car emissions,[56] its higher solubility in water became a major environmental issue[57] and no process is known for its direct production from renewable resources.[58] However, the development of biomass-based methanol and *iso*-butylene technologies could open the door to its application again. Similarly, biomass-based ethyl *tertiary*-butyl ether (ETBE), which has been introduced as a gasoline additive in EU countries,[59] requires the development of biomass based *iso*-butylene technology.

14.4.6 Gamma-Valerolactone

Gamma-valerolactone (GVL), a naturally occurring chemical in fruits and a frequently used food additive, is proposed to exhibit the most important characteristics of an ideal sustainable liquid, which could be used for the production of both energy and carbon-based consumer products.[16] GVL is renewable, easy and safe to store and move globally in large quantities, has low melting (−31 °C), high boiling (207 °C) and open cup flash (96 °C) points, a definitive but acceptable smell for easy recognition of leaks and spills, and is miscible with water, assisting biodegradation. A comparative evaluation of GVL and ethanol as fuel additives, performed on a mixture of 10 v/v% GVL or EtOH and 90 v/v% 95-octane gasoline, showed very similar properties. Since GVL does not form an azeotrope with water, the latter can be readily removed by distillation, resulting in a less energy-demanding process for the production of GVL than that of absolute ethanol.

The biomass-based synthesis of GVL could start with the dehydration of carbohydrates to 5-hydroxymethyl-2-furfural (HMF) followed by its catalytic hydrolysis to levulinic and formic acids, and completed by the reduction of levulinic acid.[60]

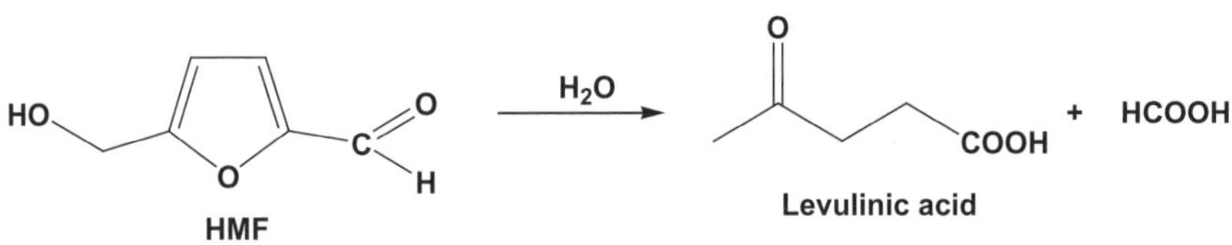

Levulinic acid has been produced from agricultural wastes or cellulose and hemicellulose by the same technology since the 1940s.[61] The initial hydrolysis with strong acids (HCl or H_2SO_4) around 100 °C to produce hexoses is followed by a treatment with dilute HCl at 200 °C. The catalytic hydrogenation of levulinic acid has been extensively studied using heterogeneous[62] and homogeneous[63] catalysts. The complete transfer-hydrogenation of levulinic acid with formic acid to GVL has also been demonstrated.[64]

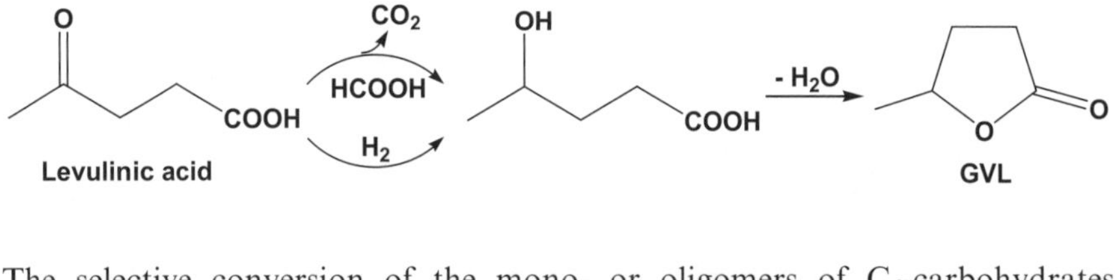

The selective conversion of the mono- or oligomers of C_6-carbohydrates directly to levulinic acid and formic acid is a particularly attractive approach. Since the formic acid can be converted to carbon dioxide and hydrogen, the latter can be used for the hydrogenation of levulinic acid to GVL, making the process hydrogen independent and readily useable even in remote parts of the world. Alternatively, formic acid could be used in the transfer hydrogenation of levulinic acid to GVL. It should be noted that the hydrogenation of GVL could lead to the formation of 1,4-pentanediol, which can be readily converted to 2-MeTHF under hydrogenation conditions.[65] The formation of even a few hundred ppm of 2-MeTHF could be a safety issue, as it will readily form peroxides under ambient conditions and any liquid containing more than 80 ppm peroxides is considered hazardous.[66]

14.4.7 2-Methyl-Tetrahydrofuran

2-Methyl-tetrahydrofuran (2-Me-THF) has been listed as a renewable component of the alternative P-fuel[17] and can be synthesized by the catalytic hydrogenation of levulinic acid and/or GVL.[65]

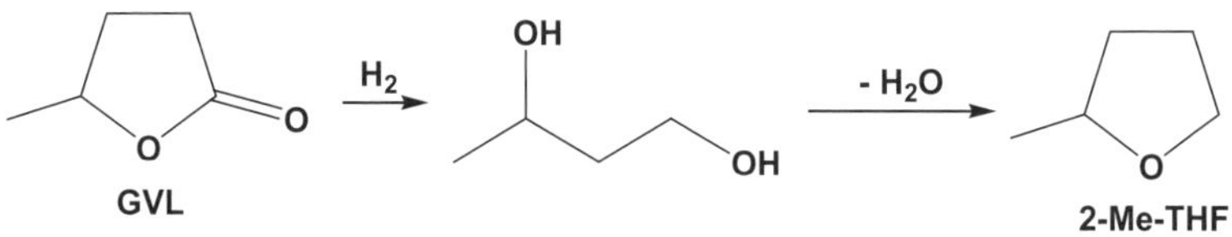

14.4.8 2-Furfural

2-Furfural is the only heterocyclic compound that is produced in large scale (300 000 tons/year) from xylose, the most common five-carbon sugar (pentose) in the world and the main dehydration product of hemicellulose, comprising almost one-third of the carbohydrates contained in woody plants.[67] Many agricultural residues (corncobs, bagasse, cornstalks, rice hulls, *etc.*) can be used to produce 2-furfural by acid-catalyzed dehydration at 200–250 °C. Sulfuric acid is the most frequently used acid with the highest possible yield being only 70% due to the formation of reactive intermediates.[68] Furfural could be used directly as a solvent, and it is an important starting material for the synthesis of a series of derivatives including furfuryl alcohol, furoic acid, furfan, tetrahydrofuran, and 2-methyl-tetrahydorfuran.

5-(Ethoxymethyl)furfural (EMF) and 5-methyl furfural (MF) have been proposed as biofuel candidates, however, both EMF and MF cannot be synthesized directly, but are prepared through an intermediate, 5-(chloromethyl)furfural.[69] It should be noted that 5-hydroxymethyfurfural can be converted into a range of substituted furans and tetrahydrofurans.[70]

14.5 Conclusions – Do We Have Enough?

The total energy consumption of the USA in the forms of coal, crude oil, and natural gas was 2050.20 million tones oil equivalent in 2008.[3] After a simple calculation one could easily conclude that about 23 000 million barrels of ethanol (based on 3.73 GJ energy per barrel ethanol) would be required to replace all fossil resources. In sharp contrast, the estimated production level of biomass-based fuel ethanol was about 220 million barrels in 2008[71]– a drop in the bottom of an empty barrel.

The various biomass-production technologies and conversion schemes therefore are facing difficult sustainability questions with serious global and societal implications. How much of the arable land should or could be used for edible or nonedible biomass production? Do we have enough land and freshwater locally or regionally to secure continuous or semi-continuous production? Will the emerging biomass production and conversion technologies bring unexpected and unknown environmental problems?[72] While the production of "biofuels" has been rapidly expanding, many of these questions have remained unanswered, probably significantly increasing the risks by moving towards an uncertain future.

References

1. J. Tarr and C. MacShane, in *The Making of Urban America*, R. Mohl, (ed.), SR Publishers, NY, 1997, pp. 105–130.
2. K. S. Deffeyes, *Beyond Oil: The View from Hubbert's Peak,* Farrar, Straus and Giroux, New York, 2005.
3. BP Statistical Review of World Energy 2009 (www.bp.com).
4. D.J.C. MacKay, Sustainable Energy – Without the Hot Air, Kindle, UIT Cambridge Ltd., Cambridge, 2009.
5. (a) J.-E. Moser, *Nature Mater.*, 2005, **4**, 723; (b) National Research Council, The hydrogen economy: opportunities, costs, barriers and R & D needs, National Academies Press, Washington D. C., 2004.
6. D. L. Klass, *Biomass for Renewable Energy, Fuels and Chemicals*, Elsevier, Amsterdam, 1998.
7. F. W. Lichtenthaler, *Acc. Chem. Res.*, 2002, **35**, 728.
8. J. F. Robyt, *Essentials of Carbohydrate Chemistry,* Springer, New York, 1998.
9. E. Khor, *Chitin: Fulfilling a Biomaterials Promise,* Elsevier, Amsterdam, 2001.
10. V. Oreopoulou and R. Winfried, *Utilization of Byproducts and Treatment of Waste in the Food Industry,* Springer, New York, 2007.
11. (a) Photosynthesis in Algae, *Advances in Photosynthesis and Respiration, Vol. 14*, S.E. Douglas, A.W.D. Larkum, J.A. Raven, ed., Kluwer Academic Publishers, Dordrecht, 2004; (b) L.E. Graham, J.E. Graham and L.W. Wilcox, Algae, Benjamin Cummings, San Francisco, 2008.
12. (a) E. W. Becker, *Biotechnol. Adv.*, 2007, **25**, 207; (b) G. H. Wikfors, J. W. Twarog and R. Ukeles, *Biol. Bull.*, 1984, **167**, 251.
13. F. Asinger, *Methanol – Chemie- und Energierohstof,* Springer-Verlag, Berlin, 1986.
14. Ford Predicts Fuel from Vegetation in *New York Times*, 1925, Sept. 20, p. 24.

15. T. C. Ezeji, N. Qureshi and H. P. Blaschek, *Curr. Opin. Biotech.*, 2007, **18**, 220.
16. I. T. Horváth, H. Mehdi, V. Fábos, L. Boda and L. T. Mika, *Green Chem.*, 2008, **10**, 238.
17. (a) S. F. Paul, US patent 5,697,987 (1996). (b) DOE, Alternative Fuel Transportation Program; P-series fuels (Proposed Rules), *Federal Register*, 1998, **63**, 40202.
18. G. W. Huber, S. Iborra and A. Corma, *Chem. Rev.*, 2006, **106**, 4044.
19. A. Corma, S. Iborra and A. Velty, *Chem. Rev.*, 2007, **107**, 2411.
20. G. A. Olah, A. Goeppert and G. K. S. Prakash, *Beyond Oil and Gas: The Methanol Economy,* Wiley-VCH, Weinheim, 2006.
21. G. A. Olah, *Angew. Chem. Int. Ed.*, 2005, **44**, 2636.
22. I. L. Bennett Jr, F. H. Cary, G. L. Mitchell Jr and M. N. Cooper, *Medicine (Baltimore)*, 1953, **32**, 431.
23. S. Lee and R. Iredell, *Methanol Synthesis Technology,* CRC Press, Boca Raton, 1990.
24. http://www.ethanolrfa.org/industry/statistics/#E
25. J. Goldemberg and P. Guardabassi, *Biofuels Bioprod. Bioref.*, 2010, **4**, 17.
26. (a) I. C. Macedo, J. E. A. Seabra and E. A. R. Silva, *Biomass Bioenerg.*, 2008, **32**, 582; (b) Worldwatch Institute, State of the World 2006 – Special Focus: China and India, Earthscan, London, 2006.
27. E. C. Petrou and C. P. Pappis, *Energy Fuels*, 2009, **23**, 1055.
28. R. Doornbosch and R. Steenblik, *Biofuels: is the cure worse than the disease?,* OECD – Roundtable on sustainable development, Paris, 2007.
29. O. Henniges, J. Zeddies, FO Licht's World Ethanol and Biofuels Report 2004, 2004, **2**, 374.
30. A. Aden, *Southwest Hydrology*, Sept./Oct. 2007, p. 22.
31. M. Balat, H. Balat and C. Oz, *Prog. Energy and Comb. Sci.*, 2008, **34**, 551.
32. L. Luo, E. Van der Voet and G. Huppes, *Renew. Sustain. Energy Rev.*, 2009, **13**, 1613.
33. A. N. Glazer and H. Nikaido, *Microbial Biotechnology: Fundamentals of Applied Microbiology,* W. H. Freeman, New York, 1995.
34. A. E. Farrell, R. J. Plevin, B. T. Turner, A. D. Jones, M. O'Hare and D. M. Kammen, *Science*, 2006, **311**, 506.
35. S. Banerjee, S. Mudliar, R. Sen, B. Giri, D. Satpute, T. Chakrabarti and R. A. Pandey, *Biofuels Bioprod. Bioref.*, 2010, **4**, 77.
36. C. Sheridan, *Nature Biotechnol.*, 2009, **27**, 1074.
37. Y. Sun and J. Cheng, *Biores. Technol.*, 2003, **83**, 1.
38. J. Sheehand and M. Himmel, *Biotechnol. Progr.*, 1999, **15**, 817.
39. O. P. Ward, *Fermentation Biotechnology: Principles, Processes and Products,* Open University Press, Milton Keynes, 1989.
40. O. P. Ward and A. Singh, *Adv. Appl. Microbiol.*, 2002, **51**, 53.
41. C. N. Hamelinck, G. van Hooijdonk and A. P. C. Faaij, *Biomass Bioenergy*, 2005, **28**, 384.
42. M. Balat, H. Balat and C. Öz, *Progr. Energy Combustion Sci.*, 2008, **34**, 551.

43. http://www.bp.com and http://www2.dupont.com
44. http://www.gevo.com/faqs.php
45. N. Qureshi and T. C. Ezeji, *Biofuels Bioprod. Bioref.*, 2008, **2**, 319.
46. http://www.butanol.com
47. P. Dürre, *Biotechnol. J.*, 2007, **2**, 1525.
48. S. Pukalskas, Z. Bogdanovièius, E. Sendžikienë, V. Makaraevièienë and P. Janulis, *Transport*, 2009, **24**, 301.
49. T. Ezeji, N. Qureshi and H. P. Blaschek, *Biotechnol. Bioeng.*, 2007, **97**, 1460.
50. R. Cascone, *Chem. Eng. Progr.*, 2008, **104**, S4.
51. S. Atsumi, W. Higashide and J. C. Liao, *Nature Biotechnol.*, 2009, **27**, 1177.
52. J. Nölling, G. Breton, M.V. Omelchenko, K.S. Makarova, Q. Zeng, R. Gibson, H. M. Lee, J. Dubois, D. Qiu, J. Hitti, GTC Sequencing Center Production, Finishing, and Bioinformatics Teams, Y. I. Wolf, R. L. Tatusov, F. Sabathe, L. Doucette-Stamm, P. Soucaille, M. J. Daly, G. N. Bennett, E. V. Koonin and D. R. Smith, *J. Bacteriology* 2001, **183**, 4823.
53. S. Y. Lee, J. H. Park, S. H. Jang, L. K. Nielsen, J. Kim and K. S. Jung, *Biotechnol. Bioeng.*, 2008, **101**, 209.
54. Y.-N. Zheng, L.-Z. Li, M. Xian, Y.-J. Ma, J.-M. Yang, X. Xu and D.-Z. He, *J. Ind. Microbiol. Biotechnol.*, 2009, **36**, 1127.
55. R. Gheshlaghi, J. M. Scharer, M. Moo-Young and C. P. Chou, *Biotechnol. Adv.*, 2009, **27**, 764.
56. Section 211(k) and Section 211(m) of the Clean Air Act, USA.
57. A. Keller, J. Froines, C. Koshland, J. Reuter, I. Suffet, J. Last, Health and Environmental Assessment of MTBE, Report to the Governor and Legislature of the State of California as Sponsored by SB 521, Volume I, Summary and Recommendations, University of California, 1998.
58. S. H. Hamid and A. M. Ashraf, *Handbook of MTBE and Other Gasoline Oxygenates,* Marcel Dekker Inc., New York, 2004.
59. Directive 2003/30/EC, *Official Journal of the European Union* 2003, **L 123**, pp. 42–46.
60. H. Mehdi, V. Fábos, R. Tuba, A. Bodor, L. T. Mika and I. T. Horváth, *Top. Catal.*, 2008, **48**, 49.
61. (a) W.W. Moyer, US patent 2,270,328 (1942); (b) S.W. Fitzpatrick, WO 8,910,362 (1989); (c) S.W. Fitzpatrick, WO 9,640,609 (1996).
62. (a) H. S. Broadbent, G. C. Campbell, W. C. Bartley and J. H. Johnson, *J. Am. Chem. Soc.*, 1959, **24**, 1847; (b) H. S. Broadbent and T. G. Selin, *J. Am. Chem. Soc.*, 1963, **28**, 2343; (c) R.M. Bullock, M. Schlaf and E.M. Hauptman, US patent 6,462,206 (2002); (d) R. A. Bourne, J. G. Stevens, J. Ke and M. Poliakoff, *Chem. Commun.*, 2007, 4632; (e) Z. Yan, L. Lin and S. Liu, *Energy & Fuels*, 2009, **23**, 3853.
63. (a) F. Joó, Z. Tóth and M. T. Beck, *Inorg. Chim. Acta*, 1977, **25**, L61; (b) F. Joó, L. Somsak and M. T. Beck, *J. Mol. Catal.*, 1984, **24**, 71; (c) K. Osakada, T. Ikariya and S. Yoshikawa, *J. Organomet. Chem.*, 1982, **231**, 79.

64. I.T. Horváth, H. Mehdi, V. Fábos and N. Kaposy, Hungarian Patent 08 00662 (2008).
65. (a) K. Folkers and H. Atkins, *J. Am. Chem. Soc.*, 1932, **54**, 1145; (b) D.C. Elliott and J.G. Frye, US patent 5,883,266 (1999).
66. V. Fábos, G. Koczó, H. Mehdi, L. Boda and I. T. Horváth, *Energy & Environ. Sci.*, 2009, **2**, 767.
67. D. Montane, J. Salvado, C. Torras and X. Farriol, *Biomass Bioenergy*, 2002, **22**, 295.
68. M. J. Antal, T. Leesoboom, W. S. Mok and G. N. Richards, *Carbohydr. Res.*, 1991, **217**, 71.
69. M. Mascal and E. B. Nikitin, *Angew. Chem. Int. Ed.*, 2008, **47**, 7924.
70. Y. Roman-Leshkov, C. J. Barrett, Z. Y. Liu and J. A. Dumesic, *Nature*, 2007, **447**, 982.
71. Fuel Ethanol Overview, 1981–2008, Energy Information Administration, Annual Energy Review, 2008.
72. I. T. Horváth, *Chem. Rev.*, 1995, **95**, 1.

CHAPTER 15

Biodiesel Production and Properties

JON H. VAN GERPEN AND BRIAN HE

University of Idaho, Department of Biological and Agricultural Engineering, Moscow, Idaho 83844, USA

15.1 Introduction

Diesel engines are used around the world in applications demanding reliability, long operational life and excellent fuel economy. Diesel engines are used almost universally in the long-haul trucking, construction, agricultural, mining, and logging industries. The majority of new automobile sales in Europe are now diesel powered. In the US, inroads to the passenger car market have been mostly limited to light-duty pick-up trucks.

One distinguishing feature of diesel engines is their use of a fuel produced from a higher boiling point fraction of petroleum than that used for gasoline. This portion of the petroleum, known as middle distillate, or diesel fuel, was historically a residual product from gasoline production and available at a substantial discount to gasoline prices. Currently, the diesel market has established itself as an internationally traded commodity with strict quality standards, and diesel fuel is frequently priced higher than gasoline.

Biodiesel is a liquid fuel with properties that make it suitable for use as a direct substitute for petroleum diesel fuel. According to ASTM Standard D6751,[1] biodiesel (B100) is defined as a "fuel comprised of mono-alkyl esters of long chain fatty acids derived from vegetable oils or animal fats". Although vegetable oils or animal fats are specified as the feedstocks for biodiesel

RSC Energy and Environment Series No. 1
Thermochemical Conversion of Biomass to Liquid Fuels and Chemicals
Edited by Mark Crocker

Published by the Royal Society of Chemistry, www.rsc.org

production, other feedstocks, such as algal oil, can also be used. The fuel properties of biodiesel, therefore, are affected by the chemical composition, or fatty-acid profiles, of the feedstock oils or fats.

The terms "renewable diesel" and "green diesel" are being used for alternative diesel fuels other than alkyl esters. Because there are currently no specifications for these fuels, there is considerable variation in how the term is used. One class of renewable diesel fuels are produced by bringing the vegetable oils and animal fats into a petroleum refinery, removing oxygen, adding hydrogen and producing a fuel rich in normal alkanes.[2] The fuel may be coprocessed with a conventional petroleum stream[3] or may be processed separately and maintained as an identifiable stream.[4] Since *n*-alkanes are substantial constituents of conventional diesel fuel, if the renewable diesel is blended with conventional diesel fuel, it may be difficult to distinguish it from the fossil fuel. Another class of renewable diesel fuels involves alkane-rich fuels produced from carbohydrates following processes described by Dumesic and Huber.[5] These fuels are alkane-based like renewable diesel, but the feedstock is a simple sugar and not an oil or fat.

Currently, biodiesel consisting of alkyl esters of oils and fats is the only commercially available biobased alternative fuel for diesel engines, so it will be the primary focus of this chapter.

15.2 Biodiesel Chemistry

15.2.1 Biodiesel Feedstocks

Almost any vegetable oil or animal fat can be used for biodiesel production. Currently, rapeseed oil is the dominant feedstock used in Europe, while soybean oil is the major feedstock in the United States.

Rapeseed is the traditional name for the *Brassica* oilseed crops that adapt well to a wide variety of climate and soil conditions. It is related to mustard, cabbage, broccoli, cauliflower and turnip. Canola is the Canadian modified cultivar of rapeseed that is very low in erucic acid and glucosinolates, a group of chemicals whose hydrolyzed products are toxic to animals and human beings. Low erucic acid, low glucosinolates varieties of rapeseed are also the most common varieties in Europe. High erucic acid rapeseed oil is still used for nonedible purposes such as lubricants, hydraulic fluids and plastics. Rapeseed adapts well to a wide variety of climate and soil conditions. Its yield in Europe is in the range of 2000–3500 kg/ha (equivalent to 1800–3100 lb/acre). Some reports have shown that biodiesel made from rapeseed oil has better cold-flow properties.[6] Mustard oils (yellow, white, and oriental mustard), also from the *Brassica* family, are suitable for making biodiesel as well.

Despite the low oil content of soybeans, soybean oil is the main biodiesel feedstock in the United States, because soybean meal is in high demand as an animal feed. Approximately 3 billion bushels (or approximately 80 billion kg) of soybeans were produced in 2008.[7] The soybean is an annual legume native to East Asia. The crop is well adapted to many countries around the world. The United States, Brazil, Argentina, Australia, China, and India are the major soybean producers. Yield varies widely, and is in the range of 2500–4000 kg/ha (2200–3600 lb/acre). Soybean seeds contain about 18 to 20% oil, and the oil is typically extracted either by mechanical pressing or by solvent extraction. Advanced genetic technologies have not only brought the unit yield of soybean to a new record high, but have also given soybean oil the desired fatty-acid profile for biodiesel production.

Other vegetable oils, such as palm, coconut, safflower, sunflower, and flaxseed, can also be used for producing biodiesel. New, high oil yield plant oils, such as camelina, jatropha, and algae oils, have also been researched and used for biodiesel production.

Oil palms are of great interest due to high oil yield per acre. The oil palms include two species, the African Oil Palm and the American Oil Palm. Palm trees are single-stemmed, and can grow up to 20 m tall. The palm fruit takes 5 to 6 months to mature; it comprises an oily outer layer and a single seed, both rich in oil. Therefore, palm oil can be extracted from both the palm fruit meat and fruit kernels. Palm oil is a very common cooking ingredient in the regions where it is produced, such as Southeast Asia and Central America. Palm oil has the highest oil yield per unit area of any plant, approximately 5000–8000 L/ha or 530–850 gal/acre yearly. However, because palm oil is high in saturated fats and is solid at room temperature, biodiesel made from palm oil tends to freeze at a relatively high temperature, making it unsuitable for use in cold climates. Due to this fact, palm oil is now used as a biodiesel feedstock mostly in Southeast Asia, for example, in Malaysia and Indonesia, where palm oil is readily available and the weather is warm.

Coconut, another abundant tropical plant, also yields a large quantity of oil. Like palm oil, coconut oil is a fat consisting of about 90% saturated fat, predominantly medium-chain triglycerides. Unrefined coconut oil is semi-solid at room temperature. Coconut oil is slow to oxidize (due to its high saturation level) and thus resistant to rancidity. The drawback for use in biodiesel production is its less desirable cold-flow properties.

Jatropha is a perennial shrub. It is an uncultivated nonfood wild species, and can grow up to 5 m high. It is originally from Central America and has been spread to other tropical and subtropical areas. Jatropha seeds contain about 30% to 35% oil. It is a poisonous plant and its oil is nonedible. Jatropha can tolerate harsh climates and harsh soil conditions, and is resistant to a high degree of aridity. It can be planted in areas where it won't compete for resources needed to grow food crops. Jatropha is considered one of the best candidates for future biodiesel production. It can yield several times more oil per acre than soybean. However, due to the variations in species and the lack of research data, reports on jatropha productivity vary widely, from 200 gal/acre to 1000 gal/acre. Harvesting jatropha seeds is very challenging, involving labor-intensive hand picking. Machine harvesting is still under investigation. Meanwhile, the long-term impact of large-scale jatropha plantations on soil quality and the environment is also not very well known and requires further investigation.

Camelina is a plant of the *Brassica* family. Among its many species is *Camelina sativa*, which is the most important oil plant of the genus. Camelina is native to Northern Europe and to Central Asia, and has been introduced to North America. Traditionally, camelina has been cultivated as an oilseed crop in eastern and central Europe, and continues to be cultivated nowadays. Camelina plants are heavily branched, growing up to three feet. Camelina can grow on marginal agricultural lands and thus does not compete with food crops. It adapts well to dry and cooler climate conditions. The yield of camelina in North America typically ranges from 800 lb/acre to 1400 lb/acre. When mature, camelina ripens rapidly. Timely harvesting is important for good yield. Camelina seeds are small and the plant is very woody when mature. These properties create challenges during harvesting. The oil content of the camelina seed is relatively high, ranging from 30 to 40%. Camelina contains exceptionally high levels of omega-3 fatty acids, up to 45%. Cold-pressed camelina oil contains more than 50% polyunsaturated fatty acid. The major fatty acids in camelina are oleic (~15%), linoleic (~17%), linolenic (~35%), and eicosenoic (~12%). This fatty-acid profile makes camelina oil a good candidate for producing biodiesel with desirable cold-flow properties.

Algal oil is considered a promising feedstock for biodiesel production because it can be potentially "industrialized". The term "algae" used here refers to all types of algae that potentially produce lipids or oil. Algae can be cultivated in open ponds and lakes, and also in bioreactors. The cultivating conditions, such as nutrient supply, water temperature, and proper subculturing, need to be well controlled. A sufficient supply of light is necessary for healthy growth of some algae. Once ready, algae can be harvested by various mechanical means. Sometimes chemical agents are used to increase the sizes of algal clusters before harvest, such as in flocculation processes. Algal fats can be obtained from dewatered algae through mechanical crushing. The oil yield per unit area of algae cultivating facility has been estimated to be 4–17 kg/m^2 (0.8–3.4 lb/ft^2) per year although the upper end of this range may not be achievable in practice. The difficulty in producing oil efficiently from algae lies in screening and/or genetically modifying algal strains that have high fat content and fast growth rates. Also, cost-effective cultivation and harvesting systems are needed before algae for oil production can be commercialized. Currently, producing oil from algae for biodiesel is at the research and development stage.

Other types of plant oils including safflower, sunflower, and flaxseed, are also suitable for making biodiesel. However, due to their limited supplies and high-value uses in other areas, such as edible pigments and flavor, food, bird feed, and as a source of omega-3 fatty acids, these feedstocks are not generally used for commercial biodiesel production.

In developing energy crops for biodiesel production, the unit yield per acre is one of the most important factors. Figure 15.1 shows the relative production rates of different oil plants. Although jatropha oil and algae oil are the top

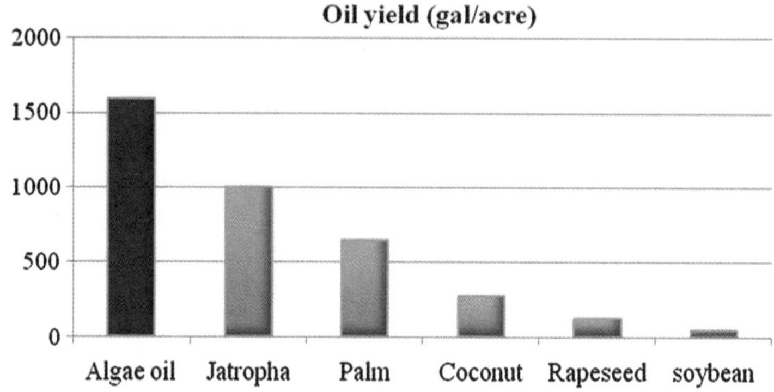

Figure 15.1 Estimated productivity of plant oils per unit land.

candidates in terms of oil production per unit of land, currently, rapeseed and soybean oils are the main feedstocks.

Animal fats and greases can also be used as biodiesel feedstocks. Like vegetable oils, animal fats and greases are chemically composed of triglycerides. They are typically solid or semi-solid at room temperature due to the high degree of saturation in their chemical bonds.

In the six years between 2001 and 2007, the US produced approximately 3.8 million tons of tallow, 0.6 million tons of lard, and 0.5 million tons of butter yearly. Tyson Foods produces about 2.3 billion pounds of chicken fat annually from its poultry-processing facilities. Animal fats are considered inexpensive and logical feedstocks for biodiesel production, and are currently used commercially for biodiesel feedstocks. However, because each of these fats has its own established market, such as the animal-feed industry, reallocating animal fats from their existing markets to biodiesel has caused concerns that biodiesel production will result in a shortage of animal fats for traditional uses.

"Grease" is the general term for animal fats and/or used vegetable oils that are semi-solid at room temperature. The term "waste vegetable oil (WVO)" is also used for used vegetable oils. Greases are sometimes further classified as yellow grease, which has less than 15% free fatty acids, and brown grease, which has more than 15% free fatty acids. Yellow grease typically refers to used frying oil from deep fryers. It also refers to lower-grade tallow from rendering facilities. Brown grease is sometimes recovered from grease traps of restaurants and food processing plants.

Greases frequently contain large amounts of free fatty acids. To avoid a large quantity of soap formation in transesterification, a two-step, acid- and then alkaline-catalyzed process is needed to convert greases to biodiesel effectively. Due to the nature of its collection, greases usually contain relatively high water content that also makes the conversion process more difficult. Another thing that concerns some biodiesel producers is that trap grease is usually contaminated with cleaning agents that may cause problems in terms of biodiesel quality control.

15.2.2 Fatty-Acid Profiles of Feedstocks for Biodiesel Production

Fat is a generic term for a class of lipids in biochemistry. It refers to the greasy, solid materials found in animal tissues and in some plants. Vegetable oil is the fat extracted from plant sources, mainly from plant seeds. Chemically, plant oils and animal fats are comprised of a family of chemicals called triglycerides.

Triglycerides (also called triacylglycerols) are fatty-acid esters of glycerol: a glycerol combined with three fatty acids on the glycerol's hydroxyl (-OH) groups. Triglycerides are the dominate components in all naturally formed plant oils and animal fats. The chemical formula of a triglyceride is shown below (Figure 15.2), where R_1, R_2 and R_3 are longer alkyl chains. The chain lengths of the alkyl chains in naturally occurring triglycerides can be of varying length, from C_{10} to C_{30}, but C_{16} to C_{18} are the most common ones. The long, nonpolar alkyl chain is an important counter balance to the polar acid functional group.

Glycerol is also known as glycerin (alternative spelling: glycerine). Glycerol is a poly hydric alcohol or triol that has three hydrophilic alcoholic hydroxyl groups (-OH). It is a colorless, odorless, hygroscopic, and sweet-tasting viscous liquid. In triglycerides, glycerol serves as the "backbone" to link the three fatty acids.

A fatty acid is a carboxylic acid of long carbon chains. In the structure of triglycerides, three fatty acids R′-COOH, R″-COOH and R‴-COOH are linked chemically to the glycerol. These three fatty acids can be the same as or different from each other. Almost without exception, only even numbers of carbon atoms are found in natural fatty acids due to the way they are biologically synthesized from acetyl Co-A, a two-carbon chemical species. Under certain conditions, fatty acids in glycerides may dissociate and become "free" fatty acids (FFAs).

Fatty acids possess different chemical and physical properties. Generally, the longer the alkyl chain of fatty acids, the higher the melting point. For fatty acids of the same length, the more double bonds a fatty acid contains, the lower its melting point. Chain length and the number of double bonds of the fatty acids are important factors affecting biodiesel fuel properties, which will be discussed in next section.

Fatty-acid (FA) profiles of vegetable/plant oils and animal fats vary widely. Even with the same type of oil or fat, its FA profile may vary from location to

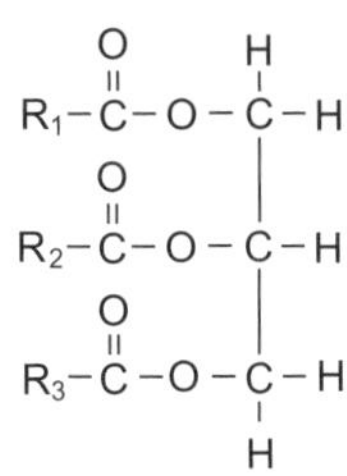

Figure 15.2 General structure of a triglyceride.

location, or even from batch to batch. Tables 15.1, 15.2, and 15.3 summarize some commonly known fatty acids and their properties, and the fatty-acid profiles of some plant oil and animal fats.

15.2.3 Biodiesel Production

Biodiesel is produced by chemically reacting a vegetable oil or animal fat with a simple alcohol, such as methanol. An alkali catalyst is used.

Figure 15.3 shows the global transesterification reaction for a generalized triglyceride with methanol to form methyl esters. Other alcohols such as ethanol[8] and isopropanol[9] can also be used as long as the resulting alkyl esters have properties that satisfy the relevant fuel standard.

The industrial process to convert triglycerides to alkyl esters can be considered a competition between the transesterification reaction, which is the desired path, and saponification, which is undesirable. Each of these reactions will be discussed, and then the requirements to direct the reaction in the preferred direction will be presented.

The transesterification process proceeds through a chain reaction as the triglyceride (TG) is converted to a diglyceride (DG), then to a monoglyceride (MG), and finally to free glycerin. Each step in the reaction scheme is catalyzed by an alkoxide ion, which when methanol is used, is methoxide, $^{-}OCH_3$. The methoxide ion can be prepared by dissolving sodium metal in methanol, or by adding sodium or potassium hydroxide to methanol. The reaction mechanism for the first step of converting the triglyceride to a diglyceride is shown in Figure 15.4.[10]

As shown, the methoxide ion attacks the triglyceride and forms a tetrahedral intermediate, which subsequently separates into a diglyceride and a methyl ester. The methoxide ion, which is consumed in the process, is regenerated when the diglyceride ion strips a proton from an available methanol molecule. The diglyceride and monoglyceride reactions follow similar pathways. In each case, the methoxide ion initiates the reaction, becomes part of the headgroup on the alkyl ester, and is then regenerated at the end by stripping a proton from the alcohol so the reaction can continue.

Saponification is the production of salts of alkali metals with carboxylic acids, generally known as *soaps*. The reaction of animal fats with lye and water to produce household soap has been practiced for thousands of years. In a typical biodiesel production reactor, the transesterification and saponification reactions are occurring simultaneously. The significance of each reaction depends on the relative amounts of methoxide ion $^{-}OCH_3$ and hydroxide ion ^{-}OH in the reaction mixture. These relative amounts are determined by the following equilibrium (shown for a sodium-based catalyst):

$$CH_3OH + Na^+ + {}^{-}OH \leftrightarrow Na^+ + {}^{-}OCH_3 + H_2O \qquad (15.1)$$

Equation 15.1 clearly illustrates the process when a hydroxide catalyst is used and how it produces the necessary methoxide ions when methanol is present. Note that this reaction produces a molecule of water for each molecule of

Table 15.1 Fatty acids commonly found in plant oils and animal fats and their properties.[1]

Abb.	*Common name*	*Chemical names*	*Formula*	*Structure*	*MW* [2]	*m.p.* [3] *(°C)*	*b.p.* [4] *(°C)*	*Density* [5] *(g/cm^3)*	*ND* [6]	*HHV*[7] *(MJ/mol)*	*HHV*[7] *(MJ/kg)*
$C_{12:0}$	lauric acid	dodecanoic acid	$C_{12}H_{24}O_2$	$CH_3(CH_2)_{10}COOH$	200.3	44	298.9	0.880	1.447	7.38	36.85
$C_{14:0}$	myristic acid	tetradecanoic acid	$C_{14}H_{28}O_2$	$CH_3(CH_2)_{12}COOH$	228.4	58	242–249	0.862	1.451	8.68	38.02
$C_{16:0}$	palmitic acid	hexadecanoic acid	$C_{16}H_{32}O_2$	$CH_3(CH_2)_{14}COOH$	256.5	63–64	351–352	0.892	1.453	9.98	38.94
$C_{18:0}$	stearic acid	octadecanoic acid	$C_{18}H_{36}O_2$	$CH_3(CH_2)_{16}COOH$	284.5	69	383	0.888	1.455	11.29	39.68
$C_{18:1}$	oleic acid	cis-9-octadecenoic acid	$C_{18}H_{34}O_2$	$CH_3(CH_2)_7CH{:}CH(CH_2)_7COOH$	282.5	13–14	360	0.895	1.466	11.12	39.38
$C_{18:2}$	linoleic acid	cis,cis-9,12-octadecadienoic acid	$C_{18}H_{32}O_2$	$CH_3(CH_2)_3(CH_2CH{:}CH)_2(CH_2)_7COOH$	280.5	– 5	229	0.911	1.478		
$C_{18:3}$	linolenic acid	cis,cis,cis-6,9,12-octadecatrienoic	$C_{18}H_{30}O_2$	$CH_3(CH_2CH{:}CH)_3(CH_2)_7COOH$	278.5	– 10	230/1 mm-Hg	0.924	1.490		
$C_{20:0}$	arachidic acid	eicosanoic acid	$C_{20}H_{40}O_2$	$CH_3(CH_2)_{18}COOH$	312.6	75	328	0.824			
$C_{20:1}$	eicosenoic acid	cis-11-eicosadienoic acid	$C_{20}H_{38}O_2$	$CH_3(CH_2)_7CH{:}CH\ (CH_2)_9COOH$	310.5	23–24		0.883	1.461		
$C_{20:5}$	eicosapentaenoic	icosa-5,8,11,14,17-pentaenoic	$C_{20}H_{30}O_2$	$CH_3(CH_2CH{:}CH)_5(CH_2)_3COOH$	302.5						
$C_{22:1}$	erucic acid	cis-13-docosenoic acid	$C_{22}H_{42}O_2$	$CH_3(CH_2)_7CH{:}CH(CH_2)_{12}COOH$	338.6	34	381 (decomp.)	0.860			

[1] Combined from various sources
[2] MW = molecular weight, g/mol
[3] m.p. = melting point, °C
[4] b.p. = boiling point, °C
[5] Density at 25 °C
[6] ND = refractive index
[7] HHV = high (gross) heating value

Table 15.2 Properties of methyl esters of fatty acids commonly found in plant oils and animal fats.[1]

Abb.	*Common name*	*Formula*	*Structure*	*MW* [2] (g/mol)	*m.p.*[3] (°C)	*b.p.*[4] (°C)	*Density*[5] (g/cm^3)	*HHV*[6] (MJ/kg)
$C_{12:0}$	methyl laurate	$C_{13}H_{26}O_2$	$CH_3(CH_2)_{10}COOCH_3$	214.3	4–5	261–262	0.869	37.89
$C_{14:0}$	methyl myristate	$C_{15}H_{30}O_2$	$CH_3(CH_2)_{12}COO\ CH_3$	242.4	18–20	323	0.866	38.93
$C_{16:0}$	methyl palmitate	$C_{17}H_{34}O_2$	$CH_3(CH_2)_{14}COO\ CH_3$	270.5	31–32	196	0.865	39.47
$C_{18:0}$	methyl stearate	$C_{19}H_{38}O_2$	$CH_3(CH_2)_{16}COO\ CH_3$	298.5	38–39	215	0.863	40.10
$C_{18:1}$	methyl oleate	$C_{19}H_{36}O_2$	$CH_3(CH_2)_7CH{:}CH(CH_2)_7$-COO CH_3	296.5	– 20	190	0.850	39.93
$C_{18:2}$	methyl linoleate	$C_{19}H_{34}O_2$	$CH_3(CH_2)_3(CH_2CH{:}CH)_2(CH_2)_7COO\ CH_3$	294.5	– 35	373–374	0.884	39.72
$C_{18:3}$	methyl linolenate	$C_{19}H_{32}O_2$	$CH_3(CH_2CH{:}CH)_3(CH_2)_7$-COO CH_3	292.5	– 52	364–365	0.895	39.37
$C_{20:0}$	methyl arachidate	$C_{21}H_{42}O_2$	$CH_3(CH_2)_{18}COO\ CH_3$	326.6	53–55	374–376	0.862	
$C_{20:1}$	methyl eicosenoate	$C_{21}H_{40}O_2$	$CH_3(CH_2)_7CH{:}CH(CH_2)_9COO\ CH_3$	324.5				
$C_{20:5}$	methyl eicosapentaenoate	$C_{21}H_{38}O_2$	$CH_3(CH_2CH{:}CH)_5(CH_2)_3$-COO CH_3	332.6				
$C_{22:1}$	methyl erucate	$C_{23}H_{44}O_2$	$CH_3(CH_2)_7CH{:}CH(CH_2)_{12}COO\ CH_3$	352.6	4	220/5.67 kPa	0.870	41.01

[1] Combined from various sources
[2] MW = molecular weight, g/mol
[3] m.p. = melting point, °C
[4] b.p. = boiling point, °C
[5] Density at 25 °C
[6] HHV = high (gross) heating value

methoxide produced. Caldin and Long,[11] in a study of the equilibrium between hydroxide and ethoxide in ethanol, found that the equilibrium strongly favors the formation of alkoxide with more than 96% of the base present as ethoxide. Assuming methanol and other alcohols behave similarly to ethanol, this equilibrium explains why hydroxides can be effective at producing the methoxide ions needed for transesterification. It also explains why hydroxides are observed to provide more soap formation.[12] The water produced by the conversion of hydroxide to methoxide, along with water that may have been introduced with the original oil and alcohol, causes the equilibrium to favor a higher concentration of hydroxide ions and more soap production. This problem is particularly acute when free fatty acids (FFAs) are present, because the saponification of FFAs regenerates water and thus the hydroxide ions needed for additional soap production.

Under the conditions that are appropriate for biodiesel production, the transesterification reaction precedes saponification, so the saponification is more likely to involve a methyl ester than a glyceride.[13] Bender and Glassom[14] have shown that the reaction rates for attack on an ester by ^{-}OH, $^{-}OCH_3$ and $^{-}OCH_2CH_3$ are similar, so the predominant factor in determining the significance of saponification compared with transesterification is the relative concentration of hydroxide and alkoxide.

The predominance of alkoxide provided by the equilibrium of eqn 15.1 indicates that methyl ester formation is not inhibited by saponification early in the reaction; later, however, the methyl esters will be subject to saponification along with residual TGs, DGs, and MGs, depending on the level of hydroxide in the mixture, which in turn is primarily dependent on the amount of water. The reaction mechanism for saponification of a methyl ester is shown in Figure 15.5. The reaction is initiated when a hydroxide ion, ^{-}OH, attacks the methyl ester in the same manner as the methoxide ion in the transesterification reaction.

Again, similar to the transesterification reaction, a tetrahedral intermediate is formed, which separates into methanol and a molecule of soap, which consists of a negatively charged fatty-acid chain ionically bonded to the positively charged metal ion from the original strong base used as a catalyst. Saponification is irreversible with the catalyst being consumed according to equation 15.1 to compensate for the loss of the hydroxide ion. Saponification also converts TGs, DGs, and MGs to soap following a mechanism that is similar to the mechanism for methyl esters shown in Figure 15.5.

If free fatty acids are present in the oil, the mechanism is slightly different, as shown in Figure 15.6. In this mechanism, a hydroxide ion, ^{-}OH, strips a proton from the fatty acid, leaving the negatively charged fatty acid, which ionically bonds to the metal ion of the catalyst to form soap. In contrast to the saponification reactions of glycerides and methyl esters, which produce other glycerides and methanol as by-products, the saponification of fatty acids produces water, which, following eqn 15.1, will be a source for new hydroxide ions.

The competition between transesterification and saponification explains the crucial importance of minimizing water in the biodiesel reactor. Methanol is

Table 15.3 Fatty acid composition and density of plant oils and animal fats.

No.	Oils/fats	Density [1] (g/cm^3) at °C	Fatty-acid profiles [2,3]										
			$C_{12:0}$	$C_{14:0}$	$C_{16:0}$	$C_{18:0}$	$C_{18:1}$	$C_{18:2}$	$C_{18:3}$	$C_{20:0}$	$C_{20:1}$	$C_{20:5}$	$C_{22:1}$
1	Algae oil	0.920		12–15	10–20		4–19	1–2	5–8			35–48	
2	Butter	0.860		7–10	24–26	10–13	28–31	1–3	0–1				
3	Camelina	(0.922–0.930)/20 [7]					12–15	15–20	30–40		12–15		2–3
4	Canola	(0.914–0.917)/20 [6]			1–3	2–3	50–60	15–25	8–12				
5	Canola, high oleic				1–3	2–3	70–80	12–15	1–3				
6	Coconut oil	(0.908–0.921)/40	45–53	16–21	7–10	2–4	5–10	1–2.5					
7	Corn	(0.917–0.925)/20		1–2	8–16	1–3	20–45	34–65	1–2				
8	Cottonseed	(0.918–0.926)/20		0–2	20–25	1–2	23–35	40–50					
9	Grape seed oil	(0.920–0.926)/20			5–11	3–6	12–28	58–78					
10	Jatropha	0.914/20 [8]			11–16	6–15	34–45	30–50		3–5			
11	Lard	0.908		1–2	25–30	10–20	40–50	6–12	0–1				
12	Flax/Linseed oil	0.922/30 [5]			4–7	2–4	25–40	35–40	25–60				
13	Mustard seed oil	(0.910–0.921)/20				1–2	8–23	10–24	6–18		5–13		20–50
14	Olive	(0.915–0.918)/15.5 [4]			9–10	2–3	72–85	10–12	0–1				
15	Palm oil	(0.891–0.899)/50		0.5–2	39–48	3–6	36–44	9–12					
16	Palm kernel oil	(0.899–0.914)/40	45–55	14–18	6–10	1–3	12–19						
17	Peanut	(0.912–0.920) /25 [4]			8–9	2–3	50–65	20–30					
18	Rapeseed, high erucic	(0.910–0.920)/20			1–3	0–1	10–15	12–15	8–12	7–10			45–60

19	Rapeseed, high oleic	(0.914–0.920)/20		1–5	1–2	60–80	16–23	10–15
20	Safflower, high linoleic	(0.922–0.927)/20		3–6	1–3	7–10	80–85	
21	Safflower, high oleic	(0.913–0.919)/20		1–5	1–2	70–75	12–18	0–1
22	Sesame oil	(0.915–0.924)/20		8–12	4–7	35–45	37–48	
23	Soybean oil	(0.919–0.925)20		6–10	2–5	20–30	50–60	5–11
24	Soybean, high oleic			2–3	2–3	80–85	3–4	3–5
25	Sunflower	(0.918–0.923)/20		5–8	2–6	15–40	30–70	
26	Sunflower, high oleic	(0.909–0.912)/25		0–3	1–3	80–85	8–10	0–1
27	Tallow	0.890/30 [5], 0.774/25 [10]	3–6	22–32	10–25	35–45	1–3	
28	Tung oil	(0.931–0.973)/30 [9]		3–4	0–1	4–15		75–90
29	Yellow greases	0.777/25 [10]	0–2	12–18	8–13	45–55	8–20	1–2

[1] From CODEX-STAN 210-1999 otherwise indicated.
[2] Composed from various sources.
[3] The rest are other fatty acids, making up to 100%.
[4] (Hodgman and Lange, 1924)
[5] (Acosta *et al.*, 1996)
[6] (Przybylski, 2009)
[7] (Oil by Nature, 2009)
[8] (Veny *et al.*, 2009)
[9] (Singleton *et al.*, 1952)
[10] (Spataru, 2009)

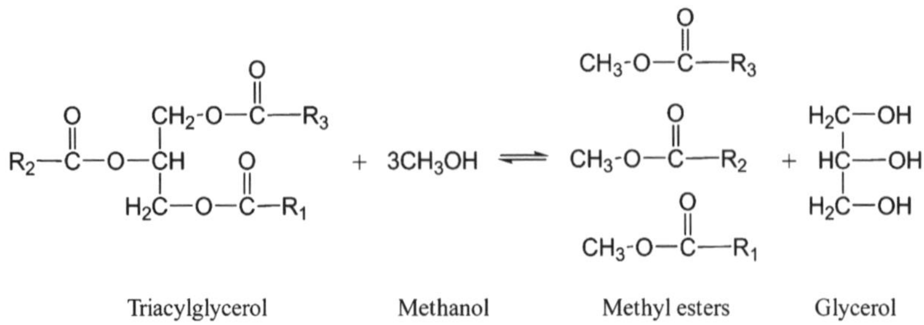

Figure 15.3 Global transesterification reaction.

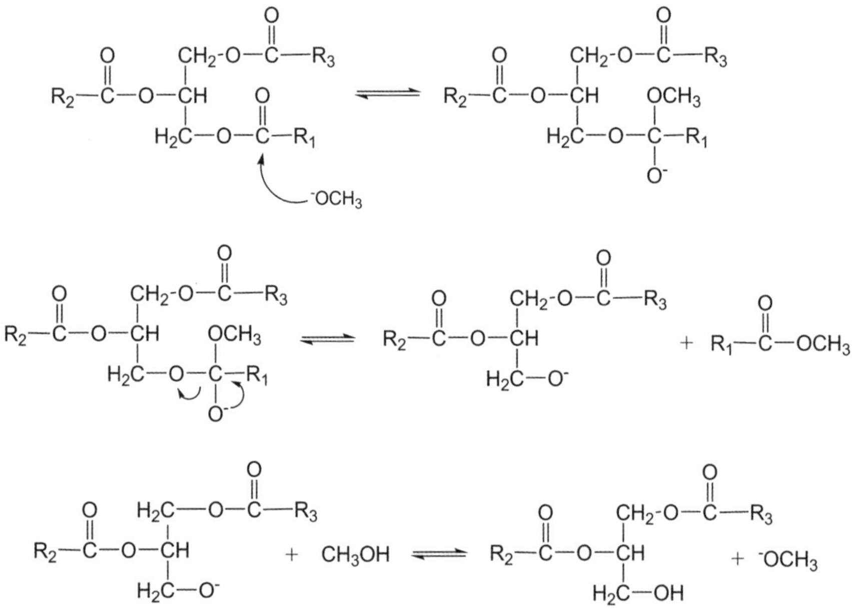

Figure 15.4 Transesterification reaction mechanism.

commonly available with water content of less than 0.05%. Oils and fats should be dried to less than 0.1% water, and preferably to 0.05%. Water-free solutions of 25 wt% or 30 wt% sodium methoxide in methanol are commercially available.

15.2.4 Effect of Catalyst

As was shown earlier, the catalyst species for transesterification with methanol is the methoxide ion, $^{-}OCH_3$. The sodium and potassium hydroxides and methoxides most often used for biodiesel production are strong bases and can

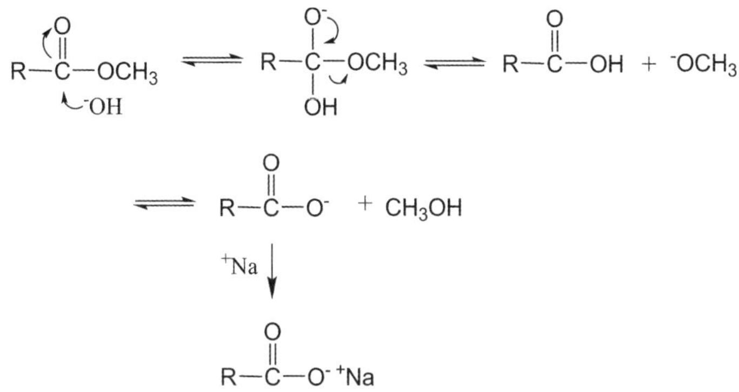

Figure 15.5 Saponification of a methyl ester.

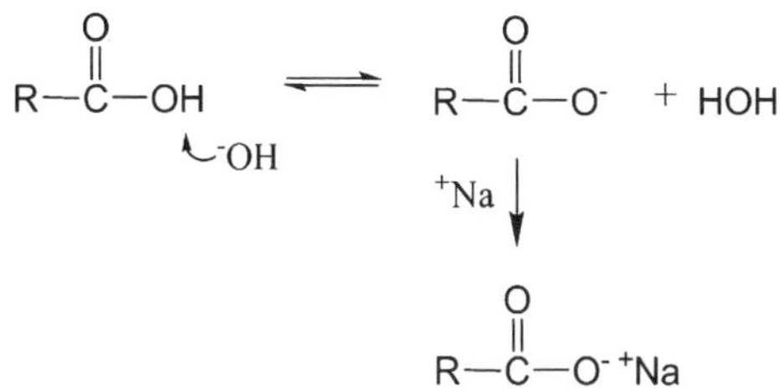

Figure 15.6 Saponification of fatty acid.

be assumed to dissociate completely in the methanol solution. Vicente *et al.*[15] showed that biodiesel yields of 99.33 and 98.46 wt% were obtained for sodium and potassium methoxides, respectively. In contrast, the yields for sodium and potassium hydroxides were only 86.71 and 91.67 wt%, respectively. The lower yield was caused by the higher soap levels associated with the higher hydroxide ion concentrations. Some of the yield loss was due to direct loss of triglyceride to soap formation which, due to its polarity, partitions with the glycerin; and some was attributed to the fact that soap causes greater methyl ester solubility in the glycerin. Singh *et al.* also found that yields with methoxide catalysts were higher than for hydroxide catalysts.[16]

15.2.5 Mass-Transfer Limitations

A fundamental problem with transesterification of triglycerides is the low solubility of short-chain alcohols in the oils or fats. Figure 15.7 shows the solubility of methanol and ethanol in edible beef tallow, which is low in FFAs. Since the normal amount of methanol used for biodiesel production is 16 to 22 wt%, based on the oil, the reaction mixture would be expected to contain

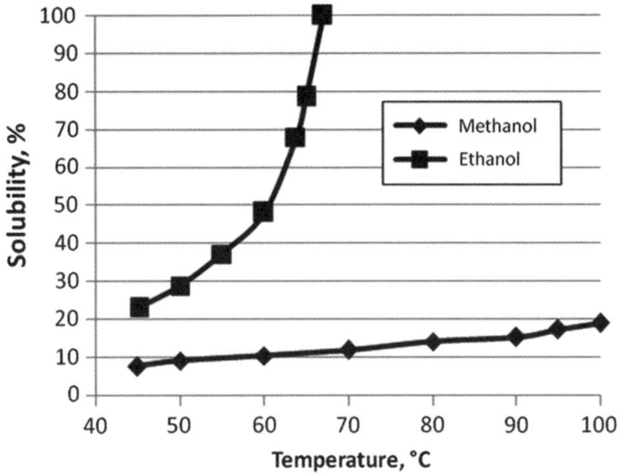

Figure 15.7 Solubility of methanol and ethanol in beef tallow (adapted from Ma *et al.*[22]).

two phases: a nonpolar phase consisting mostly of triglyceride, and a polar phase consisting mostly of methanol.

Typical reaction mixtures for ethanol contain 45% more weight of alcohol than for methanol, reflecting its greater molecular weight. At reaction temperatures near its boiling point of 78 °C, ethanol transesterification reactions could be single phase.

The high polarity of the methanol phase means that the catalyst will almost entirely reside in this phase.[17] This imposes a mass-transfer limitation on the reaction so that the reaction rate is limited by the rate at which triglyceride can diffuse into the methanol droplets and the methyl esters diffuse away. Researchers have found that the reaction rate is strongly influenced by the level of agitation in the mixture[18] and several recent developments using ultrasonic mixing[19] and hydrodynamic cavitation[20] utilize this concept.

Frascari *et al.* found that once a fine methanol/oil dispersion had been produced by an initial "high-energy" mixing event, the level of agitation could be reduced or even stopped while still providing a complete reaction.[21] Ma *et al.* suggest that after the initial stage of the reaction, monoglycerides and diglycerides are produced that act as emulsifiers and the oil and alcohol form an emulsion. They indicate that once this emulsion has formed the stirring can be stopped and that this will shorten the time required for the glycerol and ester phases to separate.[22] While this approach has the advantage that the settling of glycerin that occurs as the reaction approaches completion will help to shift the equilibrium towards more complete reaction, the disadvantage is that the glycerin extracts the catalyst and makes it unavailable for the reaction. This can make it difficult to achieve the essentially complete reaction required by the biodiesel quality specifications.

Under conditions of no cosolvent and modest levels of agitation, the transesterification rate typically starts out slow, then accelerates to a rapid conversion of the bulk of the triglyceride to methyl ester and then slows again toward the end of the reaction. The initial period is slow due to the mass-transfer limitations caused by the low solubility of methanol in the oil. During this period, the catalyst, which is mostly in the alcohol phase, acts on the triglycerides that are dissolved in the alcohol and at the interface between the alcohol droplets and the oil. As the reaction proceeds, as noted earlier, significant amounts of DGs and MGs are formed that act as solublizers or emulsifiers to allow the mixture to approach single-phase behavior with an accompanying acceleration in the reaction rate.[22] When the reaction proceeds to a point where free glycerin is produced, then the glycerin forms small droplets. As the droplets form, most of the catalyst will sequester into the glycerin droplets, which slows the reaction and inhibits it moving to completion.

Table 15.4 shows the distribution of ester, glycerin, methanol and catalyst for a typical reaction mixture.[23] As noted earlier, the catalyst partitions primarily into the glycerol-rich phase for both methanol and ethanol. There is a significant difference in the extent to which glycerol partitions between methanol and ethanol. In the case of methanol, only 2.3% of the glycerol will remain in the ester-rich phase. For ethanol, 19.3% of the glycerol will remain in this phase.

Boocock *et al.* have addressed the immiscibility of the oil and alcohol by introducing a cosolvent, which means the reaction occurs as a single phase immediately.[24] Doell *et al.* have presented kinetics data for the homogeneous (single phase) reaction of methanol and soybean oil with tetrahydrofuran used as a cosolvent. These data can be used to determine the actual chemical reaction rate that is the limiting rate achievable by enhanced mixing techniques such as

Table 15.4 Phase composition of transesterification products (adapted from Zhou and Boocock 2006).[23]

	Methanolysis mixture		*Ethanolysis mixture*	
	Ester-rich phase	*Glycerol-rich phase*	*Ester-rich phase*	*Glycerol-rich phase*
Composition of each phase (%)				
Glycerol	0.21 ± 0.01	60.3 ± 1.0	1.60 ± 0.01	70.5 ± 1.0
Alcohol	4.3 ± 0.1	33.1 ± 1.0	12.7 ± 0.1	27.5 ± 1.0
Ester	95.5 ± 1.0	< 1.0	84.3 ± 1.0	< 1.0
Catalyst	0.05 ± 0.01	5.57 ± 0.01	N/A	N/A
Distribution between phases (%)				
Glycerol	2.3	97.7	19.3	80.7
Alcohol	42.0	58.0	75.4	24.6
Ester	99.0	1.0	99.0	1.0
Catalyst	5.8	94.2	10.1	89.9

high shear mixers or ultrasonic mixing.[25] The basic reaction system is shown in eqn. 15.2.

$$\begin{aligned} &TG + {}^{-}OCH_3 \xrightarrow{k_1} ME + DG \\ &DG + {}^{-}OCH_3 \xrightarrow{k_2} ME + MG \\ &MG + {}^{-}OCH_3 \xrightarrow{k_3} ME + \text{Glycerol} \end{aligned} \tag{15.2}$$

Note that Boocock and co-workers have not included the methanol concentration in their kinetics model as others have done.[26] Rather, the reaction rate is assumed to be first order in the methoxide ion concentration, which is appropriate considering the transesterification mechanism presented in Figure 15.4.

Doell *et al.* measured the rate constants k_1, k_2, and k_3 at temperatures from 23 °C to 50 °C and these are shown in Table 15.5. For the current work, the values of ln(k) were plotted vs $1/T$ and the regression lines calculated. These lines are shown as Arrhenius expressions in Table 15.5. If the likelihood of reaction of methoxide ion with each ester group was equal, then it would be expected that the reaction rate constants k_1, k_2, and k_3 would be in the ratio 3:2:1, reflecting the number of ester groups on each molecule. In reality, k_1 is considerably smaller than k_2 and k_3, which are similar in value. Note that the activation energies for the three Arrhenius equations are very similar, as would be expected from the similarity of the reaction pathways for the three reaction steps. The pre-exponential terms increase going from k_1 to k_3 indicating steric effects associated with fatty-acid chains on the glyceride molecules, decreasing the interaction of methoxide ion with other chains on the same molecule.[26]

The reaction rate expressions given in Table 15.5 can be used to estimate the progress of the homogeneous transesterification reaction using the following rate equations:

$$\frac{d[TG]}{dt} = -k_1[TG][OCH_3^-]$$

$$\frac{d[DG]}{dt} = k_1[TG][OCH_3^-] - k_2[DG][OCH_3^-]$$

$$\frac{d[MG]}{dt} = k_2[DG][OCH_3^-] - k_3[MG][OCH_3^-] \tag{15.3}$$

$$\frac{d[ME]}{dt} = k_1[TG][OCH_3^-] + k_2[DG][OCH_3^-] + k_3[MG][OCH_3^-]$$

$$\frac{d[Gly]}{dt} = k_3[MG][OCH_3^-]$$

Table 15.5 Forward rate constants for transesterification. $L\,mol^{-1}\,min^{-1}$.

Temp (°C)	k_1	k_2	k_3
23	6.3 ± 1.3	15.2 ± 2.8	13.9 ± 2.7
30	13.6 ± 1.7	35.6 ± 7.5	35.2 ± 10.6
40	29.1 ± 5.5	68.5 ± 13	60.6 ± 16
50	54.2 ± 13	136 ± 23	139 ± 29
Arrhenius equation	$k_1 = 7.93 \times 10^{11} \exp\left(-\frac{62,715}{RT}\right)$	$k_2 = 1.94 \times 10^{12} \exp\left(-\frac{62,687}{RT}\right)$	$k_3 = 3.81 \times 10^{12} \exp\left(-\frac{64,549}{RT}\right)$

The measure of reaction completion used by most quality standards is the total glycerin, which consists of the sum of the free glycerin and the glycerin still bound with fatty acids as TG, DG, and MG. In the final product, the free glycerin is primarily a reflection of the purification processes, while the bound glycerin reflects the extent of the reaction. To monitor the progress of the reaction, the bound glycerin (BG) was calculated according to equation 15.4:

$$\begin{aligned} BG = 0.1044\left[x_{TG} \times \frac{MW_{TG}}{MW_{Ave}}\right] + 0.1488\left[x_{DG} \times \frac{MW_{DG}}{MW_{Ave}}\right] \\ + 0.2591\left[x_{MG} \times \frac{MW_{MG}}{MW_{Ave}}\right] \end{aligned} \quad (15.4)$$

The mole fractions of TG, DG, and MG are calculated from the volume concentrations and the molecular weights are used to convert from the mole fractions into a mass fraction. The constant multipliers, 0.1044, 0.1488, and 0.2591 are the mass fractions of glycerin in TG, DG, and MG assumed by the ASTM specification.[1]

The experiments conducted by Doell *et al.* to develop the kinetic parameters used less than normal amounts of catalyst to slow down the reaction to levels that would allow accurate sampling of the reactants. Testing at higher catalyst concentrations confirmed that the rate expressions were still appropriate. Comparison of the results calculated using the model described here to experimental data from Doell *et al.*[26] show an excellent match, which is to be expected since the kinetics parameters on which the model is based were extracted from these experimental data.

Although it cannot be accomplished in practice, the model can be evaluated as a homogeneous reaction without cosolvent present to provide an estimate of the limiting case where mass-transfer limitations are not present. A large amount of research has been conducted to develop heterogeneous catalysts and aggressive agitation to accelerate the transesterification reaction. Since a homogeneous solution of methoxide ions represents the ideal source of catalytic species, all proposed heterogeneous catalysts will be less effective than this ideal case. Likewise, since a single-phase homogeneous reaction eliminates mass-transfer limitations, this case represents the ideal to be approached (but never achieved) by all enhanced agitation strategies.

Figure 15.8 shows predicted values of the time required to achieve the 0.24% bound glycerin (BG) level required by the ASTM specification,[1] calculated with eqn 15.4 as the temperature is varied.

15.3 Processing Technology

15.3.1 Purification of Biodiesel

The discussion to this point has focused on the mixing of the oil and alcohol, and ensuring a complete reaction. When the reaction is complete, the glycerin is

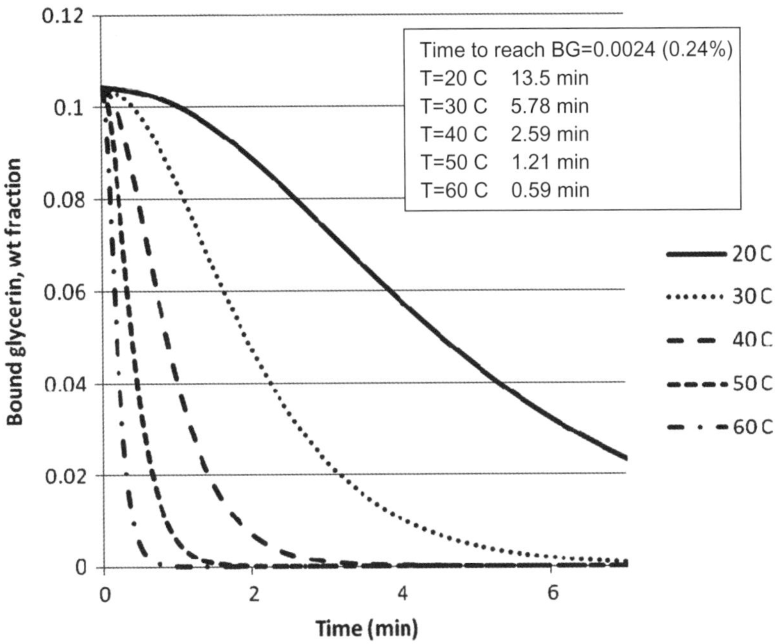

Figure 15.8 Reaction profiles of bound glycerin at various temperatures.

separated by decanting, by coalescence, or with a centrifuge. Then, the biodiesel is purified to remove contaminants such as methanol, glycerin, and soap.

Biodiesel purification involves methanol removal, usually by a vacuum flash process, with the recovered methanol condensed and recycled back to the reaction. Soap and residual free glycerin are removed by one of three methods: washing with water; using a solid adsorbent mixed with the liquid; or passing the biodiesel through a packed bed of ion-exchange resin.

Water washing produces a stream of waste water contaminated by soap and glycerin. This waste water is equal to 5 to 10% of the biodiesel production. A common strategy is to add this water to the crude glycerin emerging from the reactor. The water is then removed from the glycerin when the methanol is recovered from the glycerin, and the soap stays with the glycerin. The methanol–water mixture can be separated with a fractionating column so the methanol can be returned to the reactor, and the water can be reused for washing. This process results in zero-discharge production.

Solid adsorbents, such as magnesium silicate and ion-exchange resins, can be used in place of water. However, these processes produce solid waste streams that are usually landfilled.

Purifying biodiesel with high soap levels is problematic. If significant soap production occurs, the soap will typically be soluble in the biodiesel as long as the fuel contains more than 2% methanol. If the methanol is removed, the solubility of the soap is low, and it will precipitate out, producing a viscous phase that plugs filters and can produce a gel with the fuel. In general, purification processes will always be simpler if soap production can be kept to a minimum by keeping the levels of water and free fatty acids in the feedstock as low as possible.

15.3.2 High Free Fatty-Acid Feedstocks

Many of the cheapest feedstocks for biodiesel (such as used oil or grease) have high levels of free fatty acids (FFAs). Because the free fatty acids will react with the alkali catalyst, these feedstocks usually require pretreatment.

In some plants, the pretreatment consists of using solid adsorbents to selectively remove the fatty acids so they do not interfere with the catalyst. Steam stripping under high vacuum can also remove these FFAs and makes them available to be sold as an animal feed ingredient.

When the FFA level is high enough that stripping them is not financially viable, they can be converted to biodiesel via acid-catalyzed esterification. In this case, the high FFA material is reacted with methanol using sulfuric acid as a catalyst. When the FFAs are sufficiently converted, then the triglycerides can be converted with an alkali catalyst.[27] Solid acid catalysts that are capable of performing the esterification are now becoming available.

15.4 Biodiesel Fuel Properties as Affected by Fatty-Acid Profiles of Plant Oils and Animals Fats

15.4.1 ASTM and EN Standards of Biodiesel Specifications

In North America, biodiesel fuel quality is governed by the ASTM standard D6751, Specification for Biodiesel Fuel Blend Stock (B100) for Middle Distillate Fuels. In Europe, it is the standard EN 14214 by the European Committee for Standardization, Automotive fuels-Fatty acid methyl esters (FAME) for diesel engines-Requirements and test methods. Table 15.6 below is excerpted from the current versions of the standards, D6751-09[1] and EN 14214:2008.[28] In most major categories, the two standards are in agreement. Some specifications concern processing, and some are feedstock related.

The flash point, of importance in fuel handling and storage, is the lowest temperature at which a vapor of volatile liquid forms an ignitable mixture with air and is ready to burn when a source of ignition is present. Almost all major components in biodiesel have high flash points; therefore, biodiesel is generally categorized as a combustible, but not flammable, fuel. Biodiesel with a low flash point is generally found to have a high alcohol content.

Viscosity is a property of biodiesel that largely relates to the type of feedstock, although viscosity is also affected by other factors such as the presence of alcohol or unreacted glycerides.

Sulfur content in a biodiesel depends on its original content in feedstock. Most plant oils contain less than 15 ppm sulfur, while animal fats and used

Table 15.6 Specifications for biodiesel (B100).

No.	*Property*	*Units*	*ASTM 6751 (v.2009)* [1]	*EN14214 (v.2008)* [2]
1	Calcium and magnesium, combined	ppm (μg/g)	5 max	5 max
2	Flash point (closed cup)	°C	93 min	101 min
3	Alcohol control			
	(1) Methanol content	mass %	0.2 max	0.20 max
	(2) Flash point	°C	130 min	–
4	Water and sediment	% volume	0.050 max	0.050 max
5	Kinematic viscosity, 40 °C	mm^2/s	1.9–6.0	3.5–5.0
6	Sulfated ash	% mass	0.020 max	0.02 max
7	Sulfur (15 ppm or lower level)	% mass (ppm)	0.0015max (15)	0.0010 max
8	Copper strip corrosion		No. 3 max	No. 1
9	Cetane number		47 min	51 min
10	Cloud point	°C	Report	–
11	Carbon residue	% mass	0.050 max	0.3 max
12	Acid number	mg KOH/g	0.50 max	0.5 max
13	Cold soak filterability	seconds	360 max	
14	Free glycerin	% mass	0.020 max	0.02 max
15	Total glycerin	% mass	0.240 max	0.25 max
16	Phosphorus content	% mass	0.001 max	0.001 max
17	Distillation temperature, atmospheric equivalent temperature,90 % recovered	°C	360 max	–
18	Sodium and potassium, combined	ppm (μg/g)	5 max	5 max
19	Oxidation stability	h	3 minimum	6 minimum
20	Density (15 °C)	g/cm^3	–	0.860–0.900
21	Total contamination	mg/kg	–	24 max
22	Ester content	% mass	–	96.5 min
23	Monoglyceride	% mass	–	0.8 max
24	Diglyceride	% mass	–	0.2 max
25	Triglyceride	% mass	–	0.2 max
26	Iodine value	–	–	120 max
27	Methyl linolenate	% mass	–	12 max
28	Polyunsaturation (4 or more double bound)	% mass	–	1 max

[1] (ASTM, 2009) [2] (ECN, 2008)

vegetable oils may have higher levels.[29] Processing can also introduce additional sulfur to biodiesel such as the case of sulfuric-acid-catalyzed esterification processes. If the final sulfur content is high in biodiesel, it usually requires distillation in order to lower the sulfur and comply with the specification.

The acid number determines the level of free fatty acids or processing acids in biodiesel. High acidity may cause biodiesel to be oxidatively and thermally unstable and speed up the fuel deterioration process. Highly acid fuel can also corrode engines. Similarly, the copper strip corrosion test measures the possible corrosion of copper and brass parts of the engine and fuel system. The causes of such corrosion are mainly acids or sulfur-containing compounds. Therefore, the copper strip test relates closely to the tests of acid number and sulfur content.

To ensure the safe use of biodiesel in diesel engines, mineral ash content is strictly regulated. The ash-forming elements include calcium (Ca), magnesium (Mg), sodium (Na), and potassium (K). These ash-forming materials may be brought into biodiesel from the initial feedstock, residual catalysts, and/or processing vessels and pipelines. Other ash contributors include sulfate and sediment. Phosphorus is also regulated to a very low level because it can severely damage catalytic converters.

The test of carbon residue is an approximation of the tendency of biodiesel to form carbonaceous deposits in engines.[1] High levels of glycerin, total and/or free, lead to injector deposits and fuel-line clogging and will cause a high carbon residue level. A high level of free glycerin indicates insufficient separation of the glycerin byproduct and the fuel, while a high level of bound glycerin (*i.e.* mono- or diglycerides) is the result of incomplete conversion of oils or fats to biodiesel.

Biodiesel is vulnerable to oxidation by oxygen in the air, to hydrolysis by absorbing moisture from air, and to degradation under oxygen and sunlight. After long-time storage, biodiesel tends to go rancid, which leads to volatile impurities in the fuel. All the aforementioned involve biodiesel stability, either oxidative, thermal, or long-term storage, and biodiesel stability is closely related to the FA profiles of the feedstock. These are monitored by the acid value, water content, and oxidation stability parameters in the specification.

More recently, the issue of sterol glucosides (SG) in biodiesel has caught the industry's attention. During longer-term storage, sterol glucosides in biodiesel serve as the seeds that allow unreacted mono- and/or diglycerides to grow on their surface into larger particles. These larger particles eventually cause filter plugging in application.

Cloud point, cold-soak filterability, cetane number, and oxidative stability, are all highly related to feedstock, and are discussed in greater detail in subsequent sections of this chapter. Briefly, different feedstocks result in biodiesel with different cold-flow properties, as measured by cloud point, pour point, and cold-filter plugging point. The cetane number of the FA methyl esters is also affected by the feedstock's alkyl chain length and degree of saturation. Generally, the more double bonds a FA contains, the lower its cetane number.

15.4.2 Cold-Flow Properties

Biodiesel tends to freeze at higher temperatures than does petrodiesel. Therefore, at low temperatures, the fuel can clog filters. The low-temperature operability, or "cold-flow properties," of a diesel fuel are described by the fuel's cloud point, pour point, and cold-filter plugging point.

The cloud point (CP) of petroleum products and biodiesel fuels is the temperature of a liquid specimen when the smallest observable cluster of hydrocarbon crystals first occurs upon cooling under prescribed conditions.[30] In other words, a fuel reaches its cloud point when fine crystals are first seen as the fuel is cooled. The pour point (PP) of a fuel, either diesel or biodiesel or blends of the two, is the lowest temperature at which the fuel will maintain a flowable condition. Below its pour point, a fuel is not pumpable. However, even if above its PP, a fuel may begin to plug filters due to the accumulation of crystals on the filter. This phenomenon is described by the cold-filter plugging point (CFPP). CFPP is the highest temperature at which a given volume of fuel fails to pass through a standardized filtration device under specified testing conditions. A fuel reaches its cloud point and/or pour point due to the crystallization of its high melting point (m.p.) components. In fossil diesel, such components include long-chain, high molecular weight hydrocarbons, especially the saturated alkanes or wax. In biodiesel, such components are mainly the long-chain fatty-acid esters, and sometimes residual glycerides such as monoglycerides.

The crystallization of a compound depends on its melting temperature and solubility. Generally, the components in biodiesel have relatively higher melting temperatures than those in fossil diesel. As seen in Table 15.7, the melting point temperatures increase as the length of the fatty-acid chains elongate. For saturated pure methyl esters, the m.p. increases from about 5 °C for laurate, to 19 °C for myristate ($C_{12:0}$), 31 °C for palmitate ($C_{16:0}$), and 38 °C for stearate ($C_{18:0}$). The longer the alkyl chain, the higher the melting temperature, which then translates to less favorable cold-flow properties of biodiesel. However, the m.p. of unsaturated components is considerably less. For example, for the same chain length C18, the methyl esters of oleate ($C_{18:1}$), linoleate ($C_{18:2}$), and linolenate ($C_{18:3}$) have dramatically reduced m.p. of − 20 °C, − 35 °C, and − 52 °C, respectively. These melting points are significantly lower than that of stearate ($C_{18:0}$). Therefore, biodiesel made from more unsaturated oils has better cold-flow properties. The trade-off is that biodiesel from unsaturated oils has less oxidative stability. The higher the unsaturation, the more vulnerable the unsaturated esters are to oxidation (see Section 15.4.3 for details on biodiesel oxidative stability).

Additionally, the alcohols used in making biodiesel affect the m.p. as well. Table 15.7 below shows the m.p. of palmitate ($C_{16:0}$) and stearate ($C_{18:0}$) of

Table 15.7 Melting temperatures of selected stearate and oleate alkyl esters.[31]

		Melting point (m.p., °C)	
No.	*Alkyl*	*Palmitate ($C_{16:0}$)*	*Stearate ($C_{18:0}$)*
1	Methyl	30.5	39
2	Ethyl	19.3	31–33.4
3	Propyl	20.4	
4	Butyl	16.9	27.5
5	*iso*-Propyl	13	
6	*iso*-Butyl	22.5	

different alkyl esters.[31] The m.p. of palmitate esters decreases significantly as the alcohol used changes from methanol to other alcohols; the m.p. of *iso*-propyl palmitate is much lower than that of methyl ester. Research has shown that similar effects have been observed for biodiesel made from soybean oil and animal fat using isopropyl alcohol.[32]

However, high m.p. is not the only factor that correlates to the cloud point and/or pour point. Precipitation of a high m.p. component is also dependent on its solubility in the fuel. A high m.p. component will not crystallize or precipitate out if its concentration in the mixture is below its saturation solubility, even if the temperature is at or below its m.p. Determination of the solubility of a component is difficult due to the complexity of the composition of a given fuel. The interactions between the components may have a positive or a negative effect on the solubility of a specific component depending on the liquid–solid phase equilibrium.[33] The final CP or PP is the combined result of the properties of individual components and their liquid–solid equilibriums in the fuel mixture.

The CP and PP of biodiesel depend on the fatty-acid profiles of the feedstock. Table 15.8 below lists the cloud points of biodiesel made from feedstocks of different fatty-acid profiles. Biodiesel made from palm oil and tallow, which contain higher levels of saturated fatty acids, have much higher CP than those made from soybean and rapeseed/canola oils. The CP of biodiesel also depends on traceable matters contained initially in the feedstock and the residue components after the processing, such as monoglycerides.

Foreign matter, such as sterol glucosides, will also affect the biodiesel cold-flow properties by serving as the nuclei on which other high m.p. components can precipitate. See Section 15.4.5 for details on the effect of plant sterols on biodiesel quality.

15.4.3 Oxidative Stability

Biodiesel oxidative stability can be understood more easily by considering its oxidative instability. Biodiesel is generally considered relatively unstable compared with fossil diesel fuel. This is mainly due to the chemical characteristics of

Table 15.8 Cloud points of biodiesel made from different feedstocks as related to their fatty-acid composition (Imahara *et al.*, 2006).[40]

		Methyl ester composition (%wt)						*Cloud point*	
No.	*Oil/fat*	$C_{16:0}$	$C_{18:0}$	$C_{18:1}$	$C_{18:2}$	$C_{18:3}$	*Others*	*(K)*	*(°C)*
1	Beef tallow	23.9	17.5	43.9	2.3	0.1	12.3	286	13
2	Palm	39.5	4.1	43.2	10.6	0.2	2.4	283	10
3	Sunflower	6.1	4.2	24	63.5	0.4	1.8	274	1
4	Soybean	10.7	3.2	25	53.3	5.4	2.5	272	−1
5	Linseed	6.7	3.7	21.7	15.8	52.1	0	268	−5
6	Olive	10.7	2.6	78.7	5.8	0.7	1.5	268	−5
7	Safflower	6.4	2.2	13.9	76	0.2	1.3	267	−6
8	Rapeseed	4.3	1.9	61.5	20.6	8.3	3.1	267	−6

the biodiesel constituents. Fossil diesel is made of mixtures of straight and branched hydrocarbons, mainly saturated, and aromatics. These chemical compounds are oxidatively stable under ambient conditions for an extended period of time – even years. Biodiesel consists of long-chain fatty-acid alkyl esters. The major contributors to biodiesel instability are, therefore, twofold: the unsaturated carbon-carbon (C–C) bonds; and the carboxyl groups contained in the long, straight-chain molecules. These unsaturated C–C bonds and the carboxyl groups are vulnerable to oxidation when in contact with oxygen under ambient conditions, although the mechanism of oxidation is far more complicated than a direct oxygen attack at the unsaturated positions. Such vulnerability leads to biodiesel decomposition and polymerization. Oxidized biodiesel can cause corrosion, filter plugging, injector fouling, and malfunction of the engine and fuel systems.

15.4.3.1 Alkyl Esters and Oxidation

Biodiesel consists exclusively of long-chain fatty-acid alkyl esters, most commonly methyl esters. Figure 15.9 shows two examples of fatty acid methyl esters commonly seen in biodiesel.

Unsaturated C=C double bonds occur in most of the commonly used plant oils and animal fats, and the number of double bonds or polyunsaturation of each oil or fat varies (see Table 15.3). If three double bonds in a molecule contain a "conjugated" structure – an arrangement of adjacent carbon atoms with a C–C=C patterns – the molecule is thermochemically stable because of how the electrons are shared.[34] However, such conjugated double-bond structures are rarely found in naturally formed fatty acids. In other words, the double bonds that occur in naturally formed fatty acids don't have the conjugated structures, so that they are thermochemically vulnerable to oxidation.

15.4.3.2 Oxidation of Biodiesel

Biodiesel made from unsaturated fats is more prone to oxidation than biodiesel made from saturated fats.

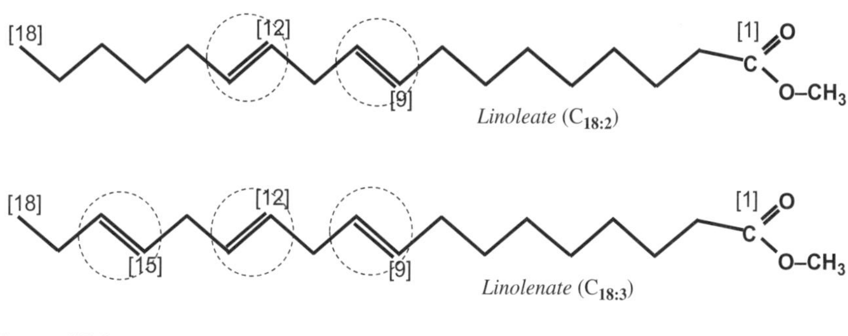

Figure 15.9 Examples of fatty acid methyl esters (methyl linoleate and methyl linoleate).

Oxidation of plant oils and animal fats, as well as biodiesel, leads to multiple families of chemicals including aliphatic alcohols, aldehydes, short-chain organic acids, small–medium-sized polymers (*i.e.* trimers and tetramers), and others in large numbers but small quantities.[35] The resulting aliphatic alcohols will reduce the flash point of biodiesel. Aldehydes cause rancidity, and short-chain fatty acids increase the acidity and speed up fuel deterioration.

During the oxidation process, intermediate products of biodiesel degradation may form larger molecules through C–O–C (in oxygen-rich environments) or C–C bonds (in oxygen-poor environments). Oxidized biodiesel with these large polymers has a higher viscosity and forms varnish deposits. Therefore, one indicator of biodiesel oxidative polymerization is the increased viscosity of the stored fuel, although polymerization can also be caused by microbial growth. The severe consequences of polymerization are sludge formation and filter plugging.

Besides the level of unsaturation, there are several other factors that affect the oxidative stability of biodiesel. Elevated acidity (high acid number) in the finished biodiesel is the most important factor leading to the oxidative instability of fuel. As discussed above, acids are the initiators and catalysts of fuel deterioration. The presence of acids will accelerate the oxidative decomposition of biodiesel. Increased acidity of stored biodiesel is an indicator of biodiesel oxidative decomposition.

Some residual minerals, such as $Fe^{2+/3+}$ and Cu^{2+}, which are contained in the feedstock and/or brought in from rusty reactors and storage vessels or pipelines, serve as catalysts for a series of fuel decomposition reactions and thus considerably decrease the oxidative stability of biodiesel.

Another factor leading to oxidation is the fact that biodiesel absorbs 15 to 25 times more moisture than petrodiesel.[36] Excess moisture content, combined with high acidity and mineral ions, provides a favorable medium for oxidative reactions and fuel deterioration.

15.4.3.3 Preventing Biodiesel Deterioration

Biodiesel will be more likely to maintain its quality during short-term storage if the fuel meets ASTM D6751 standards. The fuel's acid number, total and free glycerol, mineral content (calcium, magnesium, sodium, and potassium), and water and sediment, need to be within the specification parameters in order to ensure biodiesel oxidative stability. Storage tanks must be thoroughly cleaned to remove any deposits and rust before initially filling with biodiesel. The fuel should be stored in such a way that it will not experience extreme temperatures and direct sunlight. Underground or insulated storage tanks are ideal. To reduce biodiesel exposure to oxygen and moisture during long-term storage, a "nitrogen blanket" in the storage headspace is recommended. Plant oils contain variable amounts of natural antioxidants, but if biodiesel will be stored for longer than six months, additional antioxidants application may be necessary. If biological contamination is evident, biocides should be applied.

15.4.4 Cetane Number

The cetane number (CN) is a measure of the combustion quality of middle distillate diesel fuels during compression ignition. It measures the ignition performance of a diesel fuel oil by comparing it to a reference fuel in a standardized engine test. In other words, the cetane number test evaluates the readiness of a fuel to autoignite when injected into the engine. The higher the cetane number, the shorter the ignition delay.

Cetane (*n*-hexadecane) is a straight-chain alkane that ignites readily under compression. Although very little cetane is found in diesel fuel, cetane is used as the standard chemical to quantify a middle distillate fuel's ignition quality. A cetane number of 100 is assigned to pure cetane. Heptamethylnonane, a chemical that is very difficult to ignite under compression, is assigned a cetane number of 15. Blends of these primary reference fuels define intermediate values of cetane number.

Testing a fuel for its cetane number requires complicated equipment and procedures. The ASTM standards ASTM D-613[37] or D-6890[38] are typically used. The International Organization for Standardization (ISO) ISO5165 is also used in other countries.[39]

According to ASTM standard D975,[40] the Number 1 and Number 2 fossil diesel fuels must have a minimum cetane number of 40. For biodiesel, a minimum cetane number of 47 is required by ASTM D6751.[1] In the United States, No. 2 diesel fuel usually has a cetane number between 40 and 45. The cetane number of biodiesel is generally observed to be higher than that of fossil diesel, ranging from 45 to 55 for soybean biodiesel and 49 to 62 for rapeseed biodiesel.[41] These ranges of biodiesel cetane numbers are in agreement with a previous summary.[42]

The cetane number is generally dependent on the composition of the fuel. For biodiesel, the cetane number is directly related to the chemical properties of individual fatty-acid esters. Depending on the level of unsaturation, these long, straight-chain esters have varying cetane numbers (see Table 15.9).

The cetane number increases as the length of the saturated alkyl chain increases, from C_{12} to C_{18}. The cetane number of methyl stearate is as high as 88.6, close to the designated cetane number of cetane itself. If the alkyl chains are the same length, more unsaturated oils and fats will produce a methyl ester with a lower cetane number, as shown by the dramatic change from methyl stearate (88.6) to methyl linolenate (32.2). When the level of saturation is the same, the longer the alkyl chain, the higher the cetane number: compare the cetane numbers of methyl palmitoleate (53.8), methyl oleate (62.4), and methyl erucate (76).

The cetane numbers of fatty-acid esters are also affected by the alcohols used, such as methyl esters *vs.* ethyl esters, but the difference is not as significant.[43] However, if the alcohols are branched and the molecular weight increased, the effect may be significant. Knothe *et al.* have reported that use of branched alcohols, especially 2-ethylhexyl alcohol, had positive effects on boosting cetane numbers.[44] This positive effect, however, may be caused by the elongated molecular chain instead of the branching effect.

Table 15.9 Cetane numbers of fatty methyl esters commonly seen in biodiesel.

No.	*Methyl esters*	*Cetane Number* [1]
1	Laurate ($C_{12:0}$)	62.0 ± 4.7 [2]
2	Myristate ($C_{14:0}$)	69.5 ± 3.8
3	Palmitate ($C_{16:0}$)	81.2 ± 7.1
4	Palmitoleate ($C_{16:1}$)	53.8 [3]
5	Stearate ($C_{18:0}$)	88.6 ± 10.2
6	Oleate ($C_{18:1}$)	62.4 ± 5.9
7	Linoleate ($C_{18:2}$)	42.1 ± 3.9
8	Linolenate ($C_{18:3}$)	32.2
9	Erucate ($C_{22:1}$)	76 [4]

[1] Average data based on the summary by Lapuerta *et al.* (2009).[43]
[2] The first part of the data is the mean and second part is the standard deviation.
[3] Average of two data points.
[4] Single data point.

The results of cetane-number testing may vary from one analytical laboratory to the next, and among technicians due to the complexity of the testing procedures and the skill of the operator. This is why testing data show large standard deviations (Table 15.9). Other factors may also affect the cetane number of biodiesel, such as the oxidative status of the fuel (the age of the fuel, storage conditions, and antioxidant presence).[50]

15.4.5 Sterol Glucosides and Their Effects on Cold Soak Filtration Test

One of the most challenging fuel-quality problems facing the biodiesel industry is fuel filter plugging. Traditionally, filter plugging has been attributed to the presence of monoglycerides, the intermediate products in converting vegetable oils and animal fats to alkyl esters (biodiesel). The industry has been able to eliminate most quality issues associated with monoglycerides by strictly enforcing the ASTM specifications. However, recently, plant sterols (specifically sterol glucosides) in biodiesel have been found to cause quality problems and fuel-filter plugging.

There is very little information currently available to determine the extent of the sterol glucosides problem and to develop effective solutions. One of the primary obstacles is the fact that there are very few research organizations that possess the capability to detect and quantify these compounds. The non-acylated forms of sterol glucosides (also known as steryl glucosides, sterol glycosides, or steryl glycosides) are thought to form very fine particles that are insoluble in biodiesel.[45] Thus, after the fuel is stored for a few days, these small particles apparently act as nucleation sites for the crystallization of other compounds, such as saturated mono- and diglycerides, and possibly high melting point saturated methyl esters. The resulting precipitate complexes are found to have relatively high melting points. The consequence is filter plugging.

Currently, many biodiesel plants have installed systems to perform "cold soak filtration" on their fuels in order to remove the plant sterol precipitates. This process involves cooling the fuel to 3–5 °C and holding it at this temperature for 12 to 24 h to allow contaminants, including the sterol glucosides, to crystallize and precipitate. Then, the fuel is filtered and warmed up to ambient temperature before transport. This is a very time- and energy-consuming process. The 2009 version of Standard D6751 includes a "cold soak filter test" that uses a similar process to quantify the precipitates found above the cloud point.[1]

15.4.5.1 Plant Sterols

Sterols are a subgroup of lipid compounds of steroids found in animal and vegetal tissues. Plant sterols, also known as phytosterols, are present in all plants, including fruits and vegetables. Plant sterols help stabilize the phospholipid bilayers of plant membranes, just as cholesterol does in animal cell membranes.[46] The basic chemical structures of plant sterols are the same as those of cholesterol, but they differ with the side chain on C_{17}. More than 200 different types of plant sterols have been reported.

Plant sterols occur in plant tissues in five forms. In addition to the free form, there are four types of conjugates: (a) fatty-acid esters (3β-OH group is esterified), (b) hydroxycinnamate sterol esters (3β-OH group is esterified to hydroxycinnamic acids), (c) sterol glucosides (3β-OH group is glycosylated with a glucose), and (d) acylated sterol glucosides (3β-OH group is esterified to an acylated glucose). It is the sterol glucosides (see Figure 15.10) that are considered to be the major problem for biodiesel fuel quality.

A recent survey of 48 plants found that acylated sterol glucoside is present in all kinds of vegetable parts (fruit, tuber, root, stem, leaf, cereals), and that the acylated form is 2 to 10 times more abundant that the nonacylated sterol glucoside.[47] It was also found that 6′-linoleoyl-β-D-glucoside of β-sitosterol is predominant (47%) in soybean extracts, although other fatty acids were also detected.[48]

Vegetable oils contain primarily steryl FA esters. The plant sterol content varies widely from one oil to the next. Soil and climate conditions can also

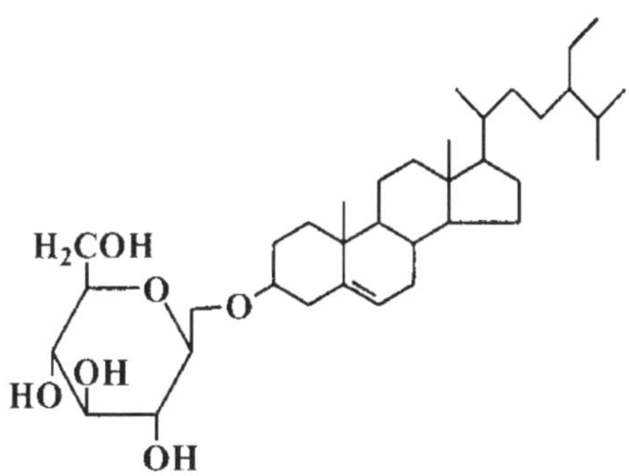

Figure 15.10 Structure of sterol glucoside.

Table 15.10 Plant sterols in vegetable oils.[49]

	Sterol	Seed oils Sunflower	Peanut	Soybean	(mg/kg oil) Cotton	Corn	Olive	Palm
1	Sitosterol	1961	1,145	1,317	3,348	9,187	732	252
2	Campesterol	242	278	563	276	2,655	18.8	88
3	Stigmasterol	236	145	564	17.3	499	0.5	42
4	Δ^5-Avenasterol	163	253	46.1	85.1	682	77.8	0.5
5	Δ^7-Stigmastenol	298	0.5	92	0.5	95.7	0.5	51
6	Δ^7-Avensterol	99.4	34.2	63.2	17.9	102	30	0.5
7	24-Methylene cycloartenol	204	0.5	53	0.5	425	580	0.5
8	Cholesterol	0.5	6.23	0.5	0.5	0.5	0.5	0.5
9	Brassicasterol	0.5	0.5	0.5	0.5	0.5	0.5	0.5
	Total=	*204.4*	1862.9	2699.3	3746.3	13 646.7	1440.6	435.5

affect plant sterol levels. Table 15.10 is an example data set of plant sterols obtained from several seed oils.[49]

Oil processing conditions, such as crude *vs.* degummed/deodorized/bleached,[50] chemically treated *vs.* physically treated,[51] *etc.*, affect the sterol content in the final product as well. Sometimes, the analytical methods employed will also lead to variations in the apparent plant sterol content.[52]

15.4.5.2 Procedure of Cold Soak Filtration Test

The 2009 version of the biodiesel specification ASTM D6751 adopts a new, mandatory test, the cold soak filtration test (CSFT) to test for insoluble substances such as plant sterols that can cause filter plugging. These substances, which can't be seen by the naked eye at room temperature, will precipitate out when cooled. The CSFT is an accelerated method of determining the tendency for these substances to plug filters. The reader is referred to the annex of ASTM D6751 for details.

15.5 Summary

Biodiesel is becoming increasingly accepted as a sustainable fuel for diesel engines. Its production is well understood and quality standards are in place. As petroleum supplies diminish, alternatives will need to be found that can take the place of conventional diesel fuel and biodiesel can play an important role in the transition process.

References

1. ASTM (American Society for Testing and Materials), D 6751-09 – Standard Specification for Biodiesel Fuel (B100) Blend Stock for Distillate Fuels, in "Annual Book of ASTM Standards", V. 05.04, ASTM International, West Conshohocken, PA, 2009.

2. E. Furimsky, *Appl. Catal. A*, 2000, **199**, 147.
3. ConocoPhillips, news release: ConocoPhillips Begins Production of Renewable Diesel Fuel at Whitegate Refinery in Cork, Ireland, 2006, http://www.conocophillips.com/newsroom/news_releases/2006news/12-19-2006.htm (accessed April 27, 2008).
4. L. Rantanen, R. Linnaila, P. Aakko and T. Harju, SAE Paper 2005-01-3771, Society of Automotive Engineers, Warrendale, PA, 2005.
5. (a) G. W. Huber, J. N. Chheda, C. J. Barrett and J. A. Dumesic, *Science*, 2005, **308**, 1446; (b) G. W. Huber, P. O'Connor and A. Corma, *Appl. Catal. A*, 2007, **329**, 120; (c) T. R. Carlson, T. P. Vispute and G. W. Huber, *Chem. Sus. Chem.*, 2008, **1**, 397; (d) R. M. West, Z. Y. Liu, M. Peter and J. A. Dumesic, *Chem. Sus. Chem.*, 2008, **1**, 417.
6. C. L. Peterson, D. L. Reece, B. L. Hammond, J. C. Thompson and S. M. Beck, *Appl. Eng. Agri.*, 1997, **13**(1), 71.
7. USDA NASS (National Agricultural Statistics Service), 2009, Soybeans – National Statistics, accessed on July 27, 2009.
8. C.L. Peterson, D.L. Reece, R. Cruz, J. Thompson, Liquid Fuels from Renewable Resources, Proceedings of an Alternative Energy Conference, American Society of Agricultural Engineers, 14–15 Dec. 1992, pp. 99.
9. P. S. Wang, M. E. Tat and J. Van Gerpen, *J. Am. Oil Chem. Soc.*, 2005, **82**(11), 845.
10. U. Schuchardt, R. Serchelli and R. M. Vargas, *J. Braz. Chem. Soc.*, 1998, **9**(1), 199.
11. E. F. Caldin and G. Long, *J. Chem. Soc.*, 1954, 3737.
12. (a) A. Singh, B. He, J. Thompson and J. Van Gerpen, *Appl. Eng. in Agri.*, 2006, **22**(4), 597; (b) G. Vicente, M. Martinez and J. Aracil, *Bioresource Technol.*, 2004, **92**, 297.
13. R. L. Glass, *Lipids*, 1971, **6**(12), 919.
14. M. L. Bender and W. A. Glasson, *J. Am. Chem. Soc.*, 1959, **81**, 1590.
15. G. Vicente, M. Martinez and J. Aracil, *Bioresource Technol.*, 2004, **92**, 297.
16. A. K. Singh, S. D. Fernando and R. Hernandez, *Energy & Fuels*, 2007, **21**, 1161.
17. W. Zhou and D. G. B. Boocock, *J. Am. Oil Chem. Soc.*, 2006, **83**(12), 1047.
18. H. Noureddini and D. Zhu, *J. Am. Oil Chem. Soc.*, 1997, **74**(11), 1457.
19. (a) C. Stavarache, M. Vinatoru, Y. Maeda and H. Bandow, *Ultrason. Sonochem.*, 2007, **14**, 413; (b) C. Stavarache C, M. Vinatoru and Y. Maeda, *Ultrason. Sonochem.*, 2007, **14**, 380; (c) J. A. Colucci, E. E. Borrero and F. Alape, *J. Am. Oil Chem. Soc.*, 2005, **82**(7), 525.
20. (a) J. Ji, J. Wang, Y. Li, Y. Yu and Z. Xu, *Ultrasonics*, 2006, **44**, e411; (b) K. S. Suslick, M. M. Mdleleni and J. T. Ries, *J. Am. Chem. Soc.*, 1997, **119**, 9303.
21. (a) D. Frascari, M. Zuccaro, D. Pinelli and A. Paglianti, *Energy & Fuels*, 2008, **22**, 1493; (b) D. Frascari, M. Zuccaro, A. Paglianti and D. Pinelli, *Ind. Eng. Chem. Res.*, 2009, **48**, 7540.
22. F. Ma, L. D. Clements and M. A. Hanna, *Ind. Eng. Chem. Res.*, 1998, **37**, 3768.

23. W. Zhou and D. G. B. Boocock, *J. Am. Oil Chem. Soc.*, 2006, **83**(12), 1047.
24. (a) D. G. B. Boocock, S. K. Konar, V. Mao and H. Sidi, *Biomass Bioenergy*, 1996, **11**(1), 43; (b) D.G.B. Boocock, S. K. Konar, V. Mao, C. Lee and S. Buligan, *J. Am. Oil Chem. Soc.*, 1998, **75**(9), 1167; (c) D. G. B. Boocock, US Patent 6,712,867, 2000.
25. R. Doell, S. K. Konar and D.G.B. Boocock, *J. Am. Oil Chem. Soc.*, 2008, **85**, 271.
26. H. Noureddini and D. Zhu, *J. Am. Oil Chem. Soc.*, 1997, **74**(11), 1457.
27. M. Canakci and J. Van Gerpen, *Trans. Am. Soc. Agri. Eng.*, 2001, **44**(6), 1429.
28. ECN (European Committee for Standardization), 2008, EN 14214 Automotive fuels – Fatty acid methyl esters (FAME) for diesel engines – Requirements and test methods, Avenue Marnix 17, B-1000, Brussels, Belgium.
29. B. He, J. Van Gerpen and J. Thompson, *Appl. Eng. Agri.*, 2009, **25**(2), 223.
30. ASTM Standard D2500-05, 2005, Standard Test Method for Cloud Point of Petroleum Products, ASTM International, West Conshohocken, PA, www.astm.org.
31. G. Knothe, *Fuel Proc. Technol.*, 2005, **86**, 1059.
32. (a) I. Lee, L. A. Johnson and E. G. Hammond, *J. Am. Oil Chem. Soc.*, 1995, **72**(10), 1155; (b) T. A. Foglia, L. A. Nelson, R. O. Dunn and W. Marmer, *J. Am. Oil Chem. Soc.*, 1997, **74**(8), 951.
33. H. Imahara, E. Minami and S. Saka, *Fuel*, 2006, **85**(12–13), 1666.
34. F. A. Carey and R. J. Sundberg, *Advanced Organic Chemistry: Part A: Structure and Mechanisms,* Springer, Dordrecht, 2008.
35. (a) K. Andersson and H. Lingnert, *J. Am. Oil Chem. Soc.*, 1998, **75**(8), 1041; (b) P. Bondioli, A. Gasparoli, L. Bella and S. Tagliabue, *Eur. J. Lipid Sci. Technol.*, 2002, **104**, 777; (c) M. Loury, *Lipids*, 1972, **7**, 671; (d) W. E. Neff, T. L. Mounts, W. M. Rinsch and H. Konishi, *J. Am. Oil Chem. Soc.*, 1993, **70**(2), 163.
36. B. He, J. Thompson, D. Routt and J. Van Gerpen, *Appl. Eng. Agri.*, 2007, **23**(1), 71.
37. ASTM Standard D613-08, 2008, Standard Test Method for Cetane Number of Diesel Fuel Oil, ASTM International, West Conshohocken, PA. www.astm.org.
38. ASTM Standard D6890-09, 2009b, Standard Test Method for Determination of Ignition Delay and Derived Cetane Number (DCN) of Diesel Fuel Oils by Combustion in a Constant Volume Chamber, ASTM International, West Conshohocken, PA. www.astm.org.
39. ISO (International Organization for Standardization), 1998, ISO5165, Determination of the ignition quality of diesel fuels - Cetane engine method, www.iso.org
40. ASTM Standard D975-07b, 2007, Standard Specification for Diesel Fuel Oils, ASTM International, West Conshohocken, PA. www.astm.org
41. M. Mittelbach and C. Remschmidt, *Biodiesel, the Comprehensive Handbook, 2004,* Boersedruck Ges.m.b.H, Vienna.

42. J. Van Gerpen, Proc. 3rd Liquid Fuel Conference, American Society of Agricultural Engineers, Nashville, TN, Sept. 15–17, 1996, pp. 197–206.
43. M. Lapuerta, J. Rodríguez-Fernándeza and E. Font de Mora, *Energy Policy*, 2009, **37**(11), 4337.
44. G. Knothe, A. C. Matheaus and T. W. Ryan III, *Fuel*, 2003, **82**, 971.
45. I. Lee, L.M. Pfalzgraf, G.B. Poppe, E. Powers and T. Haines, *Biodiesel Mag*. April 2007.
46. V. Piironen, D. G. Lindsay, T. A. Miettinen, J. Toivo and A. M. Lampi, *J. Sci. Food Agri.*, 2000, **80**, 939.
47. T. Sugawara and T. Miyazawa, *Lipids*, 1999, **34**(11), 1231.
48. L. Peng, Y. Kawagoe, P. Hogan and D. Delmer, *Science*, 2002, **295**(5552), 147.
49. P. J. Van Niekerk and A. Burger, *J. Am. Oil Chem. Soc.*, 1985, **62**(3), 531.
50. (a) R. A. Ferrari, E. Schulte, W. Esteves, L. Bruhl and K. D. Mukherjee, *J. Am. Oil Chem. Soc.*, 1996, **73**(5), 587; (b) T. Verleyen, M. Forcades, R. Verhe, K. Dewettinck, A. Huyghebaert and W. De Greyt, *J. Am. Oil Chem. Soc.*, 2002, **79**(2), 117.
51. T. Verleyen, U. Sosinska, S. Ioannidou, R. Verhe, K. Dewettinck, A. Huyghebaert and W. De Greyt, *J. Am. Oil Chem. Soc.*, 2002, **79**(10), 947.
52. M. V. Donova, D. V. Dovbnya, G. V. Sukhodolskaya, S. M. Khomutov, V. M. Nikolayeva, I. Kwon and K. Han, *J. Chem. Technol. Biotechnol.*, 2005, **80**, 55.

CHAPTER 16

Heterogeneous Catalysts for Biodiesel Production

JEAN-PHILIPPE DACQUIN, ADAM F. LEE AND KAREN WILSON

Cardiff Catalysis Instiute, School of Chemistry, Cardiff University, Cardiff, CF10 3AT, United Kingdom

16.1 Introduction

A combination of dwindling fossil-fuel reserves and major concerns over the global impact of CO_2 emissions on climate change, means that the age of cheap and abundant fossil energy is coming to an end. Oil is the most important source of energy worldwide, accounting for some 35% of primary energy consumption, followed by coal at 25% and natural gas at 21%; thus, the quest for sustainable fuel sources to meet constantly rising world energy demands is one of the main challenges for mankind this century.[1] The transition to biofuels is slow due to costly manufacture compared to fossil fuels, but through a series of tax incentives, current EU targets are aiming for 5.75% of transportation fuels to be biofuel derived by 2010.[2]

Among the renewable fuels, biodiesel is viewed as a promising, viable alternative to petroleum-based diesel used for transportation applications. Biodiesel comprises long-chain fatty-acid methyl esters (FAMEs) that are derived from vegetable oils or animal fats and contain primarily triacyl glycerides (TAGs) and free fatty acids (FFAs), and can be used either neat (100% biodiesel) or blended with petroleum diesel.[3] The high oxygen content and sulfur-free nature of biodiesel means the fuel is much cleaner burning, leading

RSC Energy and Environment Series No. 1
Thermochemical Conversion of Biomass to Liquid Fuels and Chemicals
Edited by Mark Crocker

Published by the Royal Society of Chemistry, www.rsc.org

to dramatically lower emissions of particulate matter relative to petroleum-derived diesel fuels.

The conversion of plant oils to biofuels has attracted much interest in recent years. While a number of approaches such as pyrolysis and hydrotreating/cracking methodologies remain under consideration, catalytic transesterification of TAGs to FAMEs using short-chain alcohols is by far the most energy efficient and simplest route. The objective of this review is to give an overview of current biodiesel synthesis methodologies and provide fundamental insight into the production of biodiesel by heterogeneously catalyzed routes. In an effort to identify the ideal catalyst characteristics for biodiesel synthesis, this review also strives to compare the applicability and efficiency of a number of solid acid and base catalysts, and the choice of chemical reactor design.

16.1.1 Current Processes for Biodiesel Production

The improved viscosity and volatility of FAMEs relative to the parent triglyceride means that their synthesis, particularly from vegetable oils, offers a promising alternative to diesel fuel.[4] Indeed, FAMEs typically have one-eighth the viscosity of the parent vegetable oil and thus improved pumping and low-temperature fuel characteristics. Optimization of the transesterification process, both in terms of catalyst selection and reaction conditions, is therefore a key research goal. The transesterification of oil seed TAGs to methyl (or ethyl) esters can be conducted using either acid or base catalysts (Scheme 16.1). Current manufacturing routes favor soluble alkaline catalysts,[5] such as sodium hydroxide/methoxide or potassium hydroxide, as these exhibit higher activities and are less hazardous to store and handle than mineral acids.

Unfortunately, these homogeneous base catalysts still corrode reactors (and engine manifolds), and their removal from the resulting biofuel is particularly problematic and energy intensive, requiring aqueous quench and neutralization steps, which themselves result in stable emulsions and soap formation. Furthermore, the glycerin byproduct, of significant potential value to the pharmaceutical and cosmetic industries, is heavily contaminated by aqueous waste salt streams and therefore is of limited resale value. Many plant oils also

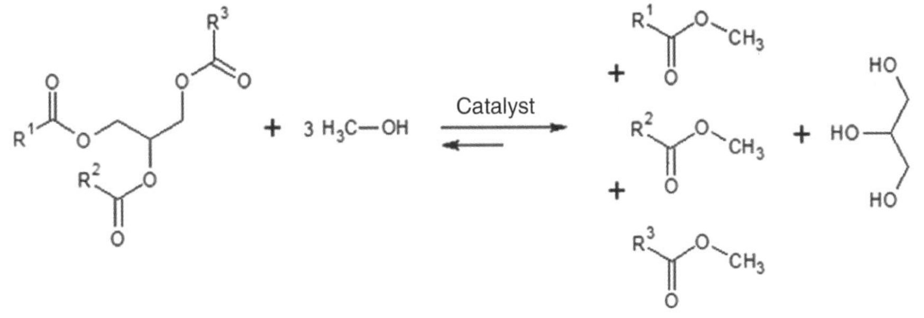

Scheme 16.1 Transesterification of TAG to FAME.

Scheme 16.2 Esterification of FFA to FAME.

contain significant levels of FFAs (Scheme 16.2), which must be removed by a preliminary acid-catalyzed esterification step before the oil is contacted with the soluble base catalyst.

In order to reduce the number of processing steps, it is often more economically viable to convert high fatty acid content oils, such as waste cooking oil, into biodiesel via batchwise esterification employing H_2SO_4. Acid-catalyzed esterification is popular within the cosmetics, pharmaceutical and food industry for manufacturing tailored fatty esters. Although industrial reactor technology is already in place for the latter batch processes, they do not exist on the very large scale necessary for synthesizing transportation fuel, especially when coupled to additonal neutralization and separation steps required for homogeneous approaches. Complete sulfuric-acid removal postreaction is also imperative due to stringent EU restrictions on the sulfur content of fuels. Consequently, the last decade has seen many industrial processes switch to solid acid catalysts for esterification.[6]

The choice of oil source for sustainable biodiesel production must also be given careful thought from both environmental and technical viewpoints. The ester chains from palm, rapeseed, sunflower or soybean oils used in biodiesel are typically C_{16}–C_{18} in length, but with two oxygen atoms retained in the FAMEs they contain ~10 wt% oxygen that gives rise to cleaner and more efficient combustion. From an application perspective, it is also important to consider the oil composition, since fatty-acid chains contain varying degrees of unsaturation that can improve their cloud points and miscibility with petroleum diesel,[7] but undergo problematic oxidation during storage. Furthermore, incomplete combustion of trace amounts of longer chain FAMEs can result in long-term residue accumulation within engines. Despite the promise of these "first generation" fuels, alternative nonfood-based oil sources are now sought to eliminate competition between land use for fuel crops versus traditional agricultural cultivation. Of equal concern are deforestation practices underway, notably in Indonesia, wherein vast tracts of rainforest and peatland are being cleared to support palm-oil plantations.[8] Alternative crops such as *Jatropha curcas*, which grow readily on arid wasteland and require minimal farming, are particularly attractive feedstocks with high yielding seeds containing ~40 wt% oil. The combination of nonfood oil seed crops, and a heterogeneously catalyzed continuous esterification/transesterification process, could dramatically improve both the environmental impact and energy efficiency of biodiesel manufacture. In addition to overcoming limitations with exisiting technologies, such new routes would dramatically improve public perception and the future development of biodiesel.

16.2 Catalysts for Triglyceride Transesterification

16.2.1 Solid Base Catalysts

Even though soluble strong base catalysts are commonly used for biodiesel production, their utilization results in undesired soap formation, requiring costly and technically difficult purification steps.[9–11] Moreover, the use of homogeneous catalysts restricts the possiblity of continuous processing methodologies.[12] Finally, such soluble bases are very sensitive to the presence of water and free fatty acids, and consequently need a high quality feedstock (refined vegetable oils) to avoid undesired side reactions (hydrolysis and saponification), or additional steps to first convert/eliminate free-fatty-acid impurities. Biodiesel synthesis using a solid base catalyst would facilitate separation of both catalyst (via a continuous-flow, packed-bed arrangement) and glycerol byproduct from the spent reaction mixture, thereby reducing production costs and enabling catalyst reuse.[13]

A variety of solid base catalysts are known, including alkali or alkaline-earth oxides, supported alkali metals, basic zeolites and clay minerals (such as hydrotalcites) and supported organic bases.[14] The origin of basicity in alkaline earth oxides has been previously reviewed, and is generally believed to arise from the presence of M^{2+}–O^{2-} ion pairs in different coordination environments.[15] The strongest basicity occurs at low coordination sites that exist at defects, corners, edges or on high Miller index surfaces as indicated in Scheme 16.3. The base strength of Group 2 oxides and hydroxides increases in the order Mg < Ca < Sr < Ba.[16]

Organic bases are another interesting class of solid base that can be synthesized by modifying either polymeric materials or inorganic porous supports to introduce primary, secondary or tertiary amine groups. Frequently investigated examples include aminopropyl functionalized silica[17] and more exotic, stronger bases such as guanidine-derivatized silica,[18] whose synthesis is outlined in Schemes 16.4(a) and (b).

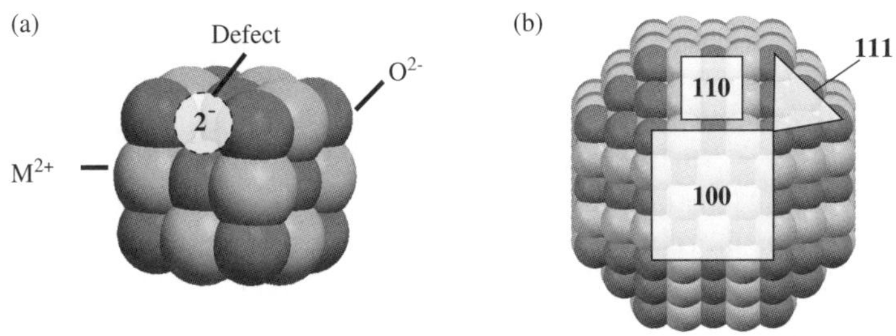

Scheme 16.3 Origin of base sites in alkaline earth oxides: (a) generation of cationic defect sites; (b) surface Mg^{2+}–O^{2-} ratio varies with crystallographic termination leading to increasing base strength across the series (100) < (110) < (111).

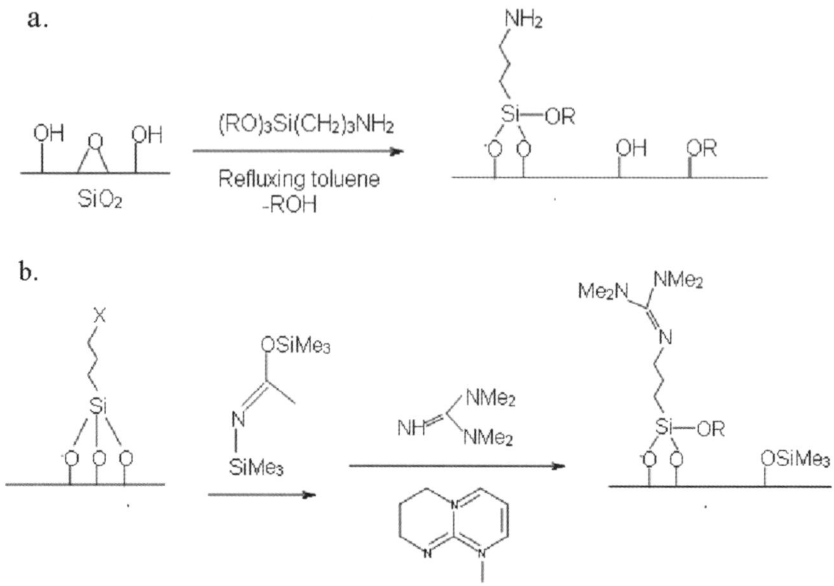

Scheme 16.4 Preparation of some common heterogenized organic bases: (a) aminopropyl silica; (b) guanidine-derivatized silica (where X is a halogen).

One of the first reported applications of heterogeneous catalysts for transesterification of vegetable oils was that of Schuchardt *et al.*[19,20] In this study, guanidines, usually employed in homogeneous catalysis, were heterogenized on organic polymers (*e.g.*, substituted polystyrenes) and tested for soybean oil transesterification with methanol over consecutive catalytic cycles. The resulting catalysts were nearly as active as their homogeneous anologs, however guanidines slowly leached from the polymers under recycling due to attack of the methoxide anions.

Classic heterogeneous base catalysts, mainly metal oxides, have been extensively tested for triglyceride transesterification.[21] La_2O_3, MgO and ZnO have been used for the solventless transesterification of glycerol with stoichiometric amounts of methylstearate[22] and it was concluded that the activity depended only on the intrinsic basicity of the strongest basic sites in the order $La_2O_3 > MgO > ZnO$. All these catalysts showed good activity and similar selectivities to the mono-, di- and triesters as observed using homogeneous basic catalysts (*e.g.*, 40% monoester at 80% conversion). Several authors[23–25] have applied CaO to catalyze the transesterification of rapeseed or sunflower oil by methanol. The morphologies of such alkaline-earth oxide catalysts and their corresponding basicities are very sensitive to their preparative route.[15] Since the active surface sites of fresh CaO catalysts are unavoidably poisoned by exposure to atmospheric H_2O and CO_2, a thermal activation treatment is required to remove surface $Ca(OH)_2$ and $CaCO_3$ in order to improve activity. Consequently, reported activites for transesterification by bulk CaO are highly variable with turnover frequencies (TOF) ranging between 2.5 to

45 $g_{(oil)}h^{-1}g_{(cat)}^{-1}$ for sunflower-oil transesterification depending on sample pretreatment and microcrystallinity.[25] CaO is generally found to be more active than MgO for rapeseed transesterification, with respective TOF of 1.66 $g_{(oil)}h^{-1}g_{(cat)}^{-1}$ versus 0.14 $g_{(oil)}h^{-1}g_{(cat)}^{-1}$ at 65 °C.[26] However, care must be taken with such comparisons, as there is concern in the literature[27,28] over partial leaching of Ca^{2+} from CaO under reaction conditions, and associated homogeneous catalytic contributions, which could limit its utility in biodiesel production. Alkali-doped CaO and MgO have also been investigated for transesterification.[29,30] These exhibit enhanced basicity that is generally associated with the formation of O^- centers when M^{2+} is replaced by M^+, resulting in charge imbalance and associated defect generation. In Li/CaO, the optimum activity occurs for dopant levels sufficient to create a saturated Li^+ monolayer, however, care must be taken during catalyst synthesis to ensure that leaching of promoters does not occur.[31]

Mixed CaO:MgO oxides prepared via thermal decomposition of natural dolomite have also shown promising activity and stability in the transesterification of a range of C_4–C_{18} triglycerides under mild conditions.[32] Dolomitic rock comprises alternating layers of $Mg(CO_3)$–$Ca(CO_3)$ and is structurally very similar to calcite ($CaCO_3$). Calcination at 900 °C transforms the mineral into a material comprising MgO nanocrystallites dispersed over larger (> 60 nm) CaO particles, Scheme 16.5. This calcined material exhibits very high activity, with a TOF of 2.9 $g_{(oil)}h^{-1}g_{(cat)}^{-1}$ towards the transesterification of olive oil at only 60 °C.

The literature suggests that such mixed MgO:CaO systems also show improved resistance to deactivation by CO_2 compared to CaO alone.[33] While the origin of this synergy requires further investigation, it may go some way to explaining the higher activity observed for calcined dolomite versus separate CaO or MgO components in plant-oil transesterification. The genesis of highly dispersed MgO nanocrystallites on the surface of CaO may also contribute to the performance of calcined dolomite, since recent reports reveal that the activity of MgO in TAG transesterification shows a strong size dependency; nanocrystalline MgO significantly outperforms commercial microcrystalline forms.[34,35] It is likely that this dependence reflects the correspondence between particle size

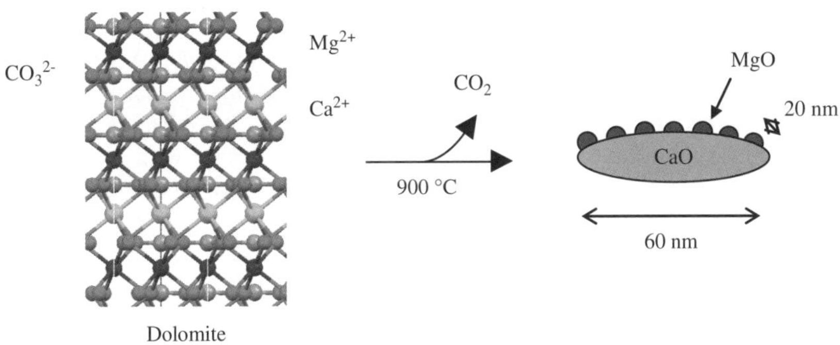

Scheme 16.5 Thermal decomposition of dolomite to MgO/CaO.

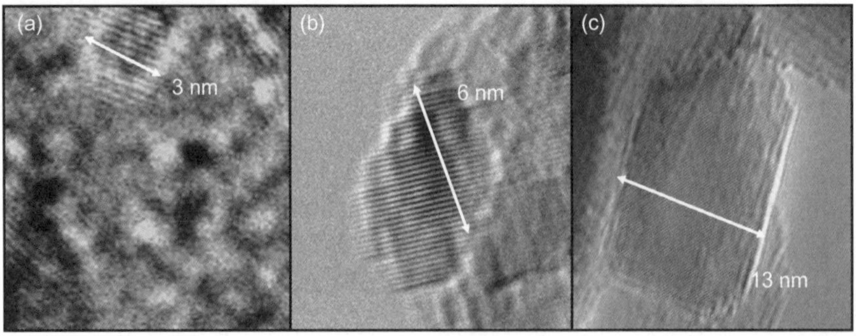

Figure 16.1 HRTEM of nanoparticulate MgO following thermal sintering, (a) as-prepared; (b) calcined at 500 °C; (c) calcined at 700 °C.

and preferential surface termination; larger crystallites ($\geq$ 7 nm) favor (111) and (110) over (100) facets.[35] Such restructuring (Figures 16.1(a)–(c)) exposes higher surface densities of polarizable, electron-donating O^{2-} centers, and creates bulk and surface defects, postulated as superbasic sites in solid base catalysts. This study highlights the importance of correlating the surface structure and basicity of solid base materials with their catalytic activity; the absence of systematic materials characterization has to date hampered catalyst optimization and thus the implementation of new heterogeneous processes.

Most of the mixed oxides described in the previous section are nonporous with low surface areas or microporous, rendering the majority of base sites inaccessible to bulky triglycerides. Further catalyst development and activity enhancement should prove possible through utilizing mesoporous solids as high area supports for active basic metal oxides and/or metals. These may confer improved stability and more facile oil diffusion.[36,37] Since the discovery of the M41S class of mesoporous materials by Mobil, there has been an explosion of interest in the preparation of such templated porous solids, and these have been extensively reviewed.[38–41] In the case of silicates, the most widely investigated mesoporous inorganic supports, network morphology and stability reflect the templating conditions (ionic or neutral surfactant), silica precursor (*e.g.*, fumed silica, tetraethylorthosilicate (TEOS), Ludox or sodium silicate) and whether hydrothermal synthesis or cosolvents are employed. A general preparative route is shown in Scheme 16.6, in which a micellar array of surfactant is used to direct the crystallization of a SiO_2 framework. Subsequent calcination, to burn out the organic template, yields materials with well defined mesostructured pores of 2–10 nm and surface areas up to 1000 $m^2 g^{-1}$.

Materials prepared using neutral polyethylene-oxide-type surfactants, such as Pluronic 123, Triton X-114 and Brij 76, enable larger pore materials (up to 10 nm) and have the added advantage that they operate via a true liquid crystal-templating mechanism (TLCT). TLCT permits final pore structures to be predicted from the liquid crystalline phase of the template, which can be rapidly identified by optical microscopy.[42] Other oxide supports, including Al_2O_3,

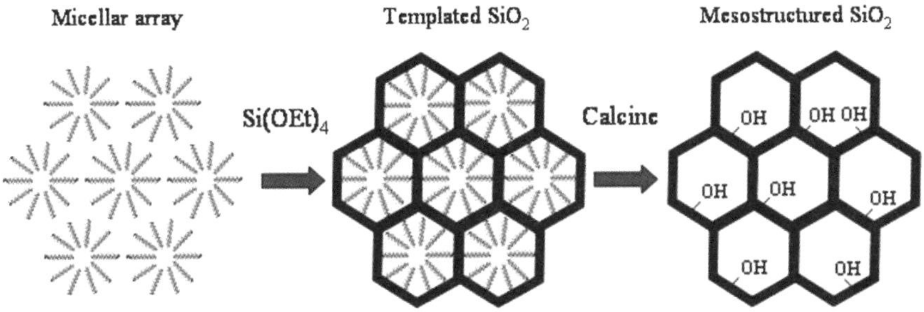

Scheme 16.6 Micelle templating of mesoporous silica.

ZrO_2 and TiO_2 and mixed metal-oxide variants can also be prepared via such templating methods. However, these typically exhibit poorer long-range mesopore order compared with that of silicates. The synthesis of robust supports with thick crystalline walls and stable mesopores with long-range crystallinity remains a challenge for materials science.

For biodiesel production, SBA-15 silicas are favored over MCM-41 as support materials due to their larger pore diameter and thicker walls that confer greater thermal stability. These mesoporous silicas, used as support hosts, have been recently investigated for various biodiesel applications such as rapeseed oil,[43] waste soybean frying oil,[44] palm oil[45] and blended vegetable oil[46] transesterification reactions. Albuquerque *et al.*[47] studied a series of SBA-15 supported CaO materials with different alkaline earth loadings, prepared via incipient wetness impregnation, for the transesterification of ethyl butyrate with methanol. They showed that SBA-15 stabilized CaO across the silica surface, and these materials exhibited negligible Ca leaching into the reaction medium compared to commercial CaO. A high loading (14 wt%) CaO/SBA-15 catalyst showed the best activity (95% conversion at 60 °C after 5 h reaction) benchmarked against soluble NaOH under the same conditions.

Rudolph *et al.*[46] screened different functionalized mesoporous catalysts (MgO/MCM-41, MgO/Kit-6, MgO/SBA-15) for the transesterification of blended vegetable oil with ethanol. The reaction was carried out at 220 °C for 5 h under continuous stirring in a batch reactor, and MgO/SBA-15 was found to be the most effective catalyst in the study, reaching 96% conversion. Although the role of support material was inconclusive from this work, catalyst activity was related to surface Mg concentration.

Amongst solid base catalysts, hydrotalcites have received considerable attention in recent years because of their high activity and robustness in the presence of water and free fatty acids.[48,49] For instance, triglyceride transesterification has been reported in the literature over hydrotalcites using both poor- and high-quality oil feeds;[50] in particular, Mg–Al–CO_3 hydrotalcite offers high transesterification rates in refined and acidic cottonseed oil, and in a high water content animal fat feed. For acidified cottonseed oil (9.5 wt% FFA) and animal fat oil (45 wt% water), 99% triglyceride conversion was achieved within 3 h at 200 °C in the presence of methanol (6:1 mole ratio of alcohol: triglyceride).

There remains some debate over the optimum Mg:Al ratio, which is an important factor when preparing hydrotalcites by precipitation from Na or K hydroxide/carbonate solutions. Complete removal of trace alkali residue from the hydrotalcite surface is very difficult, resulting in uncontrolled and unquantified homogeneous activity from leached Na or K. We recently reported on structure–reactivity correlations within thermally activated Mg-Al hydrotalcites[51] synthesized via an alkali-free precipitation route. $[Mg_{(1-x)}Al_{x}(OH)_2]^{x+}\ (CO_3)^{2-}{}_{x/n}$ materials were prepared with compositions spanning $x = 0.25$–0.55. Transesterification activity increased with Mg content in these Mg–Al hydrotalcites, which in turn tracked the intralayer electron density, and associated surface basicity. Davis *et al.*[52] investigated the influence of water on the activity and stability of activated Mg–Al hydrotalcites (Mg/Al molar ratio of 4) for the same chemistry. Brønsted base sites were active in the presence of water, but significant hydration caused rapid catalyst deactivation, presumably through ester hydrolysis to butyric acid that subsequently irreversibly neutralized surface base sites.

16.2.2 Solid Acid Catalysts

Transesterifications can be also acid catalyzed, but despite the wide range of solid acids commercially available, the corresponding direct transesterification of oils into biodiesel by solid acid catalysts has not been heavily explored. This in part reflects the lower activities for acid-catalyzed transesterification compared with base-catalyzed routes,[53] which in turn necessitates higher reaction temperatures to achieve acceptable reaction rates. However, while their activities are generally low, solid acids are less sensitive to contaminants then their base anologues and are able to function well with unrefined feedstocks containing 3-6 wt% FFAs. In contrast to solid bases that require pretreatments to remove free fatty acids, solid acids are able to esterify FFAs impurities through to FAME (see section 16.3) while simultaneously transesterifying the major triglyceride oil components without risk of soap formation.

Vicente *et al.*[54] compared different types of acid/base catalysts (both homogeneous (NaOH) and heterogeneous (Amberlyst A26, A27)) for FAME synthesis from sunflower oil at 60 °C. Unfortunately, under these conditions the heterogeneous acid catalysts were almost inactive in comparison with NaOH. However, Lopez *et al.*[55] showed that various solid acid catalysts performed favorably in triacetin transesterification, a model C_2 triglyceride, relative to H_2SO_4. In the latter case, the solid acids (Amberlyst-15, sulfated zirconia, Nafion NR 50 and tungstated zirconia) showed reasonable activities at 60 °C, indicating that they could act as alternatives to homogeneous acid catalysts, without corrosion or saponification problems. The authors also highlighted the importance of internal mass-transfer limitations, which impaired the performance of microporous heterogeneous catalysts such as ETS-10 (H) and zeolite Hβ.

Organic polymers[56] were more recently evaluated for soybean-oil transesterification with methanol or ethanol at mild temperature (64–75 °C). The sulfonated polystyrene compounds were used at 20 mol% of $-SO_3H$ groups in relation to the oil mass, with alcohol:oil molar ratios between 3:1 and 100:1. The best sulfonated polystyrene sample gave 85% conversion of refined soybean oil after 18 h of reaction.

Several studies have focused on sulfated zirconia in biodiesel synthesis, due to its strong (super) acidity and regeneration capacity in liquid-phase chemistry.[57–59] Suwannakarn *et al.* studied the activity and stability of a commercial sulfated zirconia catalyst for tricaprylin transesterification at high temperature and pressure (120 °C, 6.8 atm). While SO_4/ZrO_2 is very active towards tricaprylin, reaching 84% conversion in only two hours, subsequent reaction cycles evidence catalyst deactivation due to sulfate loss into solution. The applicability of sulfated zirconia is also of concern due to the energy input (high temperature and pressure) required for both reaction and catalyst reactivation.

Low-temperature transesterification requires robust solid acid catalyts with strong acidity. Heteropoly acids are an interesting class of well-defined superacids (materials with $pK_H+ > 12$),[60] possessing flexible structures and tuneable (super)acidity, but are unsuitable for biodiesel applications in native form due to their high solubility in polar media.[61] Although heteropoly acid dispersion onto high area supports can enhance the number of available acid sites,[62–64] this cannot overcome the solubility issue. Ion-exchanged Keggin-type phospho- and silicotungstic acids[65] can circumvent this problem, *e.g.*, Cs^+- and NH_4^+-doped variants are insoluble in water.[66] Cs salts of the general formulae $Cs_xH_{(3-x)}PW_{12}O_{40}$ and $Cs_yH_{(4-y)}SiW_{12}O_{40}$ have the added benefit of dramatically boosted surface areas relative to their fully protonated forms (Scheme 16.7), enabling transesterification at ~60 °C.[67,68] Materials with Cs contents of $x = 2.1–2.4$ or $y = 2.8–3.4$ are efficient in both esterification and transesterification reactions. This optimal composition was rationalized in

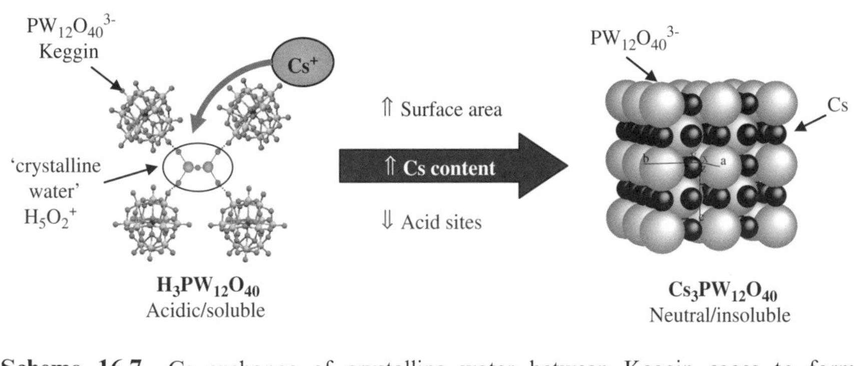

Scheme 16.7 Cs exchange of crystalline water between Keggin cages to form $Cs_xH_{(3-x)}PW_{12}O_{40}$ moieties insoluble in polar media. Complete exchange yields neutral $Cs_3PW_{12}O_{40}$ salt.

terms of a corresponding maximum in the density of *accessible* surface acid sites. In the case of $Cs_yH_{(4-y)}SiW_{12}O_{40}$, wherein C_4 and C_8 TAG transesterification were compared,[68] the absolute reaction rates were faster for the lighter triglyceride, which may evidence either poorer miscibility, or slower in-pore diffusion of the longer chain TAG in MeOH in accordance with previous observations of C_8 transesterification by solid base catalysts.[32] Furthermore, the TOF for the best Cs-doped catalyst was greater than that of highly soluble $H_4SiW_{12}O_{40}$, which operates homogeneously. This latter observation may reflect the hydrophilic nature of the parent heteropoly acid, thereby hindering ester activation of the more lipophillic C_8 TAG (the first step in acid-catalyzed transesterification) relative to more hydrophobic $Cs_xSiW_{12}O_{40}$ salts. Reactant/product polarity and associated mass transport to and from the active acid centers likely play an important role in controlling reactivity, even under homogeneous conditions, and molecular modelling may lead to a better understanding of such phenomena.

16.3 Catalysts for Free-Fatty-Acid Esterification

16.3.1 Effect of Free Fatty Acids on Biodiesel Production

Free fatty acids in plant oils and animal fats are problematic for conventional biodiesel manufacturing routes.[69,70] The negative influence of high FFA feedstocks on base-catalyzed triglyceride transesterification is well known,[4,71,72] with saponification resulting in viscous gels that hamper esterification and increase product separation costs. Consequently, there are tight specifications on feedstock compositions for transesterification, requiring FFA contents below $<0.5\%$.[12] Moreover, anhydrous alcohols (and catalysts) are required in order to avoid hydrolysis of the desired alkyl esters into FFAs and associated soap formation. There is great variability in the FFA content of different feedstocks, as shown in Table 16.1, hence highly refined vegetable oils (whose price can account for 60–75% of the overall cost of biodiesel production) are currently favored by commercial technologies.

The use of liquid inorganic or mineral acid catalysts for FFA removal via esterification adds costly neutralization/separation steps to the biodiesel manufacturing process, hence it is desirable to perform such pretreatments with a heterogeneous acid catalyst to reduce the overall number of processing steps.[73]

Table 16.1 Free-fatty-acid content of common biodiesel feedstocks.

Feedstock	*Amount of free fatty acid (wt%)*
Refined vegetable oils	<0.05%
Crude soybean oil	0.3–0.7%
Restaurant waste grease	2–7%
Animal fat	5–30%
Trap grease	75–100%

16.3.2 Heterogeneous Catalysts for Free-Fatty-Acid Esterification

Zeolites have been widely employed in catalytic esterification due to their structure, pore sizes and framework Si/Al ratios versatility. During the early 1990s zeolites such as modified H-Y, H-Beta and H-ZSM-5 were explored for the esterification of oleic, phenylacetic and benzoic acids. In order to expel the water formed, hydrophobicity of the zeolite lattice is expected to be an important parameter. The eventual suitability of a zeolite for a specific reaction depends on the polarity and miscibility of the acid and alcohol reactants.[74] More recently, comparative experiments for esterification of acetic acid with butanol have been carried out using various commercial zeolites.[75] The specific activity of the heterogeneous catalysts decreases in the following order: H-USY-20 > H-BETA-12.5 > H-MOR-45 > H-ZSM-5. It has been found that the low activity of ZSM-5 comes from internal diffusion limitations in the medium sized pores contrarily to H-BETA-12.5 and H-USY zeolite that show the best activities with bigger channel sizes. Overall, zeolite activity for esterification was strongly affected by acid strength (Si/Al ratio),[76–78] water tolerance,[13] pore size and dimensionality of the channel system.[12]

Mesoporous silicas from the MCM and SBA families have also been examined for FFA esterification. The intrinsic weak acidity of these materials has been enhanced via grafted sulfonic acid groups or surface coatings of sulfated zirconia to create stable and more active catalysts.[79–82] Synthesis of phenyl and propyl sulfonic acid SBA-15 catalysts is illustrated in Scheme 16.8.

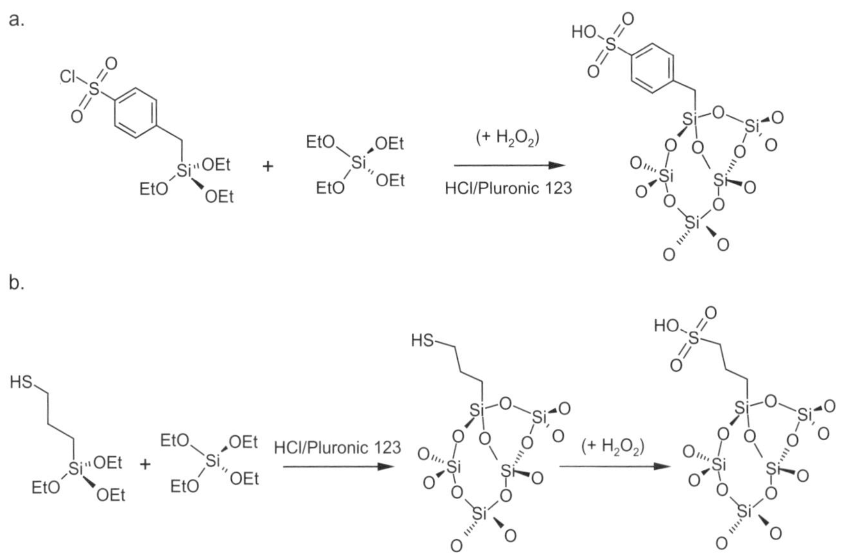

Scheme 16.8 Synthesis of sulfonic acid SBA-15 solid acids from either a) phenyl or b) propyl thiol precursors; aromatic sulfonic acid centers provide greater acidity.

These have been screened in palmitic acid esterification with methanol, and exhibit comparable activities to Nafion and Amberlyst acidic resins.[80] Phenyl sulfonic acid functionalized catalysts were more active than the corresponding propyl systems, correlating with their respective acid strengths.

In addition to good mass-transport characteristics, MCM-41 offers the possibility of metal incorporation within the oxide framework, which can improve hydrothermal stability and reactivity towards esterification. For instance, several authors[83–85] have recently studied the influence of Al doping on alkyl acid esterification reaction with several alcohols, and achieved better performance than with zeolite H-Y and H-mordenite. Modified resins have also been used for dodecanoic acid[86] and palmitic acid[87] esterification. Unfortunately, such resin catalysts are unsuitable for FFA esterification because of their poor thermal stability, which restricts their use to ~150 °C, and their relatively weak acidity.[88]

Finally, palmitic acid esterification with methanol has been conducted over Cs doped[67,68] and dispersed heteropolyacids, employing solid supports such as ZrO_2 [89] or activated carbon.[90,91] Reactant size plays an important role in regulating reactions catalyzed by carbon-supported heteropolyacids,[92] and care must also be taken to avoid leaching of $H_3PW_{12}O_{40}$ forms that are very soluble in polar media. Nevertheless, ion-exchanged heteropoly salts such as $Cs_{2.5}H_{0.5}PW_{12}O_{40}$ are promising water-tolerant catalysts for FFA esterification.[93] Structure–activity relations were recently mapped for Cs-doped $H_4SiW_{12}O_{40}$ catalysts for palmitic acid esterification,[68] which show a high activity for methylpalmitate production, associated with accessible H^+ sites within the stable mesopores.

16.4 Process Considerations for Biodiesel Production

Long-chain triglycerides in fats and plant oils are immiscible with methanol under normal reaction conditions, which limits methanol transfer into the oil phase. The solubility of TAGs in methanol decreases with increasing chain length and degree of saturation of the fatty-acid chain,[94] hence mixing efficiency is highly dependent on the oil source. For homogeneously catalyzed routes, several methods of improving miscibility have been investigated, including the use of ultrasonic or sonochemical batch reactors,[95,96] addition of a cosolvent such as tetrahydrofuran (THF),[97] or operation under supercritical methanol[98] to enhance mixing. While these approaches are successful, they all involve additional costs for either downstream separation (*e.g.*, removal of cosolvent) or high energy inputs. Membrane reactors have been proposed as a lower cost alternative to facilitate biodiesel separation.[99] Using a 0.05-μm carbon membrane, oil can be contained within the reactor zone by controlling the droplet size of the resulting oil/methanol emulsion. The membrane pore size was selected to permit facile permeation of reactively formed glycerol and FAME into the methanol zone on the other side of the membrane. Although this enables reaction equilibria to be shifted by continous FAME removal from the reaction zone, it does not overcome the problem of homogeneous catalyst separation.

Continuous processing is both preferable in terms of meeting a flexible biodiesel demand, and improving manufacturing efficiency. One of the first commercial heterogeneously catalyzed biodiesel processes, Esterfip-H^{TM}, was introduced by IFP in 2006,[5] and is based around a dual, fixed-bed reactor in which glycerol is removed after the first reactor to perturb the equilibrium and boost TAG conversion. Other approaches to intensify biodiesel production include reactive distillation,[6] continuous-flow centrifugation[100] and novel flow technologies such as the high shear spinning tube in tube (STT)[101] or oscillatory baffled reactor.[102]

The STT reactor is a commercial unit for high shear mixing that relies on thin-film formation in an annular zone between a rotor and stator tube. While this reactor is compatible with any mixture of gas, liquid and solid suspension, it is important that the feed enters the reactor zone in a homogeneous form, and that particulates are $<25\%$ of the gap dimension between interpenetrating tubes. The STT reactor has been demonstrated in biodiesel synthesis with homogeneous catalysts, however the requirement for mixing within the annular zone may result in attrition of solid catalysts. In contrast, reactive distillation involves no moving parts, and hence avoids problems associated with attrition-related deactivation. This type of reactor integrates reaction and separation into the same unit, simplifying process design, although the short residence times do necessitate a very active catalyst in order to achieve acceptable conversion. Reactive distillation has been applied to solid acid-catalyzed FFA esterification,[6] with continuous H_2O removal both assisting product separation and enhancing conversion by shifting the equilibria towards desired ester products.

The slow reaction rate of transesterification over solid acid and base catalysts means that conventional plug-flow reactors are unsuitable for biodiesel synthesis, as they necessitate very high length:diameter ratios to achieve the requisite mixing. Such designs are problematic due to their vast footprints, large pumping duties, and complex feedback control. An elegant solution to this problem is provided by the oscillatory flow (or baffled) reactor (OFR/OBR), which circumvents these issues achieving both good mixing and plug flow by superimposing an oscillatory motion onto a net slow flow of the process fluid. Orifice plate baffles equally spaced along the reactor tube enhance mixing characteristics, and can even be used to entrain solid material (heterogeneous catalysts).[103] The OBR has been successfully applied for process intensification of NaOH-catalyzed transesterification, and equivalent heterogeneously catalyzed biodiesel production is currently under investigation.

16.5 Future Perspectives and Concluding Remarks

This review highlights the significant progress made in recent years towards developing heterogeneous catalysts for biodiesel synthesis from plant oils that themselves are complex mixtures of triglycerides, free fatty acids and trace components such as phospholipids and sterols. The development of commercial heterogeneously catalyzed processes clearly requires a better understanding of how each reactant interacts with the active catalyst phase, particularly high molecular weight sterols that burn inefficiently in vehicle engines and should be removed or perhaps converted into lighter (fuel compatible) components. It should be noted that concerns remain over long-term biodiesel use and associated engine wear in high-performance diesel vehicles,[104] however, longer chain ($\sim C_{18}$) FAMEs should be easier to implement in heavy-duty diesel engines. Crucially, widespread uptake and development of next-generation biodiesel fuels requires progressive government policies and incentive schemes to place biodiesel on a comparative footing with cheaper fossil-based fuels.[105] Genetic modification, though controversial, may also assist by enabling plant biologists to engineer new higher yielding seed crops,[106] that are also rich in the optimum triglyceride chain length and degree of saturation and matched to the catalyst of choice, and desired fuel properties.

The decision whether to use a single solid acid catalyst operating at high temperature, or a dual system coupling a solid acid FFA pretreatment followed by a lower temperature solid base-catalyzed transesterification, remains to be explored, and is inextricably coupled with the particular oil composition. Solid materials capable of simultaneous esterification and transesterification under mild conditions present a future challenge for catalyst scientists, although superacids may be one solution. Alternatively, solid acids may be employed to first hydrolyze TAGs, and then esterify the resulting FFAs to FAME.[107]

Plant-oil viscosity and poor miscibility with light alcohols continues to hamper the use of new heterogeneous catalysts for continuous biodiesel production from both materials and engineering perspectives. Process optimization needs collaboration between catalyst chemists, chemical engineers and experts in molecular simulation to take advantage of innovative reactor designs; the future of biodiesel requires a concerted effort from chemists and engineers to develop catalysts and reactors in tandem. Current political concerns over the "food versus fuel" debate require urgent development of nonedible oil feedstocks as well as technical advances in order to ensure that biodiesel remains a key player in the renewable energy sector for the 21st century.

Acknowledgements

Financial support from the EPSRC under grants EP/F063423/1 and EP/G007594/1 and BP Biofuels is gratefully acknowledged. We are also grateful to Janine Montero for providing HRTEM images.

References

1. B. Walter, J.F. Gruson and G. Monnier, *Oil Gas Sci. Technol.*, 2008, **233**, 79.

2. *Directive 2003/30/EC* of the European parliament and of the council of 8 May 2003, on the promotion of the use of biofuels or other renewable fuels for transport.
3. S. Romano, *Vegetable oils: A new alternative*, *Proc. Int. Conf. on Plant and Vegetable Oils as Fuels*, 1982, 106–116. ASAE, St. Joseph, Michigan, USA.
4. B. Freedman, E.H. Pryde and T.L. Mounts, *J. Am. Oil Chem. Soc.*, 1984, **61**, 1638.
5. M. Bloch, L. Bournay, D. Casanave, J.A. Chodorge, V. Coupard, G. Hillion and D. Lorne, *Oil Gas Sci. Technol.*, 2008, **4**, 405.
6. A.A. Kiss, F. Omota, A.C. Dimian and Gadi Rothenberg, *Top. Catal.*, 2006, **40**, 141.
7. H. Imahara, E. Minami and S. Saka, *Fuel*, 2006, **85**, 1666.
8. F. Danielsen, H. Beukema, N.D. Burgess, F. Parish, C.A. Brühl, P.F. Donald, D. Murdiyarso, B. Phalan, L. Reijnders, M. Struebig and E.B. Fitzherbert, *Conserv. Biol.*, 2009, **23**, 348.
9. F.R. Ma and M.A. Hanna, *Bioresour. Technol.*, 2006, **37**, 29.
10. A. Demirbas, *Energy Convers. Manage.*, 2003, **44**, 2093.
11. A. Demirbas, *Energy Policy*, 2007, **35**, 4661.
12. E. Lotero, Y. Liu, D.E. Lopez, K. Suwannakarn, D.A. Bruce and J.G. Goodwin Jr, *Ind. Eng. Chem. Res.*, 2005, **44**, 5353.
13. H. E. Hoydonckx, D.E. De Vos, S. A. Chavan and P. A. Jacobs, *Top. Catal.*, 2004, **27**, 83.
14. K. Tanabe, M. Misono, Y. Ono and H. Hattori, *Stud. Surf. Sci. Catal.*, 1989, 51.
15. H. Hattori, *Chem. Rev 1995,* **95**, *537.*
16. N.N. Greenwood and A. Earnshaw, *Chemistry of the Elements,* Pergamon Press, Oxford, 1989.
17. D. Brunel, *Microporous Mesoporous Mater.*, 1999, **27**, 329.
18. A.C. Blanc, D.J. Macquarrie, S. Valle, G. Renard, C.R. Quinn and D. Brunel, *Green Chem.*, 2000, **2**, 283.
19. U. Schuchardt, R.M. Vargas and G. Gelbard, *J. Mol. Catal. A: Chem.*, 1996, **109**, 37.
20. U. Schuchardt, R. Sercheli and R.M. Vargas, *J. Braz. Chem. Soc.*, 1998, **9**, 199.
21. M.C.G. Albuquerque, D.C.S. Azevedo, C.L. Cavalcante Jr, J. S. Gonsalez, J.M.M. Robles, R.M. Tost, E.R. Castellon, A.J. Lopez and P.M. Torres, *J. Mol. Catal. A: Chem.*, 2009, **300**, 19.
22. S. Bancquart, C. Vanhove, Y. Pouilloux and J. Barrault, *Appl. Catal. A: Gen.*, 2001, **218**, 1.
23. G.R. Paterson and W.P. Scarrah, *J. Am. Oil Chem. Soc.*, 1984, **61**, 1593.
24. S. Gryglewicz, *Bioresour. Technol.*, 1999, **79**, 249.
25. M.L. Granados, M.D.Z. Poves, D.M. Alonso, R. Mariscal, F.C. Galisteo, R.M. Tost, J. Santamaria and J.L.G. Fierro, *Appl. Catal. B: Env.*, 2007, **73**, 317.
26. S. Yan, H. Lu and B. Liang, *Energy & Fuels*, 2008, **22**, 646.

27. M.L. Granados, D.M. Alonso, I. Sadaba, R. Mariscal and P. Ocon, *Appl. Catal. B: Env.*, 2009, **89**, 265.
28. A. Demirbas, *Energy Convers. Manage.*, 2007, **48**, 937.
29. R.S. Watkins, A.F. Lee and K. Wilson, *Green Chem.*, 2004, **6**, 335.
30. C.S. MacLeod, A.P. Harvey, A.F. Lee and K. Wilson, *Chem. Eng. J.*, 2008, **135**, 63.
31. D.M. Alonso, R. Mariscal, M.L. Granados and P.M. Torres, *Catal. Today*, 2009, **143**, 167.
32. K. Wilson, C. Hardacre, A.F. Lee, J.M. Montero and L. Shellard, *Green Chem.*, 2008, **10**, 654.
33. R. Philipp and K. Fujimoto, *J. Phys. Chem.*, 1992, **96**, 9035.
34. M. Verziu, B. Cojocara, J. Hu, R. Richards, C. Ciuculescu, P. Filip and V.I. Parvulescu, *Green Chem.*, 2008, **10**, 373.
35. J.M. Montero, P. Gai, K. Wilson and A.F. Lee, *Green Chem.*, 2009, **11**, 265.
36. P. Shah, A.V. Ramaswamy, K. Lazar and V. Ramaswamy, *Appl. Catal. A*, 2004, **1–2**, 239.
37. M.C. Gaudino, R. Valentin, D. Brunel, F. Fajula, F. Quignard and A. Riondel, *Appl. Catal. A*, 2005, **2**, 157.
38. J. Y. Ying, C. P. Mehnert and M. S. Wong, *Angew. Chem. Int. Ed.*, 1999, **38**, 56.
39. T. Linssen, K. Cassiers, P. Cool and E. F. Vansant, *Adv. Colloid Interface Sci.*, 2003, **103**, 121.
40. A. Davidson, *Curr. Opin. Colloid Interface Sci.*, 2002, 92.
41. A. Galarneau, J. Iapichella, K. Bonhomme, F. Di Renzo, P. Kooyman, O. Terasaki and F. Fajula, *Adv. Funct. Mater.*, 2006, **16**, 1657.
42. G.S. Attard, J.C. Glyde and C.G. Goltner, *Nature*, 1995, **378**, 366.
43. K.G. Georgogianni, A.K. Katsoulidis, P.J. Pomonis, G. Manos and M.G. Kontominas, *Fuel Proc. Tech.*, 2009, **90**, 1016.
44. K.G. Georgogianni, A.P. Katsoulidis, P.J. Pomonis and M.G. Kontominas, *Fuel Proc. Tech.*, 2009, **90**, 671.
45. A.Z. Abdullah, N. Razali and K.T. Lee, *Fuel Proc. Tech.*, 2009, **90**, 958.
46. E. Li and V. Rudolph, *Energy & Fuels*, 2008, **22**, 145.
47. M.C.G. Albuquerque, I.J. Urbistondo, J.S. Gonzalez, J.M.M. Robles, R.M. Tost, E.R. Castellon, A.J. Lopez, D.C.S. Azevedo, C.L. Cavalcante Jr and P.M. Torres, *Appl. Catal. A: Gen.*, 2008, **334**, 35.
48. M. Di Serio, M. Ledda, M. Cozzolino, G. Minutillo, R. Tesser and E. Santacesaria, *Ind. Eng. Chem. Res.*, 2006, **45**, 3009.
49. Siano D, Nastasi M, Santacesaria E, Di Serio M, Tesser R, Minutillo G, Ledda M and Tenore T, 2006, *PCT application no. WO2006/050925.*
50. N. Barakos, S. Pasias and N. Papayannakos, *Bioresour. Technol.*, 2008, **99**, 5037.
51. D.G. Cantrell, L.J. Gillie, A.F. Lee and K. Wilson, *Appl. Catal. A: Gen.*, 2005, **287**, 183.
52. Y. Xi and R.J. Davis, *J. Catal.*, 2008, **254**, 190.
53. H. Fukuda, A. Kondo and H. Noda, *J. Biosci. Bioeng*, 2001, **92**, 405.

54. G. Vicente, A. Coteron, M. Martinez and J. Aracil, *Ind. Crops Prod.*, 1998, **8**, 29.
55. D.E. Lopez, J.G. Goodwin, D.A. Bruce and E. Lotero, *Appl. Catal. A: Gen.*, 2005, **295**, 97.
56. R.A. Soldi, A.R.S. Oliveira, L.P. Ramos and M.A.F. César-Oliveira, *Appl. Catal. A: Gen.*, 2009, **361**, 42.
57. T.A. Peters, N.E. Benes, A. Holmen and J.T.F. Keurentjes, *Appl. Catal. A*, 2006, **297**, 182.
58. K. Suwannakarn, E. Lotero, J.G. Goodwin Jr and C. Lu, *J. Catal.*, 2008, **255**, 279.
59. A.A. Kiss, A.C. Dimian and G. Rothenberg, *Adv. Synth. Catal.*, 2006, **348**, 75.
60. N. Mizuno and M. Misono, *Chem. Rev.*, 1998, **98**, 199.
61. T. Okuhara, N. Mizuno and M. Misono, *Adv. Catal.*, 1996, **41**, 113.
62. M. Misono, *Catal. Rev.-Sci. Eng.*, 1987, **29**, 269.
63. A.D. Newman, D.R. Brown, P. Siril, A.F. Lee and K. Wilson, *PCCP*, 2006, **8**, 2893.
64. A.D. Newman, A.F. Lee, K. Wilson and N.A. Young, *Catal. Lett.*, 2005, **102**, 45.
65. T. Okuhara, T. Nishimura, H. Watanabe and M. Misono, *J. Mol. Catal.*, 1992, **74**, 247.
66. T. Okuhara, T. Arai, T. Ichiki, K.Y. Lee and M. Misono, *J. Mol. Catal.*, 1989, **55**, 293.
67. K. Narasimharao, D.R. Brown, A.F. Lee, P.F. Siril and K. Wilson, *J. Catal.*, 2007, **248**, 226.
68. L. Pesaresi, D.R. Brown, A.F. Lee, J.M. Montero, H. Williams and K. Wilson, *Appl. Catal. A.*, 2009, **360**, 50.
69. M. Canakci and J. V. Gerpen, *Trans. ASAE*, 1999, **42**, 1203.
70. F. Ma and M. A. Hanna, *Bioresour. Technol.*, 1999, **70**, 1.
71. M.P. Dorado, E. Ballesteros, J.A. Almeida, C. Shellet, H.P. Lohrlein and R. Krause, *Trans ASAE*, 2002, **45**(3), 525.
72. R. Turck, Method for producing fatty acid esters of monovalent alkyl alcohols and use thereof, US patent, 2002, **0156305B**.
73. G.R. Peterson and W.P. Sacarrah, *J. Am. Oil Chem. Soc.*, 1984, **61**, 1593.
74. N. Sanchez, M. Martinez, J. Aracil and A. Corma, *J. Am. Oil Chem. Soc.*, 1992, **69**, 1150.
75. T.A. Peters, N.E. Benes, A. Holmen and J.T.F. Keurentjes, *Appl. Catal. A: Gen.*, 2006, **297**, 182.
76. A. Corma, H. Garcia, S. Iborra and J. Primo, *J. Catal.*, 1989, **120**, 78.
77. J. Aracil, M. Martinez, N. Sanchez and A. Corma, *Zeolites*, 1992, **12**, 233.
78. N. Sanchez, M. Martinez, J. Aracil and A. Corma, *J. Am. Oil Chem. Soc.*, 1992, **69**, 1150.
79. W.D. Bossaert, D.E. De Vos, W. Van Rhijn, J. Bullen, P.J. Grobet and P.A. Jacobs, *J. Catal.*, 1999, **182**, 156.
80. K. Wilson, A.F. Lee, D.J. Macquarrie and J.H. Clark, *Appl. Catal. A: Gen.*, 2002, **228**, 127.

81. X.R. Chen, Y.H. Ju and C.Y. Mou, *J. Phys. Chem. C*, 2007, **111**, 18731.
82. I.K. Mbaraka, D.R. Radu, V.S.-Y. Lin and B.H. Shanks, *J. Catal.*, 2003, **219**, 329.
83. S. Shylesh, P.P. Samuel and A.P. Singh, *Micro. Meso. Mater.*, 2007, **100**, 250.
84. S. Ajaikumar and A. Pandurangan, *J. Mol. Catal. A: Chem.*, 2007, **266**, 1.
85. A.C. Carmo Jr, L.K.C. de Souza, C.E.F. da Costa, E. Longo, J.R. Zamian and G.N. da Rocha Filho, *Fuel*, 2009, **88**, 461.
86. S. Steinigeweg and J. Gmehling, *Ind. Eng. Chem. Res.*, 2003, **42**, 3612.
87. C.S. Caetano, L. Guerreiro, I.M. Fonseca, A.M. Ramos, J. Vital and J.E. Castanheiro, *Appl. Catal. A: Gen.*, 2009, **359**, 41.
88. P.F. Siril, H. Cross and D.R. Brown, *J. Mol. Catal. A: Chem.*, 2008, **279**, 63.
89. K.N. Rao, A. Sridhar, A.F. Lee, S.J. Tavener, N.A. Young and K. Wilson, *Green Chem.*, 2006, **8**, 790.
90. Y. Izumi and K. Urabe, *Chem. Lett.*, 1981, 663.
91. M.G. Kulkarni, R. Gopinath, L.C. Meher and A.K. Dalai, *Green Chem.*, 2006, **8**, 1056.
92. L. R. Pizzio, P. G. Vazquez, C. V. Caceres, M. N. Blanco, E. N. Alesso, M. I. Erlich, R. Torviso, L. Finkielsztein, B. Lantano, G. Y. Moltrasio and J. M. Aguirre, *Catal. Lett.*, 2004, **93**, 67.
93. T. Okuhara, *Chem. Rev.*, 2002, **102**, 3641.
94. L.K. Arnold, R.B.R. Choudhury and A. Guzman, *J. Am. Oil Chem. Soc.*, 1963, **40**, 33.
95. C. Stavarache, M. Vinatoru, R. Nishimura and Y. Maeda, *Ultrason. Sonochem.*, 2004, **12**, 367.
96. Mandar, A. Kelkar, P.R. Gogate and A.B. Pandit, *Ultrason. Sonochem.*, 2008, **15**, 188.
97. D.G.B. Boocock, S.K. Konar, V. Mao, C. Lee and S. Buligan, *J. Am. Oil Chem. Soc.*, 1998, **75**, 1167.
98. D. Kusdiana and S. Saka, *Fuel*, 2001, **80**, 693.
99. P. Cao, M.A. Dube and A.Y. Tremblay, *Biomass Bioenergy*, 2008, **32**, 1028.
100. C.L. Peterson, J.L. Cook, J.C. Thompson and J.S. Taberski, *Appl. Eng. Agric.*, 2002, **18**, 5.
101. Kreido-biofuels; *www.kreido.com*, 2009.
102. A.P. Harvey, M.R. Mackley and T. Seliger, *J. Chem. Technol. Biotechnol.*, 2003, **78**, 338.
103. M.H.I. Baird, A.P. Harvey, M.R. Mackley, X. Ni, N.V. Rama Rao and P. Stonestreet, *Trans. I. Chem. E. Part A*, 2003, **81**, 373.
104. G. Knothe, *Energy Environ. Sci.*, 2009, **2**, 759.
105. S. Pinzi, I. L. Garcia, F. J. Lopez-Gimenez, M. D. Luque de Castro, G. Dorado and M. P. Dorado, *Energy Fuels*, 2009, **23**, 2325.
106. G. Knothe, *Fuel Proc. Tech.*, 2005, **86**, 1059.
107. K. Suwannakarn, E. Lotero, K. Ngaosuwan and J.G. Goodwin Jr, *Ind. Eng. Chem. Res.*, 2009, **48**, 2810.

CHAPTER 17

Catalytic Conversion of Glycerol to Valuable Commodity Chemicals

JORGE N BELTRAMINI[a] AND CHUN-HUI (CLAYTON) ZHOU[b]

[a] ARC Centre of Excellence for Functional Nanomaterials, The University of Queensland, St. Lucia, QLD 4072, Australia; [b] School of Chemical Engineering and Materials Science, Zhejiang University of Technology, 310032, P. R, Hangzhou, Zhejiang, China

17.1 Introduction

Increasingly scarce oil resources and the deterioration of the global environment have led governments to actively seek suitable alternative energy and green chemical production technologies. Much attention has been devoted to the conversion of biorenewable feedstocks to commodity chemicals and clean fuels. One example is glycerol, which is a nontoxic, edible, biosustainable and biodegradable compound[1,2] that is widely available and can be obtained as a byproduct from a variety of processes such as soap manufacturing, fatty-acid production, fatty-ester production[1] and microbial fermentation,[1] and can be synthesized from propylene oxide.[1] Glycerol can also be commercially produced by the fermentation of sugars such as glucose and fructose, either directly[2] or as a byproduct of the industrial conversion of lignocellulose into ethanol.[3] Technical and economic analyses show that the coproduction of

RSC Energy and Environment Series No. 1
Thermochemical Conversion of Biomass to Liquid Fuels and Chemicals
Edited by Mark Crocker

Published by the Royal Society of Chemistry, www.rsc.org

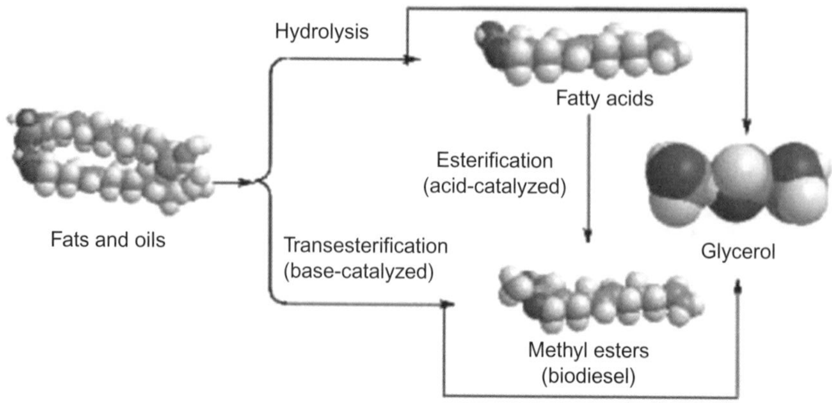

Figure 17.1 Pathways for production of biodiesel from vegetable oils.

glycerol with ethanol lowers its production costs and increases profit margins.[4,5]

More importantly, glycerol can be obtained in large quantities *via* the direct transformation of vegetable oils and animal fats into methyl esters. This reaction has been known for more than 100 years, however, it is only in the last 10 years that the transesterification of triglycerides, using vegetable oils and animal fat, has gained significance for its role in the manufacture of high-quality biodiesel fuel (Figure 17.1),[6–8] in which 1 mole of glycerol is formed for every 3 moles of methyl esters synthesized, amounting to approximately 10 wt% of the total product.[9]

Biodiesel, being renewable and clean, has proved its value as a fuel for diesel engines in the USA and in Europe,[10,11] although its life-cycle energy balance is still in dispute.[12] Recently the EU has stipulated that by the end of 2010, traffic fuels should contain at least 5.75% of renewable biocomponents.[13] If the target of this directive is to be achieved, European biodiesel demand could increase to 10 million tons per year by 2010, producing an additional 1 million tons of glycerol as coproduct.[14] At the same time, if the United States achieves its goal of replacing 2% of the on-road diesel with biodiesel by 2012, almost 363 million kg of new glycerol would be added to the market.[15] It is estimated that 5.8 billion liters of biodiesel can be produced worldwide by the end of 2015, mainly from feedstocks such as soybean, corn, trap grease and inedible tallow. On the other hand, biodiesel production would be an even more attractive fuel alternative for poor countries that are cash strapped and heavily dependent on petroleum imports.

A large quantity of impure glycerol that is produced from biodiesel processes is available at very low prices and as a result could reduce the commodity price of pure glycerol and at the same time destabilize the market for oleochemicals. Studies have shown that the current glycerol price could decline by more than 60% from its current value by 2010. To address this situation a recent study from the US Department of Energy clearly identified glycerol as a major

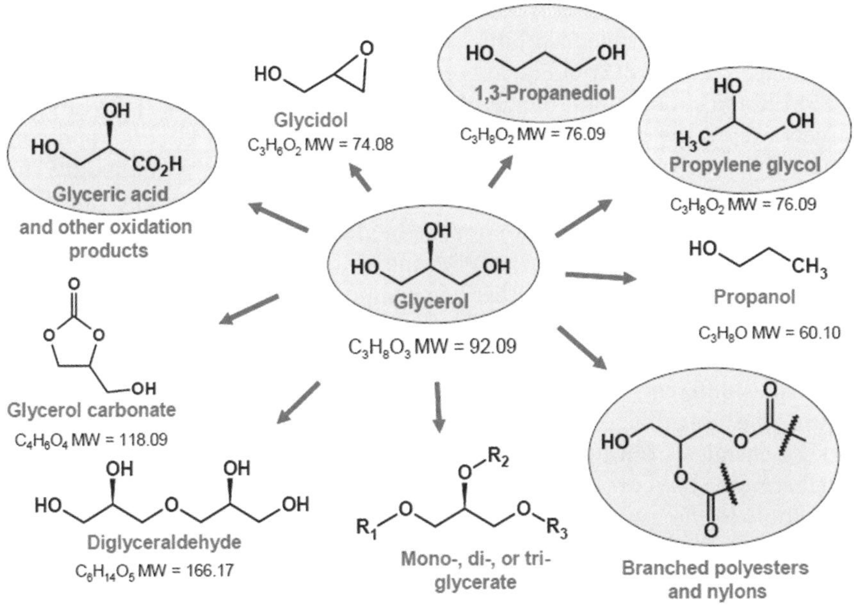

Figure 17.2 Chemical derivatives from glycerol.[16]

building block platform chemical for the future (Figure 17.2).[16] Economic forecasts demonstrated that the net production costs of B100-type biodiesel can be reduced from US$ 0.63/liter to U$S 0.35/liter by adding value to the glycerol coproduct.[15,16] However, glycerol must be purified before any attempt to convert it into valuable products.

At present, most biodiesel plants use homogeneous processes that are operated in either batch or continuous mode and utilize typically sodium hydroxide or other alkali-metal hydroxides or alkoxides as catalysts.[17,18] The resulting glycerol coproduct is of low value as it typically contains an estimated 50% impurities such as methanol, water, inorganic salts (catalyst residue), free fatty acids, unreacted mono-, di- and triglycerides, methyl esters, and a variety of other organic materials in varying quantities depending on the biodiesel process.[19] Its disposal is also difficult as the methanol content of the glycerol renders it hazardous waste.

Further treatments to improve the glycerol coproduct quality are needed and generally involve neutralization and/or removal of the excess methanol, the catalyst and soap[19] using a vacuum flash process. Depending on the type of separation process used, the final glycerol purity is about 80–95%.[20] As such, the crude glycerol can be sold to glycerol refiners at a low cost.[21] After refining, glycerol becomes a highly pure product (97% plus purity) and further refining can take the purity up to 99.5% to 99.7% using vacuum distillation or ion-exchange processes.[22] However, these processes are expensive and economically prohibitive for small biodiesel plants that prefer to discard the glycerol coproduct as waste.

New biodiesel plant designs based on the use of heterogeneous catalysts have shown that is it possible to produce biodiesel and a 98% pure glycerol that is limpid, free of any salt contaminants and colorless.[20–23] Despite these advances, the projected production volume of crude glycerol over the next 5 years will largely exceed the present commercial demand for purified glycerol, and consequently, purifying the glycerol coproduct to be sold as a commodity for traditional applications such as pharmaceuticals, cosmetics and the food industry, is not a practical option for the biodiesel industry. Therefore, the low cost of glycerol could open the way for the synthesis of value-added chemicals such as polymers, ethers, and other fine compounds that are of great industrial importance and can at the same time open new markets. From a technical standpoint, glycerol's multifunctional structure and properties can be modified by several different reaction pathways as shown in Figure 17.3. In the near future, the potential conversion of renewable resources into valuable commodity chemicals can facilitate the replacement of petroleum-based products. Furthermore, the conversion of glycerol to hydrogen or syngas can further contribute to the use of clean, renewable-energy sources. It is clear that a very large number of chemicals could be derived from glycerol. In this respect,

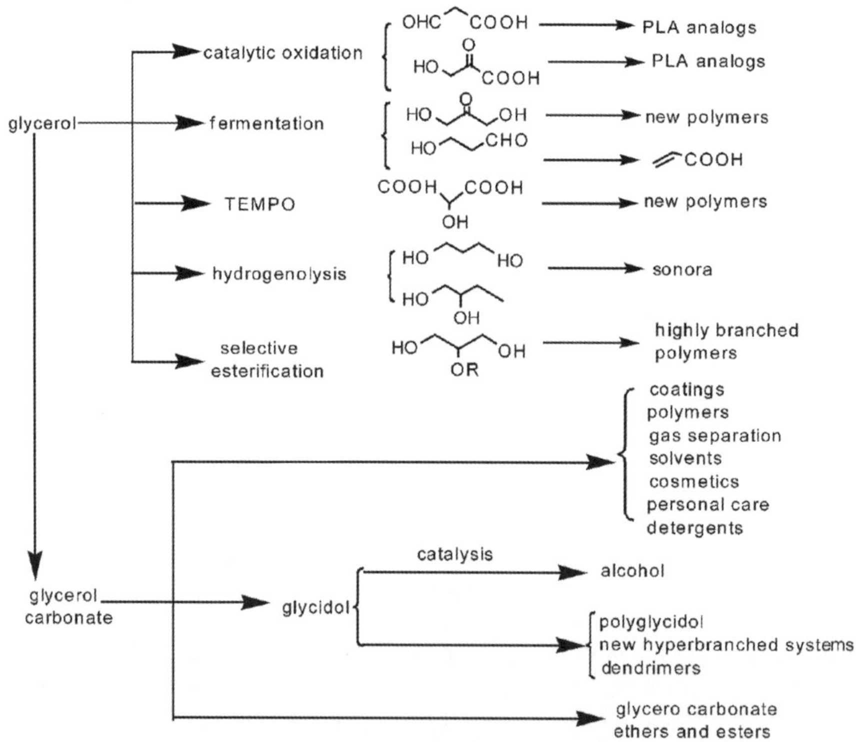

Figure 17.3 Processes for catalytic conversion of glycerol into useful chemicals.

catalysis represents a critical approach to green chemical technology in the activation and utilization of glycerol.

Recently, a series of novel catalytic conversion processes that transform glycerol into useful chemicals have been reported in the literature. This chapter discusses the most recent developments in the catalytic conversion of glycerol to promising commodity chemicals and fuels. Previous work has shown a great potential for the establishment of a variety of new catalytic processes and products from glycerol. As a result, new opportunities and challenges exist for research and industry to improve both the catalytic materials and technologies needed for the efficient transformation of glycerol into valuable chemicals. In this chapter the following sections will be considered:

- the selective oxidation of glycerol, a complex pathway of reactions that leads to a large number of products as useful intermediates or valuable fine chemicals;
- the selective etherification of glycerol to polyglycerols and fine chemicals which can be used as biodegradable surfactants, lubricants, cosmetics, food additives, *etc.*;
- the selective transesterification and esterification of glycerol;
- the selective hydrogenolysis of glycerol;
- the selective dehydration of glycerol;
- the gasification of glycerol.

17.2 Selective Oxidation of Glycerol

Glycerol selective oxidation leads to a rich reaction pathway with many possible products (Figure 17.4). Six potential C_3 oxygenated products – dihydroxyacetone, hydroxypyruvic acid, mesoxalic acid, glyceraldehydes (GLYALD), glyceric acid (GLYAC), tartronic acid together with C_2 (oxalic acid, hydroxyethanoic acid) and C_1 products (formic acid, CO_x) – can be obtained. Selective *versus* nonselective oxidation is, therefore, the big challenge associated with these catalytic oxidation reactions. The first important step in the industrialization of these new processes requires the design of new, effective heterogeneous catalysts to control the chemistry and selective orientation of the glycerol oxidation reaction towards either the oxidation of the primary alcohol functions, to give glyceric acid, or the oxidation of the secondary alcohol function, to produce dihydroxyacetone and hydroxypyruvic acid. Particularly important are tartronic and mesoxalic acids, two potentially valuable chelating agents that can be used as intermediate compounds for the synthesis of fine chemicals and novel polymers. To date, these products have a limited market because they are either produced using costly and polluting stoichiometric oxidation agents such as potassium permanganate, nitric acid or chromic acid or by the use of low-productivity fermentation processes.[24,25]

Glycerol has a highly functionalized and unique molecular structure that allows heterogeneous catalytic oxidative reactions to be carried out using inexpensive

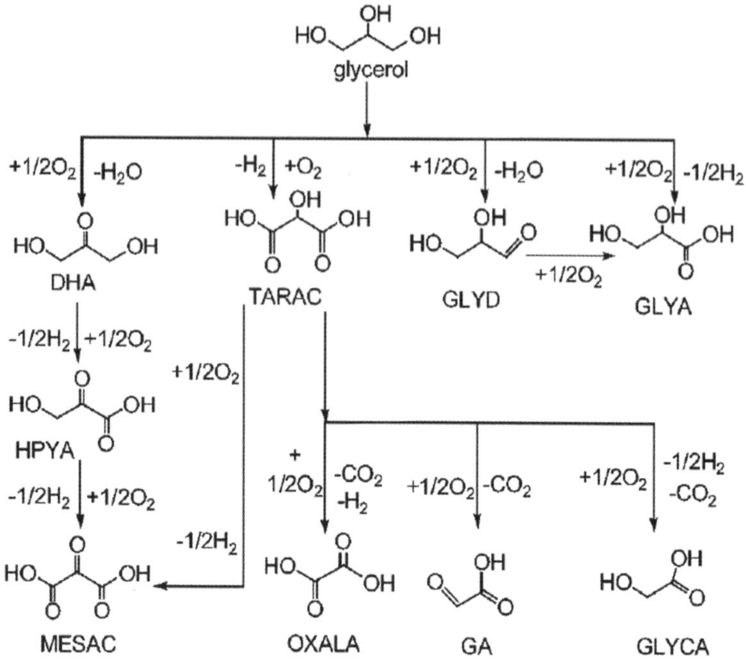

Figure 17.4 Possible reaction pathways to oxygenated derivatives of glycerol.

clean oxidizing agents such as air, oxygen, and hydrogen peroxide, instead of costly stoichiometric oxidants, leading to an economic way of producing a new class of chemical derivatives using an environmentally friendly alternative.

Table 17.1 lists some typical results for the catalytic oxidation of glycerol using supported noble metal nanoparticles such as Pd, Pt and Au as catalysts, as reported by several groups.[26–37] Given its high boiling point, the selective oxidation of glycerol with air or oxygen is normally carried out in the liquid phase using water as the solvent. It has been found that when using Pd and Pt catalysts in a basic medium, glyceric acid was obtained as the main product and tartronate and oxalate as the main overoxidation byproducts. Glyceric acid oxidation proceeded rapidly, converting into hydroxypyruvic acid. Addition of Bi to Pt catalysts greatly improves the selectivity towards the secondary alcohol.[26,27] Moreover, this reaction mainly depends upon both reaction conditions and the catalyst nature. Indeed, the products of glycerol oxidation seem to depend on how the reaction is conducted: a batch reactor working at pH 11 appears to produce more DHA than a fixed-bed reactor. However, a 20% yield of DHA was obtained in a batch reactor at 30% glycerol conversion using a 1%Bi-5%Pt/C catalyst, and a 30% DHA yield at 40% glycerol conversion was obtained in a fixed-bed catalytic reactor with a 0.6%Bi-3%Pt/C catalyst. The results suggested that, in a fixed-bed reactor loaded with Bi/Pt/C catalyst, glycerol conversion into DHA increases. It seems that Bi modifies the Pt sites, controlling the glycerol orientation towards DHA formation.[28]

Table 17.1 Some typical results from the catalytic oxidation of glycerol investigated by several groups.

Catalysts	*Oxidants*	*pH*	*Other reaction conditions*	*Glycerol conversion*	*Selectivity or yield*	*Researchers and year*	*Ref.*
Pt/charcoal	Air (0.1 MPa)	2–4	10%GLY, 323 K, 4 h	37%	4% (Y_{DHA})[b]	Kimura, 1993	26,27
Bi-Pt charcoal	Air (0.1 MPa)	2–4	10%GLY, 323 K, 4 h	30%	20%(Y_{DHA})	Kimura, 1993	26,27
Pt-Bi/charcoal[a]	Air (0.1 MPa)	–	50% GLY, 323 K, O_2/GLY = 2 (mol/mol)	80%	80%(S_{DHA})	Kimura, 1993	26,27
5% Pd/C	Air (0.1 MPa)	11	10%GLY, 333 K, 30% NaOH, 5 h	100%	8%(S_{DHA}), 70% (S_{GLYAC})	Garcia *et al.*, 1995	28
Bi-Pt/C	Air (0.1 MPa)	2	10%GLY, 333 K, 5 h	75%	50%(S_{DHA})	Garcia *et al.*, 1995	28
5%Bi-5%Pt/C	O_2 (0.02 MPa)	5.5	8 mmolGLY/60 mlH_2O, 338 K, 3 h	-	25%(Y_{DHA}), 20% (Y_{HYPAC})	Abbadi *et al.*, 1996	29
Ti-Si cogel	H_2O_2 (10% w/v)	7	10 g GLY, 353 K, 24 h	22%	37%(S_{GLYALD})	McMorn *et al.*, 1999	30
1%Au/charcoal	O_2 (0.3 MPa)	basic	12 mmolGLY/20 mlH_2O 333 K, 3 h, 12 mmol NaOH	56%	100%(S_{GLYAC})	Carrettin *et al.*, 2002	31
5%Pt/C	Air (0.1 MPa)	11	1 M GLY, 333 K, 21 h, 30%NaOH	60%	47.5%(S_{GLYAC})	Carrettin *et al.*, 2003	32
1% Au/C	O_2 (0.3 MPa)	basic	0.3 M GLY, NaOH/GLY = 4 (mol/mol), 303 K, 20 h	100%	92%(S_{GLYAC})	Porta *et al.*, 2004	33
1% Pd/graphite	O_2 (0.3 MPa)	basic	0.3 M GLY, NaOH/GLY = 4 (mol/mol), 50 °C, 1 h	90%	62.4%(S_{GLYAC})	Dimitratos *et al.*, 2005	34,35
1% (Pd + Au)/ graphite	O_2 (0.3 MPa)	basic	0.3 M GLY, NaOH/GLY = 4 (mol/mol), 323 K, 2 h	100%	39.1%(S_{GLYAC})	Dimitratos *et al.*, 2005	34,35
1% (Au@Pd)/ graphite	O_2 (0.3 MPa)	basic	0.3 M GLY, NaOH/GLY = 4 (mol/mol), 323 K, 2 h	100%	45.5%(S_{GLYAC})	Dimitratos *et al.*, 2005	34,35
1% Pt/C	O_2 (0.3 MPa)	basic	0.3 M GLY, NaOH/GLY = 4 (mol/mol), 50 °C, 4 h	81.6%	50%(S_{GLYAC})	Dimitratos *et al.*, 2006	36
1%(Au + Pt)/C	O_2 (0.3 MPa)	basic	0.3 M GLY, NaOH/GLY = 4 (mol/mol), 323 K , 4 h	69.3%	58.3%(S_{GLYAC})	Dimitratos *et al.*, 2006	36
Au/C	O_2 (0.1 MPa)	12	1.5 M GLY, NaOH/GLY = 2(mol/mol), 333 K, 3 h	30%	75%(S_{GLYAC})	Dimirel, 2005	37
Au/C	O_2 (0.1 MPa)	12	1.5 M GLY, NaOH/GLY = 2(mol/mol), 333 K, 1.5 h	50%	26%(Y_{DHA}), 44% (Y_{HYPAC})	Dimirel *et al.*, 2007	37
Au-Pt/C	O_2 (0.1 MPa)	12	1.5 M GLY, NaOH/GLY = 2(mol/mol), 333 K, 1.5 h	50%	36%(Y_{DHA}), 30% (Y_{HYPAC})	Dimirel *et al.*, 2007	37

[a]Fixed-bed reactor,
[b]selectivity to the product given in parentheses

When studying the effect of an acidic environment on the Bi promoted Pt catalyst, it was found that an 80% initial selectivity to DHA can be obtained, but as reaction time progressed the catalyst deactivated due to the increase in the number of acids strongly adsorbed on the surface.[29]

It is generally agreed[34,35] that Pd is more selective for the formation of glyceric acid than Pt. When using a basic reaction solution, the oxidation of the primary alcohol function is promoted, whereas acidic conditions promote the oxidation of the secondary alcohol function.[38,39] An improvement in the activity, selectivity and stability can be achieved in the presence of a second *p*-electron metal, particularly the heavy metal atoms of Group IV (*e.g.*, lead) and V (*e.g.*, bismuth).[26] Therefore, the nature of the individual metal can promote the reaction selectivity towards dihydroxyacetone and glyceraldehydes.[28,29] The pH of the solution can govern the nature of the complex formed between the promoter and the substrate, which constitutes the basis of the selectivity of the bismuth-modified platinum catalyst. The main disadvantage of the use of supported Pt and Pd catalysts during glycerol oxidation is their deactivation at increasing reaction time. In particular, catalysts that are based on the platinum-group metals suffer oxygen poisoning that is proportional to the oxygen partial pressure.[40,41] This is the principal reason why, when using this type of catalyst, a low partial pressure of oxygen must be used to limit oxygen dissolution. However, gold catalysts appear to be more resistant to oxygen poisoning than platinum-based catalysts, allowing the use of higher oxygen partial pressure.[37] Biella *et al.* even found that pure oxygen at 0.3 MPa pressure can be present during the reaction without any sign of catalyst deactivation.[42]

Other factors such as catalyst preparation techniques, particle size, and reaction temperature have an important effect on the catalytic conversion and selectivity during glycerol oxidation. Porta *et al.*[43] found that noble-metal catalysts prepared using sol-gel immobilization techniques performed better than catalysts prepared by impregnation or incipient wetness methods. Consequently, the overall selectivity of the reaction derives from a combination of factors such as initial selectivity of the catalyst, base-catalyzed interconversion, and stability of the products. The optimum reaction conditions were found to afford 92% glycerate selectivity and 100% glycerol conversion. Furthermore, Dimitratos *et al.* showed that by using bimetallic catalysts (Au–Pd, Au–Pt) supported on carbon, the distribution of the products could be controlled. Moreover, bimetallic catalysts were more active than monometallic catalysts[44] and selectivity could be tuned by using different preparation methods.[45] The addition of Au to Pd or Pt metals also enhanced the resistance to oxygen or byproduct poisoning with respect to monometallic catalysts[37] and resulted in superior performance during the selective oxidation of glycerol when compared with the monometallic Pd/AC, Au/AC and the physically mixed Pd–Au catalyst. This behavior can be attributed to the synergistic effect of the metal alloy formation, as the change in interatomic metal distance on the surface could have both geometric and electronic effects on the catalyst properties.

Nanosized gold particles supported on different carbons (*i.e.* carbon black, activated carbon and graphite) and oxides (TiO_2, MgO and Al_2O_3) show high activity for the heterogeneously catalyzed liquid-phase oxidation of glycerol under atmospheric pressure conditions.[46,47] Further reaction tests using Au/C catalysts with gold particle sizes in the range from 2 to 45 nm confirmed that the reaction is structure sensitive. Moreover, the selectivity of the Au/C catalyst depends on the base concentration and the reaction time. The presence of Pt as a promoter increases not only the catalyst activity but also the selectivity to DHA from 26% (Au/C) to 36% (Au–Pt/C).[37]

Under alkaline conditions the intrinsic catalyst selectivity was influenced by the nature of the catalyst (metal,[48] particle dimensions, and support[49]). At the same time, the stability of the reaction intermediates and their interconversion was found to be influenced by experimental conditions such as temperature and base concentration. Although Au appears to be an active catalyst (activity being dependent upon the particle size[50]), a problem is that the presence of a base leads to an intermediate product interconversion that can mask the true selectivity of the catalyst and allows only the production of the salt of glyceric acid. Therefore, a precise control of the preparation method and reaction conditions is needed in order to obtain active and selective catalysts.

In addition, a range of transition-metal-containing silicates and alumino-phosphate catalysts was tested in the presence of hydrogen peroxide as oxidant for the partial oxidation of glycerol.[51] It was found that despite examining a range of reaction conditions, as well as catalyst compositions such as silicalite containing Ti, V, Fe or AlPO-5 containing Cr, V, Mn, Co, no glycerol partial oxidation products were obtained. Instead, formic acid and a monoformate ester of glycerol were found to be the major products together with a complex mixture of acetals. Accordingly, it was established that increasing the pore size from *ca.* 0.5 nm to 15 nm on a titania–silica cogel catalyst significantly increased the formation of partial oxidation products of glycerol, namely glyceraldehyde, dihydroxyacetone and glyceric acid, as the diffusion of products within the narrow pores possibly leads to their facile overoxidation, resulting in the formation of formic acid. As the pore size increases, the diffusion limitations decrease and the products of partial oxidation can be easily observed. It is therefore clear that, for the oxidation of glycerol, it is important to use only macroporous catalysts to overcome any diffusion-related problems that result in the products of nonselective oxidation.

17.3 Selective Etherification of Glycerol to Fine Chemicals and Polyglycerols

As can be seen in Figure 17.5, glycerol can also be processed into branched oxygen-containing components by catalytic etherification with either alcohols (*e.g.*, methanol or ethanol) or alkenes (*e.g.*, isobutene). Among these, *tert*-butyl ethers have potential to be used as diesel fuel additives in gasoline and offer an alternative to oxygenates such as methyl *tert*-butyl ether (MTBE) and ethyl

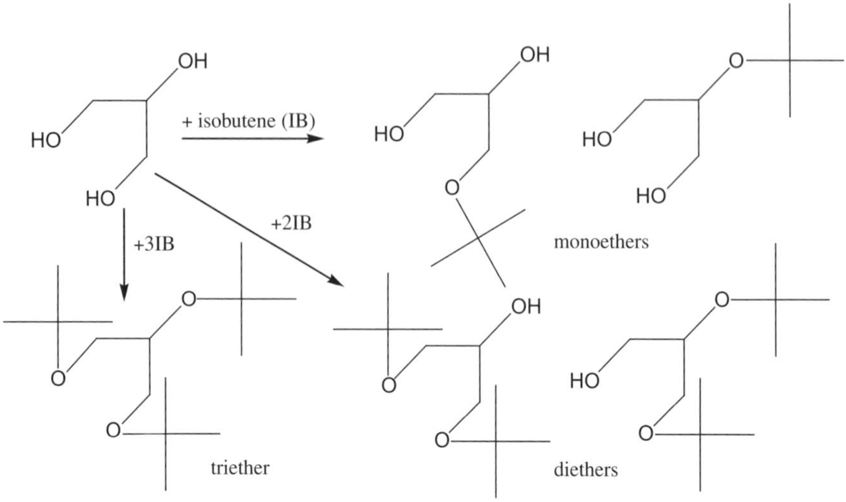

Figure 17.5 Reaction scheme for the etherification of glycerol with isobutene. Reprinted with permission from Elsevier.[57]

tert-butyl ether (ETBE), which are currently added to fuels.[21] As a consequence, the market need for *tert*-butyl ethers from glycerol with a high content of di-ethers and especially tri-ethers is expected to grow significantly in the next 5 years.[52] The etherification of glycerol can be catalyzed using acidic homogeneous catalysts such as *p*-toluenesulfonic acid and methanesulfonic acid and also preferentially by heterogeneous acid catalysts such as zeolites and strong acid ion-exchange resins.[53,54]

A comprehensive study was conducted on the etherification of glycerol by isobutylene and by *tert*-butyl alcohol over commercial strong acid ion-exchange resins and with two large pore size zeolites H-Y and H-Beta (Klepacova *et al.*[55,56]). When comparing the conversions and selectivity of glycerol to di- and tri-*tert*-butyl ethers, it was found that acid macroreticular resins in a dry form are very active catalysts for the etherification reaction with isobutylene because of a large pore diameter, and it was also stated that *tert*-butyl alcohol as an alkylation agent is not suitable because it deactivates the catalysts as a consequence of the formation of water. On the other hand, zeolites and gel-type polymer catalysts are not effective for this etherification reaction as they have small pore diameters. Karinen *et al.*[57] have also reported that for the liquid-phase etherification of glycerol with isobutene in the presence of an acidic ion-exchange resin catalyst, up to five product ethers were obtained and, as a side reaction, isobutene reacted to give C_8–C_{16} hydrocarbons. The authors reported that the optimal selectivity conditions for ether formation were found to correspond to an isobutene/glycerol molar ratio of 3 at 353 K. They also showed that the extent of the etherification reaction, and, thus the main ether products, can be changed by varying the reaction conditions.

Glycerol has traditionally played a role in the production of several types of polymers, some of which are available commercially. Selective etherification reactions can convert glycerol into polyglycerol (PG) and polyglycerol esters, which can be used as biodegradable surfactants, lubricants, cosmetics, food additives, *etc.*[58] The polyglycerols, which are ethers, have many of the properties of glycerol and offer greater flexibility and functionality than glycerol. For example, polyglycerol and polyglycerol methacrylates are used as treatments for wood to improve its stability.[59] In comparison, glycerol-based polymers offer additional benefits including a reduced tendency for leaching into the environment. This application would compete with the more widely used polyethyleneglycols.

Polyglycerols and polyglycerol esters are obtained from the oligomerization of glycerol and the esterification or transesterification of the oligomers with fatty acids or methyl esters. Normally, the reactions are performed in the presence of homogeneous catalysts so that a mixture of polyglycerols (Figure 17.6) as well as a mixture of esters can be synthesized. In addition to linear polyglycerols, branched polyols as well as oxygenated heterocyclic compounds can be obtained from cyclization reactions of glycerol and acrolein by glycerol dehydration, using either solid catalysts such as zeolites or alkaline catalysts such as calcium hydroxide and sodium carbonate.[60]

Several bases have been tested as catalysts, including hydroxides, carbonates and metal oxides.[61] It has been found that the carbonates are more active than hydroxides, despite the weaker base character of the former. This is attributed to the better solubility of carbonates in the glycerol and in the polymeric product at elevated temperatures. Oxides like MgO, CaO and ZnO are less active because of lack of solubility. Condensation of glycerol and glycidol catalyzed by acids produced polyglycerols with a high degree of polymerization and low color values.[62] Practically, the use of different filtration steps as well as neutralization and purification steps to eliminate solvents and homogeneous catalysts are necessary.[63] Over sodium hydroxide at a temperature of 503 K, only a 12.5% weight fraction of diglycerol is obtained.[64] Two distillation steps (under vacuum) are necessary to obtain 98% purity of diglycerol. Development of novel catalysts with high selectivity is still a high priority to lower processing costs.

Theoretically, porous solid catalysts could exert some shape-selective effect on the course of the polymerization reaction. In this way, zeolites NaX, NaA,[65] and zeolite beta have been used as catalysts for the selective production of diglycerol. It was reported that NaX and CsHY zeolites, which are less active at the beginning of the reaction, favor the formation of di- and triglycerol without any formation of acrolein. Indeed, the selectivity to diglycerol is higher than 90% over a CsHY or a NaX zeolite. In this particular case, it seems that the pore size of the catalyst slightly increases the selectivity of the reaction, especially when glycerol conversion is less than 80%. Furthermore, the effect of alkaline promoters on the activity and selectivity was demonstrated; the best results were obtained with the most basic promoter. In contrast, medium-pore Cs-containing zeolites like ZSM-5 are less active and selective.

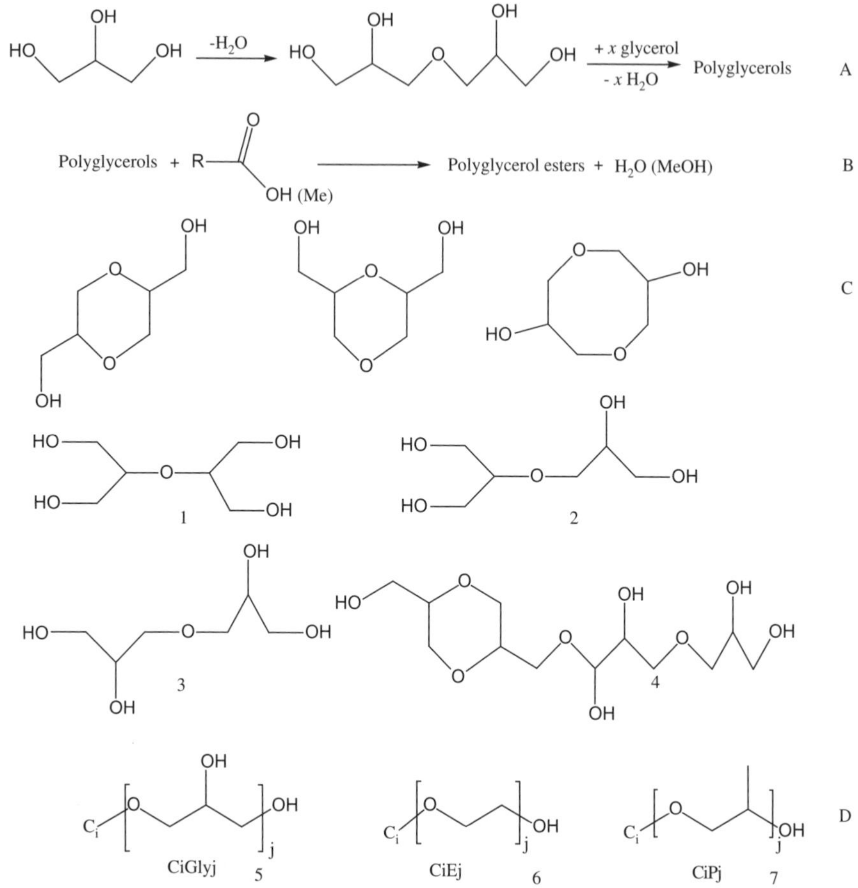

Figure 17.6 (A) glycerol oligomerization, (B) polyglycerol esterification, (C) examples of di- and triglycerol isomers (1, 2 and 3 being the linear diglycerol isomers), (D) glycerol 1-monoethers (6,7 being the ethylene glycol ether (CiEj) and propylene glycol ether (CiPj), respectively). (A)–(C) and (D) reprinted with kind permission from Springer Science + Business Media.[66]

The influence of catalyst pore size and the basicity on the polymerization reaction are confirmed by the use of basic MCM-41 catalysts. Barrault et al.[66] discovered that, if well chosen, basic mesoporous materials can be quite selective catalysts for the direct synthesis of di- and triglycerol from glycerol. They found that when basic mesoporous materials are used there is a significant increase in the selectivity and yield of di- and triglycerols without formation of cyclic compounds or acrolein. The Cs-impregnated material provided the best results, whereas Mg- and La-containing catalysts favored glycerol dehydration and the formation of acrolein. The selectivity of these mesoporous catalysts approaches that of the Cs-exchanged faujasite X. In the presence of lanthanum-

or magnesium-containing catalysts, the glycerol dehydration to acrolein is significant, whereas this unwanted product is not formed when cesium is used as a promoter during the impregnation step. It was further shown that the impregnation method results in the highest catalyst activity. Concerning the selectivity of the modified mesoporous catalysts, the best values to di- and tri-glycerol (>90%) are obtained over solids prepared by the impregnation or grafting methods. Finally, as far as catalyst leaching and stability is concerned, the best results are obtained with catalysts prepared using grafting techniques, which retain their structure and their specific area after the promoter addition to the solid support in comparison with results observed over impregnated catalysts.

It is still rather difficult to selectively obtain one type of polyglycerol or to control the mixture and the quality of the product. If an individual polyglycerol ester needs to be obtained, new catalytic methods or novel catalysts need to be found. The polyol product distribution is important as it should meet certain specifications in order to be used as a food additive. For example, according to European Union regulations, the majority of the content should be di-, tri- and tetraglycerol, whereas the content of polyglycerols equal to or higher than heptaglycerol must be low.[67] Food and Drug Administration regulations allow the use of polyglycerols up to and including decaglycerol.[68] Theoretically, low amounts of catalyst (< 1 mol%) and relatively short reaction times are required to exert some control on the polymerization degree in order to favor a mixture of products having, on average, a low polymerization degree.

17.4 Selective Glycerol Transesterification and Esterification

Monoglycerides (MG) and polyglycerol esters (PEG) consist of a hydrophilic head and a hydrophobic tail, which give them detergent characteristics. The importance of monoglycerides and their derivatives as emulsifiers in the food, pharmaceutical, and cosmetic industries is well known.[69] As such, industrial projects are under discussion to convert glycerol into ester or ether derivatives that can be processed into new marketable products, *e.g.*, biodegradable surfactants.[70] Monoglycerides are generally obtained by: (I) the hydrolysis of triglycerides, (II) the glycerolysis of triglycerides, *i.e.* base-catalyzed transesterification of triglycerides with glycerol at elevated temperature (*e.g.*, 528 K, eqn. 17.1), or the direct esterification of glycerol with fatty acids (eqn. 17.2).

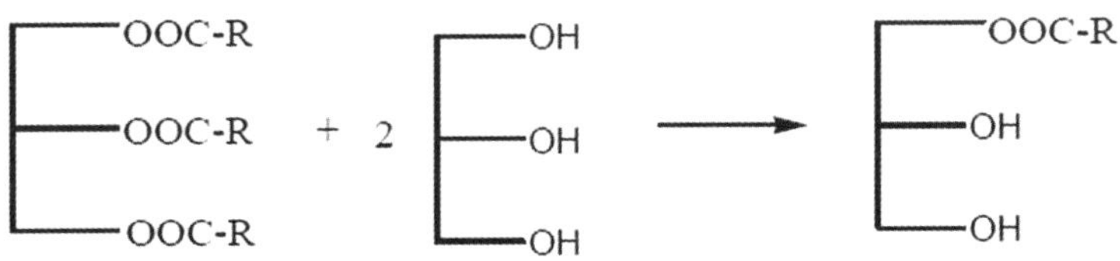

(17.1)

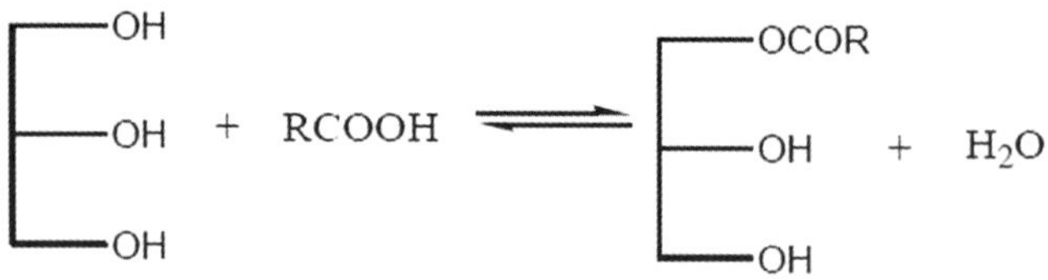

(17.2)

However, as the three hydroxyl groups in glycerol do not strongly differ in reactivity, the product of the direct esterification or transesterification of glycerol with acid and/or base catalysis is a mixture of mono-, di- and some triglyceride, and dissolved glycerol, which typically contains 40–60% monoglyceride and 35–45% diglyceride. Furthermore, the proportion of 2-acyl or β-monoglyceride remains below 10% at low temperature but increases with the reaction temperature. In order to lower the required reaction temperature for the direct esterification of glycerol with fatty acids, normally an acid catalyst is required, *e.g.*, sulfuric acid, phosphoric acid, or organic sulfonic acids.

On the other hand, basic catalysts such as KOH and $Ca(OH)_2$ can be used during the transesterification processes. After reaction, they should be neutralized with phosphoric acid, and then the formed salts must be removed. Therefore, a lot of undesired waste chemicals can be generated. Furthermore, these processes suffer from another drawback – the requirement of high temperature. In parallel, techniques for purification of monoglycerides, *e.g.*, distillation, are limited to food applications as such process steps are expensive. It is therefore highly desirable to find a greener catalytic process that can improve the yield of desired products (*e.g.*, monoester) using more economic reaction conditions on an appropriate solid catalyst amenable to regeneration. The basic features of these processes and the progress towards modifying the classical process using solid catalysts have been previously reviewed.[71,72]

Solid basic materials such as metal oxides, MgO, CeO_2, La_2O_3 and ZnO; Al–Mg hydrotalcites; Cs-exchanged sepiolite; and mesoporous MCM-41 have been tested as potential catalysts for glycerol transesterification with triglycerides.[73] Table 17.2 summarizes some recent results concerning monoglyceride synthesis over various solid catalysts.[74–78] Researchers found that, due to the negative influence of the low solubility of the hydrophilic glycerol in the fat phase, the use of organic solvents has a beneficial effect on the overall reaction but, for safety reasons, it is avoided when producing monoglycerides that are then used as food additives. In other work, Corma et al.[79] studied the transesterification of oleic acid methyl ester with glycerol and found that Brønsted-base catalysts have a higher activity and a higher specific rate than the Lewis-basic hydrotalcite catalyst. The selectivity to monoglycerides of the former is higher because of the lower deactivation of the catalyst, which allows the transesterification of diglycerides to occur at longer reaction times. Calcined Li–Al hydrotalcites result in higher activity than MgO or Al–Mg hydrotalcites because of the stronger Lewis basicity of the former. All of the solid Lewis-basic catalysts, regardless of the base strength, have the same selectivity to monoglyceride, which is lower than that of Brønsted-basic solid catalysts.

In a different approach, a variety of solid acids such as 12-membered-ring structures, like faujasite,[80] USY[81] and zeolite beta;[82] 10-membered-ring structures, like ZSM-5; or unidirectional 12-membered ring materials like mordenite,[83] have been investigated for the catalytic esterification of glycerol with fatty acids. The authors stated that, in the presence of porous zeolites, the relatively small pore opening of these materials would have the potential to reduce the formation of the bulky di- and triesters, therefore increasing the selectivity to monoglycerides. However, as di- and even traces of triglycerides are still readily observed at conversions higher than 10%, a significant contribution of the external surface to the overall reaction cannot be totally excluded. Even though such catalysts exhibit high monoester selectivity, the activity, and consequently the monoglyceride yield, was low. Mesoporous silicas are easily accessible for large reactants such as fatty acids and their esters and offer the possibility of overcoming the pore-size limitation characteristic of traditional zeolite materials. Unfortunately, the strength of the acid sites present in conventional MCM-41 materials is much lower than microporous zeolites, and, consequently, their activity and monoglyceride selectivity is low.[84] An alternative approach is to employ postsynthesis modification in order to obtain specific properties such as acidity or basicity. Perez-Pariente and coworkers recently investigated the behavior of mesoporous materials containing R-SO_3H groups in the esterification of glycerol with fatty acids[85]. The catalytic activity of Al-MCM-41 (Si/Al = 15) in the esterification of glycerol with oleic acid is much lower than that of zeolite USY with similar Si/Al ratios (Table 17.2), whereas the selectivity to mono-olein increases substantially. Nevertheless, the activity of this material is still too low. A strategy to increase the catalytic activity while keeping the benefits of the large pore diameter could be to introduce, in a controlled manner, new acid groups on the catalyst surface. The superior performance of mesoporous materials containing organosulfonic functional groups synthesized by cocondensation over those obtained *via* silylation of the calcined materials was demonstrated for the esterification of glycerol with oleic acid. The higher activity of the sample synthesized by cocondensation is most probably due to its higher content of sulfonic acid. Despite improvements in the selectivity to monoglyceride when using mesoporous acid catalysts, a monoglyceride yield of at least 90% is still needed to avoid the costly distillation of the ester mixtures.

Among different alternatives, the use of glycerol-based additives to improve the properties of biodiesel is being explored. For instance, Melero *et al.*[86] recently described the esterification of glycerol with acetic acid to yield glycerine acetates, such as diacetylglycerol (DAG) and triacetylglycerol (TAG, also called triacetin), which have been shown to be valuable fuel additives leading to either enhanced cold and viscosity properties when blended with diesel fuel or antiknocking properties when added to gasoline. The activities and selectivities achieved using sulfonic acid functionalized mesostructured materials (SBA-15) as catalysts are comparable or even superior to those displayed by conventional acid catalysts, providing values of up to 90% for the glycerol conversion and over 80% for the combined selectivity toward DAG

Table 17.2 Typical results for the synthesis of monogycerides over various solid base or acid catalysts.

Catalysts	*Reactants*	*Glycerol / fatty comp. (molar ratio)*	*Solvent*	*Reaction temp (K)*	*Time/h*	*Conversion*[a] *(%)*	*Monoglyceride selectivity (%)*	*Ref.*
ZnO	glycerol + methyl stearate	1	no	493	6	18	80	74
MgO	glycerol + methyl stearate	1	no	493	6	83	42	74
La_2O_3	glycerol + methyl stearate	1	no	493	6	97.0	28	74
CeO_2	glycerol + methyl stearate	1	no	493	6	4	100	74
ZnO	glycerol + stearic acid	1	diglyme	160	16	63.3	82.5	75
ZnO	glycerol + lauric	1	diglyme	433	16	56.1	73.3	75
ZnO	glycerol + oleic	1	diglyme	433	16	45.1	90.7	75
MgAl-MCM-41	glycerol + lauric acid	3	no	493	20	80	70	76
ZnO	glycerol + myristic	3	no	493	33	80	62	76
ZnO	glycerol + stearic acid	3	no	493	44	80	50	76

Calcined Al-Mg hydrotalcite	glycerol + methyl steararte	6	no	473	8	95	67	77
Calcined-rehy-drated(12h) Al-Mg hydrotalcite	glycerol + methyl steararte	6	no	473	8	98	80	77
KF/Al_2O_3	glycerol + methyl steararte	6	no	473	2	68	69	77
USY (Si/Al = 14.2)	glycerol + oleic acid	1	no	373	24	8.2	55	78
Beta (Si/Al = 13)	glycerol + oleic acid	1	no	373	24	8.8	64	78
Al-MCM-41 (Si/Al = 15)	glycerol + oleic acid	1	no	373	24	5.8	96	78
MCM-41-F[b]	glycerol + oleic acid	1	no	373	24	10.9	68	78
MCM-41-C[b]	glycerol + oleic acid	1	no	373	24	24.2	69	78
Phenyl-MCM-41	glycerol + oleic acid	1	no	393	8	25	67	78
Methylsulfonic/phe-nylsulfonic-MCM-41	glycerol + oleic acid	1	no	393	8	39	69	78

[a]Conversion of fatty aids or fatty-acid monoesters.
[b]Where MCM-41-F means postsynthesis functionalization with sulfonic acid and MCM-41-C stands for cocondensation procedure.

and TAG after 4 h of reaction. Within the studied range, optimal conditions were found at a temperature of 125 °C and an acetic acid to glycerol molar ratio of 9:1. The acid strength of the sulfonic acid site has also been found to be an important factor affecting the catalytic performance of these materials.

17.5 Selective Hydrogenolysis of Glycerol

Through the selective hydrogenolysis of glycerol in the presence of metallic catalysts and hydrogen, as shown in Figure 17.7, 1,2-propanediol (1,2-PD), 1,3-propanediol (1,3-PD), ethylene glycol (EG) and eventually propane can be obtained. Whether the primary or secondary hydroxyl group is more easily reduced depends on the catalyst system and reaction conditions. Catalytic -OH cleavage with noble metals under reductive conditions usually favors reduction of primary hydroxyl groups over secondary groups. Tertiary hydroxyls are cleaved faster than secondary hydroxyls. The alternating primary-secondary-primary hydroxyl reduction pathway is obviously unlikely to occur.

Propylene glycol, *i.e.* 1,2-PD, a three-carbon diol with a steriogenic center at the central carbon atom, is an important medium-value commodity chemical with a 4% annual growth in market size. It is used for polyester resins, liquid detergents, pharmaceuticals, cosmetics, tobacco humectants, flavors and fragrances, personal care, paints, animal feed, antifreeze, *etc.* Traditionally, it is produced by the hydration of propylene oxide derived from propylene by either the chlorohydrin process or the hydroperoxide process. There has been a rapid expansion of the market for 1,2-PD as antifreeze and deicing agents because of growing concern about the toxicity of ethylene glycol-based products to humans and animals. 1,3-PD is also a high-value specialty chemical that is mainly used in specialty polyester fibers, films, and coatings. 1,3-PD is copolymerized with terephthalic acid to produce the polyester SORONA® from DuPont, or CORTERRA® from Shell, which has unique properties in terms of chemical resistance, light stability, elastic recovery, and dye ability.[87] 1,3-PD is currently catalytically produced from petroleum derivatives such as ethylene oxide (Shell route) or acrolein (Degussa-DuPont route). The low conversion efficiency of the acrolein process, as well as the hazardous nature of acrolein, has spurred a great deal of interest in producing 1,3-PD from other chemical sources, especially glycerol. Another important diol, ethylene glycol, mainly derived from ethylene, is also a raw material for synthetic fibers and explosives, *etc.*[88]

Traditional hydrogenation practices have indicated that alcohol groups are stable and do not readily react at normal hydrogenating reaction conditions. Moreover, alcohols are also known as excellent solvents for hydrogenation reactions. Hence, conventional hydrogenation catalysts such as nickel, ruthenium, and palladium are not very effective when used for hydrogenation and catalytic -OH cleavage, *i.e.* as in the hydrogenolysis of glycerol. Recently, the Shell Oil Company developed a process that uses a homogenous palladium complex in a water–sulfolane mixture containing methanesulfonic acid. After a

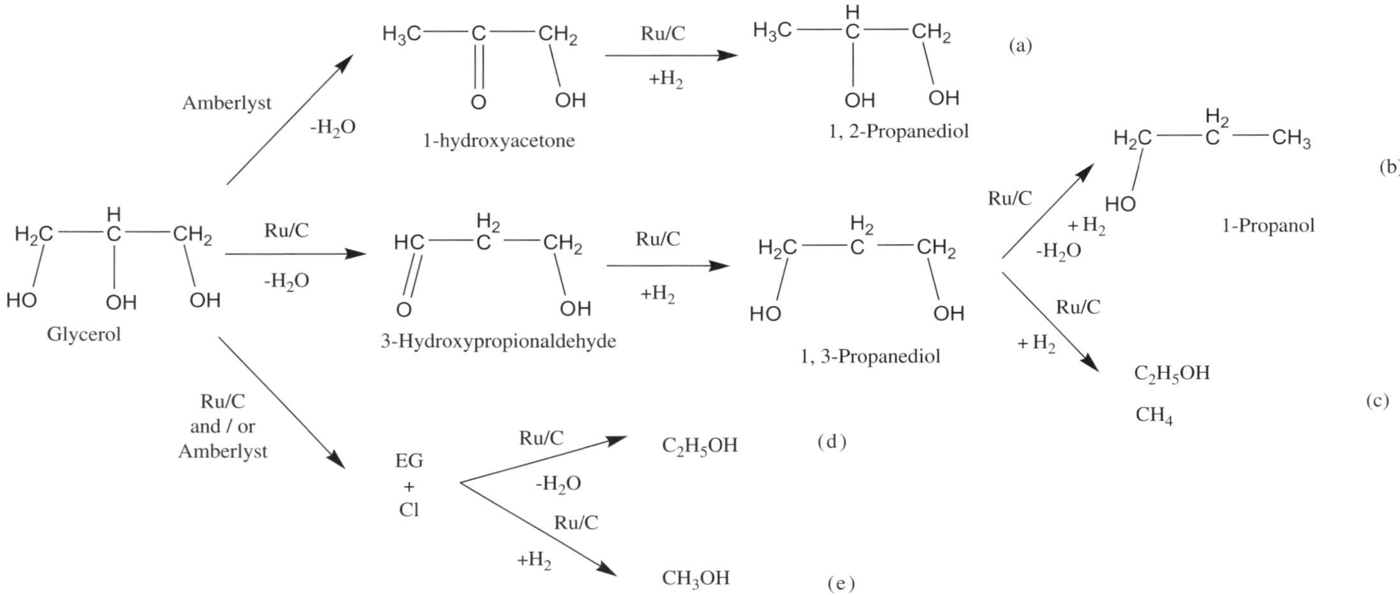

Figure 17.7 Reaction scheme for glycerol hydrogenolysis and degradation reactions. Reprinted with permission from Elsevier.[97]

10-h reaction, 1-propanol, 1,2-propanediol and 1,3-propanediol were detected in a ratio of 47 : 22 : 31.[89]

Therefore, if the catalytic hydrogenolysis of glycerol could be carried out over solid catalysts without the presence of dangerous solvents, it would become economically and environmentally attractive. Montassier *et al.*[90] carried out the hydrogenolysis of glycerol at 30 MPa H_2 at 533 K in the presence of Raney Ni, Ru, Rh and Ir catalysts. They found that mainly methane was produced, but when Raney Cu was used as a catalyst, 1,2-PD was the main reaction product. Raney Cu is known for its poor hydrogenolytic activity towards C–C bonds but it is an efficient catalyst for C–O bond hydrogenation and dehydrogenation. Casale and Gomez[91] reported the hydrogenolysis of glycerol at 15 MPa of pressure and between 513–543 K using a mixture of copper and zinc metals, as well as sulfided ruthenium supported on an activated carbon. In contrast, Ludvig and Manfred[92] studied the production of propanediols using a catalyst containing cobalt, copper, manganese, molybdenum, and an inorganic polyacid that achieved a 95% yield of 1,2-PD at 25 MPa and 523 K. Haas *et al.*[93] reported a two-stage process for simultaneous production of 1,2-PD and 1,3-PD from gaseous glycerol solutions at a temperature of 573 K. It involves: (a) dehydration of a gaseous 10–40% glycerol–water mixture at 523 K to 613 K over a solid acid catalyst, and (b) catalytic hydrogenation of the reaction mixture to yield 1,2-PD and 1,3-PD.

Analyzing the above cases we find that in order to obtain optimum selectivity values, very complex conditions and catalytic systems are needed. A more preferable technology would convert crude, natural glycerol at moderate temperatures and pressures. Hydrogen pressures of around 6–10 MPa and reaction temperatures of 453–513 K for glycerol hydrogenolysis have been reported using supported-metal catalysts.[94,95] Chaminand *et al.*[96] addressed the complexity of using heterogeneous Cu, Pd, and Rh catalysts supported on ZnO, C, and alumina for the hydrogenolysis of glycerol carried out at 8 MPa pressure and at 453 K in the presence of solvents (H_2O, sulfolane, dioxane) and proposed a general mechanism that can be seen in Figure 17.8. Additives such as tungstic acid (H_2WO_4) that act as a modifier to improve the selectivity toward 1,3-PD were added to the reaction medium. Figure 17.8 shows that tungstic acid can favor the dehydration route (route A, E1-like mechanism) *via* protonation of the hydroxyl groups with loss of water. The keto group formed as an intermediate can be easily reduced under these reaction conditions. However, when using acid (*i.e.* HCl) a low conversion was obtained, suggesting that the acidity of H_2WO_4 does not exert a dominant effect on the reaction mechanism. At the same time, the formation of a Rh–W alloy cannot be excluded and may affect the selectivity and the rate of the glycerol hydrogenolysis reaction. From these findings, we conclude that further work is needed to establish the exact reaction mechanism and to optimize the reaction pathways.

Recently, Tomishige *et al.*[97] demonstrated that when active Ru supported on carbon is used in combination with a cation-exchange resin such as Amberlyst 15, it can exhibit higher activity in glycerol hydrogenolysis under mild reaction

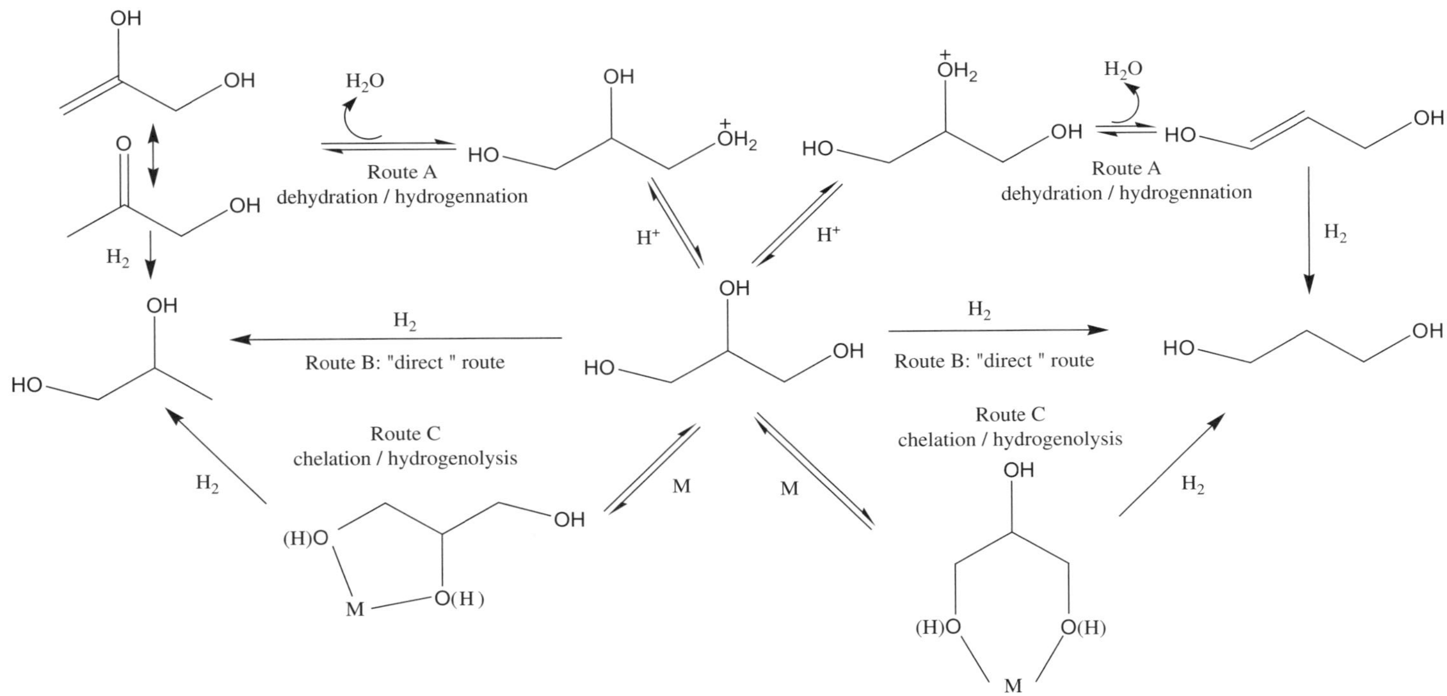

Figure 17.8 Possible reaction routes for catalytic hydrogenolysis of glycerol proposed by Chaminand et al.[96]

conditions (393 K and 4 MPa or 8 MPa H_2) than other metal–acid bifunctional catalyst systems such as zeolites, sulfated zirconia, H_2WO_4, and liquid H_2SO_4. The Ru/C catalyst showed higher conversion than Rh/C, Pd/C and Pt/C. However, the selectivity to cracking products was rather high over Ru/C, with the dehydration of glycerol to 1-hydroxyacetone being catalyzed by the acid catalysts. The subsequent hydrogenation of 1-hydroxyacetone on the metal catalysts gives 1,2-propanediol. Thus, the addition of solid acid catalysts to Ru/C was effective in increasing the conversion and hydrogenolysis selectivity. During the hydrogenolysis reaction, the activity of the metal catalyst when combined with the cation-exchange resin, can be related to that of 1-hydroxyacetone hydrogenation over the metal. In addition, the adsorbed OH group on Ru/C can also catalyze the dehydration of glycerol to 3-hydroxypropionaldehyde, which ultimately can be converted into 1,3-PD and other degradation products through subsequent hydrogenation. In this study Ru/C + H_2SO_4 showed lower activity than combined Ru/C + Amberlyst, suggesting that the solid acid was more effective for the hydrogenolysis of glycerol.

More interestingly, Perosa *et al.*[98] showed that Raney Ni is an active catalyst for the hydrogenolysis of glycerol at low hydrogen pressure (1 MPa) when the reaction is conducted in the presence of liquid phosphonium salt and without added solvents. Selectivity toward 1,2-PD is high, with the only byproducts being ethanol and CO_2. The reaction does not require solvents, promoter or harsh conditions. The addition of a liquid phosphonium salt, at the reaction temperature, improved the reaction rate and selectivity to a small extent but did not facilitate the separation of the final reaction mixture. Despite several research efforts, the potential importance of the glycerol hydrogenolysis reaction is limited to the laboratory scale as the common drawbacks of high temperature and pressure, dilute solutions and the low selectivity towards propylene glycol still require further investigation.

Interestingly, Wang *et al.*[99] have described a new approach to the selective production of 1,3-PD from glycerol. The idea is to selectively transform the second hydroxyl group of glycerol into a tosyloxyl group (tosylation) and then to remove the transformed group by catalytic hydrogenolysis (detosyloxylation). This new approach involves three steps, namely, acetalization, tosylation, and detosyloxylation. A 1,3-PD yield as high as 72%, which roughly corresponds to the overall yield of this new glycerol hydrogenolysis process, was achieved for the 2-phenyl-5-tosyl-1,3-dioxane (PTD) detosyloxylation reaction, as can be seen in Figure 17.9.

17.6 Catalytic Dehydration of Glycerol

Acrolein is an important and versatile chemical intermediate for the production of acrylic acid esters, superabsorbent polymers and detergents. A sustainable and cost-effective acid dehydration of glycerol to acrolein could offer an alternative to the currently used catalytic petrochemical process based on

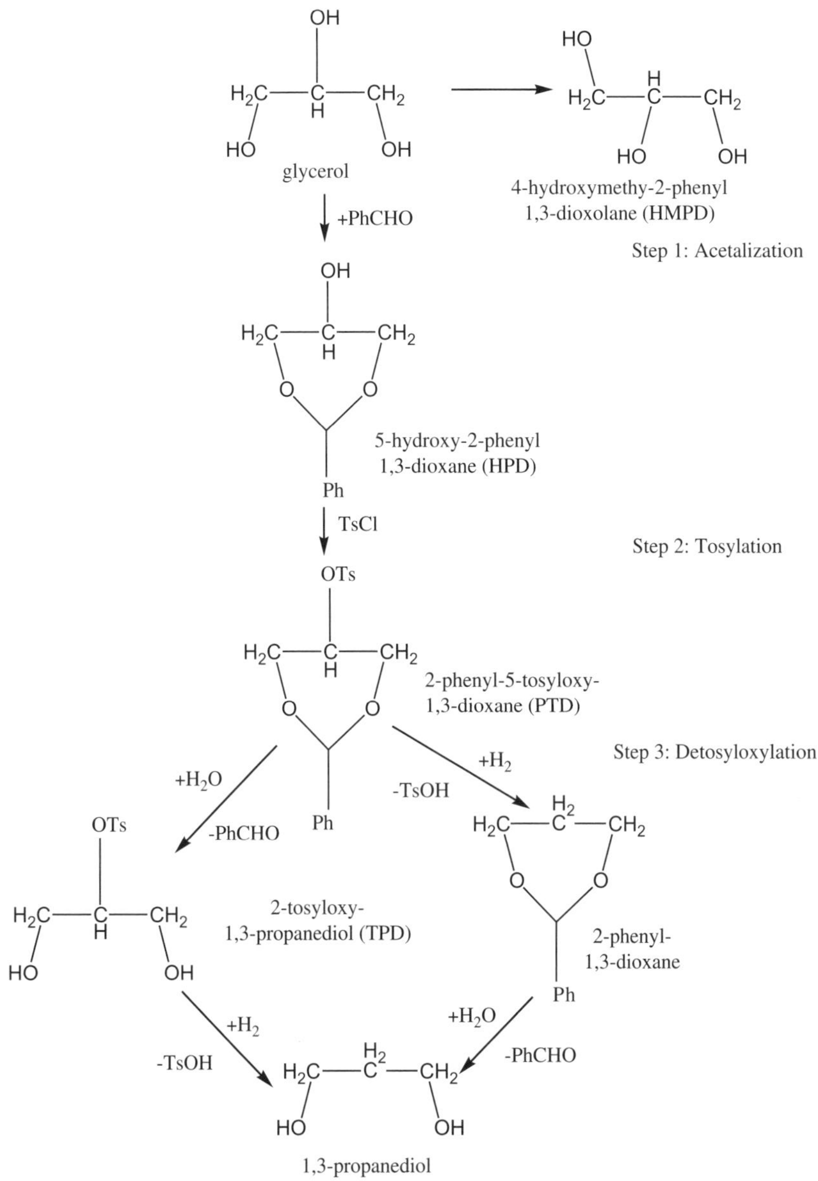

Figure 17.9 Conversion of glycerol to 1,3-propanediol *via* selective dehydroxylation in three steps. Reprinted with permission.[99] Copyright 2003 American Chemical Society.

propylene oxidation over a Bi/Mo-mixed oxide catalyst. Figure 17.10 shows a schematic reaction path when using the catalytic dehydration of glycerol.

In 1950 Waldmann and Frantisek[100] obtained 13.5 g of acrolein, distilled from a reaction mixture of 50 g glycerol, 100 g phthalic anhydride and 3 g

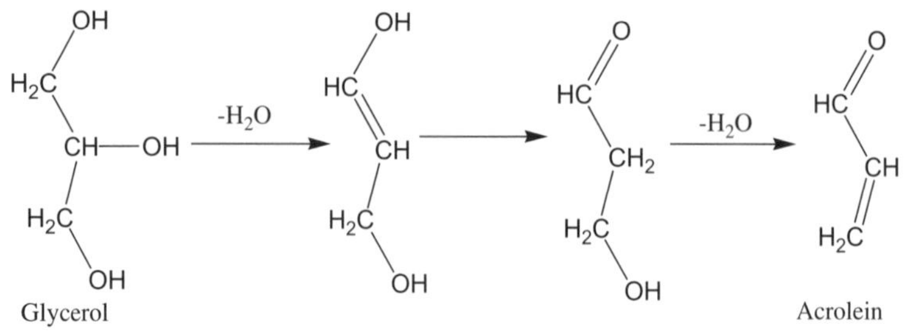

Figure 17.10 A schematic reaction path for the catalytic dehydration of glycerol.

benzenesulfonic acid. Ramayya *et al.*[101] also reported that an 84% acrolein selectivity can be obtained at 40% conversion of glycerol by adding a 5 mM H_2SO_4 solution at 573–623 K and 34.5 MPa. Antal *et al.*[102] showed that a high reaction temperature is needed for the dehydration of glycerol in water. A low 1% glycerol conversion was obtained without the formation of acrolein when the reaction was conducted at 523 K and 34.5 MPa pressure. There are several patents concerning the gas-phase dehydration of glycerol. Scheering-Kahlbaum AG[103] reported an acrolein yield of up to 80 mol% using phosphates of Cu or Li as catalysts at a temperature of 573–873 K. A more recent patent by Degussa AG[22] claimed that the reaction can be conducted at 513–623 K in the presence of an acid catalyst. On the other hand, Biswas *et al.*[104] suggested the use of pervaporation, a membrane-based separation technique, as a potential replacement process for the vacuum distillation process, which is normally used for the dehydration of glycerol–water mixtures.

Sub- and supercritical water (SCW) reactions can be also used for the production of acrolein from glycerol. Bühler *et al.*,[105] using flowing hot compressed water (HCW) equipment, found only low glycerol conversions (31 mol%) and low acrolein selectivity ($S=37$ mol%) in pure water when the system operated in a temperature range of 573–747 K and pressure of 25–45 MPa with a residence time in the range of 16–100 s. At glycerol concentrations ranging from 0.19–0.57 M, the authors suggested that the glycerol reactions in HCW progress competitively through both ionic and radical reaction mechanisms. The predominance of the ionic or radical mechanisms could be controlled by temperature and pressure: the ionic reaction was preferred in subcritical water below the critical temperature (liquid state) and the reaction *via* radical formation was favored in the supercritical region. In addition, Watanabe *et al.*[106] revealed that the yield of acrolein can be enhanced using a higher concentration of glycerol and H_2SO_4 in combination with higher pressure. Approximately 80% selectivity to acrolein was obtained at 90% glycerol conversion in the presence of H_2SO_4 at supercritical conditions (673 K and 34.5 MPa). The rate constant of acrolein decomposition was always higher than that of acrolein formation in the absence of an acid catalyst, but the rate constant of acrolein formation could overcome that of acrolein decomposition by the addition of an acid in supercritical conditions.

In another interesting study, Ott *et al.*[107] found that zinc sulfate is an effective catalyst for acrolein synthesis from glycerol in HCW (573–663 K, 25–34 MPa, 10–60 s), achieving an acrolein selectivity of 75% at 50% glycerol conversion. It turned out that the addition of zinc sulfate causes a decrease of the activation energy from 150 kJ mol^{-1} to (140 ± 2) kJ mol^{-1}. Previous research results[105–107] confirmed that the formation of acrolein from glycerol was controlled by ionic species (such as protons) and can be increased by the presence of an acid and HCW conditions. However, the current use of SCW is only promising if acid is added. As water itself induces corrosion under these conditions, the presence of an acid compound intensifies the corrosive effect, with a salt catalyst having the same corrosion potential in SCW as mineral acids. As such, another less-corrosive anion to be used under these reaction conditions is needed if acrolein synthesis from glycerol is to become an attractive commercial process in the near future.

17.7 Gasification of Glycerol

Glycerol gasification in the presence of steam yields liquid fuels at low temperatures (673–873 K) and gaseous products at high temperatures (> 1023 K).[108] Major gaseous reaction products include CO, H_2, C_2H_4, and CH_4 (in decreasing order of yield). Almost no CO_2 is detected. At initial decomposition rates mainly CO, acetaldehyde and acrolein are found. Acetaldehyde and acrolein further decompose to produce primarily carbon monoxide, ethylene, methane and hydrogen. Recently, glycerol was investigated as a source of hydrogen, and, in this regard, steam reforming of glycerol could be a suitable reaction. Xu *et al.*[109] showed that glycerol is easily and completely gasified to a hydrogen-rich gas in supercritical water without a catalyst after 44 s at 873 K and 34.5 MPa. The presence of a catalyst has little effect on the gas composition. During their experiments they observed a very low CO yield in contrast to the high H_2 yield. These results may be in disparity with those observed by Stein *et al.*[110] Evidently, the supercritical condition results in considerably different gasification chemistry than that observed at atmospheric pressure.

Using a tubular reactor system in the temperature range of 622–748 K, with pressures of 25, 35, or 45 MPa, and reaction times from 32 to 165 s, Buhler *et al.*[105] investigated the decomposition of glycerol at different initial concentrations using near- and supercritical water. A conversion between 0.4 and 31% was observed with the main products of the glycerol degradation being methanol, acetaldehyde, propionaldehyde, acrolein, allyl alcohol, ethanol, formaldehyde, carbon monoxide, carbon dioxide, and hydrogen. A non-Arrhenius behavior was found for the overall degradation rate, as well as for the pressure dependence of the reaction rate. The product distribution indicates the occurrence of two competing reaction pathways, as depicted in Figure 17.11:[102] (1) ionic reaction steps, which are preferred at higher pressures

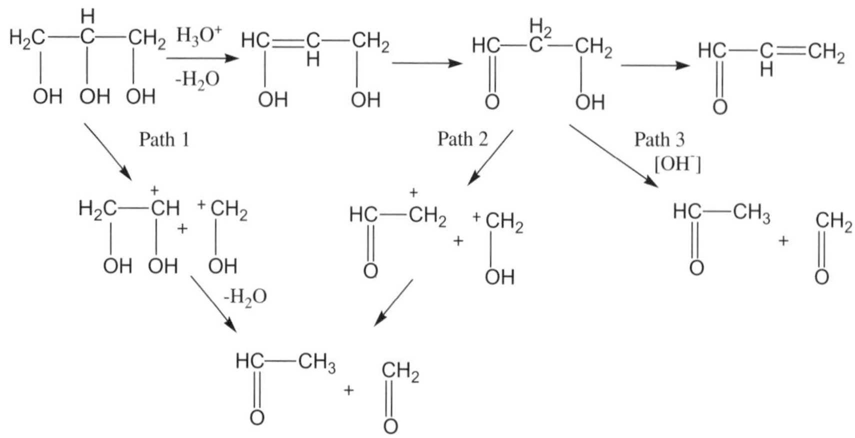

Figure 17.11 Formation of acetaldehyde, acrolein and formaldehyde as suggested by Antal et al. Reprinted with permission from Elsevier.[102]

and/or lower temperatures and (2) a free-radical degradation that dominates at lower pressures and/or higher temperatures.[101–110]

Shabaker *et al.*,[111] using Raney-NiSn catalysts at temperatures of 498 K for the aqueous-phase reforming of sorbitol, glycerol and ethylene glycol solutions, produced an effluent gas stream composed of 66 mol% H_2, 30 mol% CO_2, and 4.2 mol% methane and lower alkanes (dry basis). Although the advantage of low catalyst deactivation has been reported for the aqueous-phase reaction, it is unavoidable that the reforming of glycerol must be conducted under high pressure. On the other hand, reforming in the gas phase can be conducted under atmospheric pressure with a conventional fixed-bed flow reactor. Dauenhauer and coworkers[112] reported studies of the autothermal reforming of pure glycerol in aqueous solutions over platinum- and rhodium-based catalysts supported on alumina foams at a contact time of ~10 min. They found that rhodium catalysts, with the addition of ceria on a γ-Al_2O_3 washcoat layer, exhibited the best combination of high conversion and high selectivity to H_2 (near equilibrium). On the other hand, steam addition increased H_2 selectivity to 79%. We believe their results can be explained as occurring primarily through a surface reaction mechanism initiated by adsorption on metals through hydroxyl oxygen lone pairs to form surface alkoxides, which decompose almost exclusively to H_2 and C_1 carbon compounds. Hirai[113] developed a novel, efficient 3 wt%Ru/Y_2O_3 catalyst for glycerol steam reforming in the gas phase. The Ru/Y_2O_3 catalyst displayed very high activity in a prolonged run and is considered to be resistant to deactivation caused by carbon deposition.

More recently, Soares *et al.*[114] investigated a low-temperature catalytic route for converting glycerol into H_2/CO gas mixtures that are suitable for combination with Fischer–Tropsch and methanol syntheses. Interestingly, the authors proposed that the endothermic enthalpy change of this reaction (350 kJ mol^{-1}) corresponds to about 24% of the heating value of the glycerol

(– 1480 kJ mol^{-1}). The heat generated by the Fischer–Tropsch conversion of the CO and H_2 into liquid alkanes such as octane (– 412 kJ mol^{-1}) corresponds to about 28% of the heating value of the glycerol. Thus, combining these two reactions results in an exothermic process, with an enthalpy change (– 63 kJ mol^{-1}) that is about 4% of the heating value of the glycerol. The process pathways that can generate liquid fuels and chemicals from glycerol are outlined in Figure 17.12. Accordingly, Soares and coworkers designed Pt–Ru and Pt–Re alloy catalysts that possess the catalytic properties of Pt with respect to selective cleavage of C–C *versus* C–O bonds, but that have less exothermic enthalpy changes for CO adsorption. Results showed that the percentage conversion of glycerol and the product-gas ratios remained constant for at least 72 h time-on-stream at temperatures as low as 200 °C. The main condensable organic compound in the effluent stream of the reaction was unconverted glycerol, with smaller quantities of ethylene glycol, methanol, hydroxypropanone, and ethanol.

Generally, the generation of H_2 and CO by liquid-phase reforming at low temperatures is accompanied by selectivity challenges, as the reaction of H_2

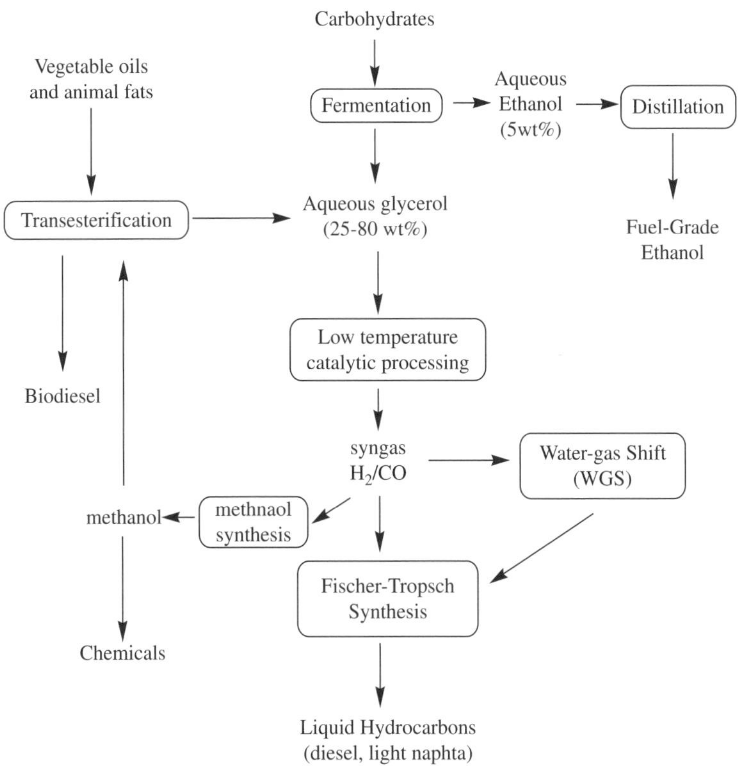

Figure 17.12 Schematic of liquid fuel and chemical production *via* catalytic processing of glycerol. Reprinted with permission from Elsevier.[115]

with CO or CO_2 to form alkanes (C_nH_{2n+2}) is highly favorable at these low temperatures. Thus, an effective catalyst for the production of H_2 by aqueous-phase reforming of glycerol must break C–C, O–H, and C–H bonds in the reactant, and the catalyst must facilitate the water-gas shift to remove adsorbed CO from the surface. However, breaking C–O bonds leads to alkane formation and the degradation of hydrogen production. Aqueous-phase reforming can also lead to the formation of oxygenated intermediates by several reactions, including dehydration and isomerization.[115] Therefore, an efficient catalyst for the steam reforming of glycerol still needs to be found.

17.8 Concluding Remarks

This chapter describes many potential routes for the conversion of glycerol into useful chemicals and materials. Some new processes, such as the oxidation of glycerol, have been extensively studied in recent years, especially in academic research institutions. In most cases, good selectivity to the desired products at high glycerol conversion is still difficult because of the functionalization of the triol glycerol molecule with hydroxyl groups of similar reactivity, and because of a lack of optimal catalysts or optimized reaction conditions. Comparatively, much research has been carried out on the oxidation, transesterification and esterification of glycerol to more valuable products, while less is being focused on novel processes such as hydrogenolysis. Some processes have been used industrially for decades, such as the production of fatty esters of glycerol, although the catalysts used are generally homogenous. For the catalytic dehydration of glycerol to acrolein, mineral acids or their salts are used as catalysts in homogeneous catalytic processes. These homogenous processes cause severe pollution, corrosion and high production costs. New solid catalysts that are stable under the reaction conditions are highly desirable.

So far there have been extensive studies on the development of new catalytic processes for glycerol conversion and some encouraging results, though only a small number of catalysts have been tested and the reaction conditions have not yet been fully optimized for specific catalysts. The challenges of increasing selectivity and yield present excellent opportunities for researchers in catalysis and nanomaterials, specifically with the aim of finding high activity catalysts for glycerol conversion to useful products. Careful design of the catalyst to control the selectivity to the targeted product is highly desired. Homogeneous catalysts need to be replaced or substituted to avoid problems arising from product separation and purification.

Traditionally, the high price of glycerol limited the economic viability of such transformations, except for products used in the high-value niche markets, *i.e.* cosmetic additives, tanning agent components, amino acid precursors, or selective metal chelating agents. As the price of glycerol decreases, new products, especially polymers, will be increasingly derived from glycerol.

Furthermore, conversion of glycerol to fuels can also be expected to grow. For example, converting glycerol into H_2 or syngas by steam reforming is very promising. The use of glycerol-based additives to improve petroleum fuel properties is also one of the possibilities currently being explored to utilize this renewable feedstock. Using glycerol to produce fuel additives, *e.g.*, *tert*-butyl ethers of glycerol and acetylated glycerol derivatives as diesel fuel or gasoline additives, could also offer an important market for biomass resources.

Finally, we would like to remark that the utilization of glycerol for the synthesis of value-added chemicals is a topic of great industrial interest because glycerol can be formed in large amounts during the production of biodiesel from natural triglycerides, and represents a waste that must be used. In particular, the following challenges need to be addressed: (1) new applications and products for directly using crude glycerol need to be found; (2) cost-effective purification processes need to be developed to purify raw glycerol from biodiesel processes; and (3) the combination of separation and purification of crude glycerol with catalytic conversion should be explored. The effective utilization of glycerol will be a key factor that can promote biodiesel commercialization and further development. As a biomass-derived chemical feedstock, its utilization is also a part of the global challenge aiming at the production of marketable chemicals *via* the catalytic transformation of biosustainable resources employed as substitutes for fossil fuels and fine chemicals.

Acknowledgements

The authors wish to acknowledge the financial support from the ARC Centre of Excellence for Functional Nanomaterials during the preparation of this review.

References

1. C. Zhou, J. N. Beltramini, Y. Fan and G. Q. Lu, *Chem. Soc. Rev.*, 2008, **37**, 527.
2. Z. J. Wang, J. Zhuge, H. Fang and B. A. Prior, *Biotechnol. Adv.*, 2001, **19**(3), 201.
3. P. L. Rogers, Y. J. Jeon and C. J. Svenson, *Process Saf. Environ. Prot.*, 2005, **83**, 499.
4. C. E. Wyman, *Biotechnol. Prog.*, 2003, **19**, 254.
5. R. Landucci, B. J. Goodman and C. E. Wyman, *Appl. Biochem. Biotechnol.*, 1994, **45–46**, 677.
6. L. Minyan, Z. Chunhui, J. N. Beltramini, Y. Weihua and F. Yogxian, *Prog. Chem.*, 2008, **20**, 1474.
7. F. Karaosmanoglu, K. B. Cigizoglu, M. Tuter and S. Ertekin, *Energy & Fuels*, 1996, **10**(4), 890.
8. L. C. Meher, D. Vidya Sagar and S. N. Naik, *Renew. Sustain. Energy Rev.*, 2006, **10**, 248.
9. M. McCoy, *Chem. Eng. News*, 2006, **84**, 7.
10. E. K. Wilson, *Chem. Eng. News*, 2002, **80**(27), 46.

11. S. T. Bagley, L. D. Gratz, J. H. Johnson and J. F. McDonald, *Environ. Sci. Technol.*, 1998, **32**(9), 1183.
12. M. Frondel and J. Peters, *Energy Policy*, 2007, **35**, 1675.
13. The Promotion of the Use of Biofuels or Related Other Renewable Fuels for Transport, EU Directive 2003/30/EC of the European Parliament and of the Council of 8 May 2003.
14. Biofuels for Transport. An International Perspective; International Energy Agency, Paris, France, 2004, p. 169.
15. K.S. Tyson, J. Bozell, R. Wallace, E. Petersen and L. Moens, Biomass Oil Analysis, Research Needs and Recommendations, *NREL/TP-510-34796*, 74–75, 2004.
16. T. Werpy and G. Petersen, *Top value-added chemicals from biomass, Vol. 1, 2004,* US Department of Energy (USDOE).
17. G. Vicente, M. Martínez and J. Aracil, *Bioresour. Technol.*, 2004, **92**(3), 297.
18. H. Fukuda, A. Kondo and H. Noda, *J. Biosci. Bioeng.*, 2001, **92**(5), 405.
19. C. W. Chiu, M. J. Goff and G. J. Suppes, *AIChE J.*, 2005, **51**(4), 1274.
20. L. Bournay, D. Casanave, B. Delfort, G. Hillion and J. A. Chodorge, *Catal. Today*, 2005, **106**, 190.
21. http:..beag.ag,utk,edu/pp/biodiesel.pdf.
22. J. Van Gerpen, *Fuel Proc. Technol.*, 2005, **86**(10), 1097.
23. M. Di Serio, R. Tesser, M. Dimiccoli, F. Cammarota, M. Nastasi and E. Santacesaria, *J. Mol. Catal. A*, 2005, **239**, 111.
24. K. Miltenberger, in *Ullmann's Encyclopedia of Industrial Chemistry*, B. Elvers, S. Hawkins, M. Ravenscroft, J.F. Rounsaville and G. Schulz, (ed.), **Vol. 13**, VCH, Weinheim, 1989, p. 507.
25. R. Bauer and D. Hekmat, *Biotechnol. Prog.*, 2006, **22**(1), 278.
26. H. Kimura, K. Tsuto, T. Wakisaka, Y. Kazumi and Y. Inaya, *Appl. Catal. A*, 1993, **96**(2), 217.
27. H. Kimura, *Appl. Catal. A*, 1993, **105**(2), 147.
28. R. Garcia, M. Besson and P. Gallezot, *Appl. Catal. A*, 1995, **127**, 165.
29. A. Abbadi and H. van Bekkum, *Appl. Catal. A*, 1996, **148**, 113.
30. P. McMorn, G. Roberts and G. J. Hutchings, *Catal. Lett.*, 1999, **63**, 19.
31. S. Carrettin, P. McMorn, P. Johnston, K. Griffin and G. J. Hutchings, *Chem. Commun.*, 2002, 696.
32. S. Carrettin, P. McMorn, P. Johnston, K. Griffin, G. J. Hutchings and C. J. Kielly, *Phys. Chem. Chem. Phys.*, 2003, **5**, 1329.
33. F. Porta and L. Prati, *J. Catal.*, 2004, **224**, 397.
34. N. Dimitratos, F. Porta, L. Prati and A. Villa, *Catal. Lett.*, 2005, **99**(3–4), 181.
35. N. Dimitratos, C. Messi, F. Porta, L. Prati and A. Villa, *J. Mol. Catal. A*, 2006, **256**, 21.
36. N. Dimitratos, F. Porta and L. Prati, *Appl. Catal. A*, 2005, **291**, 210.
37. S. Demirel, K. Lehnert, M. Lucas and P. Claus, *Appl. Catal. B*, 2007, **70**(1–4), 637.
38. R. Garcia, M. Besson and P. Gallezot, *Appl. Catal. A*, 1995, **127**, 165.

39. A. Abad, C. Almela, A. Corma and H. García, *Tetrahedron*, 2006, **62**(28), 6666.
40. P. Gallezot, *Catal. Today*, 1997, **37**, 405.
41. P. Fordham, M. Besson and P. Gallezot, *Appl. Catal. A*, 1995, **133**(2), L179.
42. S. Biella, G. L. Castiglioni, C. Fumagalli, L. Prati and M. Rossi, *Catal. Today*, 2002, **72**, 43.
43. D. Wang, A. Villa, F. Porta, D. Sua and L. Prati, *Chem. Commun.*, 2006, 1956.
44. C. L. Bianchi, P. Canton, N. Dimitratos, F. Porta and L. Prati, *Catal. Today*, 2005, **102–103**, 203.
45. G. C. Bond and D. Thompson, *Cat. Rev.-Sci. Eng.*, 1999, **41**(3–4), 319.
46. C. Bianchi, S. Biella, A. Gervasini, L. Prati and M. Rossi, *Catal. Lett.*, 2003, **85**(1–2), 91.
47. S. Demirel, K. Lehnert, M. Lucas and P. Claus, *Appl. Catal. B*, 2007, **70**(1–4), 637.
48. C. E. Wyman, *Biotechnol. Prog.*, 2003, **19**, 254.
49. C. Bianchi, S. Biella, A. Gervasini, L. Prati and M. Rossi, *Catal. Lett.*, 2003, **85**(1–2), 91.
50. G. C. Bond and D. Thompson, *Cat. Rev.-Sci. Eng.*, 1999, **41**(3–4), 319.
51. H. J. Kluytmans, A. P. Markusse, B. F. M. Kuster, G. B. Marin and J. C. Schouten, *Catal. Today*, 2000, **57**(1–2), 143.
52. R. Wessendorf, *Erdol, Kohle, Erdgas, Petrochem.*, 1995, **48**(3), 138.
53. H.S. Kesling, L.J. Karas, F.J. Liotta, US Patent 5,308,365 (1994) to ARCO Chemical Technology, L.P.
54. V.P. Gupta, US Patent 5,476,971 (1995).
55. K. Klepacova, D. Mravec and M. Bajus, *Chemical Papers-Chemicke Zvesti*, 2006, **60**(3), 224.
56. K. Klepacova, D. Mravec, E. Hajekova and M. Bajus, *Petrol. Coal*, 2003, **45**, 54.
57. R. S. Karinen and A. O. I. Krause, *Appl. Catal. A*, 2006, **306**, 128.
58. J. -M. Clacens, Y. Pouilloux and J. Barrault, *Appl. Catal. A*, 2002, **227**(1–2), 181.
59. S. Morlat, N. Cezard, B. Loubinoux, J. -L. Philippart and J. -L. Gardette, *Poly. Degrad. Stab.*, 2001, **72**(2), 199.
60. H. Stuhler, US Patent 4,551,561 (1985).
61. N. Garti, A. Aserin and B. Zaidman, *J. Am. Oil Chem. Soc.*, 1981, **58**(9), 878.
62. S. Akimoto, Japanese Patent Appl. JP58198429 (1983).
63. E.G. Harris, U. Hees, R. Bunte, J. Hachgenei, P. Kuhm, US Patent 5,349,094 (1992) assigned to Henkel.
64. L. Jeromin, B. Gutsche, R. Bunte and V. Jordan, US Pat 5,710,350 (1998) assigned to Henkel.
65. U. Hees, R. Bunte, J.W. Hachgenei, P. Kuhm and *E.G.* Harris, PCT Int. Appl. WO9325511 (1993).
66. J. Barrault, J. -M. Clacens and Y. Pouilloux, *Top. Catal.*, 2004, **27**(1–4), 137.

67. European Parliament and Council Directive No. 95/2/EC.
68. Code of Federal Regulation, Vol. 3, Title 21, Food and Drug Administration, US Government (2002).
69. H. Baumann, M. Buhler, H. Fochem, F. Hirsinger, H. Zoebelein and J. Falbe, *Angew. Chem.*, 1988, **27**(1), 41.
70. K. Gottlieb, H. Neitsch and R. Wessendorf, *Chem.-Ing.-Tech.*, 1994, **66**(1), 64.
71. N. O. V. Sonntag, *J. Am. Oil Chem. Soc.*, 1982, **59**(10), A795.
72. C. Marquez-Alvarez and E. Sastre, *J. Perez-Pariente. Top. Catal.*, 2004, **27**(1–4), 105.
73. A. Corma, S. Iborra, S. Miquel and J. Primo, *J. Catal.*, 1998, **173**(2), 315.
74. S. Bancquart, C. Vanhove, Y. Pouilloux and J. Barrault, *Appl. Catal. A*, 2001, **218**(1-2), 1.
75. Y. Pouilloux, S. Métayer and J. Barrault, *Comptes Rendus de l'Académie des Sciences-Series IIC-Chemistry*, 2000, **3**(7), 589.
76. J. Barrault, S. Bancquart and Y. Pouilloux, *C. R. Chimie*, 2004, **7**, 593.
77. A. Corma, S. B. A. Hamid, S. Iborra and A. Velty, *J. Catal.*, 2005, **234**(2), 340.
78. J. Perez-Pariente, I. Diaz, F. Mohino and E. Sastre, *Appl. Catal. A*, 2003, **254**(2), 173.
79. A. Corma, V. Fornes, M. T. Navarro and J. Perez-Pariente, *J. Catal.*, 1994, **148**(2), 569.
80. V. Gonzalez-Pena, C. Marquez-Alvarez, E. Sastre and J. Perez-Pariente, *Stud. Surf. Sci. Catal.*, 2002, **142**, 1283.
81. N. Sanchez, M. Martinez and M. J. Aracil, *Ind. Eng. Chem. Res.*, 1997, **36**(5), 1524.
82. C. Prieto, T. Blasco, M. Camblorb and J. Perez-Pariente, *J. Mater. Chem.*, 2000, **10**(6), 1383.
83. M. da Silva-Machado, J. Perez-Pariente, E. Sastre, D. Cardoso and A. Martinez-Guerenu, *Appl. Catal. A*, 2000, **203**(2), 321.
84. J. Aracil, M. Martinez, N. Sanchez and A. Corma, *Zeolites*, 1992, **12**(3), 233.
85. Y. Pouilloux, S. Abro, C. Vanhove and J. Barrault, *J. Mol. Catal. A:, Chem.*, 1999, **149**(1–2), 243.
86. J. A. Melero, R. van Grieken, G. Morales and M. Paniagua, *Energy Fuels*, 2007, **80**(1–3), 33.
87. P. N. Caley and R. C. Everett, US Patent 3,350,871 (1967) assigned to DuPont.
88. Y. C. Kim, N. V. Park, J. S. Shim, S. R. Lee and D. J. Moon, *Catal. Today*, 2003, **87**(1–4), 153.
89. E. Drent and W.W. Jager, US Patent 6,080,898 (2000) assigned to Shell Oil Co.
90. Montassier, J. M. Dumas, P. Granger and J. Barbier, *Appl. Catal. A*, 1995, **121**(2), 231.
91. B. Casale and A.M. Gomez, US Patent 5,214,219 (1993).
92. S. Ludwig and E. Manfred, US Patent 5,616,817 (1997).

93. T. Haas, A. Neher, D. Arntz, H. Klenk and W. Girke, US Patent 5,426,249 (1995).
94. T. Werpy, J. Frye, A. Zacher and D. Miller, Patent WO 03 035 582, 2002 (Michigan State University).
95. L. Peereboom, B. Koenigsknecht, M. Hunter, J. E. Jackson and D. J. Miller, *Carbon*, 2007, **45**(3), 579.
96. J. Chaminand, L. Djakovitch, P. Gallezot, P. Marion, C. Pinel and C. Rosier, *Green Chem.*, 2004, **6**, 359.
97. T. Miyazawa, Y. Kusunoki, K. Kunimori and K. Tomishige, *J. Catal.*, 2006, **240**, 213.
98. A. Perosa and P. Tundo, *Ind. Eng. Chem. Res.*, 2005, **44**(23), 8535.
99. K. Wang, M. C. Hawley and S. J. DeAthos, *Ind. Eng. Chem. Res.*, 2003, **42**(13), 2913.
100. H. Waldmann and P. Frantisek, *Chem. Ber.*, 1950, **83**, 287.
101. S. Ramayya, A. Brittain, C. DeAlmeida, W. Mok and M. J. Antal, *Fuel*, 1987, **66**, 1363.
102. M. J. Antal Jr, W. S. L. Mok and G. N. Richards, *Carbohydr. Res.*, 1990, **199**, 111.
103. E. Schwenk, M. Gehrke and F. Aichner, US Patent 1,916,743 (1933) assigned to Schering-Kahlbaum AG.
104. K. Biswas, S. Datta, S. Chaudhuri, K. Kargupta, S. Datta and S. K. Sanyal, *Separation Sci. Technol.*, 2000, **35**(9), 1391.
105. W. Bühler, E. Dinjus, H. J. Ederer, A. Kruse and C. Mas, *J. Supercrit. Fluids*, 2002, **22**(1), 37.
106. M. Watanabe, T. Iida, Y. Aizawa, T. M. Aida and H. Inomata, *Bioresour. Technol.*, 2007, **98**(6), 1285.
107. L. Ott, M. Bicker and H. Vogel, *Green Chem.*, 2006, **8**, 214.
108. T. Valliyappan, Ph.D. Dissertation, University of Saskatchewan, Saketton, Canada, 2004.
109. X. Xu, Y. Matsumura, J. Stenberg and M. J. Antal Jr, *Ind. Eng. Chem. Res.*, 1996, **35**, 2522.
110. Y. S. Stein, M. J. Antal Jr and M. Jones Jr, *J. Anal. Appl. Pyrolysis*, 1983, **4**(4), 283.
111. J. W. Shabaker, G. W. Huber and J. A. Dumesic, *J. Catal.*, 2004, **222**(1), 180.
112. P. J. Dauenhauer, J. R. Salge and L. D. Schmidt, *J. Catal.*, 2006, **244**(2), 238.
113. T. Hirai, N. O. Ikenaga, T. Miyake and T. Suzuki, *Energy & Fuels*, 2005, **19**(4), 1761.
114. R. R. Soares, D. A. Simonetti and J. A. Dumesic, *Angew. Chem. Int. ed.*, 2006, **45**, 3982.
115. A. Tanksale, J. N. Beltramini, J. A. Dumesic and G. Q. Lu, *J. Catal.*, 2008, **258**, 366.

CHAPTER 18

Renewable Diesel and Jet-Fuel Production from Fats and Oils

TOM N. KALNES,[a] MICHAEL M. MCCALL[a] AND DAVID R. SHONNARD[b]

[a] UOP LLC, A Honeywell Company, DesPlaines, Illinois, USA; [b] Department of Chemical Engineering, Sustainable Futures Institute, Michigan Technological University, Houghton, Michigan, USA

18.1 Scope of Chapter and Rationale

This chapter focuses on the conversion of biomass-derived fats and oils to hydrotreated renewable diesel (HRD) and hydrotreated renewable jet (HRJ) fuel. These high specific energy alternative fuels can be produced directly from triacylglycerol (TAG) and free fatty (FFA) acid rich feedstocks by catalytic hydroprocessing. Similar in composition to Fischer–Tropsch (FT) derived fuels, HRD and HRJ are fully deoxygenated, paraffin-rich hydrocarbon liquids that can either be used neat or blended in any proportion with existing petroleum fuels. When produced from waste streams or sustainable energy crops, HRD and HRJ exhibit a much smaller carbon footprint than their petroleum-derived counterparts.

The chapter begins with a review of the chemistry of catalytic hydroprocessing for the sequential steps of deoxygenation, isomerization and selective hydrocracking. The next section examines the critical issues of commercial scale-up: feedstock selection, process design, product properties and fuel performance in existing engine technology. Ongoing certification efforts, engine tests and fleet testing continue to document the many benefits of these fuels

RSC Energy and Environment Series No. 1
Thermochemical Conversion of Biomass to Liquid Fuels and Chemicals
Edited by Mark Crocker

Published by the Royal Society of Chemistry, www.rsc.org

including high specific energy content, good blending properties, and lower emissions. The chapter continues with life-cycle assessments of both HRD and HRJ from a variety of feedstock sources. The last section of the chapter summarizes the commercial status of HRD and HRJ production technology. The critical aspects of sustainable fuel production are also reviewed and summarized.

18.2 Overview of Hydroprocessing Chemistry

Over a century ago, a German chemist, Wilhelm Normann, found that catalytic hydrogenation could be used to convert unsaturated fatty acids or glycerides in the liquid phase into saturated ones. He was awarded a patent in Germany in 1902[1] and in Britain in 1903,[2] which was the beginning of what is now a worldwide industry. Although the use of hydrogenation in the foods industry is limited, catalytic hydrodesulfurization (HDS) of petroleum has become widespread. Over the past 60 years, a large variety of proprietary catalytic hydrogenation, hydrotreating, and hydrocracking processes were successfully developed and commercialized. Hydrotreating is a collective term for processes that remove heteroatoms such as sulfur, nitrogen and oxygen from petroleum. Today, virtually all of the petroleum refineries worldwide have one or more hydroprocessing units, thus enabling the production of low-sulfur transportation fuels from a wide variety of crude oils of varying quality and heteroatom content. An example of a typical diesel desulfurization process is depicted in the flow diagram of Figure 18.1.[3]

Hydrogenation is a class of chemical reactions in which the net result is the addition of hydrogen (H). Hydrogenolysis is a type of hydrogenation and

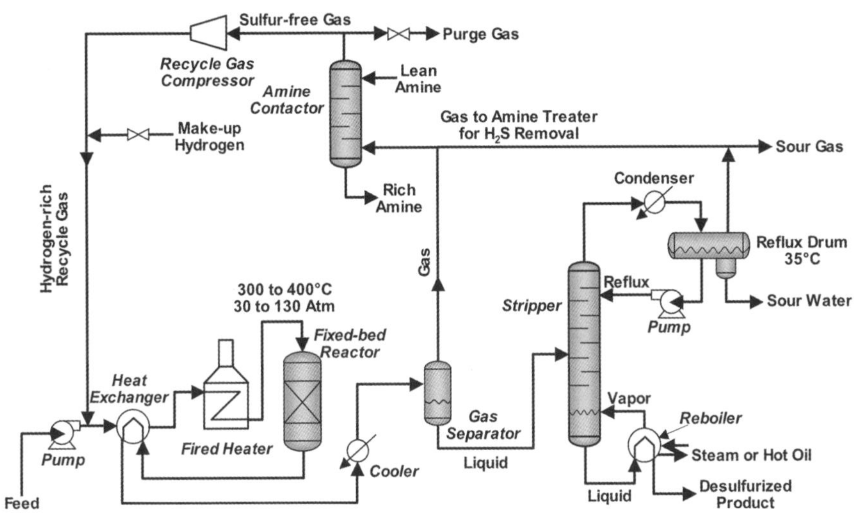

Figure 18.1 Typical diesel desulfurization (HDS) process.

results in the cleavage of the C–X chemical bond, where C is a carbon atom and X is a sulfur, nitrogen (N) or oxygen (O) atom. The net result of a hydrogenolysis reaction is the formation of C–H and H–X chemical bonds. Thus, hydrodeoxygenation is a hydrogenolysis reaction. Complete deoxygenation of fats and oils is critical to ensuring the production of transportation fuels that are chemically similar to current petroleum hydrocarbon fuels, with good storage stability and maximum specific energy.

Renewable fats and oils are primarily composed of FFAs and TAGs. The length of the fatty-acid chains ("carbon number") varies among feedstocks, as does the number and positions of double bonds ("degree of unsaturation"). Almost all fatty-acid chains have even-numbered carbon chains, although some animal fats have small amounts of odd-numbered carbon fatty acids. All feeds have high oxygen content. As shown in Table 18.1, most natural oils have 18 carbons in the majority of fatty-acid chains, with 16 carbons as the second most abundant fatty-acid chain.[4–6] Some oils also contain significant amount of C_{20} and C_{22} carbon chain lengths.[7]

When fats and oils are converted to HRD and HRJ, oxygen is completely removed through competing mechanisms: decarboxylation (DeCOx) and hydrodeoxygenation (HDO). One carbon is removed and lost in the $DeCO_x$ reaction producing CO_x and a paraffin chain with an odd number of carbon atoms. Hydrogen is required in the hydrodeoxygenation pathway (shown to the right in Figure 18.2) to produce water and a long-chain paraffin with the same even number of carbons as found in the fatty-acid chain of the feed. The glycerol portion of a TAG molecule is converted to propane. Note that since decarboxylation removes a carbon from the fatty-acid chain the liquid yield will be lower than paraffins produced through hydrodeoxygenation. For soybean oil, the theoretical hydrocarbon yield from 100% decarboxylation is about 87 wt%. The theoretical yield from 100% hydrodeoxygenation is about 91 wt%.

The hydrogenation of all double bonds occurs independently of the deoxygenation mechanism to produce a hydrogen-saturated straight-chain paraffin product. This erases one of the key differentiators of natural oils; location and number of double bonds in the fatty-acid chains. The final paraffin product has a carbon-number composition very similar to the fatty acid carbon-number distribution of the starting oil.

Deoxygenation of most plant oils and animal fats therefore produces an n-paraffin product ranging from nC_{15}–nC_{22} with a peak at nC_{17}–nC_{18}, a product in the diesel fuel range but too heavy for jet fuel, or the kerosene fuel range.

The extent to which this straight-chain paraffinic product is isomerized and hydrocracked depends on site-specific targets for fuel quality such as type of fuel, viscosity, cloud point, and cetane number. For the production of HRD, the cloud point is adjusted lower by increasing the amount of branched paraffins through hydroisomerization.

The production of HRJ requires selective hydrocracking of the n-paraffin product along with substantial isomerization to produce a synthetic paraffinic kerosene (SPK) product with a carbon-chain distribution of approximately C_9–

Table 18.1 Comparison of renewable feedstock composition.

		%Fatty Acid Composition Examples															
		8:0	*10:0*	*12:0*	*14.0*	*16.0*	*18.0*	*18:1*	*18:2*	*18:3*	*20:0*	*20:1*	*20:2*	*20:3*	*22:0*	*Other*	*Source*
Feedstock Natural Oil																	
Jatrophas Curcas						20	7	41	31								Akintayo
Jatrophas Curcas						14.1–15.3	3.7–9.8	34.3–45.8	29.0–44.2								Gubitz *et al.*
Camelina Oil						5.3–5.6	2.3–2.7	14.0–16.9	13.5 16.5	34.9–39.7	1.2–1.5	15.1–15.8	1.7–2.0	1.3–1.7	2.6–3.0		Abram *et al.*
Camelina Oil						6	2	15	16	39	0	16	2	1	4		Abram *et al.*
Algal Oil						22	5	49	10	12	1					1	UOP Data
Yellow Grease						17	10	56	10	6							UOP Data
Beef Tallow					6	26	31	31	2	0	0	0	0	0		3	Bailey's
Fatty Acid Prefix	Caproic	Caprylic	Capric	Lauric	Myristic	Palmitic	Stearic	Oleic	Linoleic	Linolenic	Arachidic	Gondoic	Eicosadienoic	Eicosatrienoic	Erucic Acid		

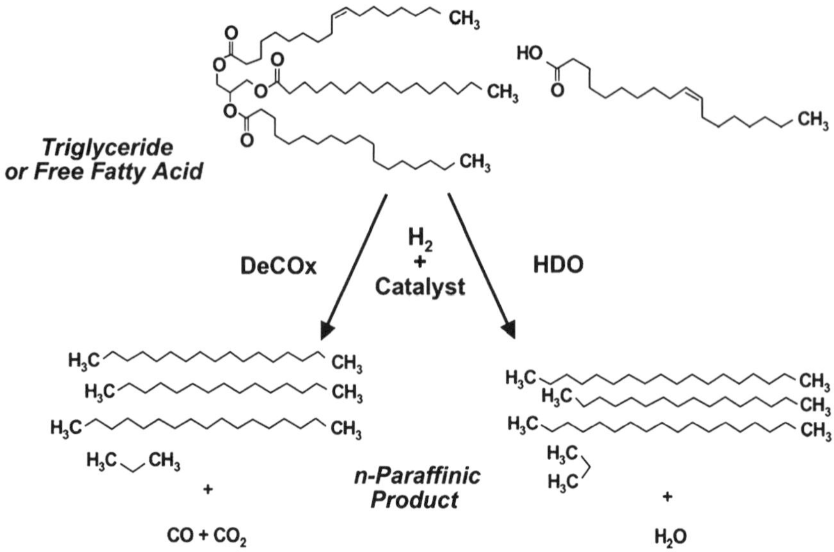

Figure 18.2 Competing mechanisms for deoxygenation of fats and oils.

C_{15}. This step is called selective cracking because overcracking will result in low yields of jet-range paraffins and high yields of light ends (C_1–C_4) and naphtha (C_5–C_8), both outside the jet-fuel range. These products of overcracking have lower economic value than diesel or jet-range paraffins. Isomerization is also required to lower the freeze point. The selective cracking and isomerization steps may be either concurrent or sequential. For example, a Brønsted-acid catalyst both isomerizes and cracks the n-paraffins through a β-scission pathway since branching is a required intermediate to cracking through this mechanism.[8] An example of the β-scission pathway for a model compound (nC_7) is shown in Figure 18.3. In this example, the paraffin is first activated through conversion to an olefin (for example by metal-catalyzed dehydrogenation). The more basic olefin is easily protonated by a catalyst acid site, followed by structural isomerization. Adjustment of the acid-site strength or process conditions will result in cracking of the iso-olefins to produce mostly branched cracked products. The choice of catalyst can also increase the amount of cracking at the end of the paraffin molecule, thereby increasing the yield of jet-range paraffins. Before desorption, the olefin is saturated by one of several mechanisms including metal catalyzed hydrogenation.

Therefore, the type of catalyst and development of the right process conditions will direct the production of either a mostly isoparaffinic diesel range fuel (HRD) or a mostly isoparaffinic jet-range fuel (HRJ SPK). There are many catalytic options to control the selectivity of the isomerization and hydrocracking steps and a complete review of these options is outside the scope of this chapter.

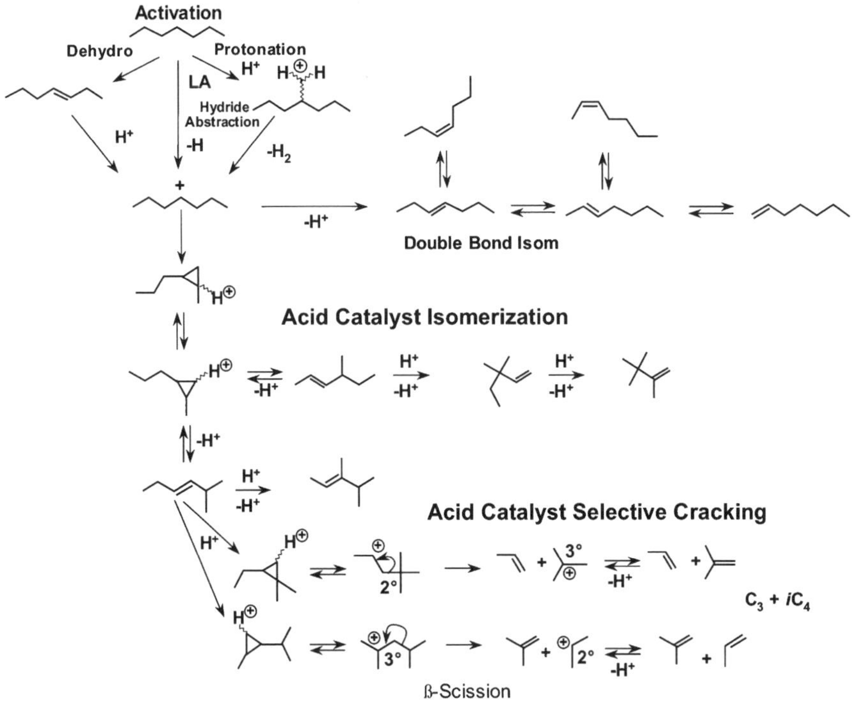

Figure 18.3 Mechanisms for isomerization and selective cracking.

As mentioned in the introduction, HRD and HRJ SPK are paraffin-rich fuels that are very similar to fuels produced *via* the Fischer–Tropsch (FT) synthesis. A comparison of the two production pathways is provided in Figure 18.4 to show why this similarity exists.

In commercial gasification – FT pathways (lower section of Figure 18.4), coal and natural gas are used as feeds to produce syngas that, after clean-up, is converted to waxy product through FT synthesis. The as-produced FT product contains mostly n-paraffins along with olefins and oxygenates. The olefins and oxygenates in FT wax need to be converted to saturated hydrocarbons, however, hydrotreating alone will produce a wide boiling point distribution with some high molecular weight paraffinic molecules or waxes that can lead to undesirable cold-flow properties when blended into diesel or jet fuel. For the production of FT-derived diesel or kerosene-range isoparaffinic product, a selective hydrocracking step is required, followed by fractionation to remove light and heavy ends.

The upper portion of Figure 18.4 shows that olefins present in (pretreated) animal fats and plant oils are also saturated in the deoxygenation stage of the hydroprocessing unit to produce an n-paraffin-rich product. Because the last processing step in both pathways is hydroprocessing of n-paraffins, it is not surprising that a high-quality paraffinic diesel or jet fuel can be made from renewable feedstocks with properties indistinguishable from paraffinic fuels

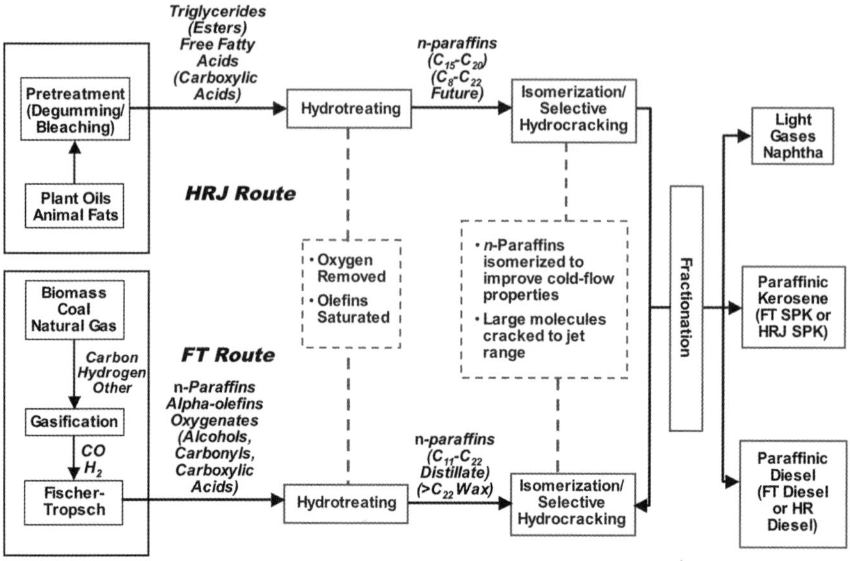

Figure 18.4 Comparison of processing steps for renewable and FT fuels.

produced by a gasification-FT (syndiesel) route. An important distinction is that the carbon distribution of the n-paraffins produced from most fats and oils is usually much narrower than that of the wax produced by FT synthesis. As a result, the deoxygenated product is already in the diesel range. Therefore, the production of HRD requires only isomerization to improve cold-flow properties, whereas the production of HRJ SPK requires selective hydrocracking coupled with isomerization. The detailed physical and chemical properties of HRD and HRJ are discussed in Section 18.3.

18.3 General Considerations for Commercial Production

Hydroprocessing for the production of renewable transportation fuels is a commercial technology. As discussed later in this chapter, more than 800 million gallons per year (~3 MMTA) of capacity has already been announced with the majority of this capacity located in existing petroleum refineries. In this section, the critical issues of commercial scale-up: feedstock selection, process design, product properties, and fuel performance in existing engine technology are discussed.

18.3.1 Feedstock Selection

The same feedstocks currently used for biodiesel (BD) production (*i.e.* rapeseed, sunflower, soybean, and palm) can be used for HRD and HRJ production.

Unlike base-catalyzed transesterification for production of BD, hydroprocessing technology is robust to feedstocks containing high concentrations of fatty acids. This flexibility has allowed several refiners to choose less-costly animal fats and used oils as feedstock for renewable diesel production. Standard pretreatment (*e.g.*, degumming) is required to reduce concentrations of phospholipids and alkali metals to an acceptable level. Less-severe pretreatment is required for refinery feedstock than for food-grade oil production. In many cases, the fat or oil is pretreated at the point of production prior to being shipped to the refinery.

The yield of HRD and HRJ is relatively insensitive to feedstock source. Figure 18.5 presents UOP pilot data to illustrate this point.[9]

In this deoxygenation catalyst stability test run, a variety of different feedstocks were processed. In all cases, close to 100 vol% yield of deoxygenated diesel range product was observed. This high yield reflects the lower density of the liquid product. Note that while deoxygenated liquid yield is relatively independent of feedstock source, less hydrogen is required when processing saturated fats than when processing unsaturated oils.

Worldwide, about 0.2 billion metric tons of renewable fats and oils are produced annually, which represents about 5% of current petroleum output. However, because more than 80% of production is currently consumed as food or animal feed, rapid growth of oilseed-based fuel production is currently constrained by feedstock availability and the potentially adverse impact of fuel use on food/feed supply and price.[10]

Long-term, expanded cultivation of new "nonfood" sources of transportation fuel feedstock is needed to enable biofuel market growth. Crops that can

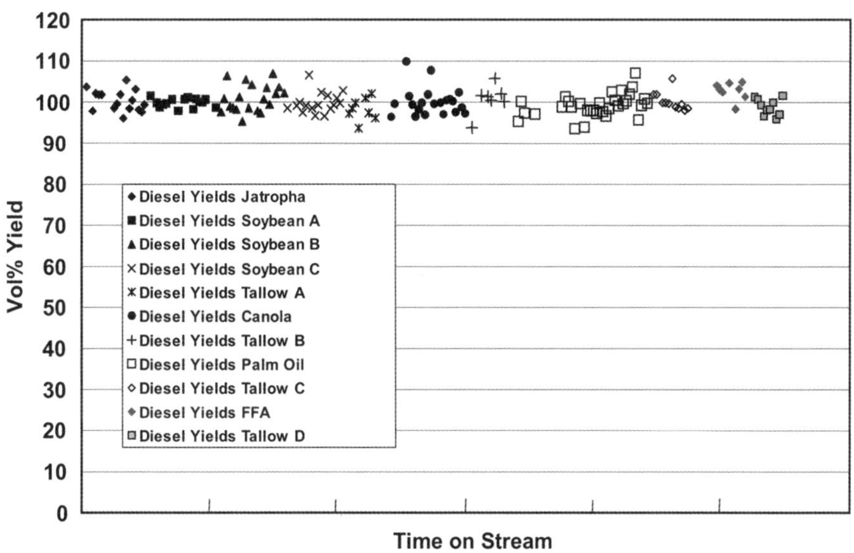

Figure 18.5 Deoxygenation yield for different feedstocks (UOP/ENI Ecofining™ process).

be grown on marginal land such as jatropha, camelina, and halophytes show promise for future expansion.[11] Large-scale production of algae oil is another area that is under rapid development.[12] Optimization of cultivation inputs (*i.e.* fertilizer and water), crop yields, oil content, harvesting and transportation cost, and markets for byproducts, are all crucial to future expansion efforts. To be sustainable, the renewable feedstock cost needs to be competitive with crude petroleum and feedstock supply should be noncompetitive with food markets. In addition, the impact on land-use change and biodiversity should be minimal.

18.3.2 Process Considerations

To a great extent, the technology for converting fats and oils to transportation fuels has evolved from the widely practiced petroleum refining technology, as previously illustrated in Figure 18.1. The most direct application of existing technology is to coprocess small amounts (*i.e.* 5–10%) of renewable fats and oils together with petroleum diesel in an existing diesel HDS unit wherein the blended feedstock is mixed with hydrogen, heated to reaction temperature and passed through a fixed-bed adiabatic reactor. Deoxygenation of the renewable feedstock will occur at operating conditions similar to those used to desulfurize petroleum diesel, however, more severe reactor conditions (*i.e.* higher temperature) may be required to maintain product sulfur targets. The deoxygenated product is corecovered with the desulfurized petroleum as a partially renewable diesel product.

Several petroleum refiners including BP, ConocoPhillips, and PetroBras have commercially practiced renewable feed coprocessing. The amount of renewable feed that can be coprocessed is limited by available hydrogen supply as well as concerns with reactor temperature control, catalyst life, diesel quality, and management of deoxygenation byproducts (*e.g.*, CO_x). In most cases, a revamp of the HDS unit is required to accommodate the blended feedstock.

As illustrated in Figure 18.6, the coprocessing of renewable feedstocks in an existing diesel HDS unit impacts diesel quality. Cetane increases due to the increase in n-paraffin content in the diesel product. However, these same n-paraffins raise concerns about the diesel cloud point.

Normal paraffins have a significantly higher cloud point than other hydrocarbons of similar carbon number. Since hydrotreating converts the fatty-acid chains into long-chain n-paraffins, it is expected that the cloud point of the hydrotreated product will increase compared to the product using the straight-run (SR) petroleum HDS unit feed.[13]

Many petroleum refiners are conservative about introducing new (non-petroleum) feed components into an existing HDS unit. Any plant upsets or deterioration in process desulfurization performance can result in shorter catalyst life, costly shutdowns and loss of operating revenue. For this reason, and because of interest in producing renewable fuels outside of the petroleum refinery, several new advanced hydroprocessing technologies have been developed to process 100% renewable feedstock. Advanced hydroprocessing

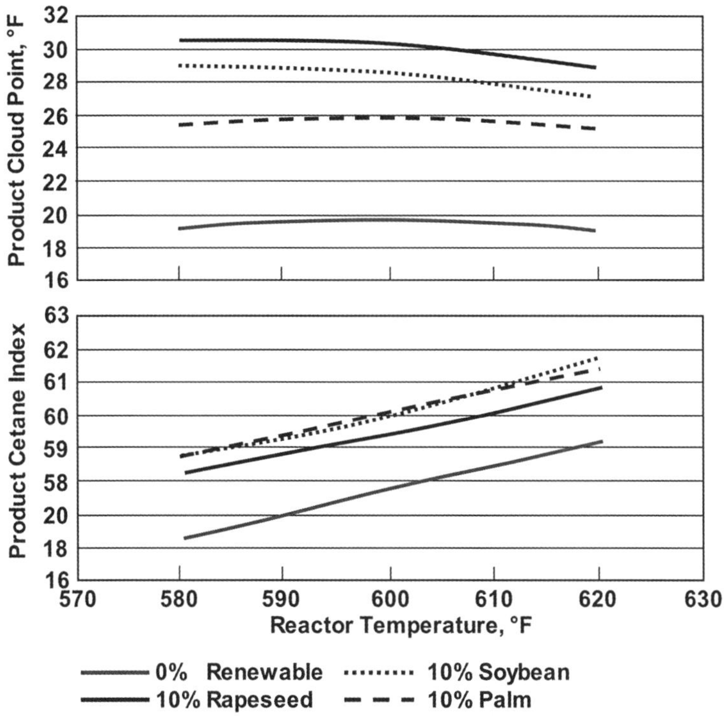

Figure 18.6 Diesel properties when coprocessing renewable oils (Grace Davison Publication, 2008).

technologies also provide flexibility to control the properties of the diesel fuel product such that cold weather (low cloud point) diesel and low freeze point jet fuels can be produced.

Processing 100% biologically derived oils offers some unique advantages, however, operating requirements differ significantly from those of a typical HDS unit. Table 18.2 shows some of the key design differences that must be accommodated in a standalone advanced hydroprocessing unit.

Whereas HDS unit feedstock is typically >99% carbon and hydrogen, typical fats and oils contain 10–12 wt% oxygen. In addition, a significant percentage of the fatty-acid chains present in renewable feedstocks are partially unsaturated. As a result, the hydrogen requirement (and heat of reaction) can be an order of magnitude higher than required for petroleum-diesel desulfurization. In addition, the advanced hydroprocessing unit design must provide for the proper management of the increased amounts of nonhydrocarbon byproducts (water and CO_x) produced during the deoxygenation step.

An example of an advanced hydroprocessing technology is the UOP/ENI Ecofining™ process. Two reaction stages are employed to produce a low cloud point HRD (Honeywell Green Diesel™). A simplified block flow diagram of the process is provided in Figure 18.7.[14]

Table 18.2 Comparison of feedstock and process requirements.

	Diesel HDS	*Tag → HRD or HRJ*
Feedstock		
Mid Boiling Point, °C (°F)	260–280 (500–536)	580–600 (1076–1112)
Carbon + Hydrogen, wt%	98–99.8	88–90
Sulfur, wt%	0.1-2	<0.05
Oxygen, wt%	<0.1	10–12
Nitrogen, wt-ppm	50–500	2–20
Olefin Content	Typically Low	Can Be High
Catalyst Poisons	Trace Si, As (Cracked Stocks)	Alkali Metals Phospholipids
Process Requirements		
H_2 Consumption, wt% Feedstock	0.2–1	2–4
Non-hydrocarbon Byproducts	H_2S, NH_3	H_2O, CO_2
Heat of Reaction	Base	5–10*Base

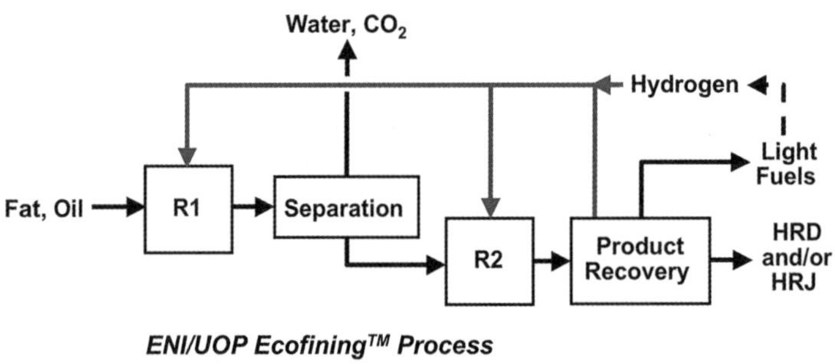

Figure 18.7 Advanced process for converting 100% fat and oil to renewable fuel.

In this process, pressurized feedstock is mixed with recycle hydrogen and then sent to a multistage adiabatic catalytic hydrodeoxygenation reactor (R1) where the biorenewable oil is saturated and completely deoxygenated. Gas recycle to R1 is set to achieve a minimum hydrogen partial pressure. Conversion of feed is complete and the volumetric yield of deoxygenated hydrocarbon products is >100%. Selectivity to diesel boiling-range paraffin is very high. The primary deoxygenation reaction byproducts are propane, water and carbon dioxide.

The effluent from R1 is immediately separated at reactor pressure to remove carbon dioxide, water and low molecular weight hydrocarbons. The resultant diesel is mixed with additional recycle gas and then routed to an integrated catalytic hydroisomerization reactor, R2, wherein a branched paraffin-rich diesel fuel is produced. In this manner, the cold-flow properties of the diesel are adjusted to meet required specifications. The isomerization reaction is also selective and, as a result, consumes very little hydrogen.

Table 18.3 Comparison of renewable diesel and biodiesel feed and products.

	Ecofining Green Diesel			*Biodiesel*	
Feeds	*Weight-%*	*Volume-%*	*Feeds*	*Weight-%*	*Volume-%*
Vegetable oil	100	100	Vegetable Oil	100	100
Hydrogen	1.5–3.8		Methanol	10	11
			Chemicals[a]	4	
Products	*Weight-%*	*Volume-%*	*Products*	*Weight-%*	*Volume-%*
Propane	5	9	FAME	96	100
Butane	0–2	0–3	Glycerol	10	7
Naphtha	<1–7	1–10			
Green Diesel	75–85	88–99			

[a]Chemicals include sodium hydroxide (NaOH) and an acid to neutralize the products

Isomerized product is separated from excess hydrogen in a conventional gas/liquid separator. After purification, the excess hydrogen is recycled back to both reactors to maintain the minimum required hydrogen partial pressure. Make-up hydrogen is added to the process to balance both chemical consumption and solution losses. The liquid product is sent to the product recovery section of the process where conventional distillation steps are employed to separate coproducts such as propane and naphtha.[15]

Table 18.3 contrasts Ecofining™ inputs and outputs to those of conventional biodiesel production. While the primary product from the Ecofining™ process is green diesel, smaller amounts of other renewable fuels are coproduced. Feeds containing less unsaturates require less hydrogen input. Higher processing severity results in a lower diesel cloud point at the expense of a yield shift from diesel to lower molecular weight fuels. The process is flexible. Operating conditions can be adjusted and optimized to accommodate changes in feedstock and seasonal variations in diesel cloud point specification.

The Ecofining process for producing HRD (Green Diesel) operates at mild processing conditions and integrates well within existing petroleum refineries. If required, a portion of the light fuel coproduct can be steam reformed to generate all of the hydrogen consumed in the process.[16]

18.3.3 Product Properties and Engine Performance

Isoparaffinic diesel and jet fuel are high-quality synthetic fuels that can be mixed in any proportion with their petroleum-derived counterparts. They have many superior properties to other alternative fuels and even exceed the quality of petroleum-derived fuels. For example, since HRD and HRJ are composed entirely of carbon and hydrogen, these hydrocarbon fuels have ultralow sulfur levels and much higher energy content than other alternative fuels such as biodiesel, a fatty-acid methyl ester fuel with high oxygen content. In fact, the higher hydrogen to carbon ratio of HRD and HRJ results in higher energy content than petroleum-derived fuels that contain higher levels of unsaturated aromatic compounds.

The properties of a HRD fuel is compared against the D 975 ASTM specification properties for diesel fuels in Table 18.4.[17] The results show that this typical neat HRD fuel meets all the specifications for a 2D diesel fuel. If fact, in many cases the properties greatly exceed the specifications, for example the very low sulfur content and the high cetane number. These excellent sulfur and cetane values make the HRD a valuable blending fuel to improve the properties of a petroleum diesel fuel.

ENI performed engine testing comparing a petroleum-derived diesel with a 27% blend of HRD with petroleum diesel.[18] The fuel was generated from a ENI/UOP joint development project that resulted in a HRD process that can be commercially licensed. Tests were completed using a direct injection, turbocharged, 1.9 liter displacement engine with an oxidation catalyst for emission

Table 18.4 Ecofining HRD quality compared to ASTM D975-08 specs for No. 2-D S15.[18]

Property	*Unit*	*Spec (No. 2-D S15)*	*HRD Actual Value*	*Test Method*
Flash point	°C	52 min	71	ASTM 93
Water and sediment by centrifuge	vol %	0.05 max.	0.0	ASTM D2709
Physical distillation, 90% recovered	°C	282 min. 338 max.	292.6	ASTM D86
Kinematic viscosity, 40 °C	cst	1.9 min. 4.1 max.	2.835	ASTM D445
Ash	wt %	0.01 max.	<0.0001	ASTM D482
Sulfur	ppm	15 max.	<3.0	ASTM D5453
Copper corrosion				
Conditions			3 h at 50 °C	ASTM D130
Rating		3 max.	1a	
Cetane number		40 min.	>66	ASTM D613
Cloud point	°C		− 8	ASTM D2500
Carbon res RCR10%				ASTM D524
Ramsbottom carbon, 10% bottoms	wt %	0.35 max.	0.02	
Lubricity by HFRR	micron	520 max.	450	
Temperature of test	°C		60	
Major axis			0.47	ASTM D6709
Minor axis			0.43	
Description of wear scar			Normal Appearance	
Electrical conductivity, portable meter*				
Conductivity	(pS/m)	25 min.	3	ASTM D2624
Test measurement	°C		20	

Table 18.5 Ecofining HRD diesel blend used for ENI engine tests.[18]

Fuel Properties		*EN 590 Diesel Fuel*	*EN590 + 27% Green Diesel*
Density@15 °C	Kg/m^3	838	822
Cetane Number		55	57
Sulfur	mg/kg	12	9
Cloud Point	°C	– 2.1	– 4.9
CFPP	°C	– 7.1	– 16
IBP	°C	201	193
5% vol	°C	220	219
50% vol	°C	276	279
95% vol	°C	356	349
FBP	°C	366	363
Total Aromatics	wt%	24.4	18.6
Poly Aromatics	wt%	3.7	2.9
Mono Aromatics	wt%	20.7	15.7

control. Table 18.5 shows the properties of the blended HRD fuel (green diesel) and the neat petroleum diesel. The HRD blend has a higher cetane number, lower sulfur content, and a lower cloud point.

Figure 18.8 shows emissions data from the ENI diesel engine tests. The hydrocarbon emissions and CO emissions were significantly reduced when HRD was blended with petroleum diesel. The emissions of NO_x and particulate matter (PM) were unaffected by blending in HRD. These data on the HRD blends demonstrate the advantages of HRD, a low sulfur – high-cetane hydrocarbon fuel that lowers emissions when blended with petroleum diesel. Lower hydrocarbon and CO emissions were also observed for Neste's NExBTL (HRD) fuel when blended with petroleum diesel.[19]

Hydroprocessed renewable jet (HRJ) is a paraffinic fuel in the kerosene boiling point range, or a synthetic paraffinic kerosene (SPK). Neat SPK meets and exceeds almost all of the specifications for turbine aviation use. In late 2008 – 2009 Boeing led a series of three test flights using 50% SPK and 50% petroleum jet fuel as summarized in Table 18.6.[20]

"UOP produced the Honeywell Green Jet Fuel™ (HRJ)" for these test flights from three different feedstocks: jatropha, algal, and camelina oils. Over 6500 gallons of HRJ was produced providing ample samples for specification testing, fit-for-purpose testing, engine tests, and the test flights themselves. The properties of the bio-SPK fuels, or HRJ, are shown in Tables 18.7 and 18.8.

Neat HRJ exceeds most of the specification properties for turbine aviation fuels, including thermal stability, heat content, and sulfur content. SPK fuels cannot meet the density requirements for most turbine aviation fuels since they do not contain many cycloparaffins and insignificant amounts of aromatics. These hydrocarbon classes give petroleum-derived kerosene the higher density. To meet density requirements for the Boeing test flights the HRJ was mixed with Petroleum Jet A in a 50% ratio. The properties of the 50:50 blended fuels

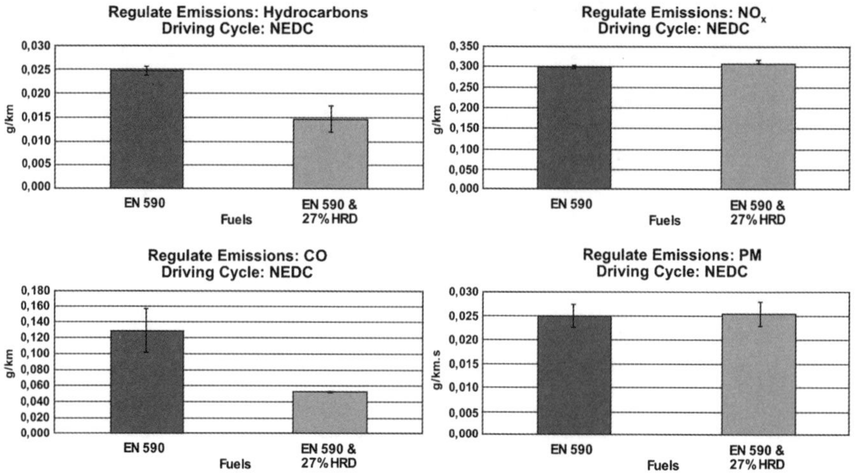

Figure 18.8 Engine test results comparing emissions from a petroleum diesel and an Ecofining HRD/ petroleum diesel blend.[9]

Table 18.6 Summary of Boeing-led test flights.[20]

Airline	*Air New Zealand (ANZ)*	*Continental Airlines (CAL)*	*Japan Airlines (JAL)*
Aircraft	Boeing 747–400	Boeing 737–800	Boeing 747–300
Engine Manufacturer	Rolls-Royce	CFM International	Pratt & Whitney
Plant Feedstock	Jatropha	95% Jatropha/ 5% algae	84% Camelina 8% Jatropha/ Algae
Fuel Provider	UOP	UOP	Nikki Universal/ UOP
Flight date	30-Dec-08	7-Jan-09	30-Jan-09

used for the test flights are shown in Table 18.9. All specification properties were met by the 50% fuels.

These 50% HRJ: JetA (A-1) fuels have been extensively tested for certification. Boeing has led fit-for-purpose testing, engine testing, and test flights. All the data will soon be compiled into an ASTM research report, a key step towards certifying the fuel. Besides the HRJ fuels produced for the test flights, this research report will include all available data from other producers of HRJ.

18.4 Life-Cycle Assessment

Although many life-cycle assessments (LCA) of plant oil-based biofuels have appeared in the literature,[21,22] only a few dealt with comparisons across

Table 18.7 HRJ Properties of 100% SPK Fuels Produced by UOP for Boeing Test Flights.

Boeing Test Flight		*ANZ*		*CAL*	*JAL*	
Feed Source		*Jatropha*	*Camelina*	*Jatropha/Algae*	*Camelina/Jatropha/Algae*	
Property	*Jet A/ Jet A-1*	*Value*	*Value*	*Value*	*Value*	*Method*
Composition						
Acidity, mg KOH/g	Max 0.10	0.002	0.002	0.001	0.003	D3242
Aromatics, volume %	Max 25	0.0	0.3	0.0		D1319
Sulfur, total mass %	Max 0.30	<0.0001	<0.0001	<0.0001	0.0001	D2622
Sulfur, mercapton, mass %	Max 0.003	0.0001	<0.0001	<0.0001	<0.0001	D3227
Volatility						
Distillation						
10% recovered (°C)	Max 205	168	162	163	163	D86
50% recovered (°C)	Report	188	186	187	183	D86
90% recovered (°C)	Report	227	226	233	226	D86
Final boiling point (°C)	Max 300	255	251	256	238	D86
Distillation residue, %	Max 1.5	1.3	1.3	1.2	1.2	D86
Distillation loss, %	Max 1.5	0.1	0.2	0.3	0.3	D86
Flash Point, °C	Min 38	46.5	42.0	41.0	42.0	D56
Density @ 15 °C, kg/L	0.776–0.840	0.749	0.753	0.748	0.752	D4052
Fluidity						
Freezing Point, °C	– 40 (Jet A) – 47 (Jet A-1)	– 57.0	– 63.5	– 54.5	– 63.5	D2386
Viscosity, – 20 °C, mm^2/s	none	3.663	3.336	3.510	3.353	D445
Combustion						
Net heat of combustion, MJ/kg	Min 42.8	44.3	44.0	44.2	44.2	D 4809
Naphthalenes, vol %	Max 3.0	<0.01	<0.01	<0.01	<0.01	D 1840
Thermal Stability						
JFTOT (2.5 h at control temperature)						D3241
Temperature °C	Min 260	340	340	340	300	D3241
Filter dP, mm Hg	Max 25					D3241
Tube Deposit Less Than	3	1	<1	1	1	D3241
Contaminants						
Existent gum, mg/100 mL	Max 7		<1	<1	<1	D381, IP 540

Table 18.8 Composition of HRJ fuels derived from different feedstocks.

HRJ Feedstock Source			*Jatropha*	*Camelina*	*Camelina/ Jatropha/ Algae*
Method	*Description*	*ASTM Pending Limit*	*Value*	*Value*	*Value*
Hydrocarbon Composition					
D2425	Cycloparaffins, mass %	Max. 15	3.8	4.1	12.8
D2425	Aromatics, mass %	Max. 0.5	0.0	0.0	0.1
D2425	Paraffins, mass %	balance	96.2	95.9	87.1
D5291	Carbon and Hydrogen, mass %	Min. 99.5	100.9	100.8	100.0
Non-Hydrocarbon Composition					
UOP 730–09	Oxygen, wt%		<0.03	<0.03	<0.03
D4629	Nitrogen, mg/kg	Max. 3	<0.4	<0.4	<1
D6304	Water, mg/kg	Max. 75	32.0	34.0	19.0
D5453	Sulfur, mg/kg	Max. 15	<0.01	<0.01	<0.01
UOP389	Metal, mg/kg (Al, Ca, Co, Cr, Cu, Fe, K, Mg, Mn, Mo, Na, Ni, P, Pb, Pd, Pt, Sn, Sr, Ti, V, Zn)	<0.1 per metal	<0.1 per total	<0.1 total	<0.1 total
D7359	Halogens, mg/kg	Max.1	<0.1	<0.1	0.2

multiple feedstocks and with several biofuel and second-generation fossil-fuel products.[15,23] In addition, there have been reports of the potential for land-use change emissions of greenhouse gases (GHG) from production of biofuels due to indirect effects on displacing food production with crop-based biofuels.[24,25] With these issues in mind, the goal of this LCA is to compare GHG emissions, as well as fossil and total energy demand, of green diesel (GD) and green jet (GJ) derived from multiple nonfood plant oil feedstocks (camelina, jatropha, and tallow) using UOP licensed technology with petroleum diesel and jet, biodiesel, and syndiesel from coal and natural gas. Note that GD is HRD from the UOP/ENI Ecofining process and that GJ is HRJ from the UOP Hydrotreated Renewable Jet process. Functional unit for this comparison is 1 MJ of energy content of final fuel product.

18.4.1 Fuel-Production Pathways and Inventory Data

The life-cycle diagrams for petroleum diesel, green diesel (GD) from camelina, and syndiesel from coal are shown in Figure 18.9 with important inputs and coproduct outputs indicated. These pathways represent the basis for the collection of inventory data at each stage of the life cycle. Similar diagrams could have been presented for the other fuels (petroleum jet, camelina HRJ, tallow GD and HRJ, jatropha GD and HRJ, camelina BD and syndiesel from natural gas), but each new diagram would be a minor variation on these three.

The sources of inventory data for each plant oil feedstock are shown in Table 18.10, where each source represents a specific geographic setting, degree of technology advancement and time relevance. For example, the cultivation of jatropha capsules and production of jatropha biofuels is assumed to occur in India using nonagricultural lands, with use of fertilizer, water, and other farm inputs provided from a single field trial. Electricity consumption assumes an India mix of energy sources and transport of finished biofuel occurs by transoceanic tanker over a 9000 km distance to Europe.

Camelina seed is assumed grown in Montana as a rotation crop with wheat. Fertilizer, fuel, and pesticide inputs are based on a forecast of improved crop yields. Chemical and energy inputs for camelina oil extraction and for local road transportation of seed are included.[26] Transportation of camelina oil from Montana to Seattle, Washington is assumed to be by rail car where upgrading to biofuel (GD and GJ) occurs. Tallow inputs for collection and conversion to a crude tallow oil are from a single study.[27] Inputs for refining of crude plant oil/tallow are also included from data in the original publications[28] (see Table 18.10), but in cases of missing data the inputs came from a CONCAWE report.[22] Emissions of N_2O from nitrogen fertilizers applied to soils during plant cultivation were estimated using an emission factor from the Intergovernmental Panel for Climate Change[29] assuming 1% of applied N in fertilizer is emitted as N_2O nitrogen. No effects of land-use change on possible GHG emissions were included in this analysis. Inputs for hydrotreating of refined oil to produce GD and GJ are from UOP LLC[30] based on engineering calculations and pilot-plant studies. In this study, the hydrogen used in the fuel-production step is assumed to be obtained 60% from a naphtha-to-gasoline reformer and 40% from steam reforming of methane.

Fuel coproducts such as naphtha, fuel gas and liquid petroleum gas (LPG) are also produced, as shown in Figure 18.9. A portion of the environmental burdens of the fuel-production processes are allocated to these coproducts. In addition, allocation to raw plant oil due to the byproduct from the oil-extraction step (seed meal, *etc.*) are also factored into the analysis using energy balances as the basis, as recommended in a recent guidance document.[31] Table 18.10 also contains all of the energy-allocation factors.

18.4.2 Impact Assessment and Comprehensive Assumptions

The software used for this LCA was SimaPro 7.1,[32] which contains a large database of inventory data for material, chemical and energy inputs. The

Table 18.9 Properties of 50% blended SPK fuels produced for Boeing test flights.

Description	*Unit*	*Jet A/A1 Specification Limit*	*ANZ 50/50*	*CAL 50/50*	*JAL 50/50*	*Method*
Composition						
Acidity	mg KOH/g	max 0.1	0.002	0.001	0.002	D3242
Aromatics	vol %	max 25	8.8	9.2	8.9	D1319
Aromatics	vol %	min 8	8.8	9.2	8.9	D1319
Sulfur, Mercapton	mass %	max 0.003	0.0004	<0.0001	0.0003	D3227
Sulfur	mass %	max 0.3	<0.015	<0.0001	0.0403	D2622
Lubricity	mm	max 0.85	0.64	0.65	0.66	D5001
Volatility						
10% recovered	°C	max 205	170.4	170.5	171.0	D86
50% recovered	°C	repot	190.3	194.0	200.5	D86
90% recovered	°C	repot	226.9	228.0	240.0	D86
Final boiling point	°C	max 300	246.8	248.5	258.0	D86
Distillation residue	%	max 1.5	1.2	1.2	1.2	D86
Distillation loss	%	max 1.5	0.4	0.2	0.2	D86
T50-T10	°C	min 15	19.9	23.5	29.5	D86
T90-T10	°C	min 40	56.5	57.5	69	D86
Flash Point, TCC	°C	min 38	45.0	45.0	44.5	D56
Density @ 15 °C	kg/m^3	775–840	779	780	789	D4052

Fluidity						
Freezing Point, °C	°C	max-40 (Jet A) max-47 (Jet A-1)	– 62.5	– 61.0	– 55.5	D2386
Viscosity, 20 °C	mm^2/s	max 8	3.606	3.817	4.305	D445
Combustion						
Net heat of combustion	MJ/kg	min 42.8	43.6	43.7	43.5	D4809
Smoke Point	mm	min 25	33.0	27.7	28.6	D1322
Naphthalenes	vol %	max 3		0.2	1.2	D1840
Corrosion						
Copper strip, 2 h @ 100 °C		max No. 1	1A	1A	1A	D130
Thermal Stability						
JFTOT (2 h at control temperature)						
Temperature	°C	min 260	300.0	300.0	300.0	D3241
Filter dP	mm Hg	max 25	3.0	0.0	0.2	D3241
Tube Deposit Less Than		less than 3	1.0	1.0	1.0	D3241
Contaminants						
Existent gum	mg/100 mL		1.0	< 1	< 1	D381
Additives						
Electrical conductivity	pS/m		123.0	< 1	< 1	D2624

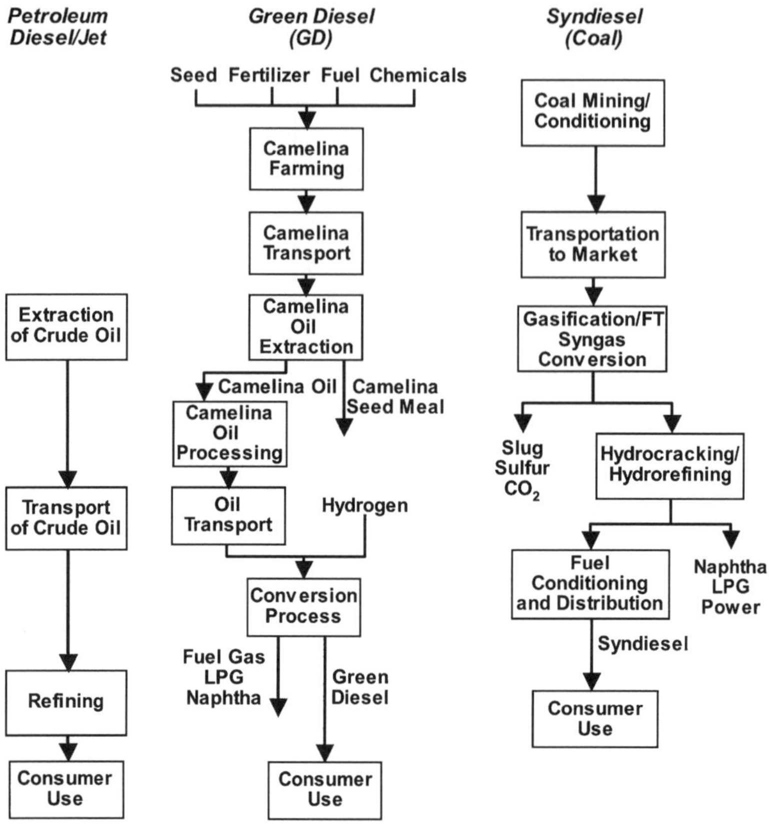

Figure 18.9 Pathway diagrams for petroleum diesel, camelina GD, and FT-syndiesel.

greenhouse-gas impact assessment method used in SimaPro was IPCC 2007 GWP 100a V1.01 method whose output is in CO_2 equivalents for all of the GHG emissions. The three primary gases that make up the CO_2 equivalence are CO_2, methane (CH_4), and nitrous oxide (N_2O), but the minor contributions from solvents and refrigerants were also included. In the IPCC 2007 GWP 100a V1.01 method, each of these gases has different GHG potentials; for example, N_2O has the GHG potential of 298 compared to CO_2, and methane has a GHG potential of 25. For the end use of the fuels, notably combustion in a vehicle, it was assumed that the fuels would completely combust, forming maximum yields of water and CO_2. This assumption is very close to reality in terms of the carbon-dioxide emissions, and therefore stoichiometric emission factors were utilized based on the carbon content of each fuel (*e.g.*, 3.17 kg CO_2/kg petroleum diesel and jet). Emission of CO_2 from combustion of biofuels does not count toward the GHG calculation because biogenic carbon is sequestered during oil seed plant growth, and combustion simply returns this carbon to the atmosphere in a closed cycle.

Table 18.10 Sources of life-cycle inventory data.[35–37]

	Energy Allocation Factors		*Inventory Data Relevance*			
Fuel	*Oil Prod.*	*Fuel Prod.*	*Geogrphic*	*Technology*	*Time*	*References*
Camelina GD, HRJ, BD	0.445	0.867, 0.635	USA	Cultivation Futuristic;	2010	(Shonnard *et al.* 2009)
Jatropha GD, HRJ	0.3355	0.8525, 0.5598	India/EU	Cultivation Field Trial	2007	(Reinhardt *et al.* 2007,2008)
Tallow GD, HRJ	1	0.8882, 0.6424	USA	Conventional	2009	(Judd, 2002)
Petroleum Diesel, Jet	NA	NA	USA	Extraction 2015 mix	2010–2015	(NETL, 2008)
FT Syndiesel; Coal, NG	NA	NA	USA	Syngas to FT Liquids	2000	(Marano *et al.* 2001)

NA; not available but allocation was included in the original publications and in the data used.

Not included within the scope of this study are the construction impacts of processing facilities, which are typically neglected in LCA studies of biofuels as they constitute a small fraction of life-cycle impacts. It is important to note here that no wastewater treatment or solid waste disposal was factored into the study, which typically constitute less than one per cent of the total. Since one percent was the cutoff criterion in these studies, neglecting these inputs will not adversely affect the accuracy of study results. Biomass feedstock transportation was included in this study for a camelina seed to oil processing facility based in Montana. In all cases, transport of the final fuel product to market was included; for camelina biofuels 150 km by road, for jatropha 500 km by pipeline; and for tallow 500 km by pipeline. These transport modes and distances are representative for serving a local market. If the transportation distance were to be increased by a factor of two or three, then the life-cycle GHG results would not change significantly, as shown in the results section below.

18.4.3 Results and Sensitivity Analysis

Figure 18.10 shows the life-cycle GHG emissions for each final fuel product based on input data from Table 18.10 for each stage in the life cycle from feedstock chemicals/production to fuels combustion. In all cases, GD, HRJ, and BD are lower in GHG emissions compared to conventional (petroleum diesel and jet) as well as advanced (FT syndiesel) fossil fuels. GD exhibits lower emissions compared to HRJ for all feedstocks and BD is higher than GD and HRJ for the camelina feedstock, which is in agreement with comparisons of BD with GD for other feedstocks.[14,15,26] Compared to conventional diesel and jet,[33] savings of GHG emissions are highest for tallow GD at 95% and lowest for

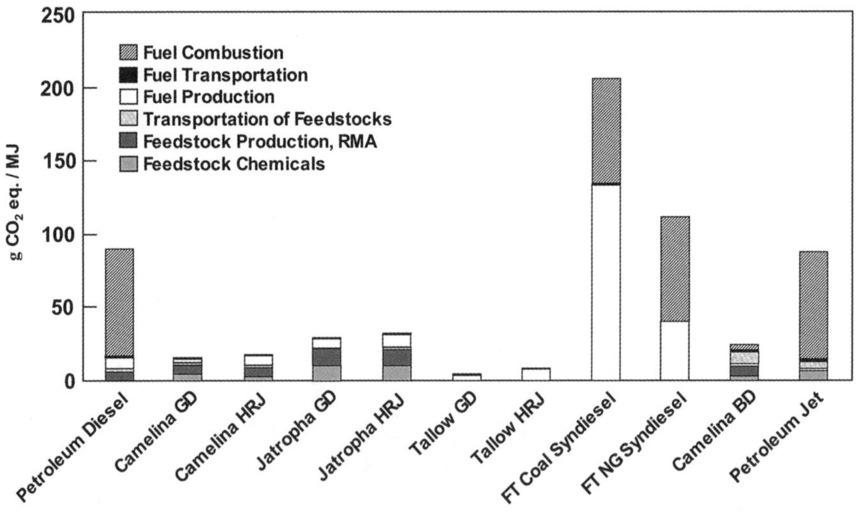

Figure 18.10 GHG emissions for GD, HRJ, petroleum fuels and FT-syndiesel.

jatropha HRJ at 62.8%, but syndiesel from coal increases GHG emissions by 129.9%. About 15-20% of petroleum fuel's GHG emissions occur during other precombustion stages of the life cycle; predominantly crude oil production and refining for fuel production. For syndiesel it was not possible to assign GHG impacts to all life-cycle stages due to limitations in the data available, but over 50% of GHG emissions occur in stages up through fuel production. In most cases, the largest contribution to HRJ and GD GHG emissions is from the combined stages of feedstock chemicals manufacture, feedstock production and fuel production. There is a small contribution to GHG emissions for fuel combustion for BD because of the presence of fossil carbon from methanol.

Figure 18.11 shows the consumption of total energy by each fuel including the energy content in the fuels by primary energy type. Overall, biofuels exhibit slightly higher total energy utilization compared to petroleum-derived diesel and jet, but significantly lower fossil energy demand over the life cycle. Savings of fossil energy compared to conventional diesel and jet are similar to those presented for GHGs, whereas savings for the petroleum component of fossil energy are even higher since fossil energy is comprised of petroleum, natural gas, and coal primary sources.

The effects of inventory uncertainty and allocation type were addressed in a prior publication, but they will be summarized here to indicate the magnitudes of these effects for camelina biofuels.[26] The most important inventory factors include nitrogen fertilizer application rate, N_2O emission rate from nitrogen fertilizer application, and source of hydrogen for GD/HRJ production. When N fertilizer application rate is doubled, GHG emissions for camelina GD increase from 16.9 to 24.6 g CO_2 eq./MJ GD; when N_2O emission rate is doubled GHG emissions increase from 16.9 to 20.7 g CO_2 eq./MJ GD; and when all H_2 is produced by steam reforming of methane GHG emissions

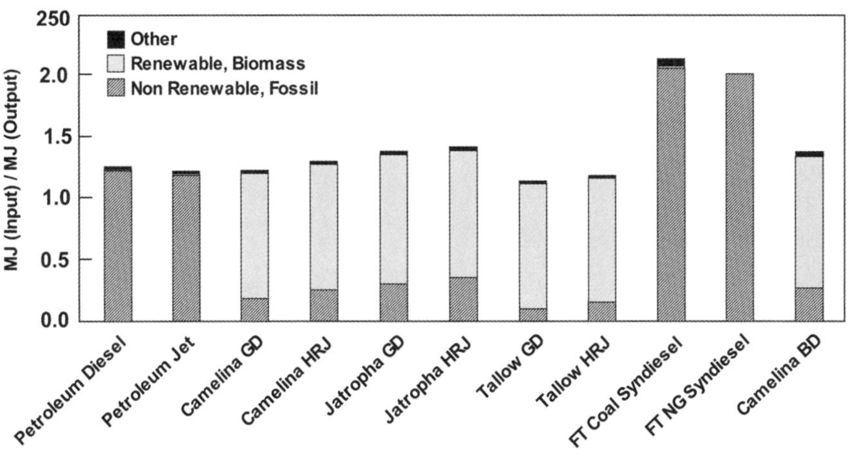

Figure 18.11 Cumulative energy for GD, HRJ, petroleum fuels and FT-syndiesel.

increase from 16.9 to 22.2 g CO_2 eq./MJ GD. When mass allocation is used in place of energy allocation, very little difference in LCA results is seen. However, when displacement allocation is employed, significant differences are observed where GHG emissions for camelina GD and HRJ are negative, indicating sequestration rather than release.

Direct land-use change (LUC) (*e.g.*, conversion of forest into agricultural land) is a large source of greenhouse gases (GHGs) and thus contributes to climate change. Indirect land-use change (iLUC) is outside the production boundary of a feedstock due to economic market forces initiated by energy-crop production on existing agricultural lands. In early 2008 two studies appeared in *Science*[24,25] indicating that biofuel production may emit more greenhouse gases than the fossil-based fuels displaced as a result of direct LUC and iLUC. The original paper estimated "payback periods" on carbon debt in the range of 100 to 1000 years. A more recent paper arrived at payback periods between 3 and 14 years[34] (Kim *et al.* 2009), indicating that this topic is still an active area of research.

18.5 Commercial Production Status and Perspectives

Several commercial facilities for HRD production are in operation or in the latter stages of design/construction. These include the NExBTL process (Neste Oy), Synfining process (Syntroleum), and Coprocessing (Petrobras and Conoco Phillips) as well as the UOP/ENI Ecofining and UOP Renewable Jet Fuel process. Table 18.11 provides a partial summary of feedstocks, plant locations, production capacity and approximate startup date. Over 800 million gallons per year (~3 MMTA) of production capacity is expected to be on-stream before 2012.

Table 18.11 Partial list of announced capacity for commercial production of HRD.

Feedstock	*Location*	*Producer*	*Technology*	*Mgals/yr*	*Start-up*[e]
Mixed	Portugal	GALP	Ecofining[d]	95	2011
Mixed	Italy	ENI Spa	Ecofining[d]	95	2012
Palm Oil	Finland	Neste	NExBTL[a]	56	2008
Palm Oil	Singapore	Neste	NExBTL[a]	264	2011
Animal Fats	Louisiana, USA	Dynamic Fuels	Bio-SynFining[b]	75	2010
Mixed/ Petroleum	Brazil	PetroBras	H-Bio[c]	NA	2006
Mixed/ Petroleum	Ireland	ConocoPhillips	ConocoPhillips[c]	14	2006
5% Tallow	Austrailia	BP	Co-processing	34	2008

[a]Neste Oil Company,
[b]Syntroleum,
[c]Both PetroBras and Conoco Phillips are co-processing vegetable oil (soy, sunflower, palm, animal fat, *etc.*)
[d]ENI/UOP,
[e]estimated from press releases

Most of the early units to come on-stream (Conoco Phillips, PetroBras, and BP Australia) coprocessed small percentages of fats and oils in existing petroleum HDS units. The remainder of the units are advanced hydroprocessing technologies designed to process feedstock that is 100% renewable fat and oil.

Tax incentives and other government subsidies for advanced biofuels are currently being used to encourage expansion of production capacity. Although the social, economic and regulatory issues associated with expanded production of HRD and HRJ are outside the scope of this chapter, it is crucial that future commercialization efforts focus on sustainable methods of producing feedstock. To support this effort, industry participants, including commercial airlines, have formed a consortium that have set criteria that are complementary to emerging internationally recognized standards such as those being developed by the Roundtable on Sustainable Biofuels: (1) Feedstock sources should be developed in a manner which is noncompetitive with food and where biodiversity impacts are minimized; in addition, the cultivation of those plant sources should not jeopardize drinking-water supplies. (2) Total life-cycle greenhouse-gas emissions from plant growth, harvesting, processing, and end use should be significantly reduced compared to those associated with jet fuels from fossil sources. (3) In developing economies, development projects should include provisions or outcomes that improve socioeconomic conditions for small-scale farmers who rely on agriculture to feed themselves and their families, and that do not require the involuntary displacement of local populations. (4) High conservation value areas and native ecosystems should not be cleared and converted for fuel plant source development.

Current and future producers are targeting sustainable production scenarios that, in addition to minimizing impact on land-use change and food and water resources, provide an energy alternative that is economically competitive with current petroleum-based fuels. Future growth will require a coordinated effort between feedstock producers, refiners, and industry regulators to ensure environmental impacts are minimized.

If done responsibly, increasing HRD and HRJ usage in the transportation sector can significantly reduce greenhouse-gas emission as well as diversify energy sources, enhance energy security and stimulate the rural agricultural economy.

References

1. DE Patent DE141029.
2. UK Patent GB190301515.
3. http://en.wikipedia.org/wiki/Hydrodesulfurization.
4. H. Abramovic and V. Abram, *Food Technol. Biotechnol.*, 2005, **43**(1), 63.
5. E. T. Akintayo, *Bioresour. Technol.*, 2004, **92**, 307.
6. G. M. Gubitz, M. Mittelbach and M. Trabi, *Bioresour. Technol.*, 1999, **67**, 732.

7. *Bailey's Industrial Oil and Fat Products*, F. Shahidi, (ed.), Volume 1, (6th edn), Wiley-Interscience, p. 4.
8. J. S. Buchanan, J. G. Santiesteban and W. O. Haag, *J. Catal.*, 1996, **158**, 279.
9. T. N. Kalnes and D. R. Shonnard, Life Cycle Analysis of Bio-renewable Jet Fuel, presented at 2008 Pacific Rim Summit on Industrial Biotechnology and Bioenergy, September 10–12, 2008, Vancouver, BC, Canada.
10. California Biodiesel Multimedia Evaluation Tier I Report, prepared by The University of California, Davis and The University of California, Berkeley for the California Environmental Protection Agency Multimedia Working Group, December 2008.
11. R. C. Hendricks and D. M. Bushnell, Halophytes, Algae, Bacteria, Weeds and Waste Fuel Feedstocks, Civil Aviation Alternate Fuels Initiative R&D Symposium, National Composite Center, Dayton, OH, 27 January, 2009.
12. L. Gouveia and A. C. Oliveira, *J. Ind. Microbiol. Biotechnol.*, 2009, **36**, 269.
13. B. Watkins, New Opportunities for Coprocessing Renewable Feeds in Refinery Process Equipment, AM-08-80, Annual Meeting of the NPRA Conference, March 9–11, 2008.
14. T. N. Kalnes, T. Marker and D. R. Shonnard, Green diesel production by hydrorefining renewable feedstock, Biofuels Technology Quarterly, www.biofuels-tech.com, Q4 2008.
15. K. P. Koers, T. N. Kalnes, T. Marker and D. R. Shonnard, *Environ. Prog. Sustainable Energy*, 2009, **28**(1), 111.
16. T. Kalnes, T. Marker and D.R. Shonnard, 2007, *Int. J. Chem. React. Eng.* 2007, **5**, A48, http://www.bepress.com/ijcre/vol5/A48
17. A. Bozzano, C. Luebke and V. Thakkar, "Hydrotreated Renewable Diesel", ASTM, June 22–25, 2009.
18. Personal communication, Emissions from engine tests of diesel /renewable diesel blends, 27% blend, Courtesy of ENI, 2008.
19. H. Aatola, M. Larmi, S. Mekkonen and T. Sarjovaara, SAE technical publication 2008-01-2500 (2008).
20. J. D. Kinder and T. Rahmes, The Boeing Company Sustainable Biofuels Research & Technology Program, "Evaluation of Bioderived Synthetic Paraffinic Kerosenes", released June 2009. Report available at: http://boeingblogs.com/randy/PDF/PAS%20biofuel%20Exec%20Summary.pdf.
21. J. Sheehan, V. Camobreco, J. Duffield, M. Graboski and H. Shapori, "A Life Cycle Inventory of Biodiesel and Petroleum Diesel for Use in an Urban Bus", National Renewable Energy Laboratory, US Department of Energy, 1998, NREL/SR-580-24089.
22. CONCAWE, Well-to-Wheels Analysis of Future Automobile Fuels and Powertrains in a European Context: Well-to-Wheels Report version 2c, March 2007, http://ies.jrc.ec.europa.eu/WTW.html.
23. M. Wang, "GREET 1.5 – Transportation Fuel Cycle Model: Volume 1: Methodology, Development, Use, and Results", Center for Transportation Research, Argonne National Laboratory, 1999, ANL/ESD-39, Vol. 1.

24. T. Searchinger, R. Heimlich, R. A. Houghton, F. Dong, A. Elobeid, J. Fabiosa, S. Tokgoz, D. Hayes and T.-H. Yu, *Science*, 2008, **319**, 1238.
25. J. Fargione, J. Hill, D. Tilman, S. Polasky and P. Hawthome, *Science*, 2008, **319**, 1235.
26. D.R. Shonnard, L. Williams and T.N. Kalnes, submitted for publication to *Environ. Prog. Sustainable Energy*, November 2009.
27. B. Judd, Biodiesel from Tallow, s.l.: Prepared for Energy Efficiency and Conservation Authority, 2002.
28. D. Zheng and M. Hanna, Preparation and Properties of Methyl Esters of Beef Tallow, Lincoln, NE: University of Nebraska Agricultural Research Division, Vol. 11010.
29. H. S. Eggleston, L. Buendia, K. Miwa, T. Ngara and K. Tanabe, IPCC Guidlines for Natinal Greenhouse Gas Inventories, Prepared by the National Greenhouse Gas Inventories Programme, Japan: IGES, 2006.
30. T. Kalnes, personal communication, Des Plaines, IL: UOP LLC, January, 2009.
31. D. T. Allen, C. Allport, K. Atkins, J. S. Cooper, R. M. Dilmore, L. C. Draucker, K. E. Eickmann, J. C. Gillen, W. Gillette, W. M. Griffin, W. E. Harrison III, J. I. Hileman, J. R. Ingham, F. A. Kimler III, A. Levy, C. F. Murphy, M. J. O'Donnell, D. Pamplin, G. Schivley, T. J. Skone, S. M. Strank, R. W. Stratton, P. H. Taylor, V. M. Thomas, M. Wang and T. Zidow, Framework and Guidance for Estimating Greenhouse Gas Footprints for Aviation Fuels, The Aviation Fuel Life Cycle Assessment Working Group, US Airforce, 2009, p. 108.
32. PRe Consultants bv. SimaPro version 7.1., Copyright 2008.
33. Development of Baseline Data and Analysis of Life Cycle Greenhouse-gas emissions of Petroleum-Based Fuels, Washington, DC: US Department of Energy, National Energy Technology Laboratory, DOE/NETL-2009/1346.
34. H. Kim, S. Kim and B. E. Dale, *Environ. Sci. Technol.*, 2009, **43**(3), 961.
35. G. Reinhardt, S. Gärtner, N. Rettenmaier, J. Munich and E. von Falkenstein, *Screening Life Cycle Assessment of Jatropha Biodiesel,* IFEU, Heidelberg, Germany, 2007.
36. G. Reinhardt, S. Gärtner, N. Rettenmaier, J. Munich, E. von Falkenstein, K. Becker, D. Chaudhary, J. Chikara, G. Francis, A. Ghosh, P. Ghosh, H. Makkar, J. Patolis, M. Reddy and S. Upadhyay, *Basic Data for Jatropha Production and Use,* IFEU, Heidelberg, Germany, June 2008.
37. J. Marano, Life Cycle Greenhouse-gas emissions Inventory for Fischer–Tropsch Fuels, prepared for US DOE National Energy Technology Laboratory by Energy and Environmental Solutions LLC, June 2001.

Footnote

Green Diesel (GD) is HRD from the UOP/ENI Ecofining Process.

CHAPTER 19

Catalytic Deoxygenation of Fatty Acids and their Derivatives for the Production of Renewable Diesel

DMITRY YU. MURZIN AND PÄIVI MÄKI-ARVELA

Åbo Akademi University, Process Chemistry Centre, FI-20500 Turku/Åbo, Finland

19.1 Technologies for the Production of Biodiesel

Based on the expectation that energy consumption will increase in the near future, especially in the transportation sector, there is a pressing need to develop alternatives to fossil fuels. The production of biodiesel has been intensively studied in recent years, in part due to political decisions to increase the use of biofuels. For instance, the EU has set a target level of 20% for the use of biofuels as a percentage of total fuel sales by 2020. The importance of renewable feeds is growing in the energy sector with one possible feedstock being naturally occurring edible oils and fats. Alternatively, there exist nonedible oil yielding crops, some of which can grow with minimal water consumption in arid areas[1–3] and that can function as an alternative feedstock for the development of biofuels.

Fatty-acid methyl esters (FAME) represent the first generation of biodiesel. Unfortunately, the use of FAME can cause problems in vehicles due to its high oxygen content, rendering it a less than ideal fuel for conventional engines; for example, the unburnt part of the biodiesel, when mixed with the lubricating oil,

RSC Energy and Environment Series No. 1
Thermochemical Conversion of Biomass to Liquid Fuels and Chemicals
Edited by Mark Crocker

Published by the Royal Society of Chemistry, www.rsc.org

can promote engine aging.[4] On the other hand, diesel-like hydrocarbons, consisting of C16 to C18 carbon atoms, provide good fuel properties, such as viscosity, cloud point and boiling point.[5] Long-chain hydrocarbons can be produced from fatty acids and their derivatives *via* hydrotreatment or catalytic deoxygenation. Second-generation biofuel (available commercially through NExBTL hydrotreating technology, developed by Neste Oil) is a diesel-like product consisting of hydrocarbons. The third method to produce diesel from biomass is based on gasification followed by conversion of the resulting syngas using Fischer–Tropsch technology (BTL).[6]

Second-generation biofuel technology (NexBTL) is used on an industrial scale in Finland. In this process, the main liquid-phase products are paraffinic hydrocarbons that are formed in a carbon-efficient way *via* hydrotreating, such that the products exhibit the same carbon number as the original feedstock. An alternative method is to use catalytic deoxygenation, which does not necessarily need hydrogen and results in a main hydrocarbon product containing one carbon less than the feedstock. In this chapter, catalytic deoxygenation of fatty acids and their derivatives is reviewed.

19.2 Catalytic Liquid-Phase Deoxygenation of Fatty Acids and their Derivatives

Catalytic deoxygenation of fatty acids in the gas phase was demonstrated as early as 1982, when octanoic and benzoic acids were decarboxylated over palladium and nickel.[7] Hydrocarbons were produced from natural oils as well,[4,8,9] but these methods were not selective towards deoxygenation. When NiO-MoO_3 and CoO-MoO_3 catalysts were used in the hydrocracking of fatty acids and triglycerides in a temperature and pressure range of 350–450 °C and 48–150 bar, respectively, the main products from palm oil were long-chain hydrocarbons (C15–C21) in a yield of 71%, in addition to substantial amounts of cracking products and wax.[4] Deoxygenation of fatty acids[10] and triglycerides[11] has also been demonstrated over activated alumina at 450 °C. In addition to hydrocarbons, the products included ketones. The treatment of triglycerides gave monoalkenes and methyl ketones, which could further be dehydrated to monoalkenes.[11] Subsequently, the liquid-phase catalytic deoxygenation of fatty acids and their derivatives was investigated using combinations of several metals and supports, with optimization of the reaction conditions. Several parameters are important in this reaction, such as catalyst selection, the nature of the feed, reactor selection and catalyst deactivation. The results from these optimization studies are discussed below.[12–23]

19.2.1 Catalyst Selection

19.2.1.1 Metal and Support Selection

In initial work, the liquid-phase deoxygenation of fatty acids and their derivatives was investigated in the presence of a variety of different catalysts.

Parameters of relevance to the catalyst selection are the metal type and support, the metal loading and the metal particle size. In the selection of a suitable support, acidity and pore-size distribution may be of importance. In order to identify the most important parameters, an intensive catalyst screening was performed and the optimal catalyst properties are reported below.

Catalyst selection for the liquid-phase deoxygenation of fatty acids was investigated using stearic acid as a feed at 300 °C under 6 bar of helium in a semi-batch reactor.[12] The following metals were studied: Pd, Pt, Mo, Ni, Ru, Rh, Ir and Os, as well as bimetallic Pd-Pt catalysts, Raney nickel and metal oxides (NiO-MoO_3/Al_2O_3). Furthermore, several supports were applied,[12] *e.g.*, Cr_2O_3, Al_2O_3, SiO_2, activated carbon and mesoporous synthetic carbon, as well as zeolites. The results revealed that the most promising catalysts were Pd and Pt supported on either activated carbon or synthetic mesoporous carbon (Table 19.1). The benefit in using a mesoporous synthetic carbon is the fact that the pore sizes are tunable, being typically about 6 nm (thereby facilitating a more stable catalyst performance; see Section 19.2.3.2), although complete conversion of stearic acid was also achieved in its catalytic deoxygenation over microporous 5 wt.% Pd/C at 300 °C (Figure 19.1). The selectivity towards the desired product, n-heptadecane, was close to 100% at a reaction time of 360 min.

Over a basic catalyst, Ru/MgO, the main product in stearic acid deoxygenation was a symmetrical ketone containing 35 carbons, obtained with a selectivity of 99% at 96% conversion at 300 °C after 360 min. Zeolite-supported-metal catalysts exhibited high conversions but very low selectivities, while NiO-MoO_3/Al_2O_3 was not active. Furthermore, when Ni-Mo catalysts were used with fatty acids, the main products were not long-chain hydrocarbons, but cracking products. Thus, it can be concluded that suitable catalysts for catalytic liquid-phase deoxygenation of fatty acids should include Pd or Pt as an active metal and activated carbon or synthetic mesoporous carbon as a support. The main liquid-phase product over Pt or Pd supported on carbon was n-heptadecane with n-heptadecene present to a

Table 19.1 Deoxygenation of stearic acid over different catalysts at 6 bar and 300 °C using 1 g catalyst with an initial stearic acid concentration of 0.16 mol/l in dodecane under helium flow.

Catalyst	*Conversion after 360 min (%)*	*Selectivity to C17 hydrocarbons (%)*	*Yield of C17 hydrocarbons (%)*	*Ref.*
5% Pd/C	100	99	99	9
1% Pd/C (Sibunit)[a]	98	99	97	17
5% Pt/C	86	95	83	9
8%Pd, 2% Pt/C	62	96	60	9
1% Pd/C	52	94	49	9
10% Pd/C	48	94	45	9

[a]The initial concentration of stearic acid was 0.1 mol/l and the reaction was performed under 17 bar argon.

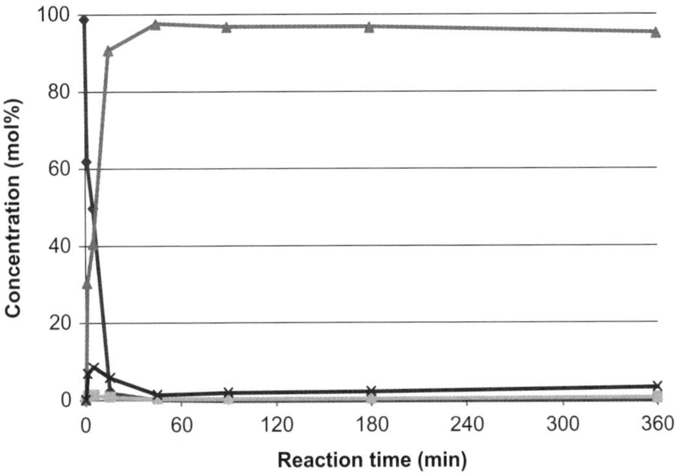

Figure 19.1 Catalytic deoxygenation of stearic acid over 5 wt.% Pd/C using 0.15 mol/l stearic acid in dodecane at 300 °C under 6 bar helium with 1 g of catalyst. Symbols: (◆) stearic acid, (▲) n-heptadecane, (×) unsaturated C17 isomers and (●) 1-heptadecene.

minor extent. Traces of aromatic compounds, formed from unsaturated hydrocarbons and containing typically C17 carbon atoms, were observed as well. The extent of dimerization was typically very minor over Pd supported on activated carbon or mesoporous carbon; this was confirmed by performing the product analysis not only with GC, but with size-exclusion chromatography as well (see Section 19.2.2.5). Relatively large amounts of dimers were, however, formed in stearic acid deoxygenation over Raney nickel, Ni/Al_2O_3, $NiO\text{-}MoO_3/Al_2O_3$, Ir/SiO_2 and Os/C.[11] Dimerization occurs *via* a Diels–Alder reaction starting from diunsaturated fatty acids, *e.g.*, linoleic acid, over the above-mentioned unselective catalysts.[24] The reaction scheme for the deoxygenation of stearic acid was developed based on the experimental results (see Figure 19.2). In the selective catalytic deoxygenation of fatty acids and their esters over Pd or Pt supported on carbonaceous materials the main gaseous products were CO and CO_2, as confirmed *via* online and offline gas analysis.[19–21] As indicated in the scheme, several reactions (decarbonylation, decarboxylation, hydrogenation, dehydrogenation, isomerization, cyclization, dimerization) are possible during the fatty-acid transformation.

19.2.1.2 Effect of Metal Dispersion

The effect of metal dispersion was systematically investigated in the deoxygenation of a mixture of stearic and palmitic acids over 1 wt.% Pd/C (Sibunit), the latter being a synthetic carbon exhibiting a specific surface area of 411 m^2/g_{cat}.[20] The metal dispersion was changed by varying the aging time during catalyst synthesis. Four different 1 wt.% Pd/C (Sibunit) catalysts were prepared

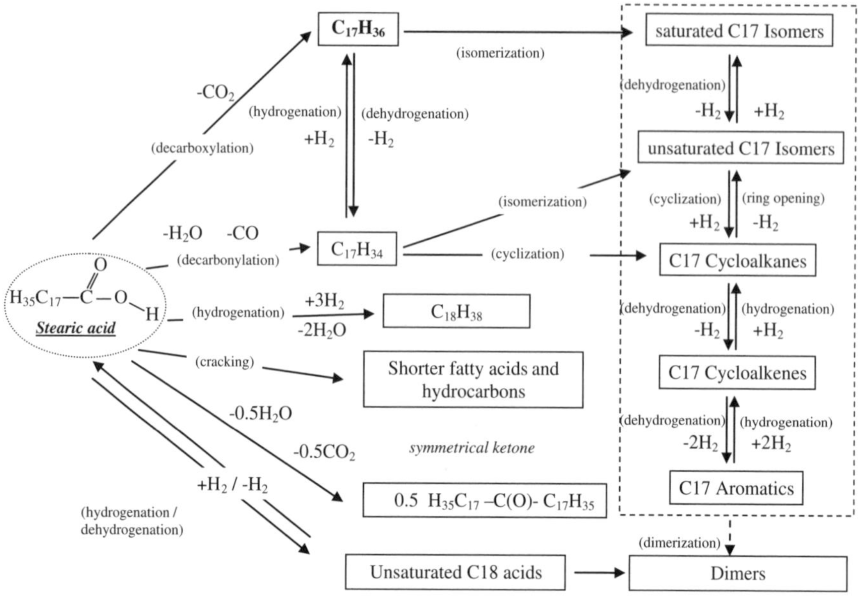

Figure 19.2 Reaction scheme for deoxygenation of stearic acid. Reprinted with permission.[12] Copyright 2006 American Chemical Society.

exhibiting Pd dispersions in the range of 18–72%. An optimum metal dispersion for the catalytic deoxygenation of fatty acids was found for two catalysts that exhibited metal dispersions of 47% and 65%. These catalysts performed equally well, giving the highest conversions in the deoxygenation of a stearic and palmitic acid mixture at 300 °C under 17.5 bar hydrogen in argon gas using 1 g catalyst and 0.1 mol/l fatty acids in dodecane as a solvent. In the case of the catalyst with the lowest metal dispersion ($D = 18\%$), the metal particles were probably too large to promote the deoxygenation effectively (*i.e.* the palladium surface area was too low) and thus extensive catalyst deactivation occurred. The metal particle-size distribution for the catalyst exhibiting the highest metal dispersion ($D = 72\%$) was broader compared to the catalyst with 65% dispersion, which was very active in the deoxygenation of fatty acids. For the catalyst with the highest metal dispersion the small Pd particles might be located in small pores, which would be the most easily blocked by coke.

19.2.2 Deoxygenation of Different Feedstocks

There are several possible feeds that can be used in the synthesis of renewable diesel *via* catalytic deoxygenation. Both saturated and unsaturated fatty acids, their esters and triglycerides have been applied.[11–19] Additionally, tall-oil fatty acids, which constitute a side stream in the pulp industry, have been shown to be amenable to deoxygenation.

19.2.2.1 Deoxygenation of Saturated Fatty Acids and their Esters

Saturated fatty acids and their esters can be deoxygenated thermally, as demonstrated for ethyl stearate in dodecane at 320 °C.[15] After 360 min of reaction the conversion of ethyl stearate was about 20%, giving stearic acid as the primary product, followed by formation of n-heptadecane in a second step. The yield of n-heptadecane was, however, only a few per cent after 360 min. In the case of catalytic deoxygenation, saturated fatty acids with carbon numbers ranging from 9 to 22 were investigated in a semi-batch reactor. As examples, the deoxygenation of stearic acid at 300 °C either over 5 wt.% Pd/C under helium in dodecane[15] or under argon using Pd/C (Sibunit) catalyst can be mentioned.[20] In the former case, complete conversion of stearic acid was achieved after 180 min with the selectivity towards n-heptadecane being 95%.[15] The deoxygenation rates of different acids were found to be the same, as might be expected, providing the reactants were of high purity (note that impurities in these acids can be a reason for catalyst deactivation).

When a saturated ester, *e.g.*, ethyl stearate, was deoxygenated at 320 °C under 17 bar of nitrogen over Pd/C catalyst, stearic acid was the primary product and n-heptadecane was formed as a secondary one. By increasing the amount of hydrogen in the gas phase, higher conversions and yields of n-heptadecane could be achieved.[15] Ethyl stearate deoxygenation was also successfully demonstrated in continuous mode (see below).

19.2.2.2 Deoxygenation of Unsaturated Fatty Acids and their Esters

Deoxygenation of unsaturated acids and their esters was investigated by using as model compounds linoleic and oleic acids, as well as methyl oleate.[17] A typical kinetic plot showing catalytic deoxygenation of oleic acid is depicted in Figure 19.3. Under hydrogen-deficient conditions, several reactions occurred in the catalytic deoxygenation of linoleic acid, namely, isomerization, hydrogenation and deoxygenation (Figure 19.4). The reaction with unsaturated feedstocks proceeds in the temperature range of 300–360 °C under 15–27 bar pressure over microporous Pd/C catalyst; in a first step, isomerization and hydrogenation occur, which is followed by decarboxylation in a second step. Under inert conditions, hydrogenation takes place over the hydrogen-pretreated catalyst, although its rate decreases with increasing reaction time since there is no constant hydrogen supply. In the case of the unsaturated ester, the corresponding saturated ester was the primary product, after which deoxygenated products were formed.

19.2.2.3 Deoxygenation of Triglycerides

Decarboxylation of a triglyceride, tristearin (glycerol tristearate), was demonstrated in a semi-batch reactor at 360 °C under 42 bar of 5 vol.% hydrogen in

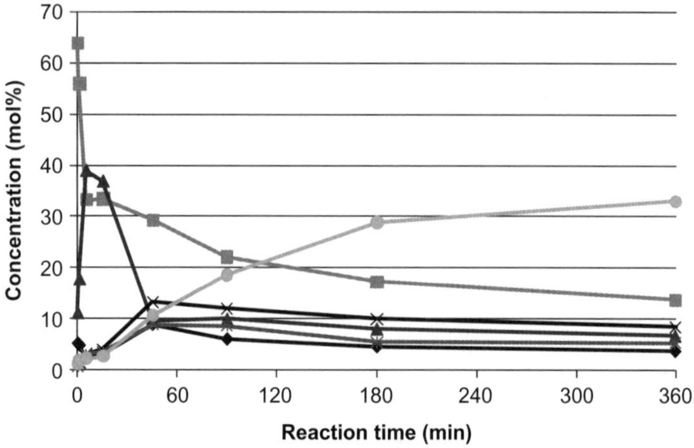

Figure 19.3 Typical reactant and intermediate concentration profiles in the catalytic deoxygenation of oleic acid over 5 wt.% Pd/C at 300 °C under 15 bar argon with the initial acid concentration $c_{\text{oleic acid}} = 3.15$ mol/l (solvent free), $m_{\text{catalyst}} = 1$ g. Symbols: (◆) linoleic acid, (▲) elaidic acid, (●) stearic acid, (*) other C18:x acids, (■) oleic acid, (×) vaccenic acid.

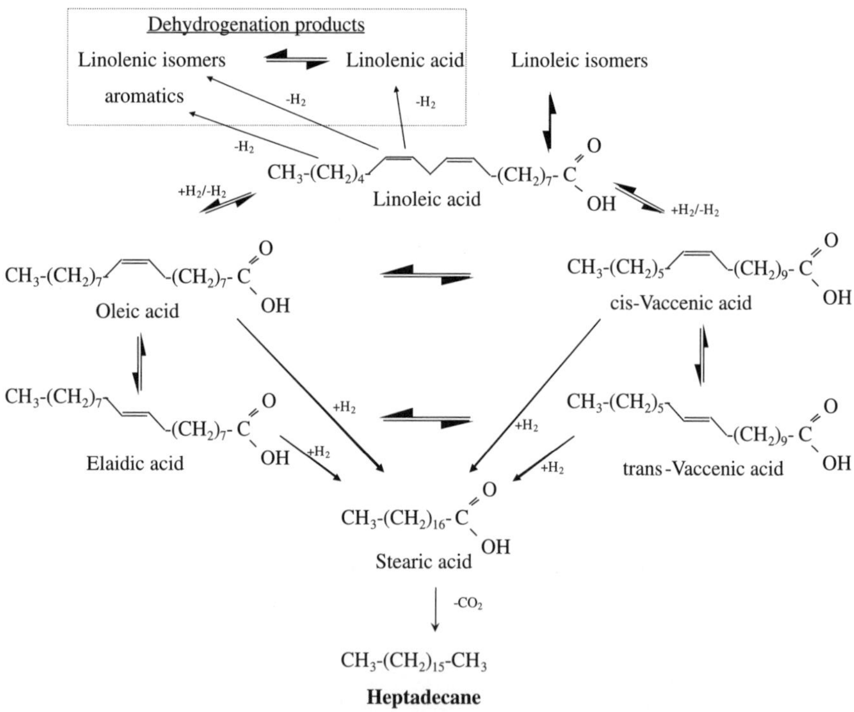

Figure 19.4 Reaction scheme for deoxygenation of linoleic acid. Reprinted with permission from Elsevier.[17]

argon over 5 wt.% Pd/C catalyst. The main product was n-heptadecane and only traces of unreacted tristearin (0.08 wt.%) were observed in the reaction mixture.[13] This result demonstrated that the catalytic deoxygenation process is also feasible for triglycerides.

19.2.2.4 Comparison of Different Feedstocks in Catalytic Deoxygenation

Results obtained in the catalytic deoxygenation of stearic acid and ethyl stearate at 300 °C, as well as ethyl stearate and tristearin at 360 °C, are compared in Figure 19.5 (5 wt.% Pd/C was used as the catalyst). Lower conversion and selectivity to n-heptadecane were achieved for the ester compared to stearic acid. This result can be explained by the fact that in the deoxygenation of an ester, the reaction proceeds *via* formation of the corresponding acid, which is the slowest step. When comparing the conversions in the catalytic deoxygenation of the ester, ethyl stearate, to that of a triglyceride, tristearin, it turned out that the latter exhibited higher selectivity to n-heptadecane at 100% conversion. This result is very important for demonstrating the industrial feasibility of the method.

19.2.2.5 Deoxygenation of Tall-Oil Fatty Acids

Tall oil is a byproduct of the pulp industry with an annual production of 1.6 million tons.[24] Typically, tall oil consists of 40–50% resin acids and 30–40% fatty acids, as well as about 10% saponifiable compounds.[26] Crude tall oil[25] and all-oil fatty-acid distillate (TOFA) have been used to prepare fatty-acid

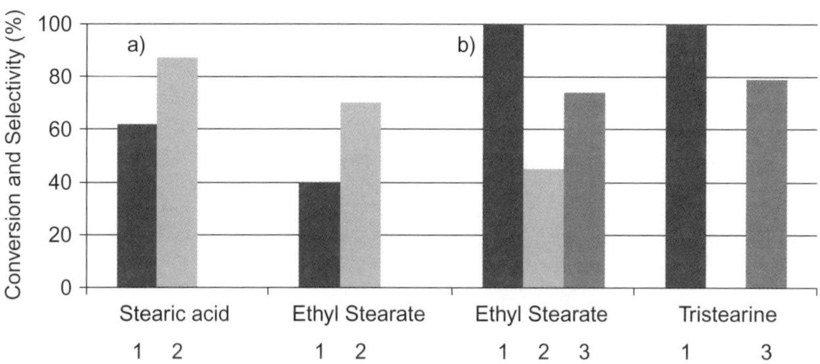

Figure 19.5 Comparison of the deoxygenation of a) stearic acid and ethyl stearate at 300 °C and 17 bar, and b) ethyl stearate and tristearin at 360 °C and 40 bar. The initial reactant concentration was 1.5 mol/l, 1 g 5 wt.% Pd/C was used as the catalyst. The reaction atmosphere was 5 vol.% hydrogen in argon. Notation: 1. Conversion after 360 min, 2. Selectivity to n-heptadecane at 40% conversion and 3. Selectivity to n-heptadecane at 100% conversion.

esters in the synthesis of biodiesel.[27] A margarine-like product was, however, reported in the hydrocracking of TOFA over CoO-MoO_3 catalyst at 350 °C.[4] Additionally, skeletal isomerization and hydrodeoxygenation have been demonstrated for tall-oil fatty acids using NiMo catalysts,[28] with the idea first to isomerize and prehydrogenate the tall-oil fatty acids, followed by deoxygenation over dried and presulfided NiO-MoO_3/Al_2O_3 catalyst. The product was typically a branched, paraffinic hydrocarbon mixture possessing a carbon chain length of one carbon less than that obtained in hydrodeoxygenation under 3 bar pressure and at 335 °C.

Recently, the selective catalytic deoxygenation of TOFA was demonstrated in the temperature range of 300–360 °C, in which TOFA dissolved in dodecane reacted in the first step to stearic acid and subsequently to C17 hydrocarbons.[22] The maximum selectivity towards the desired C17 hydrocarbons achieved at 300 °C using 1 vol.% hydrogen in argon was 84%, at an overall conversion of C18 fatty acids corresponding to 17%. Under hydrogen deficient conditions both unsaturated and saturated C17 hydrocarbons were formed as products. Higher temperatures led to higher yields of C17 hydrocarbons (Figure 19.6), but at the same time the selectivity to them decreased because higher amounts of aromatic compounds were formed. Such compounds, which partially deactivated the catalyst, were mainly obtained from unsaturated C17 hydrocarbons and consisted of undecylbenzene and related compounds. Traces of dimeric aromatic compounds were also obtained, originating from the Diels–Alder reaction.[24]

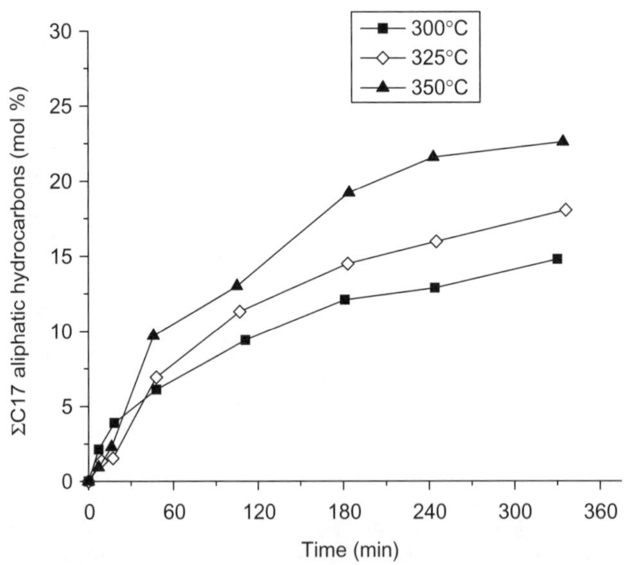

Figure 19.6 The formation of n-heptadecane and n-heptadecene over 1 wt.% Pd/C (Sibunit) at different temperatures under 17 bar 1 vol.% hydrogen in argon. The initial TOFA concentration was 0.15 mol/l.

19.2.3 Reactor Selection

19.2.3.1 Batch **versus** Semi-Batch Reactors

The use of batch and semi-batch reactors was compared in the catalytic deoxygenation of a fatty-acid ester. Ethyl stearate was deoxygenated at 300 °C with an initial concentration of 1.6 mol/l in dodecane using 5 wt.% Pd/C catalyst under 17 bar of inert gas (Figure 19.7). The results revealed that a semi-batch reactor with a purge gas flow through the reactor gave higher conversions compared to the batch operation mode. This result can be explained by the fact that the main gas-phase products, CO and CO_2, can act as catalyst poisons and thus the catalyst poisoning can be retarded by using a semi-batch reactor with purging of the product gases.[15]

19.2.3.2 Continuous Operation

The continuous catalytic deoxygenation of fatty acids and esters was investigated both in a fixed-bed reactor operating in upflow mode, as well as in trickle mode.[19–21] In the fixed-bed upflow reactor, both esters and acids were investigated. As shown in Figure 19.8, the conversion of ethyl stearate over microporous 5 wt.% Pd/C at different temperatures showed a slight decrease as

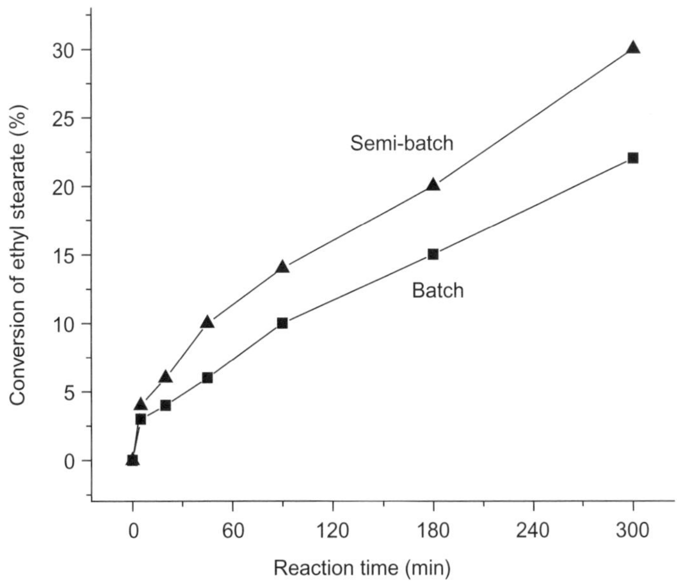

Figure 19.7 Conversion of ethyl stearate as a function of reaction time in batch and semi-batch reactors. Conditions: initial concentration of ethyl stearate 1.6 mol/l, 1 g of 5 wt.% Pd/C, temperature 300 °C, pressure 17 bar using 5 vol.% hydrogen in argon.

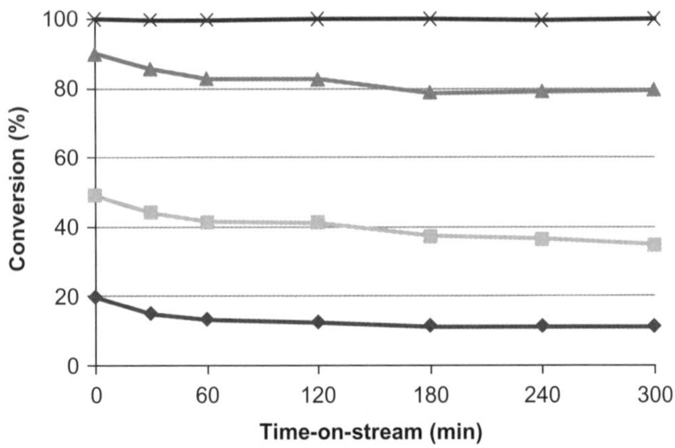

Figure 19.8 Conversion of ethyl stearate in continuous deoxygenation in a fixed-bed reactor performed using 0.16 mol/l of ethyl stearate in hexadecane, 0.4 g of 5 wt.% Pd/C, the liquid flow was 0.1 ml/min. Symbols: (×) 360 °C, $p=7$ bar, (▲) 330 °C, $p=5$ bar, (■) 300 °C, $p=3$ bar and (◆) 270 °C, pressure of argon 1 bar. Reprinted with permission from Taylor & Francis.[18]

a function of time-on-stream. The selectivity to n-heptadecane fell in the range 91–96% at temperatures corresponding to 270–360 °C. The apparent activation energy for zero order reaction kinetics was 69 kJ/mole.[18] Lauric acid was deoxygenated over the same catalyst with a volumetric flowrate of lauric acid in mesitylene of 1.0 ml/min and an initial lauric acid concentration of 0.44 mol/l.[19] At 270 °C under 10 bar of Ar the conversion of lauric acid dropped after 18 min on-stream, however, from an initial level of 68% to 10%, and thereafter it slowly declined until reaching a constant conversion level of 6%.[17] The origin of this catalyst deactivation was ascertained to be coking, resulting in blocking of active sites.

A trickle-bed reactor with down-flow operation, in combination with an inert gas flow, was applied in the deoxygenation of both lauric and stearic acids. Under solvent-free conditions, 30% conversion of lauric acid was achieved over 1 wt.% Pd/C mesoporous synthetic carbon (Sibunit) at 360 °C under 20 bar argon at a WHSV of 0.8 h^{-1}, the main products being undecane and undecene. In the case of stearic acid deoxygenation (Figure 19.9), 15% stearic acid conversion was reached at 360 °C over 1 wt.% Pd/C (Sibunit) under 5 vol.% hydrogen in argon atmosphere, giving as the main products n-heptadecane and n-heptadecene.[23] These results indicate that catalytic deoxygenation of fatty acids in the presence of a gas flow in a trickle-bed reactor is promising, giving stable conversion levels when mesoporous Pd/C is the catalyst. As discussed below, catalyst poisoning and coking are retarded when the main gaseous products, CO and CO_2, can be partially removed during the catalytic experiment. Additionally, it was confirmed by ICP-MS analysis of the liquid phase that no Pd leaching occurred during the reaction.

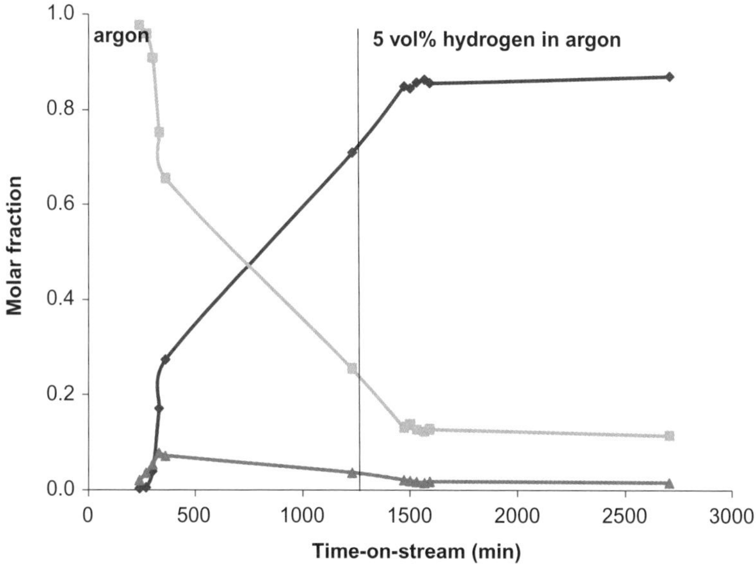

Figure 19.9 Kinetics of stearic acid deoxygenation in a trickle-bed reactor over 1 wt.% Pd/C (Sibunit) at 360 °C and 10 bar; Ar atmosphere for the first 1350 min and thereafter 5% hydrogen in argon. Symbols: (◆) stearic acid, (■) n-heptadecane and (▲) n-heptadecene. The concentration of stearic acid was 4.4 mol/l (solvent-free conditions), argon flow was 10 ml/min and the reactant flow was 0.075 ml/min. Reprinted with permission.[23] Copyright 2009 American Chemical Society.

19.2.4 Catalyst Deactivation

The origin of catalyst deactivation during catalytic deoxygenation of fatty acids and their derivatives was coking, which was confirmed by measuring the specific surface areas of the spent catalysts and comparing them with the respective values of the fresh catalysts. Additionally, aromatic compounds, which are known to adsorb strongly on the catalyst surface and block active sites, were formed (mainly from unsaturated C17 olefins). In addition, some sintering of the metal particles during the catalytic deoxygenation of fatty acids cannot be excluded.[20] Deactivation can be partially prevented by removing product gases (CO, CO_2) from the reaction mixture; consequently, higher conversion levels can be achieved in a semi-batch reactor or in a continuous trickle-bed reactor with gas flow.

19.3 Conclusions

This chapter summarizes recent work in the laboratory of the authors focused on the production of diesel-like hydrocarbons from renewable resources. The liquid-phase deoxygenation of fatty acids, esters and triglycerides was investigated over several supported-metal catalysts in a temperature and pressure ranges of 270–360 °C and 6–17 bar, respectively. In addition, tall-oil fatty-acid deoxygenation was demonstrated using Pd supported on mesoporous synthetic carbon as a catalyst. The most promising catalysts for the above-mentioned reactions were Pd and Pt supported on micro- or mesoporous carbon materials.

In stearic-acid deoxygenation, the main product was n-heptadecane, obtained in a yield of over 90%. Another product was n-heptadecene, while only traces of aromatic C17 hydrocarbons were formed. In addition to stearic acid, other fatty acids ranging from C9 to C23 were investigated, as well as selected unsaturated fatty acids and their esters. The deoxygenation rate of the different acids did not depend on the chain length. In the deoxygenation of fatty-acid esters, the formation of both the corresponding acid, and hydrocarbons possessing one carbon less than the corresponding ester, was observed. In the case of unsaturated fatty acids and esters, in addition to deoxygenation, hydrogenation, dehydrogenation and cyclization, and also isomerization of multiunsaturated acids occurred. The extent of the oligomerization of the unsaturated acids or esters was, however, very minor over the most selective catalysts. On the other hand, for nonselective catalysts, the dimerization was relatively extensive.

A semi-batch reactor exhibited superior performance compared to batch operation, since the main gaseous products, CO and CO_2, could be partially removed from the semi-batch reactor and thus the catalyst exhibited superior operational stability. The deoxygenation of fatty-acid esters was also quite successful in continuous fixed-bed operation. On the other hand, extensive catalyst deactivation occurred in lauric acid deoxygenation in the fixed-bed mode. The reason for this was coking, poisoning and pore blocking. Furthermore, metal sintering cannot be totally excluded. Metal leaching from the catalyst did not, however, occur. In trickle-bed mode, with an inert gas flow through the reactor, a stable conversion of 15% was achieved in stearic acid deoxygenation at 360 °C over a Pd catalyst supported on a mesoporous synthetic carbon. The presence of an inert-gas flow thus facilitated a more stable catalyst operation, since the gaseous products could be partially removed. Overall, these findings suggest that catalytic deoxygenation represents a viable technology for the upgrading of vegetable oils, animal fats and tall oil (a byproduct of wood pulping) to hydrocarbon fuels; as such, it offers an alternative to conventional hydrotreating.

Acknowledgements

This work is part of activities at the Åbo Akademi Process Chemistry Centre of Excellence Programmes (2000–2011) financed by the Academy of Finland.

References

1. S. V. Ghadge and H. Raheman, *Biomass Bioenergy*, 2005, **28**, 601.
2. G. M. Gübitz, M. Mittelbach and M. Trabi, *Bioresour. Technol.*, 1999, **67**, 73.
3. P. N. Giannelos, F. Zannikos, S. Stournas, E. Lois and G. Anastopoulos, *Industr. Crops Prod.*, 2002, **16**, 1.
4. N. Shinoda, E. Nagatomi, Lubricating composition for use in diesel engines compatible with biofuel, WO2009/013275 A1.
5. W. K. Craig, D. W. Soveran, Production of hydrocarbons with a relatively high cetane number, US Patent 4992605 (1991).
6. R. D. Cortright, J. A. Dumesic, Endothermic synthesis gas manufacture from biomass integrated with Fischer–Tropsch synthesis for manufacture of synthetic fuels, US Pat. Appl. Publ. 2007; US 2007225383 A1 20070927.
7. W.F. Maier, W. Roth, I. Thies and P. V. Rague Schleyer, *Chem. Ber.*, 1982, **115**, 808.
8. F. A. Twaiq, N. A. M. Zabidi and S. Bhatia, *Ind. Eng. Chem. Res.*, 1999, **38**, 3230.
9. R. O. Idem, S. P. R. Katikanieni and N. N. Bakhshi, *Fuel Process. Technol.*, 1997, **51**, 101.
10. A. Leung, D. G. B. Boocock and S. K. Konar, *Energy & Fuels*, 1995, **9**, 913.
11. E. Vonghia, D. G. B. Boocock, S. K. Konar and A. Leung, *Energy & Fuels*, 1995, **9**, 1090.
12. M. Snåre, I. Kubickova, P. Mäki-Arvela, K. Eränen and D. Yu. Murzin, *Ind. Eng. Chem. Res.*, 2006, **45**, 5708.
13. D. Yu. Murzin, I. Kubickova, M. Snåre, P. Mäki-Arvela, J. Myllyoja, Method for the manufacture of hydrocarbons, US Patent 7491858 B2 (2006).
14. I. Kubickova, M. Snåre, K. Eränen, P. Mäki-Arvela and D. Yu. Murzin, *Catal. Today*, 2005, **106**, 197.
15. P. Mäki-Arvela, I. Kubickova, K. Eränen, M. Snåre and D. Yu. Murzin, *Energy Fuels*, 2007, **21**, 30.
16. M. Snåre, I. Kubicková, P. Mäki-Arvela, K. Eränen, J. Wärnå and D. Yu. Murzin, *Chem. Eng. J.*, 2007, **134**, 29.
17. M. Snåre, I. Kubicková, P. Mäki-Arvela, C. Chichova, K. Eränen and D. Yu. Murzin, *Fuel*, 2008, **87**, 933.
18. M. Snåre, I. Kubickova, P. Mäki-Arvela K. Eränen and D. Yu. Murzin, *Catal. Org. React.*, 2006, **115**, 415.
19. P. Mäki-Arvela, M. Snåre, K. Eränen, J. Myllyoja and D. Yu. Murzin D.Yu., *Fuel*, 2008, **87**, 3543.
20. I. Simakova, O. Simakova, P. Mäki-Arvela, A. Simakov, M. Estrada and D. Yu. Murzin, *Appl. Catal. A: Gen.*, 2009, **355**, 100.
21. H. Bernas, K. Eränen, I. Simakova, A.-R. Leino, K. Kordás, J. Myllyoja, P. Mäki-Arvela, T. Salmi, D. Yu. Murzin, Deoxygenation of

dodecanoic acid under inert atmosphere, *Fuel* (2009), doi:10.1016/j.fuel.2009.11

22. P. Mäki-Arvela, B. Rozmysłowicz, S. Lestari, O. Simakova, K. Eränen, T. Salmi, D. Yu. Murzin, Catalytic Deoxygenation of Tall Oil Fatty Acid over Palladium Supported on Mesoporous Carbon, submitted, 2009.
23. S. Lestari, P. Mäki-Arvela, H. Bernas, O. Simakova, R. Sjöholm, J. Beltramini, G. Q. Max Lu, I. Simakova and D. Yu. Murzin, *Energy Fuels*, 2009, **23**, 3842.
24. R. Brütting and G. Spiteller, *Fett Wissenschaft Technologie*, 1994, **96**, 445.
25. S. G. Chatterjee, S. Omori, S. Marda, S. Shastri, Process for making biodiesel from crude tall oil, US Pat. Appl. Publ. 2007; US 2007130820 A1 20070614.
26. J. M. F. Nogueira, *Separ. Sci. Technol.*, 1996, **31**, 2307.
27. D. Altiparmak, A. Keskin, A. Koca, M. Gürü Bioresour. *Technol.*, 2007 **98**, *241*.
28. E. Koivusalmi, J. Jakkula, Catalytic process for the manufacture of branched and saturated hydrocarbons from (un)saturated fatty acid and fatty-acid esters. Eur. Pat. Appl. 2007; EP 1795576 A1 20070613.

Subject Index

Page numbers in *italics* refer to figures or tables.